Biomaterials in Tissue Engineering and Drug Delivery

Biomaterials in Tissue Engineering and Drug Delivery

Edited by **Ralph Seguin**

☐SYRAWOOD
PUBLISHING HOUSE

New York

Published by Syrawood Publishing House,
750 Third Avenue, 9th Floor,
New York, NY 10017, USA
www.syrawoodpublishinghouse.com

Biomaterials in Tissue Engineering and Drug Delivery
Edited by Ralph Seguin

International Standard Book Number: 978-1-68286-119-6 (Hardback)

Printed in the United States of America.

Contents

Preface

This book contains some path-breaking studies related to biomaterials, especially in the areas of drug delivery and tissue engineering. Concepts such as behavior of materials, design of biomaterials, preparation of nanomaterials, etc. have been discussed in a detailed manner. It aims to equip students and experts with the advanced topics and emerging concepts in biomaterial science. This book is highly recommended for graduate and postgraduate students as well as for research scholars.

Various studies have approached the subject by analyzing it with a single perspective, but the present book provides diverse methodologies and techniques to address this field. This book contains theories and applications needed for understanding the subject from different perspectives. The aim is to keep the readers informed about the progress in the field; therefore, the contributions were carefully examined to compile novel researches by specialists from across the globe.

Indeed, the job of the editor is the most crucial and challenging in compiling all chapters into a single book. In the end, I would extend my sincere thanks to the chapter authors for their profound work. I am also thankful for the support provided by my family and colleagues during the compilation of this book.

Editor

Calcium orthophosphate coatings, films and layers

Sergey V Dorozhkin

Abstract

In surgical disciplines, where bones have to be repaired, augmented or improved, bone substitutes are essential. Therefore, an interest has dramatically increased in application of synthetic bone grafts. As various interactions among cells, surrounding tissues and implanted biomaterials always occur at the interfaces, the surface properties of the implants are of the paramount importance in determining both the biological response to implants and the material response to the physiological conditions. Hence, a surface engineering is aimed to modify both the biomaterials, themselves, and biological responses through introducing desirable changes to the surface properties of the implants but still maintaining their bulk mechanical properties. To fulfill these requirements, a special class of artificial bone grafts has been introduced in 1976. It is composed of various mechanically stable (therefore, suitable for load bearing applications) biomaterials and/or bio-devices with calcium orthophosphate coatings, films and layers on their surfaces to both improve interactions with the surrounding tissues and provide an adequate bonding to bones. Many production techniques of calcium orthophosphate coatings, films and layers have been already invented and new promising techniques are continuously investigated. These specialized coatings, films and layers used to improve the surface properties of various types of artificial implants are the topic of this review.

Keywords: Calcium orthophosphates, Hydroxyapatite, Coatings, Layers, Films, Surface, Interface

Review

Introduction

All available materials have the specific characteristics of their own, namely, some of them are corrosive or biologically incompatible; some are sensitive to light or oxidation; some are hydrophilic or hydrophobic in nature, *etc*. Due to these reasons, various approaches have been already developed to modify the basic properties of diverse materials, and applying surface coatings, films or layers is a choice of option to solve some problems in a conventional form. For the particular case of artificial bone grafts, synthetic materials which are to be used in biological environments must display an adequacy of both their surface and bulk characteristics in order to fulfill the dual requirements of biocompatibility and suitable mechanical properties for the given application. Otherwise, due to a poor biocompatibility of improper compounds, fibrous tissues always encapsulate the implants made from such materials, which prolong the healing time. Considering that surface is always the first part of any insert that interacts with the host, various types of surface modifications have been developed to enhance biocompatibility and osteoconductivity of the implants (Ruckenstein and Gourisankar 1986).

On the other hand, it is well known that, due to the great chemical similarity to the inorganic part of bones and teeth of mammals, calcium orthophosphates (listed in Table 1) appear to be very friendly substances for the *in vivo* applications (Dorozhkin 2009, 2011; LeGeros 1991; Elliott 1994; Brown and Constantz 1994; Amjad 1997; Brès and Hardouin 1998; Chow and Eanes 2001; Hughes et al 2002; Dorozhkin 2012). However, since calcium orthophosphate bulk materials have a ceramic nature, they are mechanically weak (brittle); therefore, they cannot be subjected to the physiological loads as encountered in human skeletons, other than compressive ones. Therefore, for many years, the clinical applications of calcium orthophosphates alone have been largely limited to non-load bearing parts of the skeleton due to their inferior mechanical properties. Attempting to combine the advantages of various materials, which is one of the major

Correspondence: sedorozhkin@yandex.ru
Kudrinskaja sq. 1-155, Moscow 123242, Russia

Table 1 Existing calcium orthophosphates and their major properties (Dorozhkin 2009, 2011)

Ca/P molar ratio	Compound	Formula	Solubility at 25°C -(-log(K$_s$))	Solubility at 25°C (g/L)	pH stability range in aqueous solutions at 25°C
0.5	Monocalcium phosphate monohydrate (MCPM)	$Ca(H_2PO_4)_2 \cdot H_2O$	1.14	approximately 18	0.0 to 2.0
0.5	Monocalcium phosphate anhydrous (MCPA or MCP)	$Ca(H_2PO_4)_2$	1.14	approximately 17	[a]
1.0	Dicalcium phosphate dihydrate (DCPD), mineral brushite	$CaHPO_4 \cdot 2H_2O$	6.59	approximately 0.088	2.0 to 6.0
1.0	Dicalcium phosphate anhydrous (DCPA or DCP), mineral monetite	$CaHPO_4$	6.90	approximately 0.048	[a]
1.33	Octacalcium phosphate (OCP)	$Ca_8(HPO_4)_2(PO_4)_4 \cdot 5H_2O$	96.6	approximately 0.0081	5.5 to 7.0
1.5	α-Tricalcium phosphate (α-TCP)	α-$Ca_3(PO_4)_2$	25.5	approximately 0.0025	[b]
1.5	β-Tricalcium phosphate (β-TCP)	β-$Ca_3(PO_4)_2$	28.9	approximately 0.0005	[b]
1.2 to 2.2	Amorphous calcium phosphates (ACP)	$Ca_xH_y(PO_4)_z \cdot nH_2O$, $n = 3$ to 4.5%; 15 to 20% H_2O	[c]	[b]	approximately 5 to 12 [d]
1.5 to 1.67	Calcium-deficient hydroxyapatite (CDHA or Ca-def HA)[e]	$Ca_{10-x}(HPO_4)_x(PO_4)_{6-x}(OH)_{2-x}$ $(0 < x < 1)$	approximately 85	approximately 0.0094	6.5 to 9.5
1.67	Hydroxyapatite (HA, HAp or OHAp)	$Ca_{10}(PO_4)_6(OH)_2$	116.8	approximately 0.0003	9.5 to 12
1.67	Fluorapatite (FA or FAp)	$Ca_{10}(PO_4)_6F_2$	120.0	approximately 0.0002	7 to 12
1.67	Oxyapatite (OA, OAp or OXA)[f]	$Ca_{10}(PO_4)_6O$	approximately 69	approximately 0.087	[b]
2.0	Tetracalcium phosphate (TTCP or TetCP), mineral hilgenstockite	$Ca_4(PO_4)_2O$	38 to 44	approximately 0.0007	[b]

[a] Stable at temperatures above 100°C.
[b] These compounds cannot be precipitated from aqueous solutions.
[c] Cannot be measured precisely. However, the following values were found: 25.7 ± 0.1 (pH = 7.40), 29.9 ± 0.1 (pH = 6.00), 32.7 ± 0.1 (pH = 5.28). The comparative extent of dissolution in acidic buffer is ACP >> α-TCP >> β-TCP > CDHA >> HA > FA.
[d] Always metastable.
[e] Occasionally, it is called "precipitated HA".
[f] Existence of OA remains questionable.

innovations over the last approximately 40 years, researchers started to deposit biocompatible calcium orthophosphates onto the surface of mechanically strong but bio-inert or bio-tolerant materials (Ong and Chan 1999; de Groot et al. 1998; Campbell 2003; Kokubo 2008). For example, metallic implants are encountered in endoprosthesis (such as total hip joint replacements) and artificial teeth sockets because the requirements for a sufficient mechanical stability necessitate the use of a metallic body for such devices. As metals do not undergo bone bonding, *i.e.*, do not form mechanically stable links between the implant and bone tissues, they are coated by calcium orthophosphates exhibiting the bone-bonding ability between the metal and bone. After being implanted, calcium orthophosphate coatings, films and layers might be replaced by autologous bone because such coatings, films and layers participate in bone remodeling responses similar to natural bones (Ong and Chan 1999; Onoki and Hashida 2006; Kobayashi et al. 2007; Epinette and Geesink 1995; Willmann 1999; Schliephake et al. 2006; Kokubo et al. 2003; Habibovic et al. 2005; Hahn et al. 2009). Minimal requirements for HA coatings, films or layers (Table 2) have first been described in 1992 in the Food and Drug Administration (FDA) guidelines (Callahan et al. 1994), as

well as a little bit later in the ISO standards (1996). Afterwards, the FDA guidelines were updated in 1997 (U.S. FDA 1995), while the ISO standards were updated in (ISO 2000) and (ISO 2008).

General knowledge on coatings, films and layers

According to Wikipedia, the free encyclopedia, 'Coating is a covering that is applied to the surface of an object, usually referred to as the substrate. In many cases,

Table 2 FDA requirements for HA coatings (Callahan et al. 1994)

Properties	Specification
Thickness	Not specific
Crystallinity	62% minimum
Phase purity	95% minimum
Ca/P atomic ratio	1.67 to 1.76
Density	2.98 g/cm^3
Heavy metals	< 50 ppm
Tensile strength	> 50.8 MPa
Shear strength	> 22 MPa
Abrasion	Not specific

coatings are applied to improve surface properties of the substrate, such as appearance, adhesion, wettability, corrosion resistance, wear resistance and scratch resistance. In other cases, in particular, in printing processes and semiconductor device fabrication (where the substrate is a wafer), the coating forms an essential part of the finished product.' (2012a). Obviously, all the aforementioned is also valid for films. A layer is another important definition. It is determined as a single thickness of some material covering a surface or forming an overlying part or segment.

Historically, involvement with coatings, films and layers dates to the metal ages of antiquity. Consider the ancient craft of gold beating and gilding, which has been practiced continuously for, at least, 4 millennia. The Egyptians appear to have been the earliest practitioners of this art. Many magnificent examples of statuary, royal crowns and coffin cases that have survived intact attest to the level of skills achieved. For example, leaf samples from Luxor dating to the Eighteenth Dynasty (1567 to 1320 BC) appear to be approximately 0.3-μm thick. Such leaves were carefully applied and bonded to smoothed wax or resin-coated wood surfaces in a mechanical (cold) gilding process to create the earliest coatings (Ohring 2002). Concerning the subject of this review, to the best of my findings, the first research paper on calcium orthophosphate coatings was published in 1976 (Sudo et al. 1976).

In spite of the fact that the technology of coatings, films and layers appears to be simultaneously one of the oldest arts and one of the newest sciences, the distinction among the coatings, films and layers is not well established yet; moreover, it may vary depending on the field of science and/or technology. For example, in food industry, the following statement has been published: 'An edible coating (EC) is a thin layer of edible material formed as a coating on a food product, while an edible film (EF) is a preformed, thin layer, made of edible material, which once formed can be placed on or between food components (McHugh 2000). The main difference between these food systems is that the ECs are applied in liquid form on the food, usually by immersing the product in a solution-generating substance formed by the structural matrix (carbohydrate, protein, lipid or multi-component mixture), and EFs are first molded as solid sheets, which are then applied as a wrapping on the food product'. (Falguera et al. 2011). To clarify this topic further, an extensive search in the scientific databases (Scopus, ISI Web of Knowledge) has been performed, and a great number of fixed collocations have been revealed. For example, according to Scopus (as of May 2012), a combination of words 'wear-protecting + coating' in the publication titles is used more frequently if compared with that of 'wear-protecting + film' (75 and 11 publications, respectively). On the contrary, a combination of words 'ferroelectric + film' in the publication titles is used much

more frequently if compared with that of 'ferroelectric + coating' (5,861 and 28 publications, respectively). Concerning the subject of current review, a combination of words 'apatite + coating' is found in the titles of 2,635 publications, while those of 'apatite + film' and 'apatite + layer' are found in the titles of 427 and 370 publications, respectively. A similar correlation is valid for the combinations of words 'calcium + phosphate + coating', 'calcium + phosphate + film' and 'calcium + phosphate + layer' (they are found in the titles of 737, 138 and 149 publications, respectively). Therefore, both HA and all other calcium orthophosphates are most commonly associated with coatings. Perhaps, the aforementioned facts might be just a matter of terminology or even a habit for each particular sub-direction of science and technology.

Now it is necessary to classify various types of coatings, films and layers. In general, many possibilities are available. For example, they might be classified according to their structural material, such as metallic, polymeric, ceramic or composite coatings, films and layers. Furthermore, they might be classified according to their properties, such as biodegradability, edibility, transparency, reflectivity, conductivity, hardness, porosity, solubility, permeability, etc., as well as by the adhesion strength to various substrates. Besides, using a formation approach, all coatings, films and layers can be divided into two big categories: i) conversion ones, which are formed by reaction products of the base material (for example, formation of an oxide layer by surface oxidation) and ii) deposited ones. In turn, the deposited coatings, films and layers might be further classified according to the deposition techniques (Table 3). More to the point, since coatings and films may consist of either one or many individually deposited layers, all of them might be divided into monolayer coatings and films, and multilayer ones. While the former ones are produced by a single stage, the latter ones are produced by layer-by-layer deposition techniques. Furthermore, the individual layers of the multilayer coatings and films might be both indistinguishable from each other (in this case, the multilayer coatings and films behave as a thick monolayer) and distinguishable from each other. In the latter case, there might be an opportunity (sometimes, only hypothetical) to remove one or several individual layers from the surface, making coatings and films thinner. Finally yet importantly, all types of layers, coatings and films might be thin or thick. These terms appear to be relative, and the distinction between them is not well determined either; furthermore, it depends on the specific application. Nevertheless, in general, researchers consider a thin layer, film or coating as one ranging from fractions of a nanometer to several micrometers in thickness. Therefore, a thick layer, film or coating has thickness exceeding several micrometers. Interestingly that, according to the

Table 3 Various techniques to deposit bio-resorbable coatings, films and layers of calcium orthophosphates on metal implants (Sun et al. 2001; Yang et al. 2005; Narayanan et al. 2010)

Technique	Thickness	Advantages	Disadvantages
Thermal spraying	30 to 200 μm	High deposition rates; low cost	Line of sight technique; high temperatures induce decomposition; rapid cooling produces amorphous coatings; high temperatures prevent from simultaneous incorporation of biological agents
Plasma spraying	30 to 200 μm	High deposition rates; improved wear and corrosion resistance and biocompatibility	Line of sight technique; high temperatures induce decomposition; rapid cooling produces amorphous coatings; high temperatures prevent from simultaneous incorporation of biological agents
Magnetron sputtering	0.5 to 3 μm	Uniform coating thickness on flat substrates; high purity and high adhesion; dense pore-free coatings; excellent coverage of steps and small features; ability to coat heat-sensitive substrates	Line of sight technique; expensive; low deposition rates; produces amorphous coatings; high temperatures prevent from simultaneous incorporation of biological agents
Pulsed laser deposition (laser ablation)	0.05 to 5 μm	Coatings with crystalline and amorphous phases; dense and porous coatings; high adhesive strength	Line of sight technique; expensive; high temperatures prevent from simultaneous incorporation of biological agents
Ion beam deposition	0.05 to 1 μm	Uniform coating thickness; high adhesive strength	Line of sight technique; expensive; produces amorphous coatings
Dynamic mixing method	0.05 to 1.3 μm	High adhesive strength	Line of sight technique; expensive; produces amorphous coatings
Dip and spin coating	2 μm to 0.5 mm	Inexpensive; coatings applied quickly; can coat complex substrates	Requires high sintering temperatures; thermal expansion mismatch
Sol–gel technique	< 1μm	Can coat complex shapes; low processing temperatures; thin coatings; inexpensive process; can incorporate biological molecules	Some processes require controlled atmosphere processing; expensive raw materials
Electrophoretic deposition	0.1 to 2.0 mm	Uniform coating thickness; rapid deposition rates; can coat complex substrates; can incorporate biological molecules	Difficult to produce crack-free coatings; requires high sintering temperatures
Electrochemical (cathodic) deposition	0.05 to 0.5 mm	Good shape conformity; room temperature process; uniform coating thickness; short processing times; can incorporate biological molecules	Sometimes stressed coatings are produced, leading to their poor adhesion with substrate; requires good control of electrolyte parameters
Biomimetic process	< 30 μm	Low processing temperatures; can form bonelike apatite; can coat complex shapes; can incorporate biological molecules	Time consuming; requires replenishment and a pH constancy of the simulating solutions (HBSS, SBF, etc.)
Hot isostatic pressing	0.2 to 2.0 μm	Produces dense coatings	Cannot coat complex substrates; high temperature required; thermal expansion mismatch; elastic property differences; expensive; removal/interaction of encapsulation material; high temperatures prevent from simultaneous incorporation of biological agents
Micro-arc oxidation	3 to 20 μm	Simple, economical and environmentally friendly coating technique, suitable for coating of complex geometries	Except of calcium orthophosphates, coatings always contain admixture phases

HBSS, Hank's balanced salt solution; SBF, simulated body fluid.

aforementioned scientific databases, all types of coatings, films and layers are much more often 'thin' than 'thick', namely, according to Scopus (as of May 2012), a combination of words 'thin + coating' in publication titles is used more frequently if compared with that of 'thick + coating' (2,608 and 468 publications, respectively). Similarly, a combination of words 'thin + film' in publication titles is used much more frequently if compared with that of 'thick + film' (144,106 and 7,077 publications, respectively), and a combination of words 'thin + layer' in publication titles is used much more frequently if compared with that of 'thick + layer' (26,603 and 1,443 publications, respectively). Concerning the physical state

of the precursor materials, layers, coatings and films may be applied as liquids, gasses or solids, which might be used as still another classification type. The quality of coatings, films and layers is usually assessed by measuring their porosity, chemical composition, homogeneity, macro- and micro-hardness, bond strength and surface roughness.

Further, one should mention on the reasons why people apply layers, coatings and/or films to the surface of various materials. They are various, for example:

1. The core contains a material, which is toxic, provokes adverse responses, allergic reactions, etc., or has a bitter taste, an unpleasant odor, etc.;

2. Layers, coatings and/or films protect the core material from the surroundings to improve its stability and shelf life;

3. Layers, coatings and/or films develop the mechanical integrity, which means that coated products are more resistant to mishandling (abrasion, attrition, etc.);

4. To modify surface properties of the core, such as biocompatibility, light reflection, electrical conductivity, color, etc.;

5. Decoration (in the cases, when the core alone is inelegant);

6. The core contains a material, which migrates easily to stain hands, clothes and other objects;

7. To modify the release profile of active components, e.g., drugs, from the core.

Reason numbers 1, 2, 3, 4 and 7 appear to be applicable to the biomedical field in general, while reason numbers. 1, 2 and 4 are relevant to the subject of this review.

To conclude this section, one should note that, in a certain sense, all types of coated materials resemble the functionally graded ones but with an extremely high gradient in both the composition and properties at the core/coating interface.

Preparation
Brief knowledge on the important pre- and post-deposition procedures

Due to the unfavorable mechanical properties of bulk calcium orthophosphate bioceramics, an extensive research has been focused on the development of calcium orthophosphate coatings, films and layers on the surfaces of various materials. Various deposition techniques have been already proposed, which are discussed below. The major advantages and disadvantages of the available deposition techniques have been summarized in Table 3 (Sun et al. 2001; Yang et al. 2005; Narayanan et al. 2010). However, in the vast majority of the cases, prior to be coated, an object (a substrate) needs to be prepared for coating. This normally consists of some kind of cleaning (e.g., ultrasonically in an acetone or ethanol bath to remove dirt, oil and other contaminants adhering to the surface) and may include etching, tarnishing, grounding and/or application of a conversion coating (Chou and Chang 2001). Besides, various types of physical modifications of the surface, such as sand- (Cao et al. 2010a) or grit-blasting and polishing, as well as wetting or drying procedures might be applied as well. Furthermore, after calcium orthophosphate coatings, films and/or layers have been formed, various types of post-deposition treatments might be also necessary to improve their properties. For example, post-deposition heat treatment (annealing) of calcium orthophosphates leads to conversion of the deposited amorphous and non-apatite phases into HA with simultaneous increasing of coating crystallinity, enhancing corrosion resistance, as well as reducing the residual stress (Ji and Marquis 1993; Ong and Lucas 1994; Yoshinari et al. 1997; Erkmen 1999; Burgess et al. 1999; Sridhar et al. 2003; Yang et al. 2003a; Lee et al. 2005a; Johnson et al. 2006; Yang et al. 2009a). The annealing can be done by various ways, including laser treatment (Cannillo et al. 2009) or electric polarization in alkaline solutions (Huang et al. 2009). Furthermore, the presence of water during the post-deposition heat treatment also plays an important role in this conversion (Yang et al. 2009a; Cao et al. 1996; Yang et al. 2003b), namely, in comparison to heat treatments at 450°C in dry conditions, the presence of water vapor resulted in a significant increase in coating crystallinity (Yang et al. 2003b). Similar positive effect of hot water was obtained in other studies (Yang et al. 2009a; Saju et al. 2009; Ozeki et al. 2010) in which post-deposition hydrothermal treatment at 100°C to 170°C was used (Figure 1).

Thermal spraying techniques

Thermal spraying is the process in which melted, softened or heated materials are sprayed onto a surface to be deposited on it. A feedstock with a coating material or a coating precursor might be heated by various ways, such as a high temperature flame or a plasma jet, by means of which thermal spraying is classified into flame spraying and plasma spraying. The principal difference between these two techniques is the maximum temperature achievable. In general, thermal spraying provides thick (from approximately 20 μm to several millimeters, depending on the process and feedstock) coatings, films and layers over a large area at high deposition rate, as compared to other coating processes such as electroplating, physical and chemical vapor deposition (Table 3). Coating materials are fed in powder or wire form, heated to a molten or semi-molten state and accelerated towards substrates in the form of micrometer-size particles. Resulting coatings or films are formed by a continuous buildup of successive layers of liquid droplets, softened material domains and hard particles. For all types of thermal spraying techniques, the quality of coatings, layers and films is generally increased with increasing particle velocities (Fauchais et al. 2001).

Since thermal spraying occurs at very high temperatures, the substrates are heated up as well. In some cases, this might result in phase transformation and recrystallization of the near surface zones. For example, a martensitic transformation and recrystallization was found to occur in near surface of a low-modulus Ti-24Nb-4Zr-7.9Sn alloy substrate after application of a plasma-sprayed HA coating. Both phenomena were attributed to

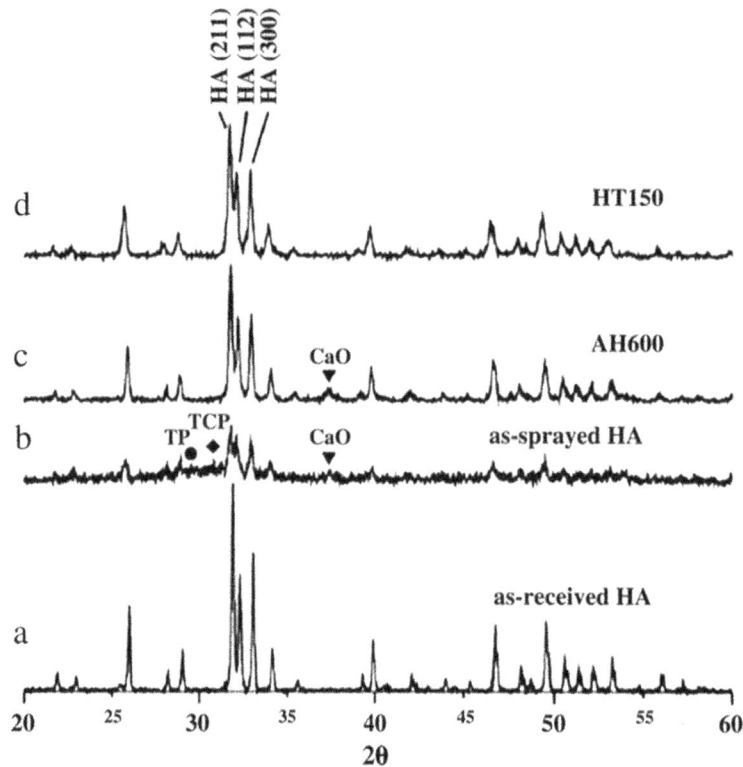

Figure 1 The XRD patterns. (a) an initial HA powder, (b) as sprayed HA coatings, (c) air heat-treated at 600°C HA coatings (AH600) and (d) hydrothermally treated at 150°C HA coatings (HT150). It is easily seen that both the crystallinity and phase purity of poorly crystalline HA coatings increased after both heat and hydrothermal treatments. Reprinted from Yang et al. (2009a) with permission.

the combination of temperature with cooling process (Zhao et al. 2011). Certainly, such phenomena introduce additional ambiguities to the mechanical and adhesive properties of the deposited coatings, films and layers.

Plasma spraying Plasma is often referred to as the fourth state of matter, as it differs from solid, liquid and gaseous states, and does not obey the classical physical and thermodynamic laws. Plasmas are used in many different processing techniques, for example, for modification and activation of various surfaces. Much research is currently being done to understand and control them (Freidberg 2007).

According to de Groot et al. (1998), a plasma spraying technique was discovered accidentally in 1970 by a student, who used the equipment to study melted and rapidly solidified aluminum oxide coatings on a metal substrate (Herman 1988). In plasma spraying, a material to be deposited (feedstock) - typically as a powder, sometimes as a liquid, suspension or wire - is introduced into a plasma jet, emanating from a plasma torch (other names: plasma arc or plasma gun) because a stream of gasses (usually, argon; however, helium, hydrogen or nitrogen might be used as well) passes through this torch. The torch turns these gasses into ionized plasma of a

very high temperature (up to approximately 20,000 K) and with a high speed of up to 400 m/s. In the jet, the material is either melted or heat softened, and these molten or softened droplets flatten and propel towards a substrate to be deposited on it. Appropriate cooling techniques keep the temperature of the substrate below 100°C to 150°C. Since the temperature of the plasma rapidly decreases as a function of distance, the droplets rapidly solidify and form deposits. Commonly, the deposits consist of a multitude of pancake-like lamellae called 'splats', formed by flattening of the liquid or softened droplets. They remain adherent to the substrate as coatings, films and layers. As the feedstock typically consists of powders with sizes from several micrometers to approximately 1 mm, the lamellae have thickness in the micrometer range and lateral dimension from several to hundreds of micrometers. That is why thick coatings, films or layers might be produced only. One pass of the plasma gun can produce a layer of about 5 to 15-lamellae thick. Once a layer has been applied to the whole substrate, the gun returns to the initial position and another layer is applied (Fauchais 2004). Typical current values that are used for spraying HA coatings range from 350 A (Cao et al. 1996) to 1,000 A (Quek et al. 1999). Good schematic setups of the plasma spraying process are available in literature (Narayanan et al. 2010; Fauchais 2004; Paital and Dahotre 2009a; Layrolle 2011; Surmenev 2012). A typical image of a plasma-sprayed HA coating is shown in Figure 2 (Layrolle 2011).

Depending on the experimental conditions, various sub-modifications of plasma spraying technique have been outlined, namely atmospheric plasma spraying (Heimann 2006), vacuum (or low-pressure) plasma spraying (Gledhill et al. 1999, 2001), powder plasma spraying, suspension plasma spraying (Jaworski et al. 2009; Gross and Saber-Samandari 2009; Podlesak et al. 2010), liquid (or solution) plasma spraying (Huang et al. 2010) and, gas plasma spraying (Morks and Kobayashi 2007; Wu et al. 2009) techniques, and all of them are used to fabricate bioactive calcium orthophosphate-based coatings, films and layers. Such modifications have some specific advantages, *e.g.*, they allow obtaining thinner coatings of 5 to 50 μm, which are a few times thinner than those obtained by dry powder processing (Table 3) (Surmenev 2012). In addition, there is a microplasma spraying technique (Dey and Mukhopadhyay 2010; Dey et al. 2011; Dey and Mukhopadhyay 2011), which is characterized by small dimensions, a low level (25 to 50 dB) of noise and hardly any dust, as well as a low power consumption. All of these make it possible to operate under normal workroom conditions. The process provides deposition of coatings, films and layers on small-sized parts and components, including those with fine sections, this being unachievable with any other methods. Due to a low heat input of the microplasma jet, overheating of the powder particles as well as excessive local overheating of the substrate is reduced. The mechanical properties of the coatings, films and layers are generally good (Dey and Mukhopadhyay 2010; Dey et al. 2011; Dey and Mukhopadhyay 2011). This

makes it possible to widen the application scales of plasma spraying and produce different functional coatings.

It is important to stress that, among the deposited lamellae, there are small voids, such as pores, cracks and regions of incomplete bonding. Due to such inhomogeneous structure, the deposits can have properties significantly different from the initial bulk materials (Fauchais 2004). In addition, due to both very high processing temperatures (leading to dehydroxylation of HA and partial decomposition of any other material) followed by rapid solidification, various admixtures and metastable phases are usually present in the deposits. For example, in the case of plasma spraying of calcium orthophosphates, complicated mixtures of various phases (high temperature ACP, α-TCP, β-TCP, HA, OA, TTCP) with other compounds, such as calcium pyrophosphates, calcium metaphosphates, CaO, etc. are obtained (Cao et al. 2010a; Zyman et al. 1993; Weng et al. 1993; Zyman et al. 1994; Tong et al. 1995; McPherson et al. 1995; Yang et al. 1995; Tong et al. 1998; Park et al. 1998; Gross et al. 1998a; Gross et al. 1998b; Heimann and Wirth 2006; Roy et al. 2011). Thus, the chemical and phase compositions of the final coatings, films and layers are dependent on the thermal history of the powder particles. This leads to variable solubility of the deposited coatings, films and layers, dictated by the amounts of more soluble phases, such as ACP. Furthermore, the distribution of by-product and metastable phases in the coatings, films and layers appears to be inhomogeneous. For example, the coating crystallinity was reported to be lower at the interface with the Ti substrate than at the surface of the coating. This happened because metals had a higher rate of thermal diffusivity than calcium orthophosphates and, thus, the cooling rate of the first layer was faster (Gross et al. 1998b). Besides, residual stresses in the plasma sprayed coatings, films and layers were found and measured (Valter et al. 1997; Tsui et al. 1998; Yang et al. 2000; 2003c; Yang and Chang 2005; Carradó 2010; Yang 2011).

A diagram for the formation of various phases during plasma spraying of HA coatings is presented in Figure 3 (Khor and Cheang 1993). According to the authors, if the outside skin of an HA particle is molten and the core remains un-molten, insufficient heat is transferred to melt the particle completely. This model is modified to a totally molten hydroxyl-rich core (with the stoichiometry of HA) with further changes depending on the heat transfer to the particle. The first condition depicts a molten droplet with a hydroxyl-depleted skin. The center containing the hydroxyl-rich molten material will crystallize upon deposition to form HA. The dehydroxylated region, which is exposed to the substrate upon droplet spreading, will form an ACP phase, but the area distant from the substrate will crystallize to form OA.

Figure 2 A scanning electron microscopy of a typical plasma sprayed HA coating on titanium implants. Bar is 10 μm. Reprinted from Layrolle (2011) with permission.

OA requires smaller atomic rearrangements to occur for crystallization from a viscous melt, and, therefore, crystallizes in preference to a mixture of TCP and TTCP. Growth of HA will begin in the hydroxyl-rich core and will finally change to OA in response to the depleted hydroxyl concentration at the top of the lamellae (Figure 3, case (i)). If the molten particle flattens to an extent where the cooling rate is increased, then the entire particle becomes amorphous. Both TCP and TTCP are observed in greater quantities when a higher heat transfer to the particle prevails. If the heat dissipation is slow through the already-solidified amorphous and crystalline layers of the coating, TCP and TTCP can be nucleated at the top surface of the lamellae (Figure 3, case (ii)). The growth of TCP and TTCP may delay the growth of OA with the latent heat of fusion. With a high level of dehydroxylation in the molten particle, lesser amounts of HA or OA will form, and so the large volume of dehydroxylated material will then mostly contain TCP and TTCP. The growth mechanism may begin within the droplet, since a more fluid droplet facilitates faster diffusion. Calcium oxide is observed when even higher heating conditions are employed. In addition to being hydroxyl deficient, the outer shell of the molten particle also becomes phosphate deficient (Figure 3, case (iii)) (Narayanan et al. 2010). In addition, a numerical simulation model of HA powder behavior in plasma jet was suggested (Dyshlovenko et al. 2004). Within this model, the authors created temperature fields inside an HA particle before impact and their transformation into crystal phases after rapid solidification and cooling (Figure 4).

There are a large number of technological parameters that influence the interaction of the particles with the plasma jet and the substrate and, therefore, the properties of final coatings, films and layers. These parameters include feedstock type, plasma gas composition, flow rate, energy input, torch offset distance, substrate cooling, etc. (Cizek et al. 2007). Furthermore, due to the very high temperatures of plasmas, the aforementioned thermodynamic instability of calcium orthophosphates at such temperatures plays an important role in the final properties of the deposited coatings, films and layers. Ideally, only a thin outer layer of each powder particle should be heated to the molten plastic state, which

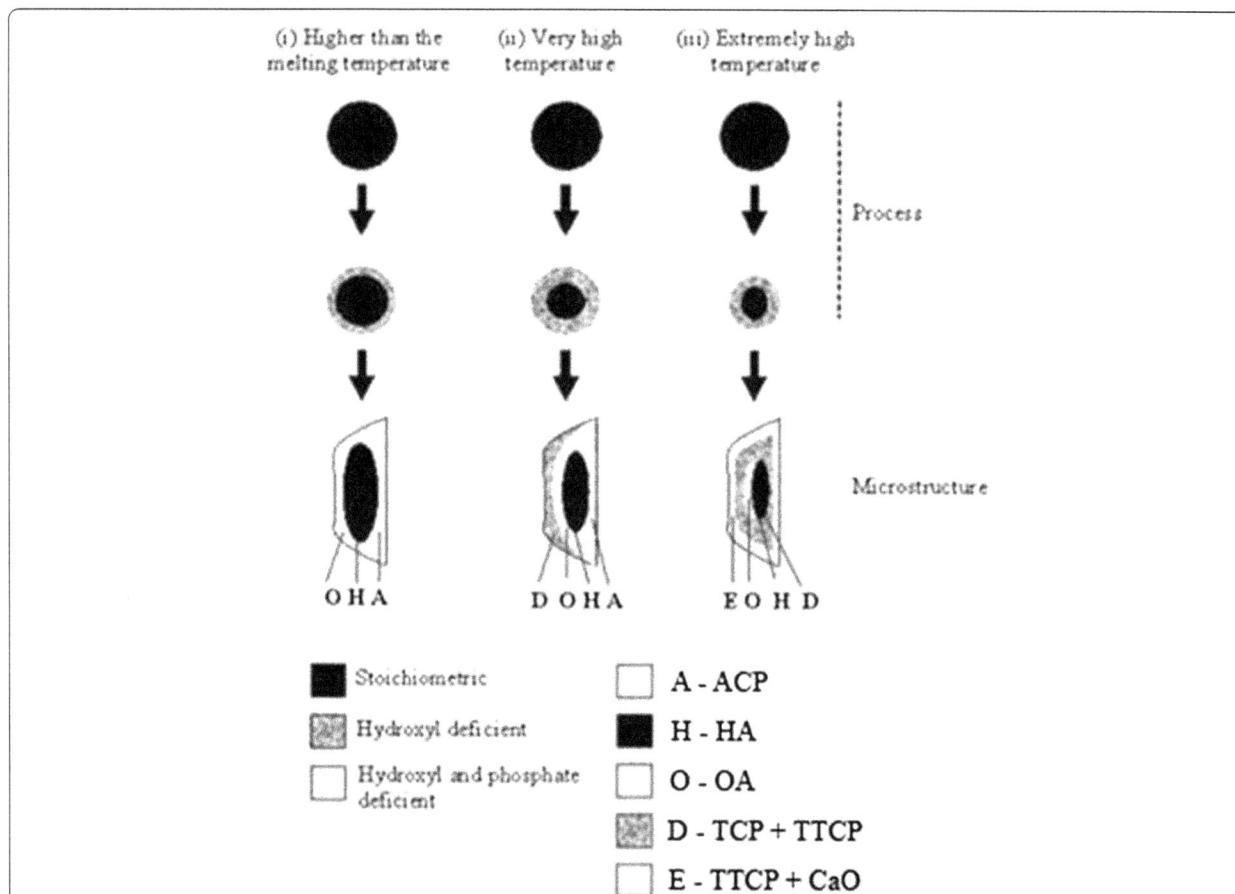

Figure 3 A proposed model for phase formation in the plasma sprayed HA coatings. The process stage depicts the various melt chemistries as a function of particle temperature. The microstructure depicts the different phases that can be formed in a lamella. Reprinted from Khor and Cheang (1993) with permission.

unavoidably undergoes both chemical transformations and phase transitions. This plastic state is necessary to ensure dense and adhesive coatings but it should comprise just a negligible volume fraction of the particles. By choosing optimum relations among particle size, type of gas, speed of the plasma and cooling process of the coated surface, one obtains calcium orthophosphate coatings films and layers with the desired thickness and crystallinity (de Groot et al. 1987; Cook et al. 1988; Stevenson et al. 1989; Wolke et al. 1992; Sun et al. 2003; Prymak et al. 2004).

The dimensions of calcium orthophosphate particles were found to affect their melting characteristics within the plasma flame, namely, large particles undergo a lesser degree of melting in the plasma flame than small particles (Cheang and Khor 1995; Kweh et al. 2000). For example, during spraying of HA particles with dimensions exceeding approximately 55 μm they were found to remain crystalline and showed little or no melting during plasma spraying. HA particles with dimensions within 30 to 55 μm were partially melted and consisted of mixtures of crystalline and amorphous phases, while HA particles less than approximately 30 μm were fully melted and contained large amounts of ACP and traces of CaO (Cheang and Khor 1995). In another study, plasma sprayed HA particles of 20 to 45 μm in size were found to produce denser lamellar coatings than the coatings obtained by plasma spraying of 45 to 75 and 75 to 125-μm HA particles. Coatings formed from 20 to 45-μm sized HA particles did not show the presence of cavities but contained a flatter smoother surface profile as a result of neatly stacked disk-like splats, while coatings formed from 45 to 75 and 75 to 125-μm sized HA particles contained numerous un melted particles, cavities and macropores (Kweh et al. 2000). Interestingly, the coating roughness might be used as a measure of the melting degree of particles within the plasma flame, namely, when the particles reach a more fluid state within the plasma

flame, they become less viscous and can be spread out to a greater degree on impact with the substrate. A smoother coating will result in this case. Partially melted particles will not be able to flatten on the coating surface. This situation will lead to large undulations and a rough coating (Gross and Babovic 2002).

Further details on the plasma spraying technique are available in excellent reviews (Surmenev 2012).

High velocity oxy-fuel spraying In 1990s, a new class of thermal spray processes called high velocity oxy-fuel (HVOF) spraying was developed (Oguchi et al. 1992; Sobolev and Guilemany 1996). A mixture of gaseous (hydrogen, methane, propane, propylene, acetylene, natural gas, etc.) or liquid (kerosene, etc.) fuel and oxygen is fed into a combustion chamber, where they are ignited and combusted continuously. The resultant hot gas at a pressure approximately 1 MPa emanates through a converging-diverging nozzle and travels through a straight section. The jet velocity (> 1,000 m/s) at the exit of the barrel exceeds the speed of sound. A powder feed stock is injected into the gas stream, which accelerates the powder up to 800 m/s. The stream of hot gas and powder is directed towards the surface to be coated. The powder partially melts in the stream and deposits upon the substrate. The resulting calcium orthophosphate coatings, layers and films have a low porosity and a high adhesion strength (Oguchi et al. 1992; Sobolev and Guilemany 1996; Haman et al. 1999; Li et al. 2000, 2002a; Khor et al. 2003a, 2003b, 2004; Hasan and Stokes 2011).

Similar to the aforementioned results on plasma spraying, in the case of HVOF spraying, larger particles of calcium orthophosphates were also found to undergo a lesser degree of melting than smaller particles (Khor et al. 2003b), namely, cross-sectional SEM investigations of the sprayed HA particles of 50 ± 10 μm in sizes revealed that they were melt only partially from the

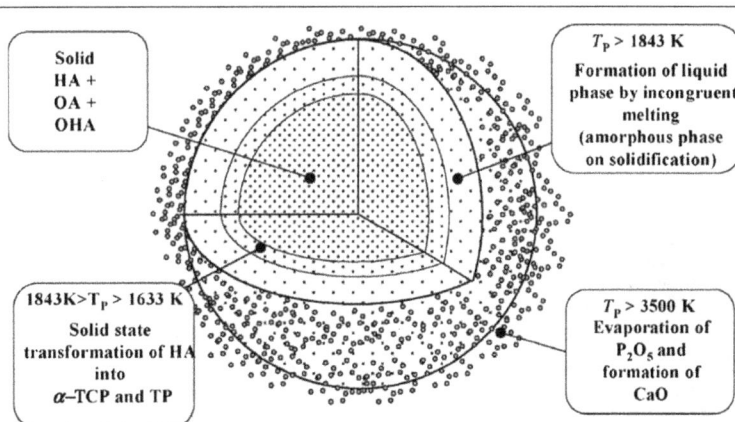

Solid
HA +
OA +
OHA

T_p > 1843 K
Formation of liquid phase by incongruent melting (amorphous phase on solidification)

1843K>T_p > 1633 K
Solid state transformation of HA into α–TCP and TP

T_p > 3500 K
Evaporation of P_2O_5 and formation of CaO

Figure 4 Temperature fields of HA powder particle at impact and assumed phase transformations. Reprinted from Dyshlovenko et al. (2004) with permission.

surface, while those for HA particles of 30 ± 10 μm in sizes revealed that they were melt almost completely. The coating morphology shown in Figure 5 further reveals the influence of the melt state on grain size of the coatings. It clearly demonstrates the interface zone between the melted and un melted parts within a HA splat. It is noted that the HA grains located in un melted part are of far larger size than those in melted part, which states the influence of rapid cooling on grain growth during coating formation (Khor et al. 2003b). Furthermore, Raman spectroscopy qualitative inspection on the sprayed HA particles (partially melted) revealed that a thermal decomposition of HA occurred within the melted part rather than the unmelted zone (Khor et al. 2004). Therefore, to both achieve high crystallinity of the coatings and reduce the amount of admixture phases, the appropriate powder size together with the apt HVOF spray parameters must be carefully selected.

Wet techniques

As follows from the definition, all types of wet deposition techniques occur from either solutions or suspensions both aqueous and non-aqueous. Furthermore, all of them occur at moderate temperatures (Nijhuis et al. 2010). Depending on the solution pH, various calcium orthophosphates might be precipitated (Table 1) and, therefore, be deposited as coatings, films and layers. In general, the deposition kinetics depends on the solution supersaturation, concentration of the reagents, temperature, presence or absence of admixtures, nucleators, inhibitors, etc. As to

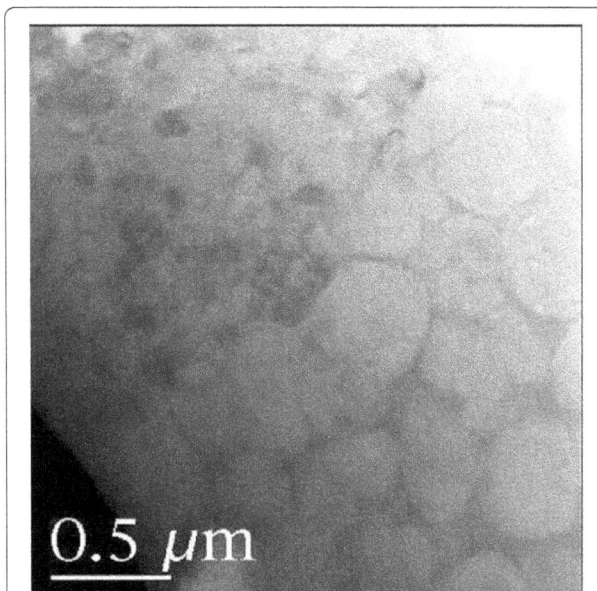

Figure 5 TEM image of as-sprayed HA coating. This show the interface between unmelted and un-melted parts within a HA splat and different grain size. Reprinted from Khor et al. (2003b) with permission.

the precipitation mechanism of calcium orthophosphates from aqueous solutions, this process appears to be rather complicated; for the biologically relevant calcium orthophosphates (OCP, CDHA and HA), the crystallization process occurs via formation of one or several intermediate and/or precursor phases, such as ACP, DCPD and/or OCP. The detailed description of the precipitation mechanisms of various calcium orthophosphates is beyond the scope of current review; the interested readers are referred to the special literature on the topic (Wang and Nancollas 2008; Wang et al. 2011).

For some types of the wet techniques, specific surface preparation techniques appear to be necessary. For example, if calcium orthophosphates need to be deposited on titanium or its alloys, a surface layer of hydrated titanium hydroxides should be created prior the deposition (Wang et al. 2008a). This can be done by various oxidation techniques, such as alkali treatment (de Andrade et al. 2002; Liang et al. 2003; Wang et al. 2000), oxidation in H_2O_2 (Wang et al. 2000), micro-arc oxidation (Song et al. 2004), pre-calcification in boiling $Ca(OH)_2$ solution (Wen et al. 1997; Chen et al. 2009a) or using water vapor treatment (Feng et al. 2002). Positive effects of the presence of hydrated silica (Li et al. 1994) and sodium (Pham et al. 2002) on the surface are known as well. Since the detailed description of the surface preparation of metals is beyond the scope of this review, the interested readers are referred to the special literature on the subject (Narayanan et al. 2010; Nanci et al. 1998; Liu et al. 2004; Rautray et al. 2010; Variola et al. 2011).

Electrophoretic deposition According to Wikipedia, the free encyclopedia. 'Electrophoretic deposition is a term for a broad range of industrial processes, which includes electrocoating, e-coating, cathodic electrodeposition and electrophoretic coating or electrophoretic painting.' (2012b). A characteristic feature of this process is that charged colloidal particles suspended in a liquid medium migrate under the influence of a direct current electric field (electrophoresis) and are deposited onto a conductive substrate of the opposite charge (Besra and Liu 2007).

Since electrophoretic deposition is designed to apply materials to any electrically conductive surface, it is used to achieve calcium orthophosphate coatings, layers or films on various metallic substrates only (Ducheyne et al. 1986, 1990; Zhitomirsky and Gal-Or 1997; Han et al. 1999a; Zhitomirsky 2000; Wei et al. 2001; Stoch et al. 2001; Wang et al. 2002; de Sena et al. 2002; Ma et al. 2003a; Mondragón-Cortez and Vargas-Gutiérrez 2004; Meng et al. 2006, 2008). This approach is especially useful for porous metallic structures. To create coatings, layers or films, calcium orthophosphate powders are suspended in water or other suitable liquids to produce a coating bath, followed by deposition onto a metallic surface.

The proper dimensions of the particles to be deposited are very important because the particles must be fine enough to remain in suspension during the coating process. Electrophoretic deposition normally involves submerging a metallic substrate into a container or vessel, which holds the coating bath, and applying direct current electricity using electrodes, where the substrate is one of the electrodes (anode or cathode). An applied electric field is the driving force of the deposition (Besra and Liu 2007). Depending on the mode and sequence of voltage applied, electrophoretic deposition of calcium orthophosphates can be carried out at either constant (Meng et al. 2006) or dynamic (Meng et al. 2008) voltage.

After deposition, an object is normally rinsed off to remove the undeposited bath, followed by sintering in a high vacuum (10^{-6} to 10^{-7} Torr) at 850°C to 950°C (Besra and Liu 2007). The resulting coatings, layers or films consist of a number of calcium orthophosphate phases plus various random admixtures. For example, in the case of electrophoretically deposited CDHA coatings, the sintering results in their transformation to biphasic (HA + β-TCP) coatings (Han et al. 1999a). Their thicknesses can be varied by changing the electrical field strength and the deposition time. Further, at the coating/substrate interface various metal-phosphorus compounds might be formed due to mutual inter-diffusion of calcium orthophosphates and atoms of the metallic substrate. Unfortunately, due to densification during sintering, shrinkage and cracking of the coatings, layers or films can occur. In addition, thermal stresses induced by the differences in thermal expansion coefficients between the core and the coating during sintering and cooling can lead to cracking (de Groot et al. 1998).

The surface morphology of the electrophoretically deposited calcium orthophosphate coatings was found to depend on applied voltage (Mondragón-Cortez and Vargas-Gutiérrez 2004), deposition time (Mondragón-Cortez and Vargas-Gutiérrez 2004) and powder concentration (Meng et al. 2006), namely, at 200 V, the deposited particles had dimensions within 0.20 to 0.35 μm; at 400 V, the particle size range increased up to 0.35 to 0.80 μm and at 800 V, the particle size range increased up to 0.80 to 1.20 μm. Furthermore, increasing voltages resulted in increasing of the amount of deposited calcium orthophosphates. Besides, porous and roughened coatings were obtained at a higher electric field, while dense coatings of finer particle size were obtained at a lower electric field (Mondragón-Cortez and Vargas-Gutiérrez 2004). Similar effect was noticed for the deposition time: the shorter the time, the smaller particles were deposited (Mondragón-Cortez and Vargas-Gutiérrez 2004). Concerning the powder concentration in suspensions, for a low HA concentration, the coatings were very rough, and a great level of agglomeration was noticed. At higher HA concentrations,

the coatings became uniform and crack free, and there was less agglomeration. At very high concentrations of HA, many cracks were found (Meng et al. 2006). These results indicate that powder concentration, deposition time and applied potential have a significant effect on the coating morphology.

Interestingly, some specific types of calcium orthophosphate bioceramics might be prepared by electrophoretic deposition (Zhitomirsky 2000; Wang et al. 2002; Ma et al. 2003b). For example, hollow HA fibers of various diameters were fabricated (Zhitomirsky 2000). In the first step, submicron HA powders were electrophoretically deposited on individual carbon fibers, carbon fibers bundles and felts. Then, they were burned out and sintered to remove the carbon substrate and leave behind the corresponding ceramic replicas (Zhitomirsky 2000). Similarly, uniform HA tubes were prepared by electrophoretic deposition of HA powders on carbon rods by repeated depositions at room temperature (Wang et al. 2002). The repeated deposition process was necessary to produce thicker multilayered coatings with no surface cracks. The green bodies were then sintered under a range of temperatures varying from 1,150°C to 1,300°C to burn out carbon and obtain HA tubes (Wang et al. 2002). Furthermore, porous calcium orthophosphate scaffolds were fabricated by electrophoretic deposition (Ma et al. 2003b).

To conclude, electrophoretically deposited calcium orthophosphate coatings on implants are commercially available. The examples include BIONIT® (DOT GmbH, Rostock, Germany) and BoneMaster® (BIOMET Corp., Warsaw, IN, USA) (Layrolle 2011). In addition, various modifications and hybrid technologies, such as plasma-assisted electrophoretic deposition (Nie et al. 2001) and a combination of micro-arc oxidation with electrophoresis (Nie et al. 2000) have been developed as well.

Electrochemical (cathodic) deposition In electrochemical deposition of calcium orthophosphates, a supersaturated or a metastable aqueous electrolyte containing calcium and orthophosphate ions is used. Various electrochemical reactions occurring in the electrolyte near electrodes induce local pH increase, and thus, calcium orthophosphate crystals are nucleated and grow on the electrodes (Manso et al. 2000; Duan et al. 2003; Lu et al. 2005). Obviously, only conductive materials might be coated by this technique. A typical setup includes a platinum electrode (anode) and a metallic implant (cathode) connected to a current generator. Since electrochemical deposition usually occurs on the negatively charged electrodes, in literature it is sometimes referred to as cathodic deposition (Zhao et al. 2003; Blackwood and Seah 2009; Roguska et al. 2011). The electrochemical reactions occurring with the ions during the deposition of calcium orthophosphates might be found in literature (Kuo and Yen 2002; Yen and Lin 2002).

Since electrochemical deposition of calcium orthophosphate coatings, films and layers occurs from aqueous solutions, it is commonly performed at ambient conditions (Rossler et al. 2003; Lin et al. 2003). However, electrochemical deposition performed in an autoclave at 80°C to 200°C is also known (Ban and Maruno 1998). The process might be performed in various electrolytes, including SBF (Wang et al. 2004; Lopez-Heredia et al. 2007). A typical example of the deposited coating is shown in Figure 6 (Layrolle 2011). A coating thickness of less than 1 μm can be achieved. Reduction of the thickness leads to an increased resistance to delamination, which is observed frequently for thicker coatings (Peng et al. 2006). Electrochemical deposition of nanosized crystals is also possible (Shirkhanzadeh 1998; Yousefpour et al. 2006; Narayanan et al. 2007, 2008a, 2008b, 2008c. Natural materials, such as shells, have been tested as the source of calcium to produce coatings by electrochemical deposition (Narayanan et al. 2006). Unfortunately, deposition of calcium orthophosphates requires a sufficient volume of electrolyte to surface ratio. Besides, approximately 20 ml of electrolyte is needed to coat 1 cm^2 of implant. Additionally, hydrogen gas production hampers the deposition due to formation of bubbles on the titanium surface, which results in non-uniform coatings (Layrolle 2011). In order to overcome the latter problem, a modulated electrochemical deposition technique has been proposed (Lin et al. 2003).

According to the literature, nucleation of calcium orthophosphate crystals during the electrochemical deposition can occur either as instantaneous nucleation or as progressive nucleation (Eliaz and Eliyahu 2007). Nucleation is said to be instantaneous whenever the formation rate of a nucleus at a given site is expected to be at least 60 times greater than the expected rate of coverage of the site by growth only. Nucleation is said to be progressive when the expected coverage of a site by growth is at least 20 times greater than the coverage of the same site by the act of nucleation. After being formed, calcium orthophosphate nuclei can grow in one, two or three dimensions resulting in different shapes of the deposits like needles, disks or hemispheres depending on deposit/substrate binding energy and their crystallographic misfit. In the electrochemical deposition of HA from aqueous electrolytes, during the first approximately 12 min, the nucleation is instantaneous and is accompanied by a two-dimensional growth. Subsequently, the nucleation becomes progressive and is accompanied by a three-dimensional growth (Eliaz and Eliyahu 2007).

In general, calcium orthophosphate coatings, layers or films obtained by the electrochemical method have a uniform structure since they are formed gradually through a nucleation and growth process at relatively low temperatures (de Groot et al. 1998). Such coatings might be porous (Duan et al. 2003). Interestingly, that in order to produce apatite coatings, non-apatitic calcium orthophosphates might be electrochemically deposited, followed by additional treatments (Redepenning et al. 1996; Han et al. 1999b, 2001; Kumar et al. 1999; Silva et al. 2001). Subsequently, the deposited calcium orthophosphate coatings, layers or films might be heat treated in water steam at 125°C (Shirkhanzadeh 1993) and/or then calcined at temperatures up to 800°C to densify and improve its bonding to the substrates.

Sol-gel deposition By definition, a sol is a two-phase suspension of colloidal particles in a liquid, while gels are regarded as composites because they consist of a solid skeleton or network that encloses a liquid phase or an excess of the solvent. Therefore, the sol-gel process, as the name implies, is a wet-chemical technique that involves transition from a liquid 'sol' into a solid 'gel' phase. Colloidal particles can be in the approximate size range of 1 to 1,000 nm; hence, gravitational forces on these particles are negligible, and interactions are dominated by both short-range forces and surface charges. To prepare sols, usually, inorganic metal salts and/or organometallic compounds such as metal alkoxides are used as precursors. Sols are formed after a series of hydrolysis and condensation reactions of the precursors. Then, the sol particles condense into a continuous liquid gel phase. Besides, a sol might be prepared by dispersion of colloidal particles in a liquid, followed by destabilization of the sol to produce a particulate gel. With further drying and heat treatment, the gel is then converted into dense ceramic or glass materials (Morris 2011). The deposited gels create coatings, films and layers. Sol-gel coatings,

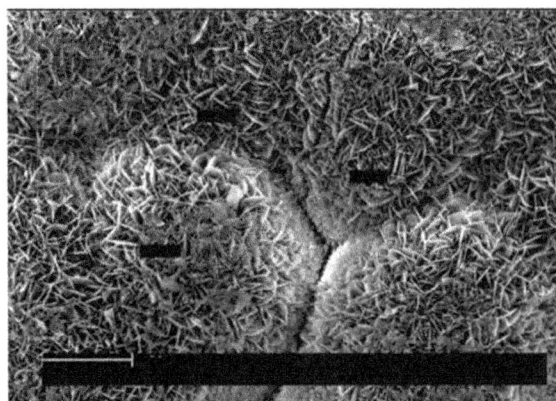

Figure 6 A scanning electron microscopy of a typical electrochemically deposited coating on titanium. Bar is 20 μm. Reprinted from Layrolle (2011) with permission.

films and layers are usually produced using spin or dip (Liu et al. 2002a) coating techniques (see below).

According to this technique, calcium orthophosphate coatings, layers or films are prepared by dipping the sample in calcium (usually, nitrate salt) and phosphorus (usually, alkyl phosphates) gels for an appropriate time at low reaction temperatures. As-formed coatings, layers or films are porous, less dense and have poor adhesion to the substrate. To improve their properties, the samples are annealed at temperatures of 400°C to 1,000°C (Gross et al. 1998c; Haddow et al. 1999; Montenero et al. 2000; Tkalcec et al. 2001; Liu et al. 2002b; Metikoš-Huković et al. 2003; Kim et al. 2004a; Gan and Pilliar 2004; Zhang et al. 2006; Stoica et al. 2008). Depending upon the temperature, different calcium orthophosphate compounds are obtained. The resulting coatings, layers or films can be extremely dense and adhere strongly to the underlying substrate (de Groot et al. 1998). Occasionally, in order to improve the bond strength between the coating and the substrate, an intervening layer of another compound might be applied prior the sol-gel deposition of a calcium orthophosphate (Kim et al. 2004b).

Biomimetic deposition Since biomimetics (synonyms: bionics, biomimicry) seeks to apply biological methods and systems found in nature, biomimetic deposition appears to be a method whereby a biologically active bone-like apatite layer is formed on a substrate surface by immersion in various simulating solutions, such as Hank's balanced salt solution (HBSS) or simulated body fluid (SBF) (Song et al. 2004; Habibovic et al. 2002; Oliveira et al. 2003; Hanawa and Ota 1992; Li et al. 1992; Leitão et al. 1995; Oliveira et al. 1999; Wang et al. 2003). This method involves a heterogeneous nucleation and growth of bone-like calcium orthophosphate crystals on the surface of implants at physiological conditions (temperatures 25°C or 37°C and solution pH within 6 to 8) for several days or even weeks. However, since all simulating solutions, such as HBSS and SBF (their chemical composition might be found in literature) contain a number of various ions, ion-substituted calcium orthophosphates might be deposited only. The thickness of such calcium orthophosphate coatings, layers or films varies within several microns (Table 3), while, according to the X-ray diffraction measurements, the majority of the biomimetic precipitates appear to be either amorphous or poorly crystalline (de Groot et al. 1998). A typical example of a biomimetically deposited calcium orthophosphate coating is shown in Figure 7 (Layrolle 2011).

The mechanism of bone-like apatite formation on an oxidized surface of titanium was investigated in details (Takadama et al. 2001; Uchida et al. 2003). Briefly, it looks as follows: First, a layer of amorphous sodium titanate is formed on the Ti surface after alkali pretreatment.

Then, immediately after immersion into SBF, the sodium titanate exchanged Na^+ ions for H_3O^+ ions in the fluid to form Ti-OH groups on its surface. Later, the Ti-OH groups incorporated calcium ions from the SBF to form a layer of amorphous calcium titanate. After longer soaking times, the amorphous calcium titanate incorporated orthophosphate ions from the SBF to form ACP coatings with a Ca/P atomic ratio of approximately 1.4. Thereafter, ACP converted into bone-like ion-substituted CDHA with a Ca/P ratio of approximately 1.65, which was close to the value of bone mineral (Takadama et al. 2001). In the next study, the authors specified that, after exchanging Na^+ ions for H_3O^+ ions, various types of titania gels might be formed but only those with the anatase or rutile structure induced apatite formation (Uchida et al. 2003). Further specific details on this topic are available in literature (Kokubo and Yamaguchi 2011).

Since biomimetic deposition of calcium orthophosphate coatings, layers or films is a slow process, ways were sought to make it faster. Using condensed versions of the simulating solutions is the most popular option. For example, time for apatite induction in the 1.5-fold SBF was significantly shortened compared to that in the standard SBF. Therefore, the concentration of SBF was increased further, namely, 2-fold (Sun and Wang 2010; Miyaji et al. 1999; Kim et al. 2000), 5-fold (Barrere et al. 2002a, 2002b, 2004) and even 10-fold (Tas and Bhaduri 2004) SBF solutions were used to accelerate precipitation and increase the amount of precipitates. However, whenever possible, this should be avoided because the application of condensed solutions of SBF leads to changes in the chemical composition of the precipitates; namely, the concentration of carbonates increases, while the concentration of orthophosphates decreases (Dorozhkina and Dorozhkin 2003).

The nucleation and growth of calcium orthophosphate coatings deposited on Ti6Al4V substrates from 5-fold SBF were investigated in details by both atomic force and environmental scanning electron microscopes (Barrere et al. 2004). Scattered calcium orthophosphate deposits of approximately 15 nm in diameter were found to appear after only 10 min of immersion in 5-fold SBF. Then, they grew up to 60 to 100 nm after approximately 4 h. With increasing immersion time, the packing of calcium orthophosphate deposits with size of tens of nanometers in diameter formed larger globules and then continuous calcium orthophosphate coatings on Ti6Al4V substrates. The coatings were composed of nano-sized deposits. A direct contact between calcium orthophosphates and the Ti6Al4V surface was observed (Barrere et al. 2004). A stable solution containing high concentrations of calcium and orthophosphate ions was prepared in another study (Li et al. 2002b). This solution became supersaturated after $NaHCO_3$ was added. A uniform coating of approximately

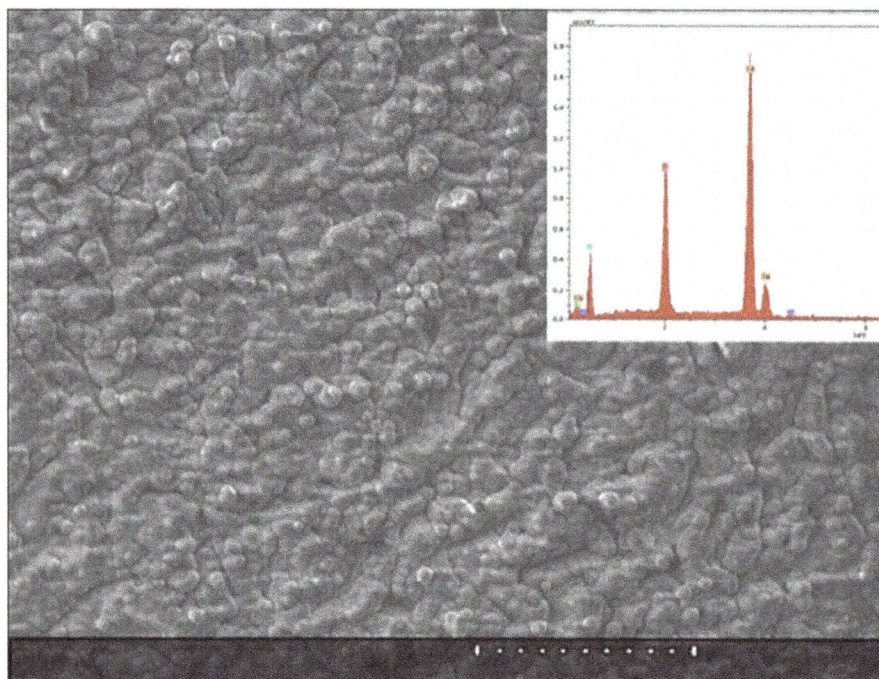

Figure 7 A scanning electron microscopy of a typical biomimetically deposited carbonated apatite coating. Inset: an EDX spectrum of the coating. Bar is 200 μm. Reprinted from Layrolle (2011) with permission.

40-μm thickness was obtained on the substrate after immersion for 24 h. The coatings contained adjustable composition from CDHA to DCPD (Li et al. 2002b).

Simplification of the ionic composition of the standard simulating solutions is still another option to increase deposition kinetics (Bigi et al. 2005a; Li 2003). For example, a fast (a few hours instead of 14 days with SBF) biomimetic deposition of CDHA coatings on Ti6Al4V substrates was obtained using a slightly supersaturated Ca/P solution with an ionic composition simpler than that of SBF. Thin film XRD indicated that the deposits obtained after approximately 3 h were poorly crystalline CDHA, and their content increased on increasing the soaking time up to 3 days (Bigi et al. 2005a). However, since adhesion of this coating to the substrate was not indicated, it is doubtful whether this coating had sufficient strength to resist dissolution inside the body.

Dip coating Dip coating is a popular way of creating coatings, films and layers for various purposes. It consists of several successive steps. A substrate is immersed into either a solution or a suspension of the coating material (in our case, calcium orthophosphate) at a constant speed. A wet coating, film or layer is deposited by itself on the substrate while it is pulled up. Usually, withdrawing is carried out at a constant speed to avoid any jitters. The speed determines the thickness (the faster, the thinner). Excess solution or suspension is drained from the surface. A solvent evaporates from the solution or suspension, forming a denser coating, film or layer. For volatile solvents, such as alcohols, evaporation starts already during the deposition and drainage steps. After being dried and sintered, a solid surface is achieved (Brinker et al. 1991). By means of dipping, uniform coatings, films and layers of calcium orthophosphates have been applied onto various substrates (Li et al. 1996; Weng and Baptisa 1998; Jiang and Shi 1998; Choi et al. 2003; Bini et al. 2009).

There are two mechanisms which govern the formation of the surface coatings, films or layers during dip coating. The first mechanism is known as liquid entrainment. It occurs when a specimen is withdrawn from slurry faster than it can drain from the surface, leaving a thin film (Pontin et al. 2005). The second mechanism is slip casting, in which the capillary suction caused by a substrate drives ceramic particles to concentrate at the substrate-suspension boundary, and a wet layer is formed (Gu and Meng 1999). The withdrawal velocity and the suspension properties (volume fraction of solids, viscosity) have influence on the liquid entrainment mechanism, while the surface microstructure of the substrate (porosity and pore diameter) together with the suspension properties have influence on the slip casting mechanism. By modifying these parameters, layers as thin as 2 μm and as thick as 0.5 mm might be formed (Pontin et al. 2005; Gu and Meng 1999).

Physical vapor deposition techniques In general, all types of physical vapor deposition (or sputtering) techniques for producing coatings, films and layers can be broadly classified into two main groups: (1) those involving thermal evaporation techniques, where a material is heated until its vapor pressure becomes greater than the ambient pressure, and (2) those involving ionic sputtering methods, where a highly energetic beam of ions and/or electrons strikes a solid target and knocks atoms off from the surface (Narayanan et al. 2010; Paital and Dahotre 2009a). Usually, physical vapor deposition occurs in vacuum; however, it might be performed in presence of some gasses. The target is the source material (in our case, a calcium orthophosphate). Substrates are placed into a chamber, and they are pumped down to a prescribed pressure. Sputtering is driven by momentum exchange between the ions and atoms in the materials due to collisions. Afterwards, the dislodged atoms or molecules are deposited on a substrate which is also placed into the same vacuum chamber. An important advantage of the sputter deposition is that even materials with very high melting points are easily sputtered. For the efficient momentum transfer, the atomic weight of the sputtering gas should be close to the atomic weight of the target, so neon or argon is preferable for sputtering of light elements, while krypton or xenon is used for heavy elements (Cuerno and Barabási 1995). However, for deposition of calcium orthophosphates, oxygen might be used as well. It has a number of features, and a better stoichiometry with respect to HA of the deposited coatings, films and layers is one of them (van Dijk et al. 1997).

To sputter calcium orthophosphates, several types of the physical vapor deposition techniques are used, such as ion beam (Stevenson et al. 1989; Barthell et al. 1989; Ong et al. 1991, 1992, 1994; Yoshinari et al. 1994; Cui et al. 1997; Kim et al. 1998; Luo et al. 1999; Choi et al. 2000; Wang et al. 2001; Hamdi and Ide-Ektessabi 2003; Lee et al. 2003; Fujihara et al. 2004; Lee et al. 2005b; Rabiei et al. 2006; Lee et al. 2007a; Blalock et al. 2007), radio-frequency (RF) magnetron (Cooley et al. 1992; Yamashita et al. 1994; Jansen et al. 1993; Wolke et al. 1994; van Dijk et al. 1995, 1996; Wolke et al. 1998, 2003; Nelea et al. 2003, 2004; Feddes et al. 2003a, 2003b; Ding 2003; Yamaguchi et al. 2006; Wan et al. 2007; Ozeki et al. 2007; Ueda et al. 2007; Snyders et al. 2008; Ievlev et al. 2008; Toque et al. 2009), pulsed laser (Nelea et al. 2000, 2002, 2004; Cotell et al. 1992, Cotell 1993; Torrisi and Setola 1993; Cotell 1993; Singh et al. 1994; Wang et al. 1997; Hontsu et al. 1997; Fernández-Pradas et al. 1998, 1999, 2001; Mayor et al. 1998; Arias et al. 1998; Craciun et al. 1999; Fernandez-Pradas et al. 1999; Zeng et al. 2000; Cleries et al. 2000a; Nelea et al. 2000; Zeng and Lacefield 2000; Fernandez-Pradas et al. 2001; Nelea et al. 2002; Socol et al. 2004; Kim et al. 2005a; Bigi et al. 2005b; Koch et al. 2007; Kim et al. 2007a; Paital and Dahotre 2008; Paital et al. 2009; Dinda et al. 2009; Tri

and Chua 2009; Sygnatowicz and Tiwari 2009), diode, direct current and reactive sputtering or deposition (Massaro et al. 2001). The physical and aggregate states of the calcium orthophosphate source might influence the deposition kinetics. For example, the deposition rate of HA was found to be much higher in a solid plate target than in a powder lump target, owing to the difference of apparent density approximately 75% and approximately 18%, respectively (Wan et al. 2007).

Depending on the type of sputtering system and parameters used for the deposition, the structure and chemical composition of the deposited coatings, layers and films may be quite different from those of the initial material used for sputtering. For example, differences in Ca/P ratios between the initial calcium orthophosphates and that in the sputtered coatings were suggested to be attributed to the preferential sputtering of calcium, probably due to a possibility of orthophosphate ions being pumped away before they are deposited at the substrate (Zalm 1989). It was also suggested that orthophosphate ions might be weakly bound to the growing coatings, layers and films, and therefore, they are sputtered away by incoming ions or electrons (van Dijk et al. 1996). Nevertheless, all sputtering techniques have the advantage of depositing thin coatings, films and layers with strong adhesion and compact microstructure.

Ion beam assisted deposition Ion beam assisted deposition (IBAD) is a vacuum technique in which ions of a material to be deposited (in our case, calcium orthophosphates) are generated by collisions with electrons. Then, the detached ions are accelerated by an electric field emanating from a grid toward a target. As the ions leave the source, they are neutralized by electrons from the second external filament and form neutral atoms. A pressure gradient between the ion source and a sample chamber is generated by placing a gas inlet at the source and shooting through a tube into the sample chamber (Ali et al. 2010). Therefore, a typical deposition system consists of two main parts: electron or ion beam bombarding and vaporizing a calcium orthophosphate bulk target to produce an elemental cloud towards the surface of a substrate and a source for simultaneous irradiation of a substrate with highly energetic inert (e.g., Ar^+) or reactive (e.g., O_2^+) gas ions to assist the deposition. Both single and dual ion beam assisted deposition systems are available. Good illustrations of both systems are presented in literature (Narayanan et al. 2010; Paital and Dahotre 2009a; Surmenev 2012).

In this approach, firstly thin (a few hundred atomic layers thick) and amorphous calcium orthophosphate coatings, layers or films are usually deposited. Then, an ion implantation technique, with ions such as argon, nitrogen and oxygen, is used to make them crystalline

(Stevenson et al. 1989; Barthell et al. 1989; Ong et al. 1991, 1992, 1994; Yoshinari et al. 1994; Cui et al. 1997; Kim et al. 1998; Luo et al. 1999; Choi et al. 2000; Wang et al. 2001; Hamdi and Ide-Ektessabi 2003; Lee et al. 2003, 2007a; Fujihara et al. 2004; Lee et al. 2005b; Rabiei et al. 2006; Blalock et al. 2007). A high bond strength associated with this deposition technique appears to be a consequence of an atomic intermixing interfacial layer, which can be up to a few microns thick. Studies revealed alterations in the chemical composition of the ion beam deposited coatings, layers or films. For example, calcium orthophosphate films were synthesized on silicon wafers by electron beam evaporation of β-TCP both with and without simultaneous Ar ion beam bombardments (Lee et al. 2003). It was observed that a simultaneous bombardment with Ar ion beam had a significant effect on both the morphology (Figure 8) and composition of the films, namely, films formed without Ar ion beam bombardment were found to have a Ca/P ratio of approximately 0.76 and reacted immediately with the moisture in the air as soon as it is removed from the chamber. In contrast, the films formed with Ar ion beam bombardment had a Ca/P ratio of approximately 0.80 with smooth and featureless surface morphology (Lee et al. 2003).

In another study, calcium orthophosphate layers on silicon substrates were prepared by using ion beam assisted simultaneous vapor deposition. The method comprised of an electron beam heater and a resistance heater vaporizing CaO and P_2O_5, respectively, while an argon ion beam was focused onto substrates to assist the deposition (Hamdi and Ide-Ektessabi 2003). All deposited layers appeared to be amorphous, regardless of the current density level of the ion beam. Therefore, a post-heat treatment was applied to crystallize the layers. The effects of ion beam current density on the phase composition of the crystallized calcium orthophosphates are shown in Figure 9. The Ca/P ratio was found to increase with increasing ion beam current density presumably due to the high sputtering rate of P_2O_5 compared to that of CaO from the layer being coated. As seen in Figure 9, biphasic (HA + TCP) formulations were found when the ion beam was either not used or used at current density of 180 mA/cm^2, while at ion beam current density of 260 mA/cm^2, only HA peaks were observed (Hamdi and Ide-Ektessabi 2003). In still another study, the X-ray photoelectron spectroscopy analysis of the deposited calcium orthophosphate coatings on titanium revealed several distinct zones: (i) the ambient-exposed surface exhibited elevated concentrations of carbon due to atmospheric contamination; (ii) the bulk zone contained relatively constant concentrations of calcium, oxygen, phosphorus and fluorine, indicating the chemistry for calcium fluoride and FA formation; (iii) while the underlaying zone exhibited elevated titanium and oxygen photoelectron peaks, suggesting the coexistence of calcium orthophosphates within titanium oxides. Furthermore, the substrate was shown to be identical to the passivated

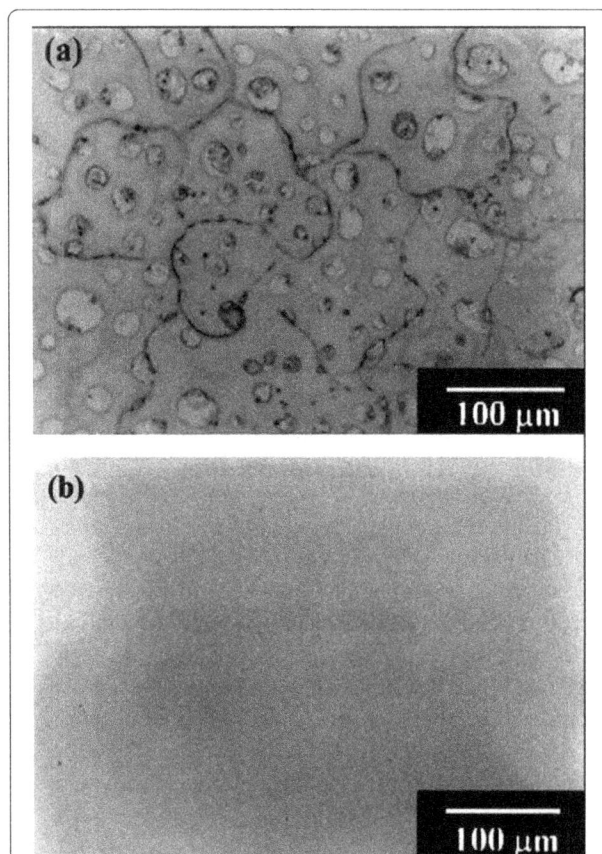

Figure 8 Optical micrographs of a calcium orthophosphate layer deposited on a Si wafer. (a) Without ion beam bombardments and (b) with Ar ion beam bombardments (120 V, 0.8 A). Reprinted from Lee et al. (2003) with permission.

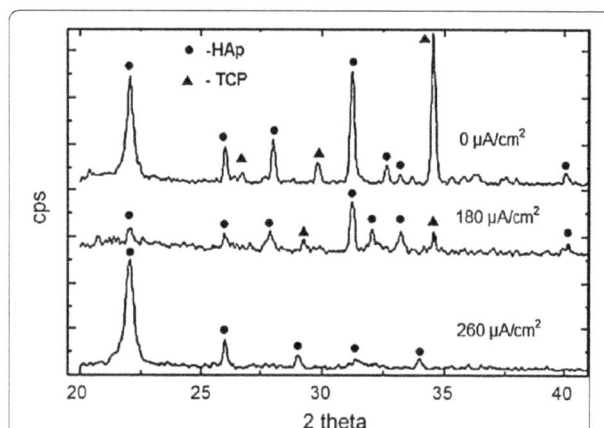

Figure 9 XRD patterns of fully crystallized (after a heat-treatment at 1200°C) calcium orthophosphate coatings. Sputtered at three different values of ion beam current density. Reprinted from Hamdi and Ide-Ektessabi (2003) with permission. HAp, hydroxyapatite; TCP, tricalcium phosphate.

titanium surface prior to deposition (Ong et al. 1991). A similar zone structure was also discovered by other researchers (Wang et al. 2001). In addition, a cross-section of functionally graded thin HA coatings on silicon substrate obtained by a dual ion beam assisted deposition and simultaneous heat treatment was investigated and the microstructural analysis of the coatings revealed a gradual decrease of the grain size and crystallinity towards the surface, leading to nano-scale grains and eventually amorphous layer at the surface (Rabiei et al. 2006).

Choi et al. (2000) deposited HA films on Ti-6Al-4 V alloy by electron beam vaporization of pure HA target and simultaneous bombardment using a focused Ar ion beam on the metal substrate to assist deposition. The effect of Ar ion beam current on the bond strength and dissolution of the coating in a physiological solution was studied. The bond strength between the coating and the substrate increased with increasing current, whereas the dissolution rate in physiological solution decreased remarkably (Choi et al. 2000). Further details on this technique are available in the aforementioned references.

Pulsed laser deposition or laser ablation deposition
Shortly after the discovery of a laser in the end of 1950s (Amy and Storb 1955), researchers began focusing their beams at materials to observe the interaction. PLD or laser ablation deposition technique for producing thin films became increasingly popular in 1970s due to the advent of lasers delivering nanosecond pulses (Singh and Narayan 1990). In this technique, a high power pulsed laser beam is focused inside a vacuum chamber to strike a target of the material (in our case, calcium orthophosphates) resulting in a gaseous cloud of various atoms, ions, molecules, molecular clusters and, in some cases, droplets and target fragments, due to a thermal decomposition of the target (Dinda et al. 2009). For sufficiently high laser energy density, each laser pulse vaporizes or ablates a small amount of the material, creating a plasma plume. The ablated material is ejected from the target in a highly forward-directed plume. The ablation plume provides the material flux, which then is deposited on a substrate. This process can occur in both ultra-high vacuum and presence of a background gas, such as oxygen, which is commonly used when depositing oxides to fully oxygenate the deposits. Argon (Bao et al. 2006) and water vapor (Fernandez-Pradas et al. 2000) can be used as well. The thorough investigation of a plasma plume expansion process during an ArF laser ablation of HA is well described elsewhere (Jedynski et al. 2008). The experimental setup of a PLD technique is available in literature (Narayanan et al. 2010; Paital and Dahotre 2009a; Surmenev 2012); it essentially consists of a laser source, an ultrahigh vacuum deposition chamber equipped with a rotating target and a fixed substrate

holder plus pumping systems. Mostly, the substrates are attached to the surface parallel to the target surface at a target-to-substrate distance of 2 to 10 cm. Usually, for ablation, ultraviolet excimer lasers with pulses of approximately 10-ns duration and power densities in the order of 10 to 500 MW/cm^2 are required. A two-laser beam technique (so-called, laser-assisted laser ablation method) is used as well (Katayama et al. 2009). In this technique, one laser beam from KrF laser, the ablation laser, is used for ablation of a HA target. The other beam from ArF laser, the assist laser, is used to irradiate a Ti substrate surface during formation of the HA coating. The assist laser plays an important role in the formation of a crystalline HA coating and improves the strength of adhesion to the Ti substrate (Katayama et al. 2009). Further details on the PLD technique might be found in literature (Willmott and Huber 2000).

A PLD process is used for forming thin (0.05 to 5 μm) calcium orthophosphate coatings, layers or films on various substrates (Nelea et al. 2000, 2002, 2004; Cotell et al. 1992, 1993; Torrisi and Setola 1993; Singh et al. 1994; Wang et al. 1997; Hontsu et al. 1997; Fernández-Pradas et al. 1998, 1999, 2001; Mayor et al. 1998; Arias et al. 1998; Craciun et al. 1999; Fernandez-Pradas et al. 1999; Zeng et al. 2000; Cleries et al. 2000a; Zeng and Lacefield 2000; Socol et al. 2004; Kim et al. 2005a, 2007a; Bigi et al. 2005b; Koch et al. 2007; Paital and Dahotre 2008; Paital et al. 2009; Dinda et al. 2009; Tri and Chua 2009; Sygnatowicz and Tiwari 2009). The process involves ablation of a calcium orthophosphate (usually, HA) target using a pulsed (usually, pulses of 30 ns and 120 mJ at a repetition of 10 Hz) KrF excimer laser beam ($\lambda = 248$ nm) in 0.3 Torr/H_2O atmosphere and deposition of the ejected HA material on a heated (400°C to 800°C) substrate. The deposition rate of PLD is about 0.02 to 0.05 nm per laser shot (de Groot et al. 1998). An investigation into the effects of high laser fluence (between 2.4 J/cm^2 and 29.2 J/cm^2) on the properties of calcium orthophosphate films was performed (Tri and Chua 2009). The films deposited at 2.4 J/cm^2 were found to be partially amorphous and had rough surfaces with many droplets, while higher laser fluences showed a higher level of crytallinity and lower surface roughness. Furthermore, higher laser fluences also decreased the ratio Ca/P of as-deposited films and, probably, increased their density (Tri and Chua 2009). The substrate heating is necessary to ensure the formation of a highly crystalline and phase pure coatings, films and layers. Besides, the substrate temperature could be varied to provide deposits with the desired fine texture and roughness, depending on their application (Saju et al. 2009; Rau et al. 2010).

Typically, during the deposition, a target should be rotated to achieve a stable ablation rate. As PLD is usually carried out at high substrate temperatures, a thin oxide layer might be formed on the substrate surface prior to the deposition of calcium orthophosphates and, thereby, it influences its adherence to the substrate (Nelea et al. 2000). The deposited coatings, layers and films frequently consist of several calcium orthophosphates (often with admixtures of other substances, such as CaO, calcium pyrophosphates, etc.) and might contain both amorphous and crystalline phases (Cleries et al. 2000a; Koch et al. 1990). Biphasic formulations, such as HA + TTCP (Kim et al. 2005b), might be deposited as well. Interestingly, TTCP in the coatings was not formed by partial conversion of previously deposited HA. Instead, it was produced by nucleation and growth of TTCP itself from the ablation products of the HA target or by accretion of TTCP grains formed during ablation (Kim et al. 2005b). Furthermore, the PLD-deposited coatings, films and layers might consist of calcium orthophosphates with different morphologies (e.g., granular and columnar), which have different resistance values to delamination (Cleries et al. 2000a). More to the point, various types of oriented textures might be created as well (Kim et al. 2005a, 2007a, 2010). A modification known as 'transmission laser coating' has been introduced (Cheng and Ye 2010).

Further details on the PLD technique might be found in literature (Surmenev 2012; Koch et al. 2007).

Magnetron sputtering A magnetron is a high-powered vacuum tube that generates microwaves using the interactions of a stream of electrons with a magnetic field. Magnetron sputtering technique has been emerged in the mid of 1960s (Gill and Kay 1965) and is considered as a high-rate vacuum coating technique for depositing metals, alloys and compounds onto a wide range of materials with thickness up to approximately 5 μm (Table 3). A sputtering system consists of an evacuated chamber, a wave generator, a magnetron, a cooling system, as well as it contains a target and a substrate. It works on the principle of applying a specially shaped magnetic field to a sputtering target. Once the substrate is placed into the vacuum chamber, air is removed, and the target material (in our case, calcium orthophosphates) is released into the chamber in the form of a gas. Powerful magnets ionize particles of the target material. Then, the negatively charged target material lines up on the substrate to form deposits (Kelly and Arnell 2000). In principle, magnetron sputtering can be done in either DC (direct current) or RF modes; however, since the DC mode might be done with conducting materials only, the RF mode is solely used to deposit calcium orthophosphates. Typically, RF magnetron sputtering employs a sinusoidal wave generator operating at 13.56, 5.28 or 1.78 MHz. The parameters that directly affect the quality and integrity of calcium orthophosphate coatings, films and layers include discharge power, gas

flow rate, working pressure, substrate temperature, deposition time, post-heat treatment and negative substrate bias (Surmenev 2012). For example, the deposition rate was found to increase with increasing argon gas pressure up to 2 Pa but decreased significantly as the pressure increased up to 5 Pa, while the Ca/P ratios of as-deposited coatings decreased significantly at the higher argon gas pressures (Boyd et al. 2007). A good schematic setup of a magnetron sputtering system is available in literature (Narayanan et al. 2010; Paital and Dahotre 2009a; Surmenev 2012).

For the first time, RF magnetron sputtering was used to prepare calcium orthophosphate coatings in 1992 (Cooley et al. 1992). Since then, it has become a convenient method for deposition of biocompatible ceramic coatings, layers and films on various substrates (Yamashita et al. 1994; Jansen et al. 1993; Wolke et al. 1994, 1998, 2003; van Dijk et al. 1995, 1996; Nelea et al. 2003, 2004; Feddes et al. 2003a, 2003b; Ding 2003; Yamaguchi et al. 2006; Wan et al. 2007; Ozeki et al. 2007; Ueda et al. 2007; Snyders et al. 2008; Ievlev et al. 2008; Toque et al. 2009). The advantages of magnetron sputtering over other sputtering processes include a high deposition rate, an excellent adhesiveness and an ability to coat implants with difficult surface geometries (Table 3). Still, several issues, such as the endurance and the Ca/P ratio, have to be solved before magnetron sputtering can be applied to deposit, on a routine basis, pure and crystalline calcium orthophosphates on implant surfaces. For example, both microstructure and mechanical properties of HA thin films, grown on Ti-5Al-2.5Fe alloys by RF magnetron sputtering, were investigated (Nelea et al. 2003). The deposition was performed from pure HA target in low pressure Ar or Ar-O_2 mixtures at substrate temperatures ranging from 70°C to 550°C. Smooth (an average roughness of approximately 50 nm) and uniform calcium orthophosphate films were fabricated. It was observed that the films grown at the substrate temperatures below approximately 300°C were prevalently amorphous (ACP) and contained a small amount of crystalline phases. On the contrary, the films obtained at a substrate temperature of 550°C or the films grown at room temperature followed by annealing at 550°C consisted of HA (Nelea et al. 2003).

The chemical composition of the deposited calcium orthophosphate coatings, films and layers might be modified by varying the RF sputtering power density (Snyders et al. 2008), namely, when the power density was increased by 240%, the Ca/P ratio increased from approximately 1.51 to approximately 1.82. X-ray diffraction indicated the phase pure HA except for the samples prepared at the highest power density values, in which the presence of CaO and TCP was also detected. Interestingly, deviations from the stoichiometric HA resulted in reduction of the elastic modulus, namely, for Ca/P approximately 1.51, the elastic modulus dropped by approximately 15%, which was attributed to Ca vacancies in the lattice, while for Ca/P approximately 1.82, the average elastic modulus decreases by approximately 10% due to formation of additional phases (Snyders et al. 2008).

Various types of calcium phosphates were magnetron sputtered from TTCP, HA, β-TCP, β-calcium pyrophosphate (CPP) and β-calcium metaphosphate (CMP) powder targets (Ozeki et al. 2007). The composition of the deposited films was changed depending on the target materials, while the Ca/P molar ratios of the films varied from 0.74 to 2.54, increasing with the Ca/P molar ratio of the target. Interestingly, the deposition rate of the aforementioned calcium phosphates was established as the following: $TTCP \approx \beta\text{-}CMP > \beta\text{-}TCP > \beta\text{-}CPP \approx HA$, which correlated well to the solubility order: $TTCP \approx \beta\text{-}CMP > \beta\text{-}TCP > \beta\text{-}CPP \approx HA$ (Ozeki et al. 2007).

RF magnetron sputtering might be combined with other deposition techniques. For example, plasma-assisted RF magnetron co-sputtering deposition method was used to deposit calcium orthophosphates on Ti6Al4V orthopedic alloy (Xu et al. 2005; Long et al. 2007). Further details on the magnetron sputtering technique are available in excellent reviews (Surmenev 2012; Shi et al. 2008).

Other deposition techniques: miscellaneous
Prior describing the below mentioned deposition techniques, one should note that they are rare and are mentioned in just a few research papers. Therefore, the detailed description is not always possible.

Hot isostatic pressing Hot isostatic pressing (HIP) is a manufacturing process used to reduce porosity and increase the density of many types of materials. The HIP process subjects a component to both elevated temperature and isostatic gas pressure in a high-pressure containment vessel. To deposit coatings, initially, solid cores are covered by a calcium orthophosphate (usually HA) powder. Both organic binders and some other additives are used to simplify deposition. A furnace is constructed within the high-pressure vessel, and the coated samples are placed inside to be pressed. Then, the specimens are heated at temperatures within 700°C to 1,200°C and pressed at pressures within 20 to 100 MPa. The obtained coatings, films and layers are usually thick (0.2 to 2.0 mm) and dense (Bocanegra-Bernal 2004).

The HIP technique was used to manufacture calcium orthophosphate coatings, films and layers on various materials (Lacefield 1988; Herø et al. 1994; Wie et al. 1998; Kameyama 1999). For example, HA granules (32 to 38 μm in diameter) were implanted into a substrate of superplastic titanium alloy. First, the HA granules were spread over this surface and, then, hot pressed at 750°C and 17 MPa for 1 h with a plunger to implant them into the substrate. After 10 min of implantation, the implantation ratio was approximately 20%, and some granules were not on the substrate. After 60 min of implantation, the implantation ratio was 100%, but the upper areas of granules were exposed (Kameyama 1999).

A variation in the HIP technique was proposed in which thin HA coatings were prepared with a curved surface at low temperatures (Onoki and Hashida 2006). The method used double-layered capsules in order to create suitable hydrothermal conditions; the inner capsule encapsulated the coating materials and a Ti substrate, while the outer capsule was subjected to isostatic pressing under the hydrothermal conditions. It was demonstrated that a HA layer of approximately 50 μm thick could be deposited on Ti cylindrical rods at 135°C under the confining pressure of 40 MPa. The deposited HA layer had a porous microstructure with the relative density of approximately 60%. According to the results of pullout tests, the shear strength was in the range of 4.0 to 5.5 MPa. These results also revealed that a crack propagation occurred within the HA coating layer but not along the HA/Ti interface. This observation suggests that the fracture property of the HA/Ti interface was higher than that of the HA ceramics only. Thus, hydrothermal HIP technique appears to be a useful method for producing bioactive HA ceramic coatings on curved prostheses surfaces (Onoki and Hashida 2006). However, the majority of the calcium orthophosphate coatings, layers or films produced by HIP technique are contaminated with metal and SiO_2 particles due to the use of a glass encapsulating tubes (de Groot et al. 1998).

Frit enameling Frit is a ceramic composition that was fused in a special fusing oven, quenched to form a glass and granulated. According to this technique, a metal bar was dipped into a HA slurry, dried and sintered at 1,100°C to 1,200°C for a minimum of 3 h in a protective Ar atmosphere (Lacefield 1988). Coatings created in this technique show very low interfacial shear strength (approximately 0.22 MPa). Probably, the furnace atmosphere used for sintering was inadequate, which resulted in excessive substrate oxide layer thickness and poor bonding (de Groot et al. 1998).

Aerosol-gel An aerosol is a colloid suspension of fine solid particles or liquid droplets in a gas. Examples are clouds and air pollution, such as smog and smoke. While gels are regarded as composites because they consist of a solid skeleton or network that encloses a liquid phase or an excess of the solvent. Therefore, the aerosol-gel process, as the name implies, is a gas-chemical technique that involves transition from a gaseous 'aerosol' into a solid 'gel' phase.

The aerosol-gel technique was also applied to produce highly porous calcium orthophosphate coatings on various materials (Manso et al. 2002a; 2002b;

Manso-Silván et al. 2003). Calcium nitrate and triethylphosphate diluted in ethanol were used as precursor solutions. After production of a steady state aerosol, the microdroplets were conducted into a deposition chamber by an air flux. After being deposited, the coatings were sintered at temperatures within 500°C to 1,000°C. The composition, structure and morphology of the final coatings were found to fit highly porous polycrystalline HA. The adhesive strength, measured by means of indentation techniques, was found to be in the order of 100 MPa, which was significantly higher than the values obtained by sol-gel deposition technique (Manso et al. 2002a; 2002b; Manso-Silván et al. 2003).

Micro-arc oxidation Micro-arc oxidation (MAO), also called plasma electrolytic oxidation, anodic spark deposition, or micro-arc discharge oxidation, is a plasma-chemical and electrochemical process. The process combines electrochemical oxidation with a high-voltage spark treatment in an aqueous electrolytic bath, which also contains modifying elements in the form of dissolved salts (*e.g.*, silicates) to be incorporated into the resulting coatings. A schematic setup of a MAO system is available in literature (Narayanan et al. 2010; Paital and Dahotre 2009a).

By means of a MAO technique, calcium orthophosphate coatings, films and layers have been prepared on various metals (Song et al. 2004; Liu et al. 2005; Sun et al. 2007; Wei et al. 2007; Han et al. 2008). For example, MAO was performed on titanium in an electrolyte containing calcium glycerophosphate and calcium acetate using a direct current power supply. The MAO technique appeared to be suitable to form porous and rough ceramic coatings containing Ca and P. Then, the coatings were hydrothermally treated in aqueous solutions with pH within 7.0 to 11.0 (adjusted by adding NaOH) at 190°C for 10 h in an autoclave. This procedure converted undisclosed calcium orthophosphates into CDHA and/or HA crystals, while the amount of precipitated HA increased with solution pH increasing (Liu et al. 2005).

Direct laser melting According to the direct laser melting technique, a starting precursor (calcium orthophosphate) powder is mixed thoroughly in a water-based organic solvent. Then, the suspension is sprayed onto the substrate surfaces to create a coating, which is then air dried to remove the moisture, followed by direct laser melting using either continuous wave or pulsed laser beams to produce strong bonds between the calcium orthophosphate coating and the substrate (Paital and Dahotre 2007, 2008, 2009a, 2009b; Kurella and Dahotre 2006. A schematic setup of a direct laser melting system is available in literature (Paital and Dahotre 2009a).

Laser cladding Calcium orthophosphate coatings, films and layers were synthesized on various substrates by laser cladding (Lusquiños et al. 2001, 2003, 2005). Cheap precursors, such as mixed powders of calcium carbonate and DCPA/DCPD, can be used to prepare coatings (Wang et al. 2008b; Zheng et al. 2010; Lü et al. 2011a, 2011b; Lv et al. 2012). The reactions between $CaCO_3$ and DCPA/DCPD can produce high crystallized HA in the coatings, as well as TTCP, α-TCP, β-TCP, $Ca_2P_2O_7$ and CaO, if the Ca/P ratio is deviated from 1.67. Since the reactions between the powders produce gaseous by-products (CO_2 and water vapor), the prepared coatings were porous. Furthermore, when deposition was performed on metals (e.g., Ti), other admixtures, such as $CaTiO_3$, were formed. As the laser power increased, the amount of TTCP, HA and CaO in the coatings decreased gradually and, finally, only α-TCP and $CaTiO_3$ remained. Nevertheless, the amount of HA could be increased greatly by heat treatment at 800°C for 5 h followed by furnace cooling, due to the total transformation of TTCP and α-TCP to HA (Wang et al. 2008b; Zheng et al. 2010; Lü et al. 2011a, 2011b; Lv et al. 2012).

Detonation gun spraying Detonation gun spraying is a high temperature and a high velocity technique which is thought to introduce a higher degree of melting to starting powder. This technique has also been explored to prepare calcium orthophosphate coatings on titanium alloys (Gledhill et al. 1999, 2001). In this process, a mixture of oxygen and acetylene is fed into a barrel together with a charge of the power. The gas is ignited, and detonation waves accelerate the powder up to about 750 m/s. The process produces a denser coating, which has a higher proportion of the amorphous phase with some evidence for the appearance of β-TCP. A lower crystallinity and higher residual stress found in the detonation gun sprayed coatings resulted in a faster dissolution rate both *in vitro* and *in vivo* (Gledhill et al. 1999, 2001).

Cold spraying technique In recent years, a new coating technology, known as cold spraying, has been developed (Gärtner et al. 2006). In this process, spraying particles (1 to 50 μm in size) experiencing both a little change in microstructure and a little oxidation or decomposition are accelerated by a supersonic jet of a compressed gas stream passing through a Laval type nozzle to a very high velocity (300 to 1,200 m/s). The deposition system consists of a gas pressure regulator, a gas heater, a powder feeder and a spray gun. In this technique, the deposited calcium orthophosphate particles are always in a solid state and at temperatures below their melting point. Thus, all phenomena occurring at high temperatures, such as thermal decomposition and phase transformations (see section Thermal spraying techniques above), are avoided. Deposition of calcium orthophosphate particles takes place through intensive plastic deformations (Zhang and Zhang 2011; Zhang et al. 2012a, 2012b). However, for successful bonding, the deposited particles have to exceed a critical velocity on impacts, which is dependent on the properties of the particular sprayed material (Gärtner et al. 2006).

Thermal substrate deposition Thermal substrate deposition technique is based on the solubility differences at low and high temperatures, namely by heating a substrate in suitable saturated aqueous solutions, coatings, films and layers can be directly deposited onto the substrate. Various heating techniques of substrates have been proposed, namely conductive substrates, such as foil or wire, can be heated by electric current through them. Non-contact techniques, such as high frequency induction, can be used to heat materials with complex shapes. In either case, the immersed metallic sample can be heated up to 160°C in solutions, giving local supersaturations to perform crystallization.

Using this approach, calcium orthophosphate coatings, films and layers were obtained on titanium (Ziani-Cherif et al. 2002; Okido et al. 2002; Kuroda et al. 2002a, 2002b, 2003, 2005). An alternating current is passed through the metallic sample immersed in an aqueous solution containing calcium and phosphorus compounds. The deposition is usually performed for 10 to 30 min at solution pH 4 to 8. The type of precipitates varies, depending on the solution pH, temperature and ion concentrations, namely, precipitates of high quality, whose predominant component was HA (at pH > 6) or DCPA (at pH = 4), were obtained on titanium substrates by this technique. The content of HA in the deposits was found to increase with increasing temperature and heating time (Ziani-Cherif et al. 2002; Okido et al. 2002; Kuroda et al. 2002a, 2002b, 2003, 2005).

Matrix-assisted pulsed laser evaporation Matrix assisted pulsed laser evaporation technique was developed as an alternative to PLD for delicate and accurate deposition of calcium orthophosphate films, coatings and layers combined with organic and/or biologic materials. The examples include deposition calcium orthophosphate-based biocomposites with sodium maleate (Negroiu et al. 2008), alendronate (Bigi et al. 2009) and silk fibroin (Miroiu et al. 2010). Thermally unstable calcium orthophosphates, such as OCP (Boanini et al. 2012), might be deposited as well. This technique provides a more gentle mechanism for transferring different compounds, including large molecular weight species, and it is expected to ensure an improved stoichiometric transfer, a more accurate thickness control and a higher uniformity of the coatings.

Electrostatic spray deposition Electrostatic spray deposition (ESD) is based on generation of an aerosol composed of organic solvents containing inorganic precursors under the influence of high voltages. According to this technique, spray droplets are generated by pumping a solution through a nozzle. Between the nozzle and substrate, a high voltage is applied. Consequently, droplets coming out the nozzle are dispersed into a spray, and this spray is deposited upon the substrate. When the solvent has evaporated, a coating is formed. The schematic setup of the ESD technique is available in literature (Leeuwenburgh et al. 2003, 2004, 2005a).

To perform ESD, a soluble calcium salt (nitrate or chloride) and phosphoric acid were dissolved in an alcohol. The obtained solutions were pumped, quickly mixed prior the nozzle and electrostatically sprayed onto a substrate, while the substrate itself might be heated to 300°C to 450°C (Leeuwenburgh et al. 2003, 2004, 2005a, 2005b, 2006a, 2006b). Besides, calcium orthophosphate powders might be suspended in alcohols, and the obtained suspensions are electrostatically sprayed (Lee et al. 2007b; Jiang et al. 2008; Iafisco et al. 2012). The chemical and morphological characteristics of the deposited calcium orthophosphate coatings, films and layers were found to be strongly dependent on both the composition of the precursor solutions (pH, absolute and relative precursor concentrations) and the deposition parameters, such as temperature, the nozzle-to-substrate distance, the liquid flow rate, as well as the geometry of the spraying nozzle. By varying these parameters, several phases and phase mixtures might be deposited by ESD technique: carbonate apatite, carbonated HA, α-TCP, β-TCP, DCPA, β- and γ-calcium pyrophosphates, calcium metaphosphate, $CaCO_3$, CaO (Leeuwenburgh et al. 2005b, 2006a, 2006b). Since ESD might be performed at ambient temperatures, thermally unstable compounds could be deposited. As seen in Figure 10, the electrostatically sprayed calcium orthophosphate coatings, layers and films might be porous (Leeuwenburgh et al. 2006a, 2006b; Lee et al. 2007b; Jiang et al. 2008; Iafisco et al. 2012; Zhu et al. 2012). Nevertheless, after the deposition, the coated samples might be annealed at high temperatures. The annealing stage is necessary to aggregate and/or melt the deposited calcium orthophosphate particles and form highly dense and homogeneous coatings.

To conclude this part, combined techniques, such as sol-gel-assisted electrostatic spray deposition (Kim et al. 2007b) and electrostatic spray-assisted vapor deposition (Hou et al. 2007), have been developed as well. Further details on the ESD techniques are available in the aforementioned references.

Spin coating Spin coating is a procedure used to apply uniform thin films to flat substrates. It is rather similar

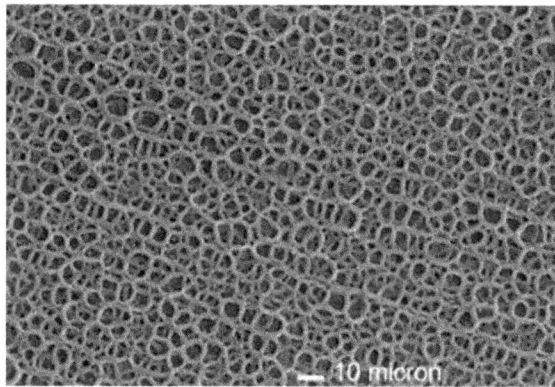

Figure 10 A scanning electron microscopy of an electrostatic spray deposited calcium orthophosphate coating. Characterized by a porous surface morphology. Reprinted from Leeuwenburgh et al. (2006a) with permission.

to dip coating. The coating process consists of four stages: deposition, spin up, spin off and evaporation. In this process, a sample is dipped in a solution or suspension and then withdrawn at a constant speed, usually with the help of a motor. Rotational draining and solvent evaporation result in the deposition of a coating, film or layer. A machine used for spin coating is called a spin coater, or simply spinner (Mennicke and Salditt 2002). Just a few publications on spin coating of calcium orthophosphates were published (You et al. 2005; Yuan et al. 2009; Carradò and Viart 2010).

Properties

Generally, for clinical applications, slowly or non-resorbable high crystalline coatings, films and layers have been recommended in order to retain the bonding strength with implants. However, this contradicts to the statement that the ideal interface between the implants and surrounding tissues should match the tissues being replaced. For example, in the case of HA, its crystallinity has been stated to be in the inverse proportion to its bioactivity (LeGeros 2008). Therefore, from the bioactivity point of view, calcium orthophosphate coatings, films and layers should be of low crystallinity and also contain various ionic substitutions, such as sodium, magnesium and carbonate. Thus, since one of the first steps in bonding involves dissolution of the coating surface, it might be suggested that coatings, films and layers prepared from less crystalline and/or more resorbable calcium orthophosphates would be more beneficial for early bone ingrowth than those prepared from high crystalline HA (Narayanan et al. 2010). However, soluble coatings, films and layers will weaken the bonding strength between them and substrates. In particular, a rapid dissolution of coatings, films and layers may loosen the bonding strength between the implant surface and the host bone. For example, a

comparative study on the biological stability and osteo-conductivity of HA coatings on Ti produced by pulsed laser deposition and plasma spraying was conducted. After 24 weeks of implantation, the plasma sprayed HA coatings showed considerable instability and reduction in thickness but no statistical difference to the uncoated Ti (the control), while the pulsed laser deposited ones remained almost intact but showed a significantly higher amount of bone apposition (Peraire et al. 2006). Thus, in that study the coating stability prevailed over its solubility. Furthermore, the excessive amount of the dissolved ions from the soluble coatings, films and layers may cause local inflammatory reactions.

Except for crystallinity and chemical composition, a number of other factors appear to influence the physical, chemical and mechanical properties of calcium orthophosphate coatings, films and layers. They include thickness (this will influence adhesion and fixation - the agreed optimum now seems to be within 50 to 100 μm), phase and chemical purity, fatigue resistance, porosity and adhesion (de Groot et al. 1998; Sun et al. 2001). Abrasion resistance might be important as well (Morks et al. 2007).

Fatigue properties

Several studies have already demonstrated that cyclic loading of the coated samples leads to fatigue failure. Further, it has been shown that a combination of an aqueous environment with stress can result in delamination or accelerated dissolution of calcium orthophosphate coatings, which can influence the long-term stability of the implants (Kummer and Jaffe 1992; Reis et al. 1994; Wolke et al. 1997). For example, calcium orthophosphate coatings were RF magnetron sputtered on Ti-6Al-4 V bars and, afterwards, some of them were annealed at 650°C to convert ACP into crystalline structure. Then, the coated samples were mechanically tested in either dry or wet (SBF solution) conditions (Wolke et al. 1997). The results of SEM demonstrated that, after cyclic loading conditions in air, the bars coated by crystalline calcium orthophosphates showed a partial coating loss. Furthermore, in wet conditions only the heat-treated sputter-coated bars appeared to be stable. On the other hand, the ACP coatings showed signs of delamination in more stressed regions only (Wolke et al. 1997). Thus, the fatigue properties of amorphous and crystalline calcium orthophosphate coatings, films and layers are different. Furthermore, the fatigue behavior shows substantial differences when tested in either dry or wet/conditions.

Thickness

Depending on the deposition technique, the thickness of the calcium orthophosphate coatings, films and layers varies from nanometric dimensions to several millimeters (Table 3), and this parameter appears to be very important, namely, if calcium orthophosphate coatings, films and layers are too thick, they are easy to break. Furthermore, the outer layers might tend to detach from the inner ones. On the contrary, if calcium orthophosphate coatings, films and layers are too thin, they are easy to dissolve because resorbability of HA, which is the second least soluble among calcium orthophosphates (Table 1), is about 15 to 30 μm per year under the physiological conditions (Gineste et al. 1999). To complicate the situation, the failure mechanisms for thinner and thicker coatings, films and layers appear to be different, namely the failure mode of thinner (50 μm) HA coatings on a Ti alloy was found to be conclusively at or near the coating/bone interface, while that of thicker (200 μm) HA coatings was found to be at the coating/bone interface, inside the HA lamellar splat layer, as well as at the coating/Ti alloy substrate interface (Wang et al. 1993; Yang et al. 1997). A similar conclusion was made in another study, in which the mechanical behavior of thin (0.1, 1 and 4 μm) calcium orthophosphate coatings was compared (Vercaigne et al. 2000a). Considering these points, commercial plasma-sprayed HA coatings, films and layers have thicknesses between 50 and 200 μm (Sun et al. 2001), though cells and tissues interact with only top surface, and thus, thickness of approximately 10 nm would be sufficient for cell activity.

Calcium orthophosphate coatings of various thickness ranging from 170 nm up to 1.5 μm were obtained depending upon the deposition times (Fernandez-Pradas et al. 2001). The coating morphology was found to be grain-like particles and droplets. During growth, the grain-like particles grew in size, partially masking the droplets, and a columnar structure was developed. The thinnest (170 nm) coating consisted mainly of ACP. The coating of approximately 350-nm thick also contained HA, whereas even thicker coatings contained some α-TCP in addition to HA. All coatings failed under the scratch test by spalling from the diamond tip; however, the failure load increased as thickness decreased until only plastic deformation and cohesive failure for the thinnest coating was observed (Fernandez-Pradas et al. 2001). Therefore, both the structure and the phase composition of calcium orthophosphate coatings, films and layers might depend on their thickness.

Adhesion

In surgical practice, failure of implants and undesirable tissue responses take place when decohesion of coatings occurs. Therefore, all types of coatings, films and layers must adhere satisfactorily to the underlying substrate irrespective of their intended functions. Generally, the bottom surfaces of the coatings, films and layers are not in the full contact with the substrates. The areas that are in contact are called 'welding points' or 'active zones'.

Voids of various shapes and dimensions are located among them. In general, the greater the contact area, the better adhesion of the coating is (Pawlowski 2008). Since the chemical interactions between deposited calcium orthophosphates and substrates are rare, mechanical anchorage is the main mechanism involved in adhesion of calcium orthophosphate coatings, films and layers, in which the substrate surface roughness is the paramount parameter to achieve good adhesion. In many cases of ceramic coatings, the adhesion strength is found to be a linear function of the average surface roughness. Therefore, substrate preparation techniques, such as grit blasting, are used to increase roughness prior to spraying and, hence, increase the adhesion strength (see section 'Brief knowledge on the important pre- and postdeposition procedures'). On the other hand, the amount of mechanical anchorage is reduced if a large amount of shrinkage occurs during solidification of the particles (Heimann 2006). Since the strength of human bones is approximately 18 MPa, all types of coatings, films and layers on the implant surface should have higher or, at least, comparable bond strength. Thus, according to the ISO requirements, the adhesion strength of calcium orthophosphate coatings, films and layers should not be less than 15 MPa (ISO 2000; ISO 2008).

Specifically, the adhesion forces of calcium orthophosphate coatings, films and layers should be high enough to maintain their bioactivity after a surgical implantation. Generally, tensile adhesion testing according to standards ASTM C633 (ASTM C633 2008) and ASTM F-1147-05 (2011) is the most common procedure to determine the quantitative adhesion values to the underlying substrates. Furthermore, fatigue (Surmenev 2012; Mukherjee et al. 2000), scratch (Cheng et al. 2009; Hamdi et al. 2010) and pullout (Cheng et al. 2009) testing, as well as wear resistance (Hamdi et al. 2010), are among the most valuable techniques to provide additional information on the mechanical behavior of calcium orthophosphate coatings, films and layers. Changes in the surface topography can give an indication of wear resistance. For example, coatings with good adherence to the substrate have shown less alteration of its surface roughness, while the study on the different parameters revealed that deposition time was the most influential factor in the wear behavior (Hamdi et al. 2010). The latter was attributed to its correlation with coating thickness. The scratch test is performed with reference to ISO 20502:2005 (ISO 20502 2005). The load at which complete removal of the coating occurs is usually taken as an indication of the adhesion strength. Further details on the mechanical testing methods of calcium orthophosphate coatings, films and layers might be found in literature (Ben-Nissan et al. 2011).

The adhesion strength of calcium orthophosphate coatings, films and layers depends on very many parameters. In the first instance, it strongly depends on the deposition technique. For example, HA coatings, obtained by PLD, showed greater adherence to a titanium alloy when compared with plasma-sprayed HA coatings (Vasanthan et al. 2008). Besides, it might depend on the coating thickness and its chemical composition, namely coatings of 50-μm thick gave higher values of the adhesion strength than those of 240-μm thick (Filiaggi et al. 1991), while scratch tests revealed that the sol-gel-fluorinated HA coating adhered to Ti-alloy substrate up to 35% better as the fluorine concentration increased in the coating (Zhang et al. 2006). Furthermore, the nature, structure and chemical composition of the substrate surface play an important role. For example, a highly roughened substrate surface exhibited higher bond strength as compared to a smooth substrate surface (Nimb et al. 1993). Besides, the adhesion strength of plasma-spayed coatings was found to decrease when either the plate power was reduced (from 28 to 22 kW) or the working distance was increased (from 90 to 130 mm) (Roy et al. 2011). Additionally, the bond strength of calcium orthophosphate coatings deposited on Ti plates pre-treated in an alkali solution followed by heat-treating (600°C for 1 h) in air had a higher value (approximately 35 MPa) if compared to those followed by heat-treating vacuum (approximately 21 MPa). This was attributed to the structural and compositional differences in the interfacial layer of sodium titanates (Wang et al. 2008c). For plasma-assisted deposition techniques of calcium orthophosphates, a good overview on the adhesion strength values of coatings, films and layers is presented in Table 3 of Surmenev (2012).

However, application of various inter-layers (synonym: buffer layers) seems to be the most important way to influence the adhesion strength of calcium orthophosphate coatings, films and layers to diverse substrates. A big number of the available deposition techniques (see 'Preparation' section above), which should be multiplied to a big selection of various substrates, result in a great number of potentially appropriate chemicals to be used as inter-layers between the substrates and calcium orthophosphates. For example, for plasma-assisted deposition methods of calcium orthophosphates, such chemicals as TiO_2 (Rajesh et al. 2011), TiN (Nelea et al. 2000; Yang et al. 2009b, 2009c; Man et al. 2009), ZrO_2 (Nelea et al. 2000) or Al_2O_3 (Nelea et al. 2000), were used as buffer layers. TiO_2 (Nelea et al. 2007; Berezhnaya et al. 2010) and TiN (Nelea et al. 2003) were also used as under-layers for RF magnetron sputtering. Similarly, formation of intermediate layers of titanium hydroxides is required for biomimetic deposition of calcium orthophosphates on Ti (Wang et al. 2008a). To complicate things even further, one should mention, that mutual inter-diffusion of atoms,

ions and molecules of calcium orthophosphates from coatings, films and layers from one side and those of a substrate from another side might occur. Especially, this is valid for high temperature deposition techniques; however, the mutual inter-diffusion might happen for any technique at the post-deposition annealing stage (see section 'Brief knowledge on the important pre- and post-deposition procedures'). Various atomic mixed inter-layers are formed as the result. For example, the width (measured by Auger electron spectroscopy) of such inter-layer between a HA coating and magnesium substrate formed by IBAD technique was found to be approximately 3 μm (Yang et al. 2008). Such inter-layers can reduce the mismatch of thermal expansion coefficients between calcium orthophosphates and substrates, or increase the surface area of the material, wettability or heat conductivity; thus, increasing the bonding strength without affecting biocompatibility. Since the subject of inter-layers appears to be very broad, additional details are not specified further.

The adhesion forces depend on various factors, namely for dense coatings under tensile loading, failure usually occurs at the coating/substrate interface because the cohesive strength is higher than the bond strength. For porous coatings, the cohesive strength is low and the fracture occurs inside them (Han et al. 2001). The amorphous coatings have a more brittle nature and less adhesion compared to the crystalline ones (Cleries et al. 2000a). In general, the bond strength of apatite layer to Ti metal substrate is reported to range from 10 to 30 MPa (Kokubo et al. 1996; Kim et al. 1997). Similar values were obtained in another study, where calcium orthophosphate coatings were deposited on Ti substrates by a biomimetic method from two types of SBF. The results indicated that both the ionic concentrations of the SBFs and the surface roughness of the substrates had a significant influence on formation, morphology and bond strength of calcium orthophosphate precipitates. The highest bond strength of the precipitated coatings was about 15.5 MPa (Chen et al. 2009b).

Biodegradation

Biodegradation (synonyms: biotic degradation or biotic decomposition) is a chemical dissolution of materials by bacteria and/or other biological means. Since chemical composition of the body fluids might be considered as constancy, biodegradation of calcium orthophosphate coatings, films and layers is controlled by the properties of calcium orthophosphates themselves, which include their chemical composition, Ca/P ratio, crystal structure, crystallinity, porosity, lattice defects, particle sizes and purity (de Groot et al. 1998). For example, the dissolution kinetics of HA layers was studied using the dual constant composition method, and dissolution rates decreased when HA crystallinity increased (Tucker et al. 1996).

Similar results were obtained in another study: after implantation, HA coatings with crystallinity of approximately 55% were found to degrade faster and possess better osteoinductivity than those with crystallinity of approximately 98% (Xue et al. 2005). Although biodegradation supposed to be *in vivo* process, various *in vitro* simulations are widely investigated. However, to be closer to the *in vivo* conditions, the biological assessments of calcium orthophosphate coatings, films and layers are performed in various simulating solutions, such as SBF (Kim et al. 2010; Chen et al. 2009b; Verestiuc et al. 2004; van der Wal et al. 2005, 2006; Heimann 2009; Łatka et al. 2010; d'Haese et al. 2010; Ntsoane et al. 2011), HBSS (Ueda et al. 2007; Man et al. 2009; Luo et al. 2000), aqueous saline solution (Surmenev et al. 2010; Ueda et al. 2009), Ringer's solution (Gross and Berndt 1994; Gross et al. 1997), phosphate buffered saline (PBS) (Ueda et al. 2007; Boyd et al. 2006; Coelho et al. 2009a), Eagle's minimum essential medium (Lim et al. 2005). Since the simulating solutions often contain dissolved ions of calcium and orthophosphates, both partial dissolution of calcium orthophosphate coatings, films and layers and their re-crystallization occurred (Verestiuc et al. 2004; Heimann 2009; d'Haese et al. 2010; Ntsoane et al. 2011; Lim et al. 2005). For example, as written in the abstract of d'Haese et al. (2010), 'The soaking in SBF homogenizes the morphology of coatings. The sintered zone disappears, and the pores get filled by the reprecipitated calcium phosphates.' One should stress that, due to the presence of other ions in the chemical composition of the aforementioned simulating solutions, in the vast majority of the cases not chemically pure but ion-substituted calcium orthophosphates are precipitated.

Usually, the biodegradation kinetics of calcium orthophosphate coatings, films and layers appears to be proportional to the solubility values of the individual ingredients, listed in Table 1. For example, both bone bonding and bone formation of HA, α-TCP and TTCP plasma-sprayed coatings were evaluated by mechanical push-out tests and histological observations after 3, 5, 15 and 28 months of implantation. Among them, α-TCP (which was the most soluble phase) showed the most significant degradation after approximately 3 months of implantation, while HA and TTCP showed significant signs of degradation only after approximately 5 months of implantation (Klein et al. 1994a). This resulted in lesser values of the mechanical push-out tests for α-TCP-coated implants if compared with those coated by HA and TTCP (Klein et al. 1991). Plasma-spray deposited coatings of HA were found to dissolve faster than the stoichiometric HA did because a high temperature melted HA powder and partly decomposed it into more soluble compounds, such as high-temperature ACP and OA (Pezeshki et al. 2010). Similarly, as-deposited magnetron spattered calcium orthophosphate coatings were almost

amorphous (i.e., ACP), and therefore, they completely dissolved after exposure to PBS for only 24 h, while the dissolution rate of the same coatings after annealing (they became crystalline) was found to be more restrained (Boyd et al. 2006). Additionally, HA coatings were found to be less stable than those of FA (Klein et al. 1994b; Dhert et al. 1992; Dhert et al. 1993; Caulier et al. 1995) and of a similar stability with magnesium-whitlockite (*i.e.*, Mg-substituted β-TCP) coatings (Dhert et al. 1992, 1993).

On the other hand, there are cases (Gineste et al. 1999; de Bruijn et al. 1994; Cleries et al. 2000b), in which the biodegradation kinetics of calcium orthophosphate coatings, films and layers appeared to be correlated imperfectly with their solubility values (see Table 1). For example, three types of calcium orthophosphate (HA, ACP and β-TCP) coatings on titanium alloy substrates, deposited by the laser ablation technique, were immersed in SBF in order to determine their behavior in conditions similar to the human blood plasma. Neither HA nor ACP coatings were found to dissolve in SBF, while a β-TCP coating slightly dissolved. Precipitation of an apatitic phase was favored onto both HA and β-TCP coatings; however, no precipitation occurred onto ACP coating (Cleries et al. 2000b). Additionally, degradation rates of dental implants with 50- and 100-micron thick coatings of HA, FA and fluorhydroxylapatite (FHA) were studied (Gineste et al. 1999). The implants were inserted in dog jaws and retrieved for histological analysis after 3, 6, and 12 months. The thickness of the calcium orthophosphate coatings was evaluated using an image analysis device. HA and FA coatings (even at 100 micron thickness) were almost totally degraded within the implantation period. In contrast, the FHA coatings did not show significant degradation during the same period (Gineste et al. 1999).

Interaction with cells and tissue responses

The interactions of calcium orthophosphate coatings, films and layers with either cells *in vitro* or surrounding tissues *in vivo* have been studied a lot (Huang et al. 2009; Wang et al. 2004; Choi et al. 2003; Ueda et al. 2007; Massaro et al. 2001; Wie et al. 1998; Maistrelli et al. 1993; Hulshoff et al. 1996a; Caulier et al. 1997a, 1997b; Antonov et al. 1998; Cleries et al. 2000c; Lo et al. 2000; Jung et al. 2001; Manso et al. 2002c; Heimann et al. 2004; Siebers et al. 2004, 2005; Manders et al. 2006; Simank et al. 2006; Mello et al. 2007; Hashimoto et al. 2008; Coelho and Lemons 2009; Sima et al. 2010; Quaranta et al. 2010; Cairns et al. 2010). The *in vitro* trials using different cell lines revealed that in the vast majority of the cases, calcium orthophosphate coatings, films and layers enhanced cellular adhesion, proliferation and differentiation, while the results of the *in vivo* studies revealed that they promoted bone regeneration. For example, a combination of surface geometry and calcium orthophosphate coatings was found to benefit the implant-bone response during the healing phase (Hayakawa et al. 2002). Calcium orthophosphate coatings on titanium implants followed by bisphosphonate-immobilization appeared to be effective in the promotion of osteogenesis on surfaces of dental implants (Yoshinari et al. 2002). A greater percent of bone contact lengths were detected for calcium orthophosphate-coated Ti implants compared with control Ti implants 3 and 12 weeks after implant placement (Ong et al. 2002). Similar results were obtained in other studies (Dalton and Cook 1995; Nguyen et al. 2004; Yan et al. 2006).

Concerning the experiments with cells, human gingival fibroblasts attachment, spreading, extracellular matrix production and focal adhesion plaque formation were investigated on commercially pure Ti, HA-coated Ti and porous TCP/HA-coated Ti. TCP/HA and HA coatings exhibited that both the attached cell number and cell spreading area were higher than that on pure Ti and focal adhesion plaque formed earlier than that of un-coated substrate. The attached cell number and type I collagen formation on TCP/HA coatings were more than that on HA ones (Zhao et al. 2005). Osteoblasts were successfully grown on the surface of OCP (Bigi et al. 2005b) and HA (Cao et al. 2010a; Bigi et al. 2005b; Ball et al. 2001); both types of coatings, layers and films were found to favor osteoblast proliferation, activation of their metabolism and differentiation. Furthermore, the *in vitro* cell-culture studies using MG63 osteoblast-like cells were performed on calcium orthophosphate coatings deposited on titanium by plasma spray, sol-gel and sputtering techniques. The study demonstrated the ability of cells to proliferate on the materials tested. The sol-gel coating was found to promote higher cell growth, greater alkaline phosphatase activity and greater osteocalcin production compared to the sputtered and plasma-sprayed coatings (Massaro et al. 2001). In another study, calcium orthophosphate coatings were found to induce significantly higher cell differentiation levels than the uncoated control (Bucci-Sabattini et al. 2010).

A study by Cairns et al. (2010) should be described especially. Calcium orthophosphate thin films were deposited onto substrates with varying topography. Then, a layer of fibronectin was deposited from solution onto each surface, and the response of MG63 osteoblast-like cells was studied. The results revealed that, in all cases, the presence of the adsorbed fibronectin layer improved cell adhesion, proliferation and promoted early onset differentiation. Moreover, the nature and scale of the response appeared to be influenced by the surface topography of the substrates. Specifically, cells on the fibronectin-coated calcium orthophosphate thin films with regular topographical features in the nanometer range showed statistically significant differences in focal adhesion assembly, osteocalcin expression and alkaline phosphase activity compared to the calcium orthophosphate films without those topographical features (Cairns et al. 2010). Therefore, both an adsorbed bioorganics and a surface topography of the substrates appear to influence cell adhesion and differentiation.

In vivo results correlate well with the *in vitro* ones. Osseointegration rates of porous-surfaced Ti6Al4V implants with control (unmodified sintered coatings) were compared to porous-surfaced implants modified through the addition of either an inorganic or organic route sol-gel-formed calcium orthophosphate films. Implants were placed in distal femoral rabbit condyle sites and, following a 9-day healing period, implant fixation strength was evaluated using a pullout test. Both types of calcium orthophosphate films significantly enhanced the early rate of bone in-growth and fixation as evidenced by higher pull-out force and interface stiffness compared with controls. However, there was no significant difference between calcium orthophosphate-coated implants prepared using the two different methods (Gan et al. 2004).

To conclude this section, one should note that the positive clinical benefits of calcium orthophosphate coatings, layers and films were not always detected. For example, a study was undertaken to evaluate the processes involved in biological responses of the Ti-6Al-7Nb alloy with and without HA coatings with both *in vitro* and *in vivo* tests. The results with HA coating appeared to be similar to those obtained on the uncoated samples (Lavos-Valereto et al. 2001). Similarly, neither positive nor negative influence of the presence of HA coatings on the surface of implants was detected during the 10-year (Lazarinis et al. 2011), 13-year (Camazzola et al. 2009), 15-year (Stilling et al. 2009) and undisclosed (Lee et al. 2000; Gandhi et al. 2009) follow-ups. Besides, there are cases in which short-term (4 weeks) advantages of the calcium orthophosphate-coated implants were found in animal studies, whereas no significant differences to the uncoated samples were found after 6 months (Gottlander et al. 1997a). Furthermore, inflammatory tissue reaction cases have been detected (Piattelli et al. 1995; Walschus et al. 2009). Interestingly, the short-term inflammatory response against a HA coating on Ti was lower in comparison to a DCPD coating on Ti. The observed differences between the Ti-DCPD implants and the Ti-HA implants were attributed to their dissolution characteristics: the HA coating on Ti showed increased stability and, hence reduced the inflammatory response (Walschus et al. 2009). Furthermore, HA coatings were found to be a risk factor for cup revision due to aseptic loosening (Lazarinis et al. 2010). Thus, precautions to prevent contamination (asepsis) and/or infection (perioperative antibiotics) appear to be more important for the calcium orthophosphate-coated implants if compared with the uncoated ones (Oosterbos et al. 2002).

Biomedical applications

Already in 1987, de Groot et al. (1987), published a paper on the development of plasma-sprayed HA coatings on metallic implants. The same year the same researchers published the results of the first clinical study (Geesink et al. 1987). Shortly afterwards, Furlong and Osborn, two leading surgeons in the orthopedics field, began implanting plasma-sprayed HA stems in patients (Furlong and Osborn 1991) followed by other clinicians (Bauer et al. 1991; Buma and Gardeniers 1995). Since then, plentiful reports have been published about the biomedical advantages of such coated implants. To summarize the available information on the biomedical and biomechanical properties of implants coated by calcium orthophosphates, one can claim the following: If compared to uncoated implants, the presence of calcium orthophosphate deposits were found to induce bone contacts to the implants (Dhert et al. 1992, 1993; Thomas et al. 1989; Jansen et al. 1991; Gottlander et al. 1997b; Hulshoff and Jansen 1997; Hayakawa et al. 2000; Mohammadi et al. 2004; Park et al. 2005; Siebers et al. 2007; Kuroda et al. 2007; Chae et al. 2008; Schwarz et al. 2009; Junker et al. 2010; Suzuki et al. 2010); improve implant fixation (Yang et al. 1997; Søballe et al. 1993; Daugaard et al. 2010); show higher torque values (Park et al. 2005; Junker et al. 2010; Granato et al. 2009) and push-out strength (Ozeki et al. 2001); facilitate bridging of small gaps between implants and surrounding bones (Søballe et al. 1991; Stephenson et al. 1991), reduce metal ion release from the metallic substrates (Surmenev et al. 2010; Ducheyne and Healy 1988; Sousa and Barbosa 1996; Ozeki et al. 2003); slow down metal degradation and/or its corrosion (Metikoš-Huković et al. 2003; Yang et al. 2008; Cheng and Roscoe 2005); accelerate bone growth (Cook et al. 1992; Wang et al. 2009), remodeling (Pilliar et al. 1991; Yoon et al. 2009) and osteointegration rate (Bigi et al. 2008; Lee et al. 2011); induce osteoconductivity (Cao et al. 2010b), improve the early bone (Yang et al. 1996; Mohammadi et al. 2003) and healing (Vercaigne et al. 2000b) responses; and result in lack of formation of fibrous tissues (Figure 11) (Layrolle 2011; Dostálová et al. 2001), as well as increase the clinical performance of orthopedic hip systems (see below). In addition, calcium orthophosphate coatings, films and layers might be used for incorporation of drugs and important biologically active compounds, such as peptides, hormones and growth factors (Siebers et al. 2006). In the case of porous implants, calcium orthophosphate coatings enhance bone ingrowth into the pores (Suchanek and Yoshimura 1998). Furthermore, studies concluded that there was significantly less pin loosening in calcium orthophosphate-coated groups (Saithna 2010). Thus, the majority of the clinical studies are optimistic about the *in vivo* performance of calcium orthophosphate-coated prostheses.

However, to be objective, one must mention on the studies in which no positive biomedical and/or biomechanical effects of calcium orthophosphate coatings, films and layers have been detected (Tieanboon et al. 2009; Coelho et al. 2009b). Besides, the presence or absence of the positive biomedical and/or biomechanical effects of calcium orthophosphate coatings, films and layers might depend on the deposition technique used (Hulshoff et al. 1996b, 1997), as well as on the coating vendor (Dalton and Cook 1995). These uncertainties might be due to several reasons, such as variability in chemical and phase composition, porosity, admixtures, *etc.*

In biomedical applications, bone grafts are usually much thicker than coatings, films or layers applied to them. Nevertheless, the coated implants combine the surface biocompatibility and bioactivity of calcium orthophosphates with the core strength of strong substrates (Figure 12). The clinical results for calcium orthophosphate-coated implants reveal that they have much longer life times after implantation than uncoated devices, and therefore, they are particularly beneficial for younger patients (Capello et al. 1997). Their biomedical properties are approaching those of bioactive glass-coated implants (Wheeler et al. 2001; Mistry et al. 2011).

Since, among calcium orthophosphates, HA is the most popular material to be deposited as coatings, films and layers, the vast majority of the clinical investigations was performed with HA. For example, HA coating as a system of fixation of hip implants *in vivo* was found to work well in the short to medium terms (2 years (Geesink 1990), 6 years (Geesink and Hoefnagels 1995), 8 years (Wheeler 1996; Chang et al. 2006), 9 to 12 years (MaNally et al. 2000), 10 years (Oosterbos et al. 2004; Trisi et al. 2005), 10 to 15.5 years (Matsumine et al. 2004), 10 to 17 years (Muirhead-Allwood et al. 2010), 13 to 15 years (Shetty et al. 2005), 15 to 21 years (Rajaratnam et al. 2008), 16 years (Buchanan 2005), 17 years (Buchanan 2006) and 19 years (Buchanan and Goodfellow 2008)). In 2004, a special book summarizing the studies with HA-coated implants and the 'state of the art' of HA coatings in orthopedics at the close of 2002 was published (Epinette and Manley 2004). Similar data for HA-coated dental implants are also available (Tinsley et al. 2001; Binahmed et al. 2007; Iezzi et al. 2007). Nevertheless, even longer-term clinical results are awaited with a great interest. The biomedical aspects of osteoconductive coatings for total joint arthroplasty have been reviewed elsewhere (Geesink 2002). Additional details on calcium orthophosphate coatings, films and layers might be found in excellent reviews (Narayanan et al. 2010; Paital and Dahotre 2009a; León and Jansen 2009).

Nevertheless, one must stress that although many experiments concerning the *in vivo* studies of calcium orthophosphate coatings, films and layers have indicated

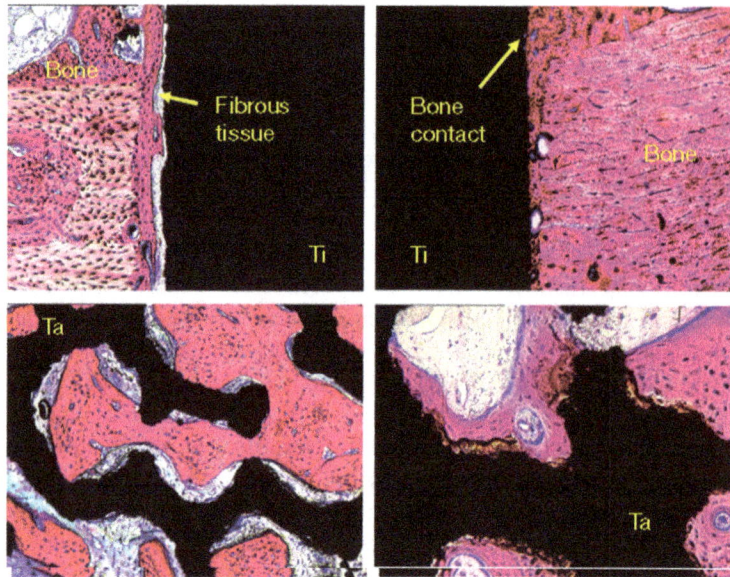

Figure 11 Comparison of bone-integrative properties. Non-coated (left) and biomimetically coated by calcium orthophosphates metal implants (right) after implantation in the femur of goats for 6 weeks. Reprinted from Layrolle (2011) with permission.

a stronger and faster fixation, as well as more bone ingrowth at the interface, the clinical performance of such coatings, films and layers is still far from the perfection. Some of the major concerns associated with the usage of calcium orthophosphate coatings, films and layers in actual body environment, with regard to their long-term stability can be listed as follows (Hulshoff et al. 1996a; Caulier et al. 1997a, 1997b):

- The degradation and resorption of calcium orthophosphate coatings, films and layers in a biological environment could lead to disintegration

of the coating, resulting in the loss of both coating-substrate bond strength and the implant fixation.
- Coating delamination and disintegration with the formation of particulate debris are also major concern
- Calcium orthophosphate coatings, films and layers may also lead to increased polyethylene wear from the acetabular cup and, thereby, alleviate the problem of osteolysis.

In spite of the long history and the aforementioned achievements, still not all concerns on the surgical applications of calcium orthophosphate coatings, films and

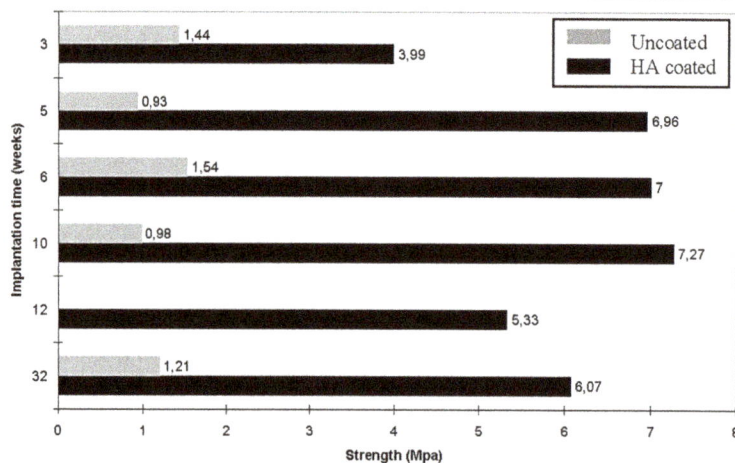

Figure 12 Time-dependent plasma-sprayed HA coating. This figure shows how a plasma-sprayed HA coating on a porous titanium (dark bars) dependent on the implantation time will improve the interfacial bond strength compared to uncoated porous titanium (light bars). Reprinted from Hench (1991) with permission.

layers have been eliminated. Still, a limited amount of the *in vivo* studies is available in the literature. The limitations to such experiments may be attributed to any of the following reasons:

- Difficulty in selection of a suitable animal model to simulate the actual mechanical loading and unloading conditions the implant might undergo in a human body environment.
- The need to sacrifice a large number of animals, since most of these experiments demand a statistical analysis to validate the results.
- A high cost and a long period of clinical testing these experiments demand.
- Lack of coordination among material scientists and biologists and thereby an insufficient understanding of this interdisciplinary subject.
- Serious ethical concerns on the use of animals for experimental studies as they are subjected to painful procedures or toxic exposures during the course of test.

To conclude this section, one must note the following: Even though the importance and the need for development of calcium orthophosphate coatings, films and layers have been recognized, it is still mostly being explored on research level, and after extensive search of open literature, these coatings appear to have made limited headway into commercialization. In spite of mention of the commercial products such as hip and dental implants produced by Zimmer Orthopedics (Freiburg, Germany), Smith and Nephew (Memphis, TN, USA), and Biomet, the science and technology related to their manufacturing is not disclosed by any one of them due to the proprietary reasons. Hence, at this point it is difficult to bring a detailed discussion on commercialization of calcium orthophosphate coatings, films and layers (Paital and Dahotre 2009a).

Future directions

A potential drawback of the majority of the deposition techniques of calcium orthophosphate coatings, films and layers is their relatively high cost for a large scale production. Therefore, to decrease processing time and make their manufacturing commercially viable, it is desirable to process the thinnest coating that would significantly increase the biological response (Coelho et al. 2009c). Much attention should be paid to functionally graded structures with an amorphous top layer and a crystalline layer underneath (Wang and Zreiqat 2010). This allows adjusting the coating resorption rates to the values at which new bone grows at early stages when it is of the most importance for the bone mineralization process. Furthermore, therapeutic capabilities of calcium

orthophosphate coatings, films and layers as templates for the *in situ* delivery of drugs and osteoinductive agents (peptides, hormones and growth factors) at the required times should be elucidated much better.

Conclusions

Solid implants prepared from various materials often possess a poor biocompatibility with a simultaneous lack of the osteogenic properties in order to promote bone healing. In addition, direct bone-to-implant contacts are desired for a biomechanical anchoring of implants rather than fibrous tissue encapsulation. All these problems might be solved by applying calcium orthophosphate coatings, films and layers. The aim is to provide the implants with surface biological properties for adsorption of proteins, adhesion of cells and bone apposition. Therefore, the available knowledge on calcium orthophosphate and, most notably, HA coatings, films and layers on various substrates has been summarized in this review. Since all available deposition techniques have both advantages and shortcomings of their own (Table 3), still there are no standard guidelines for depositing calcium orthophosphates on the implant surfaces. In general, dissolution of calcium orthophosphate coatings, films and layers improves implant osseointegration and is the basic requirement for bioactivity. However, this dissolution diminishes the stability and increases the potential for loosening of the implants. Calcium orthophosphate coatings, films and layers of lower solubility and higher stability are desirable for the long-term performance of implants because they promote faster initial bone fixation, bridge larger gaps in the misfit and degrade at a controlled rate.

Although animal and *in vitro* studies have already reported on the benefits of using calcium orthophosphate-coated implants, as well as the risks of dissolution, the short-term studies did not demonstrate that the dissolution of calcium orthophosphate coatings, films and layers led to a loss of implants. In addition, many *in vivo* and clinical studies did not consider the chemical and structural characterizations of the coatings. Under these conditions, any comparisons among various reports and studies are difficult.

New promising techniques for coating medical devices are continuously investigated. Future investigations on various coating processes will have to include clinical trials to get better understanding of bone responses to coated-implant surfaces, as well as studies on coupling of calcium orthophosphate coatings, layers and films with drugs, growth factors and cells. Although it has been generally accepted that calcium orthophosphate coatings, layers and films improve bone strength and initial osteointegration rate, the optimal coating properties required to achieve maximal bone response are yet to be reported. As such, the use of well-characterized calcium orthophosphate coatings, layers and films cell culture studies, animal studies and clinical studies should be well documented to avoid controversial results.

In addition, the clinicians need to take into consideration the enhanced bacterial susceptibility of calcium orthophosphate-coated implants if compared to the metallic ones. Besides, the clinicians need to consider possible failures of calcium orthophosphate coatings, films and layers as a result of coating-substrate interfacial fracture. It is also important that the clinical investigators be well versed with the material characterizations of the coated implants.

Competing interests
The author declares that he has no competing interests.

References
Ali MY, Hung W, Yongqi F (2010) A review of focused ion beam sputtering. Int J Precision Eng Manuf 11:157–170

Amjad Z (1997) Calcium phosphates in biological and industrial systems. Kluwer, Boston, MA, USA, p 529

Amy RL, Storb R (1955) Selective mitochondrial damage by a ruby laser microbeam: an electron microscopic study. Science 122:756–758

Antonov EN, Bagratashvili VN, Popov VK, Sobol EN, Howdle SM, Joiner C, Parker KG, Parker TL, Doctorov AL, Likhanov VB, Volozhin AI, Alimpiev SS, Nikiforov SM (1998) Biocompatibility of laser-deposited hydroxyapatite coatings on titanium and polymer implant materials. J Biomed Opt 3:423–428

Arias JL, García-Sanz FJ, Mayor MB, Chiussi S, Pou J, León B, Pérez-Amor M (1998) Physicochemical properties of calcium phosphate coatings produced by pulsed laser deposition at different water vapour pressures. Biomaterials 19:883–888

ASTM International (2008) ASTM C633 - 01 Standard test method for adhesion or cohesion strength of thermal spray coatings. http://www.astm.org/Standards/C633.htm

ASTM International (2011) ASTM F1147 - 05 Standard test method for tension testing of calcium phosphate and metallic coatings. http://www.astm.org/Standards/F1147.htm

Ball MD, Downes S, Scotchford CA, Antonov EN, Bagratashvili VN, Popov VK, Lo WJ, Grant DM, Howdle SM (2001) Osteoblast growth on titanium foils coated with hydroxyapatite by pulsed laser ablation. Biomaterials 22:337–347

Ban S, Maruno S (1998) Hydrothermal-electrochemical deposition of hydroxyapatite. J Biomed Mater Res 42:387–395

Bao Q, Chen C, Wang D, Lei T, Liu J (2006) Pulsed laser deposition of hydroxyapatite thin films under Ar atmosphere. Mater Sci Eng, A 429:25–29

Barrere F, van Blitterswijk CA, de Groot K, Layrolle P (2002a) Influence of ionic strength and carbonate on the Ca-P coating formation from SBF×5 solution. Biomaterials 23:1921–1930

Barrere F, van Blitterswijk CA, de Groot K, Layrolle P (2002b) Nucleation of biomimetic Ca-P coatings on Ti6Al4V from a SBF×5 solution: influence of magnesium. Biomaterials 23:2211–2220

Barrere F, Snel MME, van Blitterswijk CA, de Groot K, Layrolle P (2004) Nano-scale study of the nucleation and growth of calcium phosphate coating on titanium implants. Biomaterials 25:2901–2910

Barthell BL, Archuleta TA, Kossowsky R (1989) Ion beam deposition of calcium hydroxyapatite. Mater Res Soc Symp Proc 110:709–715

Bauer TW, Geesink RGT, Zimmerman R, McMahon JT (1991) Hydroxyapatite-coated femoral stems. Histological analysis of components retrieved at autopsy. J Bone Joint Surg A 73:1439–1452

Ben-Nissan B, Latella BA, Bendavid A (2011) 3.305. Biomedical thin films: mechanical properties. In: Ducheyne P, Healy K, Hutmacher DW, Grainger DW, Kirkpatrick CJ (eds) Comprehensive biomaterials. Elsevier, Amsterdam, Netherlands, pp 63–73, Vol. 3

Berezhnaya AY, Mittova VO, Kukueva EV, Mittova IY (2010) Effect of high-temperature annealing on solid-state reactions in hydroxyapatite/TiO_2 films on titanium substrates. Inorg Mater 46:971–977

Besra L, Liu M (2007) A review on fundamentals and applications of electrophoretic deposition (EPD). Prog Mater Sci 52:1–61

Bigi A, Boanini E, Bracci B, Facchini A, Panzavolta S, Segatti F, Struba L (2005a) Nanocrystalline hydroxyapatite coatings on titanium: a new fast biomimetic method. Biomaterials 26:4085–4089

Bigi A, Bracci B, Cuisinier F, Elkaim R, Fini M, Mayer I, Mihailescu IN, Socol G, Sturba L, Torricelli P (2005b) Human osteoblast response to pulsed laser deposited calcium phosphate coatings. Biomaterials 26:2381–2389

Bigi A, Fini M, Bracci B, Boanini E, Torricelli P, Giavaresi G, Aldini NN, Facchini A, Sbaiz F, Giardino R (2008) The response of bone to nanocrystalline hydroxyapatite-coated Ti13Nb11Zr alloy in an animal model. Biomaterials 29:1730–1736

Bigi A, Boanini E, Capuccini C, Fini M, Mihailescu IN, Ristoscu C, Sima F, Torricelli P (2009) Biofunctional alendronate-hydroxyapatite thin films deposited by matrix assisted pulsed laser evaporation. Biomaterials 30:6168–6177

Binahmed A, Stoykewych A, Hussain A, Love B, Pruthi V (2007) Long-term follow-up of hydroxyapatite-coated dental implants - a clinical trial. Int J Oral Max Impl 22:963–968

Bini RA, Santos ML, Filho EA, Marques RFC, Guastaldi AC (2009) Apatite coatings onto titanium surfaces submitted to laser ablation with different energy densities. Surf CoatTechnol 204:399–403

Blackwood DJ, Seah KHW (2009) Electrochemical cathodic deposition of hydroxyapatite: improvements in adhesion and crystallinity. Mater Sci Eng C 29:1233–1238

Blalock T, Bai X, Rabiei A (2007) A study on microstructure and properties of calcium phosphate coatings processed using ion beam assisted deposition on heated substrates. Surf CoatTechnol 201:5850–5858

Boanini E, Torricelli P, Fini M, Sima F, Serban N, Mihailescu IN, Bigi A (2012) Magnesium and strontium doped octacalcium phosphate thin films by matrix assisted pulsed laser evaporation. J Inorg Biochem 107:65–72

Bocanegra-Bernal MH (2004) Hot isostatic pressing (HIP) technology and its applications to metals and ceramics. J Mater Sci 39:6399–6420

Boyd AR, Meenan BJ, Leyland NS (2006) Surface characterisation of the evolving nature of radio frequency (RF) magnetron sputter deposited calcium phosphate thin films after exposure to physiological solution. Surf CoatTechnol 200:6002–6013

Boyd AR, Duffy H, McCann R, Cairns ML, Meenan BJ (2007) The influence of argon gas pressure on co-sputtered calcium phosphate thin films. Nucl Instrum Methods Phys Res B 258:421–428

Brès E, Hardouin P, (eds) (1998) Les matériaux en phosphate de calcium: aspects fondamentaux. Sauramps Medical, Montpellier, France, p 176

Brinker CJ, Frye GC, Hurd AJ, Ashley CS (1991) Fundamentals of sol–gel dip coating. Thin Solid Films 201:97–108

Brown PW, Constantz B (eds) (1994) Hydroxyapatite and related materials. CRC Press, Boca Raton, FL, USA, p 343

Bucci-Sabattini V, Cassinelli C, Coelho PG, Minnici A, Trani A, Ehrenfest DMD (2010) Effect of titanium implant surface nanoroughness and calcium phosphate low impregnation on bone cell activity in vitro. Oral Surg Oral Med Oral Pathol Oral Radiol Endodontol 109:217–224

Buchanan JM (2005) 16 year review of hydroxyapatite ceramic coated hip implants – a clinical and histological evaluation. Key Eng Mater 284–286:1049–1052

Buchanan JM (2006) 17 year review of hydroxyapatite ceramic coated hip implants – a clinical and histological evaluation. Key Eng Mater 309–311:1341–1344

Buchanan JM, Goodfellow S (2008) Nineteen years review of hydroxyapatite ceramic coated hip implants: a clinical and histological evaluation. Key Eng Mater 361–363:1315–1318

Buma P, Gardeniers JW (1995) Tissue reactions around a hydroxyapatite-coated hip prostheses: case report of a retrieveal specimen. J Arthroplasty 10:389–395

Burgess AV, Story BJ, La D, Wagner WR, LeGeros JP (1999) Highly crystalline MP-1 hydroxylapatite coating. Part I: In vitro characterization and comparison to other plasma-sprayed hydroxylapatite coatings. Clin Oral Implant Res 10:245–256

Cairns ML, Meenan BJ, Burke GA, Boyd AR (2010) Influence of surface topography on osteoblast response to fibronectin coated calcium phosphate thin films. Coll Surf B 78:283–290

Callahan TJ, Gantenberg JB, Sands BE (1994) Calcium phosphate (Ca-P) coating draft guidance for preparation of Food and Drug Administration (FDA) submissions for orthopedic and dental endosseous implants. In: Horowitz E, Parr JE (eds) Characterization and performance of calcium phosphate coatings for implants. ASTM STP 1196, Philadelphia, PA, USA, pp 185–197

Camazzola D, Hammond T, Gandhi R, Davey JR (2009) A randomized trial of hydroxyapatite-coated femoral stems in total hip arthroplasty. A 13-year follow-up. J Arthroplasty 24:33–37

Campbell AA (2003) Bioceramics for implant coatings. Mater Today 6:26–30

Cannillo V, Lusvarghi L, Sola A, Barletta M (2009) Post-deposition laser treatment of plasma sprayed titania-hydroxyapatite functionally graded coatings. J Eur Ceram Soc 29:3147–3158

Cao Y, Weng J, Chen J, Feng J, Yang Z, Zhang X (1996) Water vapor-treated hydroxyapatite coatings after plasma spraying and their characteristics. Biomaterials 17:419–424

Cao N, Dong J, Wang Q, Ma Q, Wang F, Chen H, Xue C, Li M (2010a) Plasma-sprayed hydroxyapatite coating on carbon/carbon composite scaffolds for bone tissue engineering and related tests in vivo. J Biomed Mater Res Am 92A:1019–1027

Cao N, Dong J, Wang Q, Ma Q, Xue C, Li M (2010b) An experimental bone defect healing with hydroxyapatite coating plasma sprayed on carbon/carbon composite implants. Surf CoatTechnol 205:1150–1156

Capello WD, D'Antonio JA, Feinberg JR, Manley MT (1997) Hydroxyapatite-coated total hip femoral components in patients less than fifty years old. Clinical and radiographic results after five to eight years of follow-up. J Bone Joint Surg Am 79:1023–1029

Carradó A (2010) Structural, microstructural, and residual stress investigations of plasma-sprayed hydroxyapatite on Ti-6Al-4 V. ACS Appl Mater Interf 2:561–565

Carradò A, Viart N (2010) Nanocrystalline spin coated sol–gel hydroxyapatite thin films on Ti substrate: towards potential applications for implants. Solid State Sci 12:1047–1050

Caulier H, van der Waerden JPCM, Paquay YCGJ, Wolke JGC, Kalk W, Naert I, Jansen JA (1995) Effect of calcium phosphate (Ca-P) coatings on trabecular bone response: a histological study. J Biomed Mater Res 29:1061–1069

Caulier H, van der Waerden JPCM, Wolke JGC, Kalk W, Naert I, Jansen JA (1997a) A histological and histomorphometrical evaluation of the application of screw-designed calciumphosphate (Ca-P)-coated implants in the cancellous maxillary bone of the goat. J Biomed Mater Res 35:19–30

Caulier H, Hayakawa T, Naert I, van der Waerden JPCM, Wolke JGC, Jansen JA (1997b) An animal study on the bone behaviour of Ca-P-coated implants: influence of implant location. J Mater Sci Mater Med 8:531–536

Chae GJ, Jung UW, Jung SM, Lee IS, Cho KS, Kim CK, Choi SH (2008) Healing of surgically created circumferential gap around nano-coating surface dental implants in dogs. Surf Interf Analysis 40:184–187

Chang JK, Chen CH, Huang KY, Wang GJ (2006) Eight-year results of hydroxyapatite-coated hip arthroplasty. J Arthroplasty 21:541–546

Cheang P, Khor KA (1995) Thermal spraying of hydroxyapatite (HA) coatings: effects of powder feedstock. J Mater Process Technol 48:429–436

Chen XB, Li YC, Plessis JD, Hodgson PD, Wen C (2009a) Influence of calcium ion deposition on apatite-inducing ability of porous titanium for biomedical applications. Acta Biomater 5:1808–1820

Chen X, Li Y, Hodgson PD, Wen C (2009b) Microstructures and bond strengths of the calcium phosphate coatings formed on titanium from different simulated body fluids. Mater Sci Eng C 29:165–171

Cheng X, Roscoe SG (2005) Corrosion behavior of titanium in the presence of calcium phosphate and serum proteins. Biomaterials 26:7350–7356

Cheng GJ, Ye C (2010) Experiment, thermal simulation, and characterizations on transmission laser coating of hydroxyapatite on metal implant. J Biomed Mater Res A 92A:70–79

Cheng K, Ren C, Weng W, Du P, Shen G, Han G, Zhang S (2009) Bonding strength of fluoridated hydroxyapatite coatings: a comparative study on pull-out and scratch analysis. Thin Solid Films 517:5361–5364

Choi JM, Kim HE, Lee IS (2000) Ion-beam-assisted deposition (IBAD) of hydroxyapatite coating layer on Ti-based metal substrate. Biomaterials 21:469–473

Choi J, Bogdanski D, Koller M, Esenwein SA, Muller D, Muhr G, Epple M (2003) Calcium phosphate coating of nickel-titanium shape-memory alloys. Coating procedure and adherence of leukocytes and platelets. Biomaterials 24:3689–3696

Chou BY, Chang E (2001) Interface investigation of plasma-sprayed hydroxyapatite coating on titanium alloy with ZrO₂ intermediate layer as bond coat. Scr Mater 45:487–493

Chow LC, Eanes ED (eds) (2001) Octacalcium phosphate (Monographs in oral science). vol. 18. S Karger Pub, Basel, Switzerland, p 168

Cizek J, Khor KA, Prochazka Z (2007) Influence of spraying conditions on thermal and velocity properties of plasma sprayed hydroxyapatite. Mater Sci Eng C 27:340–344

Cleries L, Martinez E, Fernandez-Pradas JM, Sardin G, Esteve J, Morenza JL (2000a) Mechanical properties of calcium phosphate coatings deposited by laser ablation. Biomaterials 21:967–971

Cleries L, Fernandez-Pradas JM, Morenza JL (2000b) Behaviour in simulated body fluid of calcium phosphate coatings obtained by laser ablation. Biomaterials 21:1861–1865

Cleries L, Fernandez-Pradas JM, Morenza JL (2000c) Bone growth on and resorption of calcium phosphate coatings obtained by pulsed laser deposition. J Biomed Mater Res 49:43–52

Coelho PG, Lemons JE (2009) Physico/chemical characterization and in vivo evaluation of nanothickness bioceramic depositions on alumina-blasted/acid-etched Ti-6Al-4 V implant surfaces. J Biomed Mater Res A 90A:351–361

Coelho PG, de Assis SL, Costa I, Thompson VP (2009a) Corrosion resistance evaluation of a Ca- and P-based bioceramic thin coating in Ti-6Al-4 V. J Mater Sci Mater Med 20:215–222

Coelho PG, Cardaropoli G, Suzuki M, Lemons JE (2009b) Early healing of nanothickness bioceramic coatings on dental implants. An experimental study in dogs. J Biomed Mater Res B Appl Biomater 88B:387–393

Coelho PG, Granjeiro JM, Romanos GE, Suzuki M, Silva NRF, Cardaropoli G, van Thompson P, Lemons JE, Coelho PG, Granjeiro JM, Romanos GE, Suzuki M, Silva NRF, Cardaropoli G, van Thompson P, Lemons JE (2009c) Basic research methods and current trends of dental implant surfaces. J Biomed Mater Res B Appl Biomater 88B:579–596

Cook SD, Thomas KA, Kay JF, Jarcho M (1988) Hydroxyapatite-coated titanium for orthopedic implant applications. Clin Orthop Rel Res 232:225–243

Cook SD, Thomas KA, Dalton JE, Volkman TK, Whitecloud TS III, Kay JF (1992) Hydroxylapatite coating of porous implants improves bone ingrowth and interface attachment strength. J Biomed Mater Res 26:989–1001

Cooley DR, van Dellen AF, Burgess JO, Windeler S (1992) The advantages of coated titanium implants prepared by radiofrequency sputtering from hydroxyapatite. Prosthetic Dent 67:93–100

Cotell CM (1993) Pulsed laser deposition and processing of biocompatible hydroxylapatite thin films. Appl Surf Sci 69:140–148

Cotell CM, Chrisey DB, Grabowski KS, Sprague JA, Gosset CR (1992) Pulsed laser deposition of hydroxylapatite thin films on Ti-6Al-4 V. J Appl Biomed 3:87–93

Craciun V, Boyd IW, Craciun D, Andreazza P, Perriere J (1999) Vacuum ultraviolet annealing of hydroxyapatite films grown by pulsed laser deposition. J Appl Phys 85:8410–8414

Cuerno R, Barabási AL (1995) Dynamic scaling of ion-sputtered surfaces. Phys Rev Lett 74:4746–4749

Cui FZ, Luo ZS, Feng QL (1997) Highly adhesive hydroxyapatite coatings on titanium alloy formed by ion beam assisted deposition. J Mater Sci Mater Med 8:403–405

d'Haese R, Pawlowski L, Bigan M, Jaworski R, Martel M (2010) Phase evolution of hydroxapatite coatings suspension plasma sprayed using variable parameters in simulated body fluid. Surf CoatTechnol 204:1236–1246

Dalton JE, Cook SD (1995) In vivo mechanical and histological characteristics of HA-coated implants vary with coating vendor. J Biomed Mater Res 29:239–245

Daugaard H, Elmengaard B, Bechtold JE, Jensen T, Soballe K (2010) The effect on bone growth enhancement of implant coatings with hydroxyapatite and collagen deposited electrochemically and by plasma spray. J Biomed Mater Res A 92A:913–921

de Andrade MC, Filgueiras MRT, Ogasawara T (2002) Hydrothermal nucleation of hydroxyapatite on titanium surface. J Eur Ceram Soc 22:505–510

de Bruijn JD, Bovell YP, van Blitterswijk CA (1994) Structural arrangements at the interface between plasma sprayed calcium phosphates and bone. Biomaterials 15:543–550

de Groot K, Geesink RGT, Klein CPAT, Serekian P (1987) Plasma sprayed coatings of hydroxylapatite. J Biomed Mater Res 21:1375–1381

de Groot K, Wolke JGC, Jansen JA (1998) Calcium phosphate coatings for medical implants. Proc Inst Mech Eng Part J Eng Med 212:137–147

de Sena LA, de Andrade MC, Rossi AM, Soares GDA (2002) Hydroxypatite deposiiton by electrophoresis on titanium sheets with different surface finishing. J Biomed Mater Res 60:1–7

Dey A, Mukhopadhyay AK (2010) Anisotropy in nanohardness of microplasma sprayed hydroxyapatite coating. Adv Appl Ceram 109:346–354

Dey A, Mukhopadhyay AK (2011) Fracture toughness of microplasma-sprayed hydroxyapatite coating by nanoindentation. Int J Appl Ceram Technol 8:572–590

Dey A, Nandi SK, Kundu B, Kumar C, Mukherjee P, Roy S, Mukhopadhyay AK, Sinha MK, Basu D (2011) Evaluation of hydroxyapatite and β-tri calcium phosphate microplasma spray coated pin intra-medullary for bone repair in a rabbit model. Ceram Int 37:1377–1391

Dhert WJA, Klein CPAT, Wolke JGC, van der Velde EA, de Groot K, Rozing PM (1992) A mechanical investigation of fluorapatite, magnesium whitlockite and hydroxylapatite plasma-sprayed coatings in goats. J Biomed Mater Res 25:1183–1200

Dhert WJA, Klein CPAT, Jansen JA, van der Velde EA, Vriesde RC, de Groot K, Rozing PM (1993) A histological and histomorphometrical investigation of fluorapatite, magnesium whitlockite and hydroxylapatite plasma-sprayed coatings in goats. J Biomed Mater Res 27:127–138

Dinda GP, Shin J, Mazumder J (2009) Pulsed laser deposition of hydroxyapatite thin films on Ti-6Al-4 V: effect of heat treatment on structure and properties. Acta Biomater 5:1821–1830

Ding SJ (2003) Properties and immersion behavior of magnetron-sputtered multi-layered hydroxyapatite/titanium composite coatings. Biomaterials 24:4233–4238

Dorozhkin SV (2009) Calcium orthophosphates in nature, biology and medicine. Materials 2:399–498

Dorozhkin SV (2011) Calcium orthophosphates: occurrence, properties, biomineralization, pathological calcification and biomimetic applications. Biomatter 1:121–164

Dorozhkin SV (2012) Calcium orthophosphates: applications in nature, biology, and medicine. Pan Stanford, Singapore, p 850

Dorozhkina EI, Dorozhkin SV (2003) Structure and properties of the precipitates formed from condensed solutions of the revised simulated body fluid. J Biomed Mater Res A 67A:578–581

Dostálová T, Himmlová L, Jélinek M, Grivas C (2001) Osseointegration of loaded dental implant with KrF laser hydroxylapatite films on Ti6Al4V alloy by minipigs. J Biomed Optics 6:239–243

Duan K, Fan Y, Wang R (2003) Electrochemical deposition and patterning of calcium phosphate bioceramic coating. Ceram Transact 147:53–61

Ducheyne P, Healy KE (1988) The effect of plasma-sprayed calcium phosphate ceramic coatings on the metal ion release from porous titanium and cobalt-chromium alloys. J Biomed Mater Res 22:1137–1163

Ducheyne P, van Raemdonck W, Heughebaert JC, Heughebaert M (1986) Structural analysis of hydroxylapatite coatings on titanium. Biomaterials 7:97–103

Ducheyne P, Radin S, Heughebaert M, Heughebaert JC (1990) Calcium phosphate ceramic coatings on porous titanium: effect of structure and composition on electrophoretic deposition, vacuum sintering and in vitro dissolution. Biomaterials 11:244–254

Dyshlovenko S, Pateyron B, Pawlowski L, Murano D (2004) Numerical simulation of hydroxyapatite powder behaviour in plasma jet. Surf CoatTechnol 179:110–117, corrigendum: Surf. Coat. Technol. 2004, 187, 408–409

Eliaz N, Eliyahu M (2007) Electrochemical processes of nucleation and growth of hydroxyapatite on titanium supported by realtime electrochemical atomic force microscopy. J Biomed Mater Res A 80A:621–634

Elliott JC (1994) Structure and chemistry of the apatites and other calcium orthophosphates. In: Studies in inorganic chemistry. Elsevier, Amsterdam, Netherlands, p 389, 18

Epinette JAMD, Geesink RGT (1995) Hydroxyapatite coated hip and knee arthroplasty. Elsevier, Amsterdam, Netherlands, p 394

Epinette JA, Manley MT (eds) (2004) Fifteen years of clinical experience with hydroxyapatite coatings in joint arthroplasty. Springer, France, p 452

Erkmen ZE (1999) The effect of heat treatment on the morphology of D-gun sprayed hydroxyapatite coatings. J Biomed Mater Res (Appl Biomater) 48:861–868

Falguera V, Quintero JP, Jiménez A, Muñoz JA, Ibarz A (2011) Edible films and coatings: structures, active functions and trends in their use. Trends Food Sci Technol 22:292–303

Fauchais P (2004) Understanding plasma spraying. J Phys D: Appl Phys 37: R86–R108

Fauchais P, Vardelle A, Dussoubs B (2001) Quo vadis thermal spraying? J Thermal Spray Technol 10:44–66

Feddes B, Wolke JGC, Jansen JA, Vredenberg AM (2003a) Radio frequency magnetron sputtering deposition of calcium phosphate coatings: Monte Carlo simulations of the deposition process and depositions through an aperture. J Appl Phys 93:662–670

Feddes B, Wolke JGC, Jansen JA, Vredenberg AM (2003b) Radio frequency magnetron sputtering deposition of calcium phosphate coatings: the effect of resputtering on the coating composition. J Appl Phys 93:9503–9507

Feng B, Chen Y, Zhang XD (2002) Effect of water vapor treatment on apatite formation on precalcified titanium and bond strength of coatings to substrates. J Biomed Mater Res 59:12–17

Fernández-Pradas JM, Sardin G, Clèrics L, Serra P, Ferrater C, Morenza JL (1998) Deposition of hydroxyapatite thin films by excimer laser ablation. Thin Solid Films 317:393–396

Fernandez-Pradas JM, Cleries L, Sardin G, Morenza JL (1999) Hydroxyapatite coatings grown by pulsed laser deposition with a beam of 355 nm wavelength. J Mater Res 14:4715–4719

Fernandez-Pradas JM, Cleries L, Martinez E, Sardin G, Esteve J, Morenza JL (2000) Calcium phosphate coatings deposited by laser ablation at 355 nm under different substrate temperatures and water vapour pressures. Appl Phys A 71:37–42

Fernandez-Pradas JM, Cleries L, Martinez E, Sardin G, Esteve J, Morenza JL (2001) Influence of thickness on the properties of hydroxyapatite coatings deposited by KrF laser ablation. Biomaterials 22:2171–2175

Filiaggi MJ, Coombs NA, Pilliar RM (1991) Characterization of the interface in plasma-sprayed HA coating/Ti–6Al–4 V implant system. J Biomed Mater Res 25:1211–1229

Freidberg JP (2007) Plasma physics and fusion energy. Cambridge University Press, Cambridge, UK, p 692

Fujihara T, Tsukamoto M, Abe N, Miyake S, Ohji T, Akedo J (2004) Hydroxyapatite film formed by beam irradiation. Vacuum 73:629–633

Furlong RJ, Osborn JF (1991) Fixation of hip prostheses by hydroxyapatite ceramic coating. J Bone Joint Surg B 73:741–745

Gan L, Pilliar R (2004) Calcium phosphate sol–gel-derived thin films on porous surfaced implants for enhanced osteoconductivity. Part I: Synthesis and characterization Biomaterials 25:5303–5312

Gan L, Wang J, Tache A, Valiquette N, Deporter D, Pilliar R (2004) Calcium phosphate sol–gel-derived thin films on porous-surfaced implants for enhanced osteoconductivity. Part II: short-term in vivo studies. Biomaterials 25:5313–5321

Gandhi R, Davey JR, Mahomed NN (2009) Hydroxyapatite coated femoral stems in primary total hip arthroplasty. A meta-analysis J Arthroplasty 24:38–42

Gärtner F, Stoltenhoff T, Schmidt T, Kreye H (2006) The cold spray process and its potential for industrial applications. J Thermal Spray Technol 15:223–232

Geesink RGT (1990) Hydroxyapatite-coated total hip prostheses; two-year clinical and roentgenographic results of 100 cases. Clin Orthop Rel Res 261:39–58

Geesink RGT (2002) Osteoconductive coating for total joint arthroplasty. Clin Orthop Rel Res 395:53–65

Geesink RGT, Hoefnagels NHM (1995) Six-year results of hydroxyapatite-coated total hip replacement. J Bone Jt Surg Br 77B:534–547

Geesink RGT, de Groot K, Klein CPAT (1987) Chemical implant fixation using hydroxyl-apatite coatings. The development of a human total hip prosthesis for chemical fixation to bone using hydroxyl-apatite coatings on titanium substrates. Clin Orthop Rel Res 225:147–170

Gill WD, Kay E (1965) Efficient low voltage sputtering in a large inverted magnetron suitable for film synthesis. Rev Scientif Instrum 36:277–282

Gineste L, Gineste M, Ranz X, Ellefterion A, Guilhem A, Rouquet N, Frayssinet P (1999) Degradation of hydroxylapatite, fluorapatite, and fluorhydroxyapatite coatings of dental implants in dogs. J Biomed Mater Res 48:224–234

Gledhill HC, Turner IG, Doyle C (1999) Direct morphological comparison of vacuum plasma sprayed and detonation gun sprayed hydroxyapatite coatings. Biomaterials 20:315–322

Gledhill HC, Turner IG, Doyle C (2001) In vitro fatigue behavior of vacuum plasma and detonation gun sprayed hydroxyapatite coatings. Biomaterials 22:1233–1240

Gottlander M, Johansson CB, Wennerberg A, Albrektsson T, Radin S, Ducheyne P (1997a) Bone tissue reactions to an electrophoretically applied calcium phosphate coating. Biomaterials 18:551–557

Gottlander M, Johansson CB, Albrektsson T (1997b) Short- and long-term animal studies with a plasma-sprayed calcium phosphate-coated implant. Clin Oral Implant Res 8:345–351

Granato R, Marin C, Suzuki M, Gil JN, Janal MN, Coelho PG (2009) Biomechanical and histomorphometric evaluation of a thin ion beam bioceramic deposition on plateau root form implants: an experimental study in dogs. J Biomed Mater Res B Appl Biomater 90B:396–403

Gross KA, Babovic M (2002) Influence of abrasion on the surface characteristics of thermally sprayed hydroxyapatite coatings. Biomaterials 23:4731–4737

Gross KA, Berndt CC (1994) In vitro testing of plasma-sprayed hydroxyapatite coatings. J Mater Sci Mater Med 5:219–224

Gross KA, Saber-Samandari S (2009) Revealing mechanical properties of a suspension plasma sprayed coating with nanoindentation. Surf CoatTechnol 203:2995–2999

Gross KA, Berndt CC, Goldschlag DD, Iacono VJ (1997) In vitro changes of hydroxyapatite coatings. Int J Oral Maxillofac Implants 12:589–597

Gross KA, Gross V, Berndt CC (1998a) Thermal analysis of amorphous phases in hydroxyapatite coatings. J Am Ceram Soc 81:106–112

Gross KA, Berndt CC, Herman H (1998b) Amorphous phase formation in plasma-sprayed hydroxyapatite coatings. J Biomed Mater Res 39:407–414

Gross KA, Chai CS, Kannangara GSK, Ben-Nissan B, Hanley L (1998c) Thin hydroxyapatite coatings via sol–gel synthesis. J Mater Sci Mater Med 9:839–843

Gu Y, Meng G (1999) A model for ceramic membrane formation by dip-coating. J Eur Ceram Soc 19:1961–1966

Habibovic P, Barre're F, van Blitterswijk CA, de Groot K, Layrolle P (2002) Biomimetic hydroxyapatite coating on metal implants. J Am Ceram Soc 85:517–522

Habibovic P, Li J, van der Valk CM, Meijer G, Layrolle P, van Blitterswijk CA, de Groot K (2005) Biological performance of uncoated and octacalcium phosphate-coated Ti6Al4V. Biomaterials 26:23–36

Haddow DB, James PF, van Noort R (1999) Sol–gel derived calcium phosphate coatings for biomedical applications. J Sol–gel Part Sci Technol 13:261–265

Hahn BD, Park DS, Choi JJ, Ryu J, Yoon WH, Kim KH, Park C, Kim HE (2009) Dense nanostructured hydroxyapatite coating on titanium by aerosol deposition. J Am Ceram Soc 92:683–687

Haman JD, Chittur KK, Crawmer DE, Lucas LC (1999) Analytical and mechanical testing of high velocity oxy-fuel thermal sprayed and plasma sprayed calcium phosphate coatings. J Biomed Mater Res (Appl Biomater) 48:856–860

Hamdi M, Ide-Ektessabi A (2003) Preparation of hydroxyapatite layer by ion beam assisted simultaneous vapor deposition. Surf CoatTechnol 163–164:362–367

Hamdi M, Toque JA, Ide-Ektessabi A (2010) Wear characteristics and adhesion behavior of calcium phosphate thin-films. Key Eng Mater 443:469–474

Han Y, Xu KW, Lu J, Wu Z (1999a) The structural characteristics and mechanical behaviors of nonstoichiometric apatite coatings sintered in air atmosphere. J Biomed Mater Res 45:198–203

Han Y, Xu K, Lu J (1999b) Morphology and composition of hydroxyapatite coatings prepared by hydrothermal treatment on electrodeposited brushite coatings. J Mater Sci Mater Med 10:243–248

Han Y, Fu T, Lu J, Xu K (2001) Characterization and stability of hydroxyapatite coatings prepared by an electrodeposition and alkaline-treatment process. J Biomed Mater Res 54:96–101

Han Y, Sun J, Huang X (2008) Formation mechanism of HA-based coatings by micro-arc oxidation. Electrochem Comm 10:510–513

Hanawa T, Ota M (1992) Characterization of surface film formed on titanium in electrolyte using XPS. Appl Surf Sci 55:269–276

Hasan S, Stokes J (2011) Design of experiment analysis of the Sulzer Metco DJ high velocity oxy-fuel coating of hydroxyapatite for orthopedic applications. J Thermal Spray Technol 20:186–194

Hashimoto Y, Kawashima M, Hatanaka R, Kusunoki M, Nishikawa H, Hontsu S, Nakamura M (2008) Cytocompatibility of calcium phosphate coatings deposited by an ArF pulsed laser. J Mater Sci Mater Med 19:327–333

Hayakawa T, Yoshinari M, Nemoto K, Wolke JGC, Jansen JA (2000) Effect of surface roughness and calcium phosphate coating on the implant/bone response. Clin Oral Implant Res 11:296–304

Hayakawa T, Yoshinari M, Kiba H, Yamamoto H, Nemoto K, Jansen JA (2002) Trabecular bone response to surface roughened and calcium phosphate (Ca-P) coated titanium implants. Biomaterials 23:1025–1031

Heimann RB (2006) Thermal spraying of biomaterials. Surf CoatTechnol 201:2012–2019

Heimann RB (2009) Characterization of as-plasma-sprayed and incubated hydroxyapatite coatings with high resolution techniques. Mater Werkst 40:23–30

Heimann RB, Wirth R (2006) Formation and transformation of amorphous calcium phosphates on titanium alloy surfaces during atmospheric plasma spraying and their subsequent in vitro performance. Biomaterials 27:823–831

Heimann RB, Schürmann N, Müller RT (2004) In vitro and in vivo performance of Ti6Al4V implants with plasma-sprayed osteoconductive hydroxylapatite-bioinert titania bond coat "duplex" systems: an experimental study in sheep. J Mater Sci Mater Med 15:1045–1052

Hench LL (1991) Bioceramics: from concept to clinic. J Am Ceram Soc 74:1487–1510

Herman H (1988) Plasma-sprayed coatings. Sci Am 9:112–117

Herø H, Wie H, Jorgensen RB, Ruyter IE (1994) Hydroxyapatite coatings on Ti produced by hot isostatic pressing. J Biomed Mater Res 28:343–348

Hontsu S, Matsumoto T, Ishii J, Nakamori M, Tabata H, Kawai T (1997) Electrical properties of hydroxyapatite thin films grown by pulsed laser deposition. Thin Solid Films 295:214–217

Hou X, Choy KL, Leach SE (2007) Processing and in vitro behavior of hydroxyapatite coatings prepared by electrostatic spray assisted vapor deposition method. J Biomed Mater Res A 83A:683–691

Huang Y, Qu Y, Yang B, Li W, Zhang B, Zhang X (2009) In vivo biological responses of plasma sprayed hydroxyapatite coatings with an electric polarized treatment in alkaline solution. Mater Sci Eng C 29:2411–2416

Huang Y, Song L, Liu X, Xiao Y, Wu Y, Chen J, Wu F, Gu Z (2010) Hydroxyapatite coatings deposited by liquid precursor plasma spraying: controlled dense and porous microstructures and osteoblastic cell responses. Biofabrication 2:045003

Hughes JM, Kohn M, Rakovan J (eds) (2002) Reviews in mineralogy and geochemistry. Phosphates: geochemical, geobiological and materials importance. vol. 48. Mineralogical Society of America, Washington, DC, USA, p 742

Hulshoff JEG, Jansen JA (1997) Initial interfacial healing events around calcium phosphate (Ca-P) coated oral implants. Clin Oral Implant Res 8:393–400

Hulshoff JEG, van Dijk K, van der Waerden JPCM, Kalk W, Jansen JA (1996a) A histological and histomorphometrical evaluation of screw-type calciumphosphate (Ca-P) coated implants; an in vivo experiment in maxillary cancellous bone of goats. J Mater Sci Mater Med 7:603–609

Hulshoff JEG, van Dijk K, van Der Waerden JPCM, Wolke JGC, Kalk W, Jansen JA (1996b) Evaluation of plasma-spray and magnetron-sputter Ca-P-coated implants: an in vivo experiment using rabbits. J Biomed Mater Res 31:329–337

Hulshoff JEG, Hayakawa T, van Dijk K, Leijdekkers-Govers AFM, van der Waerden JPCM, Jansen JA (1997) Mechanical and histologic evaluation of Ca-P plasma-spray and magnetron sputter-coated implants in trabecular bone of the goat. J Biomed Mater Res 36:75–83

Iafisco M, Bosco R, Leeuwenburgh SCG, van den Beucken JJJP, Jansen JA, Prat M, Roveri N (2012) Electrostatic spray deposition of biomimetic nanocrystalline apatite coatings onto titanium. Adv Eng Mater 14:B13–B20

Ievlev VM, Domashevskaya EP, Putlyaev VI, Tret'yakov YD, Barinov SM, Belonogov EK, Kostyuchenko AV, Petrzhik MI, Kiryukhantsev-Korneev FV (2008) Structure, elemental composition, and mechanical properties of films prepared by radio-frequency magnetron sputtering of hydroxyapatite. Glass Phys Chem 34:608–616

Iezzi G, Scarano A, Petrone G, Piattelli A (2007) Two human hydroxyapatite-coated dental implants retrieved after a 14-year loading period: a histologic and histomorphometric case report. J Periodontol 78:940–947

ISO (1996) Implants for surgery: coating for hydroxyapatite ceramics. ISO, Geneva, pp 1–8

ISO (2000) ISO 13779-2 Implants for surgery – Hydroxyapatite – Part 2: Coatings of hydroxyapatite. http://www.iso.org/iso/iso_catalogue/catalogue_tc/catalogue_detail.htm?csnumber=26841. Accessed

ISO (2005) ISO 20502. Fine ceramics (advanced ceramics, advanced technical ceramics) – determination of adhesion of ceramic coatings by scratch testing. http://www.iso.org/iso/iso_catalogue/catalogue_tc/catalogue_detail.htm?csnumber=34189. Accessed

ISO (2008) ISO 13779-2 Implants for surgery – Hydroxyapatite – Part 2: Coatings of hydroxyapatite. http://www.iso.org/iso/iso_catalogue/catalogue_tc/catalogue_detail.htm?csnumber=43827

Jansen JA, van der Waerden JPCM, Wolke JGC, de Groot K (1991) Histologic evaluation of the osseous adaptation to titanium hydroxyapatite-coated implants. J Biomed Mater Res 25:973–989

Jansen JA, Wolke JGC, Swann S, van der Waerden JPCM, de Groot K (1993) Application of magnetron sputtering for producing ceramic coatings on implant materials. Clin Oral Impl Res 4:28–34

Jaworski R, Pierlot C, Pawlowski L, Bigan M, Martel M (2009) Design of the synthesis of fine HA powder for suspension plasma spraying. Surf CoatTechnol 203:2092–2097

Jedynski M, Hoffman J, Mroz W, Szymanski Z (2008) Plasma plume induced during ArF laser ablation of hydroxyapatite. Appl Surf Sci 255:2230–2236

Ji H, Marquis PM (1993) Effect of heat treatment on the microstructure of plasma-sprayed hydroxyapatite coating. Biomaterials 14:64–68

Jiang G, Shi D (1998) Coating of hydroxyapatite on highly porous Al_2O_3 substrate for bone substitutes. J Biomed Mater Res (Appl Biomater) 43:77–81

Jiang W, Sun L, Nyandoto G, Malshe AP (2008) Electrostatic spray deposition of nanostructured hydroxyapatite coating for biomedical applications. J Manufact Sci Eng Transact ASME 130:0210011–0210017

Johnson S, Haluska M, Narayan RJ, Snyder RL (2006) In situ annealing of hydroxyapatite thin films. Mater Sci Eng C 26:1312–1316

Jung YC, Han CH, Lee IS, Kim HE (2001) Effects of ion beam-assisted deposition of hydroxyapatite on the osseointegration of endosseous implants in rabbit tibiae. Int J Oral Maxillofac Implants 16:809–818

Junker R, Manders PJD, Wolke J, Borisov Y, Jansen JA (2010) Bone-supportive behavior of microplasma-sprayed CaP-coated implants: mechanical and histological outcome in the goat. Clin Oral Implant Res 21:189–200

Kameyama T (1999) Hybrid bioceramics with metals and polymers for better biomaterials. Bull Mater Sci 22:641–646

Katayama H, Katto M, Nakayama T (2009) Laser-assisted laser ablation method for high-quality hydroxyapatite coating onto titanium substrate. Surf CoatTechnol 204:135–140

Kelly PJ, Arnell RD (2000) Magnetron sputtering: a review of recent developments and applications. Vacuum 56:159–172

Khor KA, Cheang P (1993) Characterization of plasma sprayed hydroxyapatite powders and coatings. In: Berndt CC, Bernecki TF (eds) Thermal spray coatings: research, design and applications. ASM International, Materials Park, Ohio, USA, pp 347–352

Khor KA, Li H, Cheang P (2003a) Characterization of the bone-like apatite precipitated on high velocity oxy-fuel (HVOF) sprayed calcium phosphate deposits. Biomaterials 24:769–775

Khor KA, Li H, Cheang P (2003b) Processing-microstructure-property relations in HVOF sprayed calcium phosphate based bioceramic coatings. Biomaterials 24:2233–2243

Khor KA, Li H, Cheang P (2004) Significance of melt-fraction in HVOF sprayed hydroxyapatite particles, splats and coatings. Biomaterials 25:1177–1186

Kim HM, Miyaji F, Kokubo T, Nakamura T (1997) Bonding strength of bonelike apatite layer to Ti metal substrate. J Biomed Mater Res 38:121–127

Kim TN, Feng QL, Luo ZS, Cui FZ, Kim JO (1998) Highly adhesive hydroxyapatite coatings on alumina substrates prepared by ion-beam assisted deposition. Surf CoatTechnol 99:20–23

Kim HM, Kishimoto K, Miyaji F, Kokubo T, Yao T, Suetsugu Y, Tanaka J, Nakamura T (2000) Composition and structure of apatite formed on organic polymer in simulated body fluid with a high content of carbonate ion. J Mater Sci Mater Med 11:421–426

Kim HW, Knowles JC, Salih V, Kim HE (2004a) Hydroxyapatite and fluor-hydroxyapatite layered film on titanium processed by a sol–gel route for hard-tissue implants. J Biomed Mater Res B Appl Biomater 71B:66–76

Kim HW, Koh YH, Li LH, Lee S, Kim HE (2004b) Hydroxyapatite coating on titanium substrate with titania buffer layer processed by sol–gel method. Biomaterials 25:2533–2538

Kim H, Vohra YK, Louis PJ, Lacefield WR, Lemons JE, Camata RP (2005a) Biphasic and preferentially oriented microcrystalline calcium phosphate coatings: in-vitro and in-vivo studies. Key Eng Mater 284–286:207–210

Kim H, Camata RP, Vohra YK, Lacefield WR (2005b) Control of phase composition in hydroxyapatite/tetracalcium phosphate biphasic thin coatings for biomedical applications. J Mater Sci Mater Med 16:961–966

Kim H, Camata RP, Lee S, Rohrer GS, Rollett AD, Vohra YK (2007a) Crystallographic texture in pulsed laser deposited hydroxyapatite bioceramic coatings. Acta Mater 55:131–139

Kim BH, Jeong JH, Jeon YS, Jeon KO, Hwang KS (2007b) Hydroxyapatite layers prepared by sol–gel assisted electrostatic spray deposition. Ceram Int 33:119–122

Kim H, Camata RP, Chowdhury S, Vohra YK (2010) In vitro dissolution and mechanical behavior of c-axis preferentially oriented hydroxyapatite thin films fabricated by pulsed laser deposition. Acta Biomater 6:3234–3241

Klein CPAT, Patka P, van der Lubbe HBM, Wolcke JGC, de Groot K (1991) Plasma-sprayed coatings of tetracalcium phosphate, hydroxylapatite, and α-TCP on titanium alloy: an interface study. J Biomed Mater Res 25:53–65

Klein CPAT, Patka P, Wolke JGC, de Blieck-Hogervorst JMA, de Groot K (1994a) Long-term in vivo study of plasma-sprayed coatings on titanium alloys of tetracalcium phosphate, hydroxyapatite and α-tricalcium phosphate. Biomaterials 15:146–150

Klein CPAT, Wolke JGC, de Blieck-Hogervorst JMA, de Groot K (1994b) Calcium phosphate plasma-sprayed coatings and their stability: an in vivo study. J Biomed Mater Res 28:909–917

Kobayashi T, Itoh S, Nakamura S, Nakamura M, Shinomiya K, Yamashita K (2007) Enhanced bone bonding of hydroxyapatite-coated titanium implants by electrical polarization. J Biomed Mater Res A 82A:145–151

Koch B, Wolke JGC, de Groot K (1990) X-ray diffraction studies on plasma-sprayed calcium phosphate-coated implants. J Biomed Mater Res 24:655–667

Koch CF, Johnson S, Kumar D, Jelinek M, Chrisey DB, Doraiswamy A, Jin C, Narayan RJ, Mihailescu IN (2007) Pulsed laser deposition of hydroxyapatite thin films. Mater Sci Eng C 27:484–494

Kokubo T (ed) (2008) Bioceramics and their clinical applications. Woodhead Publishing, Abington, Cambridge, UK, p 784

Kokubo T, Yamaguchi S (2011) Bioactive layer formation on metals and polymers. In: Ducheyne P, Healy K, Hutmacher DW, Grainger DW, Kirkpatrick CJ (eds) Comprehensive biomaterials, vol. 1. Elsevier, Amsterdam, Netherlands, pp 231–244

Kokubo T, Miyaji F, Kim HM (1996) Spontaneous formation of bonelike apatite layer on chemically treated titanium metals. J Am Ceram Soc 79:1127–1129

Kokubo T, Kim HM, Kawashita M (2003) Novel bioactive materials with different mechanical properties. Biomaterials 24:2161–2175

Kumar M, Dasarathy H, Riley C (1999) Electrodeposition of brushite coatings and their transformation to hydroxyapatite in aqueous solutions. J Biomed Mater Res 45:302–310

Kummer FJ, Jaffe WL (1992) Stability of a cyclically loaded hydroxylapatite coating: effect of substrate material, surface, preparation and testing environment. J Appl Mater 3:211–215

Kuo MC, Yen SK (2002) The process of electrochemical deposited hydroxyapatite coatings on biomedical titanium at room temperature. Mater Sci Eng C 20:153–160

Kurella A, Dahotre NB (2006) A multi-textured calcium phosphate coating for hard tissue via laser surface engineering. JOM 58:64–66

Kuroda K, Miyashita Y, Ichino R, Okido M, Takai O (2002a) Preparation of calcium phosphate coatings on titanium using the thermal substrate method and their in vitro evaluation. Mater Transact 43:3015–3019

Kuroda K, Ichino R, Okido M, Takai O (2002b) Effects of ion concentration and pH on hydroxyapatite deposition from aqueous solution onto titanium by the thermal substrate method. J Biomed Mater Res 61:354–359

Kuroda K, Miyashita Y, Ichino R, Okido M (2003) Hydroxyapatite coating on titanium by thermal substrate method in an aqueous solution and its behavior in SBF. Mater Sci Forum 426–432:3189–3194

Kuroda K, Nakamoto S, Ichino R, Okido M, Pilliar RM (2005) Hydroxyapatite coatings on a 3D porous surface using thermal substrate method. Mater Transact 46:1633–1635

Kuroda K, Nakamoto S, Miyashita Y, Ichino R, Okido M (2007) Osteoinductivity of hydroxyapatite films with different surface morphologies coated by the thermal substrate method in aqueous solutions. J Jpn Inst Metals 71:342–345

Kweh SWK, Khor KA, Cheang P (2000) Plasma-sprayed hydroxyapatite (HA) coatings with flame-spheroidized feedstock: microstructure and mechanical properties. Biomaterials 21:1223–1234

Lacefield WR (1988) Hydroxyapatite coatings. Ann New York Acad Sci 523:72–80

Łatka L, Pawlowski L, Chicot D, Pierlot C, Petit F (2010) Mechanical properties of suspension plasma sprayed hydroxyapatite coatings submitted to simulated body fluid. Surf CoatTechnol 205:954–960

Lavos-Valereto IC, Wolynec S, Deboni MCZ, Knig B Jr (2001) In vitro and in vivo biocompatibility testing of Ti-6Al-7Nb alloy with and without plasma-sprayed hydroxyapatite coating. J Biomed Mater Res 58:727–733

Layrolle P (2011) Calcium phosphate coatings. In: Ducheyne P, Healy K, Hutmacher DW, Grainger DW, Kirkpatrick CJ (eds) Comprehensive biomaterials, vol. 1. Elsevier, Amsterdam, Netherlands, pp 223–229, Vol. 1

Lazarinis S, Krärholm J, Hailer NP (2010) Increased risk of revision of acetabular cups coated with hydroxyapatite: a Swedish Hip Arthroplasty Register study involving 8,043 total hip replacements. Acta Orthop 81:53–59

Lazarinis S, Krrholm J, Hailer NP (2011) Effects of hydroxyapatite coating on survival of an uncemented femoral stem. Acta Orthop 82:399–404

Lee JJ, Rouhfar L, Beirne OR (2000) Survival of hydroxyapatite-coated implants: a meta-analytic review. J Oral Maxillofac Surg 58:1372–1379

Lee IS, Whang CN, Lee GH, Cui FZ, Ito A (2003) Effects of ion beam assist on the formation of calcium phosphate film. Nucl Instr Methods Phys Res B 206:522–526

Lee YP, Wang CK, Huang TH, Chen CC, Kao CT, Ding SJ (2005a) In vitro characterization of post heat-treated plasma-sprayed hydroxyapatite coatings. Surf CoatTechnol 197:367–374

Lee EJ, Lee SH, Kim HW, Kong YM, Kim HE (2005b) Fluoridated apatite coatings on titanium obtained by electron-beam deposition. Biomaterials 26:3843–3851

Lee IS, Zhao B, Lee GH, Choi SH, Chung SM (2007a) Industrial application of ion beam assisted deposition on medical implants. Surf CoatTechnol 201:5132–5137

Lee WH, Kim YM, Oh NH, Cheon YW, Cho YJ, Lee CM, Kim KB, Lee NS (2007b) A study of hydroxyapatite coating on porous Ti compact by electrostatic spray deposition. Diffusion and Defect Data B. Solid State Phenomena 124–126:1789–1792

Lee JH, Kim SG, Lim SC (2011) Histomorphometric study of bone reactions with different hydroxyapatite coating thickness on dental implants in dogs. Thin Solid Films 519:4618–4622

Leeuwenburgh S, Wolke J, Schoonman J, Jansen J (2003) Electrostatic spray deposition (ESD) of calcium phosphate coatings. J Biomed Mater Res A 66A:330–334

Leeuwenburgh SC, Wolke JG, Schoonman J, Jansen JA (2004) Influence of precursor solution parameters on chemical properties of calcium phosphate coatings prepared using electrostatic spray deposition (ESD). Biomaterials 25:641–649

Leeuwenburgh SCG, Wolke JGC, Schoonman J, Jansen JA (2005a) Influence of deposition parameters on morphological properties of biomedical calcium phosphate coatings prepared using electrostatic spray deposition. Thin Solid Films 472:105–113

Leeuwenburgh S, Wolke J, Schoonman J, Jansen JA (2005b) Influence of deposition parameters on chemical properties of calcium phosphate coatings prepared by using electrostatic spray deposition. J Biomed Mater Res A 74A:275–284

Leeuwenburgh SCG, Wolke JGC, Schoonman J, Jansen JA (2006a) Deposition of calcium phosphate coatings with defined chemical properties using the electrostatic spray deposition technique. J Eur Ceram Soc 26:487–493

Leeuwenburgh SCG, Heine MC, Wolke JGC, Pratsinis SE, Schoonman J, Jansen JA (2006b) Morphology of calcium phosphate coatings for biomedical applications deposited using electrostatic spray deposition. Thin Solid Films 503:69–78

LeGeros RZ (1991) Calcium phosphates in oral biology and medicine. In: Myers HM (ed) Monographs in oral science, vol. 15. Karger, Basel, Switzerland, p 201

LeGeros RZ (2008) Calcium phosphate-based osteoinductive materials Chem Rev 108:4742–4753

Leitão E, Barbosa MA, de Groot K (1995) In vitro calcification of orthopaedic implant materials. J Mater Sci Mater Med 6:849–852

León B, Jansen JA (eds) (2009) Thin calcium phosphate coatings for medical implants. Springer, New York, USA, p 326

Li P (2003) Biomimetic nano-apatite coating capable of promoting bone ingrowth. J Biomed Mater Res A 66A:79–85

Li P, Ohtsuki C, Kokubo T, Nakanishi K, Soga N (1992) Apatite formation induced by silica gel in a simulated body fluid. J Am Ceram Soc 75:2094–2097

Li P, Ohtsuki C, Kokubo T, Nakanishi K, Soga N, de Groot K (1994) The role of hydrated silica, titania, and alumina in inducing apatite on implants. J Biomed Mater Res 28:7–15

Li T, Lee J, Kobayashi T, Aoki H (1996) Hydroxyapatite coating by dipping method, and bone bonding strength. J Mater Sci Mater Med 7:355–357

Li H, Khor KA, Cheang P (2000) Effect of the powders' melting state on the properties of HVOF sprayed hydroxyapatite coatings. Mater Sci Eng, A 293:71–80

Li H, Khor KA, Cheang P (2002a) Properties of heat-treated calcium phosphate coatings deposited by high-velocity oxy-fuel (HVOF) spray. Biomaterials 23:2105–2112

Li F, Feng QL, Cui FZ, Li HD, Schubert H (2002b) A simple biomimetic method for calcium phosphate coating. Surf CoatTechnol 154:88–93

Liang F, Zhou L, Wang K (2003) Apatite formation on porous titanium by alkali and heat-treatment. Surf CoatTechnol 165:133–139

Lim YM, Kim BH, Jeon YS, Jeon KO, Hwang KS (2005) Calcium phosphate films deposited by electrostatic spray deposition and an evaluation of their bioactivity. J Ceram Process Res 6:255–258

Lin S, LeGeros RZ, LeGeros JP (2003) Adherent octacalciumphosphate coating on titanium alloy using modulated electrochemical deposition method. J Biomed Mater Res A 66A:819–828

Liu DM, Troczynski T, Hakimi D (2002a) Effect of hydrolysis on the phase evolution of water-based sol-gel hydroxyapatite and its application to bioactive coatings. J Mater Sci Mater Med 13:657–665

Liu D, Yang Q, Troczynski T (2002b) Sol-gel hydroxyapatite coatings on stainless steel substrates. Biomaterials 23:691–698

Liu X, Chu PK, Ding C (2004) Surface modification of titanium, titanium alloys, and related materials for biomedical applications. Mater Sci Eng R 47:49–121

Liu F, Song Y, Wang F, Shimizu T, Igarashi K, Zhao L (2005) Formation characterization of hydroxyapatite on titanium by microarc oxidation and hydrothermal treatment. J Biosci Bioeng 100:100–104

Lo WJ, Grant DM, Ball MD, Welsh BS, Howdle SM, Antonov EN, Bagratashvili VN, Popov VK (2000) Physical, chemical, and biological characterization of pulsed laser deposited and plasma sputtered hydroxyapatite thin films on titanium alloy. J Biomed Mater Res 50:536–545

Long J, Sim L, Xu S, Ostrikov K (2007) Reactive plasma-aided RF sputtering deposition of hydroxyapatite bio-implant coatings. Chem Vapor Deposition 13:299–306

Lopez-Heredia MA, Weiss P, Layrolle P (2007) An electrodeposition method of calcium phosphate coatings on titanium alloy. J Mater Sci Mater Med 18:381–390

Lu X, Zhao Z, Leng Y (2005) Calcium phosphate crystal growth under controlled atmosphere in electrochemical deposition. J Cryst Growth 284:506–516

Lü X, Lin X, Guan T, Gao B, Huang W (2011a) Effect of the mass ratio of $CaCO_3$ to $CaHPO_4 \cdot 2H_2O$ on in situ synthesis of hydroxyapatite coating by laser cladding. Rare Metal Mater Eng 40:22–27

Lü X, Lin X, Cao Y, Hu J, Gao B, Huang W (2011b) Effects of processing parameters and heat treatment on phase structure of the hydroxyapatite coating on pure Ti surface by laser cladding in-situ synthesis. Rare Metal Mater Eng 40:714–717

Luo ZS, Cui FZ, Li WZ (1999) Low-temperature crystallization of calcium phosphate coatings synthesized by ion beam-assisted deposition. J Biomed Mater Res 46:80–86

Luo ZS, Cui FZ, Feng QL, Li HD, Zhu XD, Spector M (2000) In vitro and in vivo evaluation of degradability of hydroxyapatite coatings synthesized by ion beam-assisted deposition. Surf CoatTechnol 131:192–195

Lusquiños F, Pou J, Arias JL, Boutinguiza M, Léon B, Pérez-Amor M, Driessens FCM, Merry JC, Gibson I, Best S, Bonfield W (2001) Production of calcium phosphate coatings on Ti6Al4V obtained by Nd: yttrium-aluminum-garnet laser cladding. J Appl Phys 90:4231–4236

Lusquiños F, de Carlos A, Pou J, Arias JL, Boutinguiza M, León B, Pérez-Amor M, Driessens FCM, Hing K, Gibson I, Best S, Bonfield W (2003) Calcium phosphate coatings obtained by Nd: YAG laser cladding: physicochemical and biologic properties. J Biomed Mater Res A 64A:630–637

Lusquiños F, Pou J, Boutinguiza M, Quintero F, Soto R, León B, Pérez-Amor M (2005) Main characteristics of calcium phosphate coatings obtained by laser cladding. Appl Surf Sci 247:486–492

Lv X, Lin X, Hu J, Gao B, Huang W (2012) Phase evolution in calcium phosphate coatings obtained by in situ laser cladding. Mater Sci Eng C 32:872–877

Ma J, Liang CH, Kong LB, Wang C (2003a) Colloidal characterization and electrophoretic deposition of hydroxyapatite on titanium substrate. J Mater Sci Mater Med 14:797–801

Ma J, Wang C, Peng KW (2003b) Electrophoretic deposition of porous hydroxyapatite scaffold. Biomaterials 24:3505–3510

Maistrelli GL, Mahomed N, Fornasier V, Antonelli L, Li Y, Binnington A (1993) Functional osseointegration of hydroxyapatite-coated implants in a weight bearing canine model. J Arthroplasty 8:549–554

Man HC, Chiu KY, Cheng FT, Wong KH (2009) Adhesion study of pulsed laser deposited hydroxyapatite coating on laser surface nitrided titanium. Thin Solid Films 517:5496–5501

MaNally SA, Shepperd HAN, Mann CV, Walczak JP (2000) The results at nine to twelve years of the use of a hydroxyapatite-coated femoral stem. J Bone Jt Surg Br 82B:378–382

Manders PJD, Wolke JGC, Jansen JA (2006) Bone response adjacent to calcium phosphate electrostatic spray deposition coated implants: an experimental study in goats. Clin Oral Implant Res 17:548–553

Manso M, Jimenez C, Morant C, Herrero P, Martinez-Duart JM (2000) Electrodeposiiton of hydroxyapatite coatings in basic conditions. Biomaterials 21:1755–1761

Manso M, Langletm M, Jimenezm C, Martinez-Duart JM (2002a) Microstructural study of aerosol-gel derived hydroxyapatite coatings. Biomol Eng 19:63–66

Manso M, Martínez-Duart JM, Langlet M, Jiménez C, Herrero P, Millon E (2002b) Aerosol-gel-derived microcrystalline hydroxyapatite coatings. J Mater Res 17:1482–1489

Manso M, Ogueta S, Herrero-Fernández P, Vázquez L, Langlet M, García-Ruiz JP (2002c) Biological evaluation of aerosol-gel-derived hydroxyapatite coatings with human mesenchymal stem cells. Biomaterials 23:3985–3990

Manso-Silván M, Langlet M, Jiménez C, Fernández M, Martínez-Duart JM (2003) Calcium phosphate coatings prepared by aerosol-gel. J Eur Ceram Soc 23:243–246

Massaro C, Baker MA, Cosentino F, Ramires PA, Klose S, Milella E (2001) Surface and biological evaluation of hydroxyapatite-based coatings on titanium deposited by different techniques. J Biomed Mater Res 58:651–657

Matsumine A, Myoui A, Kusuzaki K, Araki N, Seto M, Yoshikawa H, Uchida A (2004) Calcium hydroxyapatite ceramic implants in bone tumor surgery. A long-term follow-up study. J Bone Joint Surg B 86:719–725

Mayor B, Arias J, Chiussi S, Garcia F, Pou J, Fong BL, Pérez-Amor M (1998) Calcium phosphate coatings grown at different substrate temperatures by pulsed ArF-laser deposition. Thin Solid Films 317:363–366

McHugh TH (2000) Protein-lipid interactions in edible films and coatings. Nahrung 44:148–151

McPherson R, Gane N, Bastow TJ (1995) Structural characterization of plasma-sprayed hydroxylapatite coatings. J Mater Sci Mater Med 6:327–334

Mello A, Hong Z, Rossi AM, Luan L, Farina M, Querido W, Eon J, Terra J, Balasundaram G, Webster T, Feinerman A, Ellis DE, Ketterson JB, Ferreira CL (2007) Osteoblast proliferation on hydroxyapatite thin coatings produced by right angle magnetron sputtering. Biomed Mater 2:67–77

Meng X, Kwon TY, Kim KH (2006) Different morphology of hydroxyapatite coatings on titanium by electrophoretic deposition. Key Eng Mater 309–311:639–642

Meng X, Kwon TY, Kim KH (2008) Hydroxyapatite coating by electrophoretic deposition at dynamic voltage. Dent Mater J 27:666–671

Mennicke U, Salditt T (2002) Preparation of solid-supported lipid bilayers by spin-coating. Langmuir 18:8172–8177

Metikoš-Huković M, Tkalacec E, Kwokal A, Piljac J (2003) An in vitro study of Ti and Ti-alloys coated with sol–gel derived hydroxyapatite coatings. Surf CoatTechnol 165:40–50

Miroiu FM, Socol G, Visan A, Stefan N, Craciun D, Craciun V, Dorcioman G, Mihailescu IN, Sima LE, Petrescu SM, Andronie A, Stamatin I, Moga S, Ducu C (2010) Composite biocompatible hydroxyapatite-silk fibroin coatings for medical implants obtained by matrix assisted pulsed laser evaporation. Mater Sci Eng B 169:151–158

Mistry S, Kundu D, Datta S, Basu D (2011) Comparison of bioactive glass coated and hydroxyapatite coated titanium dental implants in the human jaw bone. Australian Dent J 56:68–75

Miyaji F, Kim HM, Handa S, Kokubo T, Nakamura T (1999) Bonelike apatite coating on organic polymers: novel nucleation process using sodium silicate solution. Biomaterials 20:913–919

Mohammadi S, Esposito M, Hall J, Emanuelsson L, Krozer A, Thomsen P (2003) Short-term bone response to titanium implants coated with thin radiofrequent magnetron-sputtered hydroxyapatite in rabbits. Clin Implant Dent Rel Res 5:241–253

Mohammadi S, Esposito M, Hall J, Emanuelsson L, Krozer A, Thomsen P (2004) Long-term bone response to titanium implants coated with thin radiofrequent magnetron-sputtered hydroxyapatite in rabbits. Int J Oral Maxillofac Implants 19:498–509

Mondragón-Cortez P, Vargas-Gutiérrez G (2004) Electrophoretic deposition of hydroxyapatite submicron particles at high voltages. Mater Lett 58:1336–1339

Montenero A, Gnappi G, Ferrari F, Cesari M, Salvioli E, Mattogno L, Kaciulis S, Fini M (2000) Sol–gel derived hydroxyapatite coatings on titanium substrate. J Mater Sci 35:2791–2797

Morks MF, Kobayashi A (2007) Effect of gun current on the microstructure and crystallinity of plasma sprayed hydroxyapatite coatings. Appl Surf Sci 253:7136–7142

Morks MF, Kobayashi A, Fahim NF (2007) Abrasive wear behavior of sprayed hydroxyapitite coatings by gas tunnel type plasma spraying. Wear 262:204–209

Morris RE (ed) (2011) The sol–gel process: uniformity, polymers and applications. Nova Science, Hauppauge, NY, USA, p 887

Muirhead-Allwood SK, Sandiford N, Skinner JA, Hua J, Kabir C, Walker PS (2010) Uncemented custom computer-assisted design and manufacture of hydroxyapatite-coated femoral components: survival at 10 to 17 years. J Bone Joint Surg B 92:1079–1084

Mukherjee DP, Dorairaj NR, Mills DK, Graham D, Krauser JT (2000) Fatigue properties of hydroxyapatite-coated dental implants after exposure to a periodontal pathogen. J Biomed Mater Res 53:467–474

Nanci A, Wuest JD, Peru L, Brunet P, Sharma V, Zalzal S, McKee MD (1998) Chemical modification of titanium surfaces for covalent attachment of biological molecules. J Biomed Mater Res 40:324–335

Narayanan R, Dutta S, Seshadri SK (2006) Hydroxy apatite coatings on Ti-6Al-4 V from seashell. Surf CoatTechnol 200:4720–4730

Narayanan R, Seshadri SK, Kwon TY, Kim KH (2007) Electrochemical nano-grained calcium phosphate coatings on Ti-6Al-4 V for biomaterial applications. Scripta Mater 56:229–232

Narayanan R, Kim SY, Kwon TY, Kim KH (2008a) Nanocrystalline hydroxyapatite coatings from ultrasonated electrolyte: preparation, characterization and osteoblast responses. J Biomed Mater Res A 87A:1053–1060

Narayanan R, Kwon TY, Kim KH (2008b) Direct nanocrystalline hydroxyapatite formation on titanium from ultrasonated electrochemical bath at physiological pH. Mater Sci Eng C 28:1265–1270

Narayanan R, Kwon TY, Kim KH (2008c) Preparation and characteristics of nano-grained calcium phosphate coatings on titanium from ultrasonated bath at acidic pH. J Biomed Mater Res B Appl Biomater 85B:231–239

Narayanan R, Kim KH, Rautray TR (2010) Surface modification of titanium for biomaterial applications. Nova Science, Hauppauge, NY, USA, p 352

Negroiu G, Piticescu RM, Chitanu GC, Mihailescu IN, Zdrentu L, Miroiu M (2008) Biocompatibility evaluation of a novel hydroxyapatite-polymer coating for medical implants (in vitro tests). J Mater Sci Mater Med 19:1537–1544

Nelea V, Ristoscu C, Chiritescu C, Ghica C, Mihailescu IN, Pelletier H, Mille P, Cornet A (2000) Pulsed laser deposition of hydroxyapatite thin films on Ti-5Al-2.5Fe substrates with and without buffer layers. Appl Surf Sci 168:127–131

Nelea V, Pelletier H, Iliescu M, Werckmann J, Craciun V, Mihailescu IN, Ristoscu C, Ghica C (2002) Calcium phosphate thin film processing by pulsed laser deposition and in situ assisted ultraviolet pulsed laser deposition. J Mater Sci Mater Med 13:1167–1173

Nelea V, Morosanu C, Iliescu M, Mihailescu IN (2003) Microstructure and mechanical properties of hydroxyapatite thin films grown by RF magnetron sputtering. Surf CoatTechnol 173:315–322

Nelea V, Morosanu C, Iliescu M, Mihailescu IN (2004) Hydroxyapatite thin films grown by pulsed laser deposition and radio-frequency magnetron sputtering: comparative study. Appl Surf Sci 228:346–356

Nelea V, Morosanu C, Bercu M, Mihailescu IN (2007) Interfacial titanium oxide between hydroxyapatite and TiAlFe substrate. J Mater Sci Mater Med 18:2347–2354

Nguyen HQ, Deporter DA, Pilliar RM, Valiquette N, Yakubovich R (2004) The effect of sol–gel-formed calcium phosphate coatings on bone ingrowth and osteoconductivity of porous-surfaced Ti alloy implants. Biomaterials 25:865–876

Nie X, Leyland A, Matthews A (2000) Deposition of layered bioceramic hydroxyapatite/TiO$_2$ coatings on titanium alloys using a hybrid technique of micro-arc oxidation and electrophoresis. Surf CoatTechnol 125:407–414

Nie X, Leyland A, Matthews A, Jiang JC, Meletis EI (2001) Effects of solution pH and electrical parameters on hydroxyapatite coatings deposited by a plasma-assisted electrophoresis technique. J Biomed Mater Res 57:612–618

Nijhuis AWG, Leeuwenburgh SCG, Jansen JA (2010) Wet-chemical deposition of functional coatings for bone implantology. Macromol Biosci 10:1316–1329

Nimb L, Gotfredsen K, Steen JJ (1993) Mechanical failure of hydroxyapatite-coated titanium and cobalt-chromium-molybdenum alloyimplants. An animal study Acta Orthop Belg 59:333–338

Ntsoane TP, Topic M, Bucher R (2011) Near-surface in vitro studies of plasma sprayed hydroxyapatite coatings. Powder Diffraction 26:138–143

Oguchi H, Ishikawa K, Ojima S, Hirayama Y, Seto K, Eguchi G (1992) Evaluation of a high-velocity flame-spraying technique for hydroxyapatite. Biomaterials 13:471–477

Ohring M (2002) Materials science of thin films, 2nd edn. Academic, San Diego, CA, USA, p 794

Okido M, Kuroda K, Ishikawa M, Ichino R, Takai O (2002) Hydroxyapatite coating on titanium by means of thermal substrate method in aqueous solutions. Solid State Ion 151:47–52

Oliveira AL, Elvira C, Reis RL, Vazquez B, San Roman J (1999) Surface modification tailors the characteristics of biomimetic nucleated on starch-based polymers. J Mater Sci Mater Med 10:827–835

Oliveira AL, Mano JF, Reis RL (2003) Nature-inspired calcium phosphate coatings: present status and novel advances in the science of mimicry. Curr Opin Solid State Mater Sci 7:309–318

Ong JL, Chan DCN (1999) Hydroxyapatite and their use as coatings in dental implants: a review. Crit Rev Biomed Eng 28:667–707

Ong JL, Lucas LC (1994) Post-deposition heat treatment for ion beam sputter deposited calcium phosphate coatings. Biomaterials 15:337–341

Ong JL, Harris LA, Lucas LC, Lacefield WR, Rigney D (1991) X-ray photoelectron spectroscopy characterization of ion beam sputter deposited calcium phosphate coatings. J Am Ceram Soc 74:2301–2304

Ong JL, Harris LA, Lucas LC, Lacefield WR, Rigney D (1992) Structure, solubility and bond strength of thin calcium phosphate coatings produced by ion beam sputter-deposited. Biomaterials 13:249–254

Ong JL, Lucas LC, Raikar GN, Weimer JJ, Gregory JC (1994) Surface characterization of ion beam sputter-deposited Ca-P coatings after in vitro immersion. Coll Surf A 87:151–162

Ong JL, Bessho K, Cavin R, Carnes DL (2002) Bone response to radio frequency sputtered calcium phosphate implants and titanium implants in vivo. J Biomed Mater Res 59:184–190

Onoki T, Hashida T (2006) New method for hydroxyapatite coating of titanium by the hydrothermal hot isostatic pressing technique. Surf CoatTechnol 200:6801–6807

Oosterbos CJM, Vogely HC, Nijhof MW, Fleer A, Verbout AJ, Tonino AJ, Dhert WJA (2002) Osseointegration of hydroxyapatite-coated and noncoated Ti6Al4V implants in the presence of local infection: a comparative histomorphometrical study in rabbits. J Biomed Mater Res 60:339–347

Oosterbos CJM, Rahmy AIA, Tonino AJ, Witpeerd W (2004) High survival rate of hydroxyapatite-coated hip prostheses 100 consecutive hips followed for 10 years. Acta Orthop Scandinavica 75:127–133

Ozeki K, Yuhta T, Aoki H, Nishimura I, Fukui Y (2001) Push-out strength of hydroxyapatite coated by sputtering technique in bone. Bio-Med Mater Eng 11:63–68

Ozeki K, Yuhta T, Aoki H, Fukui Y (2003) Inhibition of Ni release from NiTi alloy by hydroxyapatite, alumina, and titanium sputtered coatings. Bio-Med Mater Eng 13:271–279

Ozeki K, Fukui Y, Aoki H (2007) Influence of the calcium phosphate content of the target on the phase composition and deposition rate of sputtered films. Appl Surf Sci 253:5040–5044

Ozeki K, Aoki H, Masuzawa T (2010) Influence of the hydrothermal temperature and pH on the crystallinity of a sputtered hydroxyapatite film. Appl Surf Sci 256:7027–7031

Paital SR, Dahotre NB (2007) Laser surface treatment for porous and textured Ca-P bio-ceramic coating on Ti-6Al-4 V. Biomed Mater 2:274–281

Paital SR, Dahotre NB (2008) Review of laser based biomimetic and bioactive Ca-P coatings. Mater Sci Technol 24:1144–1161

Paital SR, Dahotre NB (2009a) Calcium phosphate coatings for bio-implant applications: materials, performance factors, and methodologies. Mater Sci Eng R 66:1–70

Paital SR, Dahotre NB (2009b) Wettability and kinetics of hydroxyapatite precipitation on a laser-textured Ca–P bioceramic coating. Acta Biomater 5:2763–2772

Paital SR, Balani K, Agarwal A, Dahotre NB (2009) Fabrication and evaluation of a pulse laser-induced Ca-P coating on a Ti alloy for bioapplication. Biomed Mater 4:015009

Park S, Condrate R, Hoelzer DT, Fischman GS (1998) Interfacial characterization of plasma-spray coated calcium phosphate on Ti-6Al-4 V. J Mater Sci Mater Med 9:643–649

Park YS, Yi KY, Lee IS, Han CH, Jung YC (2005) The effects of ion beam-assisted deposition of hydroxyapatite on the grit-blasted surface of endosseous implants in rabbit tibiae. Int J Oral Maxillofac Implants 20:31–38

Pawlowski L (2008) The science and engineering of thermal spray coatings, 2nd edn. Wiley, New York, USA, p 691

Peng P, Kumar S, Voelcker NH, Szili E, Smart RSC, Griesser HJ (2006) Thin calcium phosphate coatings on titanium by electrochemical deposition in modified simulated body fluid. J Biomed Mater Res A 76A:347–355

Peraire C, Arias JL, Bernal D, Pou J, León B, Araño A, Roth W (2006) Biological stability and osteoconductivity in rabbit tibia of pulsed laser deposited hydroxylapatite coatings. J Biomed Mater Res A 77A:370–379

Pezeshki P, Lugowski S, Davies JE (2010) Dissolution behavior of calcium phosphate nanocrystals deposited on titanium alloy surfaces. J Biomed Mater Res A 94A:660–666

Pham MT, Matz W, Grambole D, Herrmann F, Reuther H, Richter E, Steiner G (2002) Solution deposition of hydroxyapatite on titanium pre-treated with a sodium ion implantation. J Biomed Mater Res 59:716–724

Piattelli A, Cosci F, Scarano A, Trisi P (1995) Localized chronic suppurative bone infection as a sequel of peri-implantitis in a hydroxyapatite-coated dental implant. Biomaterials 16:917–920

Pilliar RM, Deporter DA, Watson PA, Pharoah M, Chipman M, Valiquette N, Carter S, de Groot K (1991) The effect of partial coating with hydroxyapatite on bone remodeling in relation to porous-coated titanium-alloy dental implants in the dog. J Dent Res 70:1338–1345

Podlesak H, Pawlowski L, D'Haese R, Laureyns J, Lampke T, Bellayer S (2010) Advanced microstructural study of suspension plasma sprayed hydroxyapatite coatings. J Thermal Spray Technol 19:657–664

Pontin MG, Lange FF, Sanchez-Herencia AJ, Moreno R (2005) Effect of unfired tape porosity on surface film formation by dip coating. J Am Ceram Soc 88:2945–2948

Prymak O, Bogdansky D, Esenwein SA, Köller M, Epple M (2004) NiTi shape memory alloys coated with calcium phosphate by plasma-spraying. Chemical and biological properties. Mater Werkst 35:346–351

Quaranta A, Iezzi G, Scarano A, Coelho PG, Vozza I, Marincola M, Piattelli A (2010) A histomorphometric study of nanothickness and plasma-sprayed calcium-phosphorous-coated implant surfaces in rabbit bone. J Periodontology 81:556–561

Quek CH, Khor KA, Cheang P (1999) Influence of processing parameters in the plasma spraying of hydroxyapatite/Ti-6Al-4 V composite coatings. J Mater Process Technol 89–90:550–555

Rabiei A, Thomas B, Jin C, Narayan R, Cuomo J, Yang Y, Ong JL (2006) A study on functionally graded HA coatings processed using ion beam assisted deposition with in situ heat treatment. Surf CoatTechnol 200:6111–6116

Rajaratnam SS, Jack C, Tavakkolizadeh A, George MD, Fletcher RJ, Hankins M, Shepperd JAN (2008) Long-term results of a hydroxyapatite-coated femoral component in total hip replacement: a 15- to 21-year follow-up study. J Bone Joint Surg B 90:27–30

Rajesh P, Muraleedharan CV, Komath M, Varma H (2011) Pulsed laser deposition of hydroxyapatite on titanium substrate with titania interlayer. J Mater Sci Mater Med 22:497–505

Rau JV, Smirnov VV, Laureti S, Generosi A, Varvaro G, Fosca M, Ferro D, Cesaro SN, Albertini VR, Barinov SM (2010) Properties of pulsed laser deposited fluorinated hydroxyapatite films on titanium. Mater Res Bull 45:1304–1310

Rautray TR, Narayanan R, Kwon TY, Kim KH (2010) Surface modification of titanium and titanium alloys by ion implantation. J Biomed Mater Res B Appl Biomater 93B:581–591

Redepenning J, Schlessinger T, Burnham S, Lippiello L, Miyano J (1996) Characterization of electrolytically prepared brushite and hydroxyapatite coatings on orthopedic alloys. J Biomed Mater Res 30:287–294

Reis RL, Monteiro FJ, Hastings GW (1994) Stability of hydroxylapatite plasma-sprayed coated Ti-6Al-4 V under cyclic bending in simulated physiological solution. J Mater Sci Mater Med 5:457–462

Roguska A, Hiromoto S, Yamamoto A, Woźniak MJ, Pisarek M, Lewandowska M (2011) Collagen immobilization on 316 L stainless steel surface with cathodic deposition of calcium phosphate. Appl Surf Sci 257:5037–5045

Rossler S, Sewing A, Stolzel M, Born R, Scharnweber D, Dard M, Worch H (2003) Electrochemically assisted deposition of thin calcium phosphate coatings at near-physiological pH and temperature. J Biomed Mater Res A 64A:655–663

Roy M, Bandyopadhyay A, Bose S (2011) Induction plasma sprayed nano hydroxyapatite coatings on titanium for orthopaedic and dental implants. Surf CoatTechnol 205:2785–2792

Ruckenstein E, Gourisankar SV (1986) Preparation and characterization of thin film surface coatings for biological environments. Biomaterials 7:403–422

Saithna A (2010) The influence of hydroxyapatite coating of external fixator pins on pin loosening and pin track infection: a systematic review. Injury 41:128–132

Saju KK, Reshmi R, Jayadas NH, James J, Jayaraj MK (2009) Polycrystalline coating of hydroxyapatite on TiAl6V4 implant material grown at lower substrate temperatures by hydrothermal annealing after pulsed laser deposition. Proc Inst Mech Eng H 223:1049–1057

Schliephake H, Scharnweber D, Roesseler S, Dard M, Sewing A, Aref A (2006) Biomimetic calcium phosphate composite coating of dental implants. Int J Oral Max Impl 21:738–746

Schwarz MLR, Kowarsch M, Rose S, Becker K, Lenz T, Jani L (2009) Effect of surface roughness, porosity, and a resorbable calcium phosphate coating on osseointegration of titanium in a minipig model. J Biomed Mater Res A 89A:667–678

Shetty AA, Slack R, Tindall A, James KD, Rand C (2005) 1 Results of a hydroxyapatite-coated (Furlong) total hip replacement. A 13- to 15-year follow-up. J Bone Joint Surg B 87:1050–1054

Shi JZ, Chen CZ, Yu HJ, Zhang SJ (2008) Application of magnetron sputtering for producing bioactive ceramic coatings on implant materials. Bull Mater Sci 31:877–884

Shirkhanzadeh M (1993) Electrochemical preparation of bioactive calcium phosphate coatings on porous substrates by the periodic pulse technique. J Mater Sci Lett 12:16–19

Shirkhanzadeh M (1998) Direct formation of nanophase hydroxyapatite on cathodically polarized electrodes. J Mater Sci Mater Med 9:67–72

Siebers MC, Walboomers XF, Leeuwenburgh SCG, Wolke JGC, Jansen JA (2004) Electrostatic spray deposition (ESD) of calcium phosphate coatings, an in vitro study with osteoblast-like cells. Biomaterials 25:2019–2027

Siebers MC, Matsuzaka K, Walboomers XF, Leeuwenburgh SCG, Wolke JGC, Jansen JA (2005) Osteoclastic resorption of calcium phosphate coatings applied with electrostatic spray deposition (ESD), in vitro. J Biomed Mater Res A 74A:570–580

Siebers MC, Walboomers XF, Leewenburgh SCG, Wolke JCG, Boerman OC, Jansen JA (2006) Transforming growth factor-β1 release from a porous electrostatic spray deposition-derived calcium phosphate coating. Tiss Eng 12:2449–2456

Siebers MC, Wolke JGC, Walboomers FX, Leeuwenburgh SCG, Jansen JA (2007) In vivo evaluation of the trabecular bone behavior to porous electrostatic spray deposition-derived calcium phosphate coatings. Clin Oral Implant Res 18:354–361

Silva MHPD, Lima JHC, Soares GA, Elias CN, de Andrade MC, Best SM, Gibson IR (2001) Transformation of monetite to hydroxyapatite in bioactive coatings on titanium. Surf CoatTechnol 137:270–276

Sima LE, Stan GE, Morosanu CO, Melinescu A, Ianculescu A, Melinte R, Neamtu J, Petrescu SM (2010) Differentiation of mesenchymal stem cells onto highly adherent radio frequency-sputtered carbonated hydroxylapatite thin films. J Biomed Mater Res A 95A:1203–1214

Simank HG, Stuber M, Frahm R, Helbig L, van Lenthe H, Müller R (2006) The influence of surface coatings of dicalcium phosphate (DCPD) and growth and differentiation factor-5 (GDF-5) on the stability of titanium implants in vivo. Biomaterials 27:3988–3994

Singh RK, Narayan J (1990) Pulsed-laser evaporation technique for deposition of thin films: physics and theoretical model. Phys Rev B 41:8843–8859

Singh RK, Qian F, Nagabushnam V, Damodaran R, Moudgil BM (1994) Excimer laser deposition of hydroxylapatite thin films. Biomaterials 15:522–528

Snyders R, Bousser E, Music D, Jensen J, Hocquet S, Schneider JM (2008) Influence of the chemical composition on the phase constitution and the elastic properties of RF-sputtered hydroxyapatite coatings. Plasma Processes Polym 5:168–174

Søballe K, Hansen ES, Brockstedt-Rasmussen HB, Hjortdal VE, Juhl GI, Pedersen CM, Hvid I, Bünger C (1991) Gap healing enhanced by hydroxyapatite coatings in dogs. Clin Orthop 272:300–307

Søballe K, Hansen ES, Brockstedt-Rasmussen HB, Bünger C (1993) Hydroxyapatite coating converts fibrous tissue to bone around loaded implants. J Bone Jt Surg 75B:270–278

Sobolev VV, Guilemany JM (1996) Dynamic processes during high velocity oxyfuel spraying. Int Mater Rev 41:13–32

Socol G, Torricelli P, Bracci B, Iliescu M, Miroiu F, Bigi A, Werckmann J, Mihailescu IN (2004) Biocompatible nanocrystalline octacalcium phosphate thin films obtained by pulsed laser deposition. Biomaterials 25:2539–2545

Song WH, Jun YK, Han Y, Hong SH (2004) Biomimetic apatite coatings on micro-arc oxidized titania. Biomaterials 25:3341–3349

Sousa SR, Barbosa MA (1996) Effect of hydroxyapatite thickness on metal ion release from Ti6Al4V substrates. Biomaterials 17:397–404

Sridhar TM, Kamachi MU, Subbaiyan M (2003) Sintering atmosphere and temperature effects on hydroxyapatite coated type 316 L stainless steel. Corr Sci 45:2337–2359

Stephenson PK, Freeman MAR, Revell PA, Germain J, Tuke M, Pirie CJ (1991) The effect of hydroxyapatite coating on growth of bone into cavities in an implant. J Arthroplasty 6:51–58

Stevenson JR, Solnick-Legg H, Legg KO (1989) Production of high adherent hydroxyapatite coatings by ion beam and plasma techniques. Mater Res Soc Symp Proc 110:715–719

Stilling M, Rahbek O, Søballe K (2009) Inferior survival of hydroxyapatite versus titanium-coated cups at 15 years. Clin Orthop Rel Res 467:2872–2879

Stoch A, Brozek A, Kmita G, Stoch J, Jastrzebski W, Rakowska A (2001) Electrophoretic coating of hydroxyapatite on titanium implants. J Mol Struct 596:191–200

Stoica TF, Morosanu C, Slav A, Stoica T, Osiceanu P, Anastasescu C, Gartner M, Zaharescu M (2008) Hydroxyapatite films obtained by sol–gel and sputtering. Thin Solid Films 516:8112–8116

Suchanek WL, Yoshimura M (1998) Processing and properties of hydroxyapatite-based biomaterials for use as hard tissue replacement implants. J Mater Res 13:94–117

Sudo SZ, Schotzko NK, Folke LEA (1976) Use of hydroxyapatite coated glass beads for preclinical testing of potential antiplaque agents. Appl Environmental Microbiology 32:428–437

Sun T, Wang M (2010) Electrochemical deposition of apatite/collagen composite coating on NiTi shape memory alloy and coating properties. Mater Res Soc Symp Proc 1239:141–146

Sun L, Berndt CC, Gross KA, Kucuk A (2001) Material fundamentals and clinical performance of plasma sprayed hydroxyapatite coatings: a review. J Biomed Mater Res (Appl Biomater) 58:570–592

Sun L, Berndt CC, Grey CP (2003) Phase, structural and microstructural investigations of plasma sprayed hydroxyapatite coatings. Mater Sci Eng, A 360:70–84

Sun J, Han Y, Huang X (2007) Hydroxyapatite coatings prepared by micro-arc oxidation in Ca- and P-containing electrolyte. Surf CoatTechnol 201:5655–5658

Surmenev RAA (2012) Review of plasma-assisted methods for calcium phosphate-based coatings fabrication. Surf CoatTechnol 206:2035–2056

Surmenev RA, Ryabtseva MA, Shesterikov EV, Pichugin VF, Peitsch T, Epple M (2010) The release of nickel from nickel-titanium (NiTi) is strongly reduced by a sub-micrometer thin layer of calcium phosphate deposited by rf-magnetron sputtering. J Mater Sci Mater Med 21:1233–1239

Suzuki M, Calasans-Maia MD, Marin C, Granato R, Gil JN, Granjeiro JM, Coelho PG (2010) Effect of surface modifications on early bone healing around plateau root form implants: an experimental study in rabbits. J Oral Maxillofac Surg 68:1631–1638

Sygnatowicz M, Tiwari A (2009) Controlled synthesis of hydroxyapatite-based coatings for biomedical application. Mater Sci Eng C 29:1071–1076

Takadama H, Kim HM, Kokubo T, Nakamura T (2001) TEM-EDX study of mechanism of bonelike apatite formation on bioactive titanium metal in simulated body fluid. J Biomed Mater Res 57:441–448

Tas AC, Bhaduri SB (2004) Rapid coating of Ti6Al4V at room temperature with a calcium phosphate solution similar to 10x simulated body fluid. J Mater Res 19:2742–2749

Thomas KA, Cook CD, Ray RJ, Jarcho M (1989) Biologic response to hydroxylapatite coated titanium hips. J Arthroplasty 4:43–53

Tieanboon P, Jaruwangsanti N, Kiartmanakul S (2009) Efficacy of hydroxyapatite in pedicular screw fixation in canine spinal vertebra. Asian Biomedicine 3:177–181

Tinsley D, Watson CJ, Russell JL (2001) A comparison of hydroxylapatite coated implant retained fixed and removable mandibular prostheses over 4 to 6 years. Clin Oral Implant Res 12:159–166

Tkalcec E, Sauer M, Nonninger R, Schmidt H (2001) Sol–gel-derived hydroxyapatite powders and coatings. J Mater Sci 36:5253–5263

Tong W, Chen J, Zhang X (1995) Amorphorization and recrystallization during plasma spraying of hydroxyapatite. Biomaterials 16:829–832

Tong W, Yang Z, Zhang X, Yang A, Feng J, Cao Y, Chen J (1998) Studies on diffusion maximum in X-ray diffraction patterns of plasma-sprayed hydroxyapatite coatings. J Biomed Mater Res 40:407–413

Toque JA, Hamdi M, Ide-Ektessabi A, Sopyan I (2009) Effect of the processing parameters on the integrity of calcium phosphate coatings produced by RF-magnetron sputtering. Int J Modern Phys B 23:5811–5818

Torrisi L, Setola R (1993) Thermally assisted hydroxyapatite obtained by pulsed-laser deposition on titanium substrates. Thin Solid Films 227:32–36

Tri LQ, Chua DHC (2009) An investigation into the effects of high laser fluence on hydroxyapatite/calcium phosphate films deposited by pulsed laser deposition. Appl Surf Sci 256:76–80

Trisi P, Keith DJ, Rocco S (2005) Human histologic and histomorphometric analyses of hydroxyapatite-coated implants after 10 years of function: a case report. Int J Oral Maxillofac Implants 20:124–130

Tsui YC, Doyle C, Clyne TW (1998) Plasma sprayed hydroxyapatite coatings on titanium substrates. Part 1: Mechanical properties and residual stress levels. Biomaterials 19:2015–2029

Tucker BE, Cottel CM, Auyeung RCY, Spector M, Nancollas GH (1996) Pre-conditioning and dual constant composition dissolution kinetics of pulsed laser deposited hydroxyapatite thin films on silicon substrates. Biomaterials 17:631–637

Uchida M, Kim HM, Kokubo T, Fujibayashi S, Nakamura T (2003) Structural dependence of apatite formation on titania gels in a simulated body fluid. J Biomed Mater Res A 64A:164–170

Ueda K, Narushima T, Goto T, Taira M, Katsube T (2007) Fabrication of calcium phosphate films for coating on titanium substrates heated up to 773 K by RF magnetron sputtering and their evaluations. Biomed Mater 2:S160–S166

Ueda K, Kawasaki Y, Narushima T, Goto T, Kurihara J, Nakagawa H, Kawamura H, Taira M (2009) Calcium phosphate films with/without heat treatments fabricated using RF magnetron sputtering. J Biomech Sci Eng 4:392–403

U.S. Food and Drug Administration (1995). http://www.fda.gov/MedicalDevices/DeviceRegulationandGuidance/GuidanceDocuments/ucm080224.htm. Accessed

Valter S, Orfeo S, Clarke DR (1997) Mechanical and chemical consequences of the residual stresses in plasma sprayed hydroxyapatite coatings. Biomaterials 18:477–482

van der Wal E, Wolke JGC, Jansen JA, Vredenberg AM (2005) Initial reactivity of rf magnetron sputtered calcium phosphate thin films in simulated body fluids. Appl Surf Sci 246:183–192

van der Wal E, Oldenburg SJ, Heij T, van der Gon AWD, Brongersma HH, Wolke JGC, Jansen JA, Vredenberg AM (2006) Adsorption and desorption of Ca and PO_4 species from SBFs on RF-sputtered calcium phosphate thin films. Appl Surf Sci 252:3843–3854

van Dijk K, Schaeken HG, Wolke JGC, Maree CHM, Habraken FHPM, Verhoven J, Jansen JA (1995) Influence of discharge power level on the properties of hydroxyapatite films deposited on Ti6Al4V with RF magnetron sputtering. J Biomed Mater Res 29:269–276

van Dijk K, Schaeken HG, Wolke JGC, Jansen JA (1996) Influence of annealing temperature on RF magnetron sputtered calcium phosphate coatings. Biomaterials 17:405–410

van Dijk K, Verhoeven J, Marée CHM, Habraken FHPM, Jansen JA (1997) Study of the influence of oxygen on the composition of thin films obtained by r.f. sputtering from a $Ca_5(PO_4)_3OH$ target. Thin Solid Films 304:191–195

Variola F, Brunski JB, Orsini G, de Oliveira TP, Wazen R, Nanci A (2011) Nanoscale surface modifications of medically relevant metals: State-of-the art and perspectives. Nanoscale 3:335–353

Vasanthan A, Kim H, Drukteinis S, Lacefield W (2008) Implant surface modification using laser guided coatings: in vitro comparison of mechanical properties. J Prosthodontics 17:357–364

Vercaigne S, Wolke JGC, Naert I, Jansen JA (2000a) A mechanical evaluation of TiO_2-gritblasted and Ca-P magnetron sputter coated implants placed into the trabecular bone of the goat: Part 1. Clin Oral Implant Res 11:305–313

Vercaigne S, Wolke JGC, Naert I, Jansen JA (2000b) A histological evaluation of TiO_2-gritblasted and Ca-P magnetron sputter coated implants placed into the trabecular bone of the goat: Part 2. Clin Oral Implant Res 11:314–324

Verestiuc L, Morosanu C, Bercu M, Pasuk I, Mihailescu IN (2004) Chemical growth of calcium phosphate layers on magnetron sputtered HA films. J Cryst Growth 264:483–491

Walschus U, Hoene A, Neumann HG, Wilhelm L, Lucke S, Luthen F, Rychly J, Schlosser M (2009) Morphometric immunohistochemical examination of the inflammatory tissue reaction after implantation of calcium phosphate-coated titanium plates in rats. Acta Biomater 5:776–784

Wan T, Aoki H, Hikawa J, Lee JH (2007) RF-magnetron sputtering technique for producing hydroxyapatite coating film on various substrates. Bio-Med Mater Eng 17:291–297

Wang L, Nancollas GH (2008) Calcium orthophosphates: crystallization and dissolution. Chem Rev 108:4628–4669

Wang G, Zreiqat H (2010) Functional coatings or films for hard-tissue applications. Materials 3:3994–4050

Wang BC, Lee TM, Chang E, Yang CY (1993) The shear strength and the failure mode of plasma-sprayed hydroxyapatite coating to bone: the effect of coating thickness. J Biomed Mater Res 27:1315–1327

Wang CK, Lin JHC, Ju CP, Ong HC, Chang RPH (1997) Structural characterization of pulsed laser-deposited hydroxyapatite film on titanium substrate. Biomaterials 18:1331–1338

Wang XX, Hayakawa S, Tsuru K, Osaka A (2000) A comparative study of in vitro apatite deposition on heat-, H_2O_2-, and NaOH-treated titanium surfaces. J Biomed Mater Res 52:172–178

Wang CX, Chen ZQ, Guan LM, Wang M, Liu ZY, Wang PL (2001) Fabrication and characterization of graded calcium phosphate coatings produced by ion beam sputtering/mixing deposition. Nucl Instr Methods Phys Res B 179:364–372

Wang C, Ma J, Cheng W, Zhang R (2002) Thick hydroxyapatite coatings by electrophoretic deposition. Mater Lett 57:99–105

Wang XX, Yan W, Hayakawa S, Tsuru K, Osaka A (2003) Apatite deposition on thermally and anodically oxidized titanium surfaces in a simulated body fluid. Biomaterials 24:4631–4637

Wang J, Layrolle P, Stigter M, de Groot K (2004) Biomimetic and electrolytic calcium phosphate coatings on titanium alloy: physicochemical characteristics and cell attachment. Biomaterials 25:583–592

Wang XJ, Li YC, Lin JG, Hodgson PD, Wen CE (2008a) Apatite-inducing ability of titanium oxide layer on titanium surface: the effect of surface energy. J Mater Res 23:1682–1688

Wang DG, Chen CZ, Ma J, Zhang G (2008b) In situ synthesis of HA coating by laser cladding. Coll Surf B 66:155–162

Wang X, Li Y, Lin J, Hodgson PD, Wen C (2008c) Effect of heat-treatment atmosphere on the bond strength of apatite layer on Ti substrate. Dental Mater 24:1549–1555

Wang C, Gross KA, Anderson GI, Dunstan CR, Carbone A, Berger G, Ploska U, Zreiqat H (2009) Bone growth is enhanced by novel bioceramic coatings on Ti alloy implants. J Biomed Mater Res A 90A:419–428

Wang LJ, Lu JW, Xu FS, Zhang FS (2011) Dynamics of crystallization and dissolution of calcium orthophosphates at the near-molecular level. Chinese Sci Bull 56:713–721

Wei M, Ruys AJ, Milthorpe BK, Sorrell CC, Evans JH (2001) Electrophoretic deposition of hydroxyapatite coatings on metal substrates: a nanoparticulate dual-coating approach. J Sol–gel Part Sci Technol 21:39–48

Wei D, Zhou Y, Wang Y, Jia D (2007) Characteristic of microarc oxidized coatings on titanium alloy formed in electrolytes containing chelate complex and nano-HA. Appl Surf Sci 253:5045–5050

Wen HB, Wolke JGC, de Wijn JR, Liu Q, Cui FZ (1997) Fast precipitation of calcium phosphate layers on titanium induced by simple chemical treatments. Biomaterials 18:1471–1478

Weng W, Baptisa JL (1998) Alkoxide route for preparing hydroxyapatite and its coatings. Biomaterials 19:125–131

Weng J, Liu X, Zhang X, Ma Z, Ji X, Zyman Z (1993) Further studies on the plasma-sprayed amorphous phase in hydroxyapatite coatings and its deamorphization. Biomaterials 14:578–582

Wheeler SL (1996) Eight-year clinical retrospective study of titanium plasma-sprayed and hydroxyapatite-coated cylinder implants. Int J Oral Maxillofac Implants 11:340–350

Wheeler DL, Montfort MJ, McLoughlin SW (2001) Differential healing response of bone adjacent to porous implants coated with hydroxyapatite and 45 S5 bioactive glass. J Biomed Mater Res 55:603–612

Wie H, Hero H, Solheim T (1998) Hot isostatic pressing-processed hydroxyapatite-coated titanium implants: light microscopic and scanning electron microscopy investigations. Int J Oral Maxillofac Implants 13:837–844

Wikipedia (2012a) Coating. http://en.wikipedia.org/wiki/Coating. Accessed May 2012

Wikipedia (2012b) Electrophoretic deposition. http://en.wikipedia.org/wiki/Electrophoretic_deposition. Accessed May 2012

Willmann G (1999) Coating of implants with hydroxyapatite – material connections between bone and metal. Adv Eng Mater 1:95–105

Willmott PR, Huber JR (2000) Pulsed laser vaporization and deposition. Rev Modern Phys 72:315–328

Wolke JGC, de Blieck-Hogervorst JMA, Dhert WJA, Klein CPAT, de Groot K (1992) Studies on thermal spraying of apatite bioceramics. J Thermal Spray Technol 1:75–82

Wolke JGC, van Dijk K, Schaeken HG, de Groot K, Jansen JA (1994) Study of the surface characteristics of magnetron-sputter calcium phosphate coatings. J Biomed Mater Res 28:1477–1484

Wolke JGC, van der Waerden JPCM, de Groot K, Jansen JA (1997) Stability of radiofrequency magnetron sputtered calcium phosphate coatings under cyclically loaded conditions. Biomaterials 18:483–488

Wolke JGC, de Groot K, Jansen JA (1998) In vivo dissolution behavior of various RF magnetron sputtered Ca-P coatings. J Biomed Mater Res 39:524–530

Wolke JGC, van der Waerden JPCM, Schaeken HG, Jansen JA (2003) In vivo dissolution behavior of various RF magnetron-sputtered Ca-P coatings on roughened titanium implants. Biomaterials 24:2623–2629

Wu GM, Hsiao WD, Kung SF (2009) Investigation of hydroxyapatite coated polyether ether ketone composites by gas plasma sprays. Surf CoatTechnol 203:2755–2758

Xu S, Long J, Sim L, Diong CH, Ostrikov K (2005) RF plasma sputtering deposition of hydroxyapatite bioceramics: synthesis, performance, and biocompatibility. Plasma Processes Polym 2:373–390

Xue W, Liu X, Zheng X, Ding C (2005) Effect of hydroxyapatite coating crystallinity on dissolution and osseointegration *in vivo*. J Biomed Mater Res A 74A:553–561

Yamaguchi T, Tanaka Y, Ide-Ektessabi A (2006) Fabrication of hydroxyapatite thin films for biomedical applications using RF magnetron sputtering. Nucl Instrum Methods Phys Res B 249:723–725

Yamashita K, Arashi T, Kitagaki K, Yamada S, Umegaki T (1994) Preparation of apatite thin films through RF-sputtering from calcium phosphate glasses. J Am Ceram Soc 77:2401–2407

Yan Y, Wolke JGC, de Ruijter A, Li Y, Jansen JA (2006) Growth behavior of rat bone marrow cells on RF magnetron sputtered hydroxyapatite and dicalcium pyrophosphate coatings. J Biomed Mater Res A 78A:42–49

Yang YC (2011) Investigation of residual stress generation in plasma-sprayed hydroxyapatite coatings with various spraying programs. Surf CoatTechnol 205:5165–5171

Yang Y, Chang E (2005) Measurements of residual stresses in plasma-sprayed hydroxyapatite coatings on titanium alloy. Surf CoatTechnol 190:122–131

Yang CY, Wang BC, Chang E, Wu JD (1995) The influences of plasma spraying parameters on the characteristics of hydroxyapatite coatings: a quantitative study. J Mater Sci Mater Med 6:249–257

Yang CY, Wang BC, Chang WJ, Chang E, Wu JD (1996) Mechanical and histological evaluations of cobalt-chromium alloy and hydroxyapatite plasma-sprayed coatings in bone. J Mater Sci Mater Med 7:167–174

Yang CY, Wang BC, Lee TM, Chang E, Chang GL (1997) Intramedullary implant of plasma-sprayed hydroxyapatite coating: an interface study. J Biomed Mater Res 36:39–48

Yang YC, Chang E, Hwang BH, Lee SY (2000) Biaxial residual stress states of plasma-sprayed hydroxyapatite coatings on titanium alloy substrate. Biomaterials 21:1327–1337

Yang Y, Agarwal CM, Kim KH, Martin H, Schul K, Bumgardner JM, Ong JL (2003a) Characterization and dissolution behavior of sputtered calcium phosphate coatings after different postdeposition heat treatment temperatures. J Oral Implantology 29:270–277

Yang Y, Kim KH, Agarwal CM, Ong JL (2003b) Effect of post-deposition heating temperature and the presence of water vapor during heat treatment on crystallinity of calcium phosphate coatings. Biomaterials 24:5131–5137

Yang YC, Chang E, Lee SY (2003c) Mechanical properties and Young's modulus of plasma-sprayed hydroxyapatite coating on Ti substrate in simulated body fluid. J Biomed Mater Res A 67A:886–899

Yang Y, Kim KH, Ong JL (2005) A review on calcium phosphate coatings produced using a sputtering process – an alternative to plasma spraying. Biomaterials 26:327–337

Yang JX, Jiao YP, Cui FZ, Lee IS, Yin QS, Zhang Y (2008) Modification of degradation behavior of magnesium alloy by IBAD coating of calcium phosphate. Surf CoatTechnol 202:5733–5736

Yang CW, Lui TS, Chen LH (2009a) Hydrothermal crystallization effect on the improvement of erosion resistance and reliability of plasma-sprayed hydroxyapatite coatings. Thin Solid Films 517:5380–5385

Yang S, Man HC, Xing W, Zheng X (2009b) Adhesion strength of plasma-sprayed hydroxyapatite coatings on laser gas-nitrided pure titanium. Surf CoatTechnol 203:3116–3122

Yang S, Xing W, Man HC (2009c) Pulsed laser deposition of hydroxyapatite film on laser gas nitriding NiTi substrate. Appl Surf Sci 255:9889–9892

Yen SK, Lin CM (2002) Cathodic reactions of electrolytic hydroxyapatite coating on pure titanium. Mater Chem Phys 77:70–76

Yoon HJ, Song JE, Um YJ, Chae GJ, Chung SM, Lee IS, Jung UW, Kim CS, Choi SH (2009) Effects of calcium phosphate coating to SLA surface implants by the ion-beam-assisted deposition method on self-contained coronal defect healing in dogs. Biomed Mater 4:044107

Yoshinari M, Ohshiro Y, Derand T (1994) Thin hydroxyapatite coating produced by the ion beam dynamic mixing method. Biomaterials 15:529–535

Yoshinari M, Watanabe Y, Ohtsuka Y, Dérand T (1997) Solubility control of thin calcium-phosphate coating with rapid heating. J Dent Res 76:1485–1494

Yoshinari M, Oda Y, Inoue T, Matsuzaka K, Shimono M (2002) Bone response to calcium phosphate-coated and bisphosphonate-immobilized titanium implants. Biomaterials 23:2879–2885

You C, Yeo IS, Kim MD, Eom TK, Lee JY, Kim S (2005) Characterization and *in vivo* evaluation of calcium phosphate coated cp-titanium by dip-spin method. Curr Appl Phys 5:501–506

Yousefpour M, Afshar A, Yang X, Li X, Yang B, Wu Y, Chen J, Zhang X (2006) Nano-crystalline growth of electrochemically deposited apatite coating on pure titanium. J Electroanal Chem 589:96–105

Yuan Q, Sahu LK, D'Souza NA, Golden TD (2009) Synthesis of hydroxyapatite coatings on metal substrates using a spincasting technique. Mater Chem Phys 116:523–526

Zalm PC (1989) Quantitative sputtering. In: Cuomo JJ, Rossnagel SM, Kaufman HR (eds) Handbook of ion beam processing technology. Noyes Publications; Park Ridge, NJ, USA, pp 78–111

Zeng H, Lacefield WR (2000) The study of surface transformation of pulsed laser deposited hydroxyapatite coatings. J Biomed Mater Res 50:239–247

Zeng H, Lacefield WR, Mirov S (2000) Structural and morphological study of pulsed laser deposited calcium phosphate bioceramic coatings: influence of deposition conditions, laser parameters, and target properties. J Biomed Mater Res 50:248–258

Zhang L, Zhang WT (2011) Numerical investigation on particle velocity in cold spraying of hydroxyapatite coating. Adv Mater Res 188:717–722

Zhang S, Xianting Z, Yongsheng W, Kui C, Wenjian W (2006) Adhesion strength of sol–gel derived fluoridated hydroxyapatite coatings. Surf CoatTechnol 200:6350–6354

Zhang L, Zhang W, Wu Z (2012a) Numerical simulation of hydroxyapatite particle impacting on Ti substrate in cold spraying. Appl Mech Mater 130–134:900–903

Zhang L, Zhang W, Li H, Geng W, Bao Y (2012b) Development of a cold spraying system for fabricating hydroxyapatite coating. Appl Mech Mater 151:300–304

Zhao Z, Li H, Huo G, Sun P, Li Y, Kuroda K, Ichino R, Okido M (2003) Cathodic deposition of hydroxyapatite without H_2 evolution. TMS Annual Meeting, pp 169–173

Zhao BH, Lee IS, Bai W, Cui FZ, Feng HL (2005) Improvement of fibroblast adherence to titanium surface by calcium phosphate coating formed with IBAD. Surf CoatTechnol 193:366–371

Zhao GL, Wen G, Song Y, Wu K (2011) Near surface martensitic transformation and recrystallization in a Ti-24Nb-4Zr-7.9Sn alloy substrate after application of a HA coating by plasma spraying. Mater Sci Eng C 31:106–113

Zheng M, Fan D, Li XK, Zhang JB, Liu QB (2010) Microstructure and *in vitro* bioactivity of laser-cladded bioceramic coating on titanium alloy in a simulated body fluid. J Alloys Compounds 489:211–214

Zhitomirsky I (2000) Electrophoretic hydroxyapatite coatings and fibers. Mater Lett 42:262–271

Zhitomirsky I, Gal-Or L (1997) Electrophoretic deposition of hydroxyapatite. J Mater Sci Mater Med 8:213–219

Zhu Y, Chen Y, Xu G, Ye X, He D, Zhong J (2012) Micropattern of nano-hydroxyapatite/silk fibroin composite onto Ti alloy surface via template-assisted electrostatic spray deposition. Mater Sci Eng C 32:390–394

Ziani-Cherif H, Abe Y, Imachi K, Matsuda T (2002) Hydroxyapatite coating on titanium by thermal substrate method in aqueous solution. J Biomed Mater Res 59:390–397

Zyman Z, Weng J, Liu X, Zhang X, Ma Z (1993) Amorphous phase and morphological structure of hydroxyapatite plasma coatings. Biomaterials 14:225–228

Zyman Z, Weng J, Liu X, Li X, Zhang X (1994) Phase and structural changes in hydroxyapatite coatings under heat treatment. Biomaterials 15:151–155

2

Polystyrene microsphere and 5-fluorouracil release from custom-designed wound dressing films

Maryam Mobed-Miremadi[1*], Raki Komarla Nagendra[2], Sujana Lakshmi Ramachandruni[2], Jason James Rook[3], Mallika Keralapura[4] and Michel Goedert[1]

Abstract

Custom-designed wound dressing films of chitosan and alginate have been prepared by a casting/solvent evaporation method for hydrophobic therapeutic agent encapsulation. In this parametric study, the propylene glycol (PG) and calcium chloride ($CaCl_2$) concentrations were varied for chitosan and alginate films, respectively. Mechanical and chemical inter-related responses under observations included thickness (*th*), elasticity (*E*), tensile strength (*TS*), sorption ability (*S%*) and kinetics of *in-vitro* drug release, specifically in terms of membrane time to burst (t_B) and duration of release (t_R). As shown by results of a one tailed t-test significance testing at the 95% confidence interval ($\alpha = 0.05$), alginate films were significantly more elastic ($p = 0.003$), thinner ($p = 0.004$) and more susceptible to osmotic burst ($p = 0.011$) and characterized by a longer duration of release ($p = 0.03$). Meanwhile chitosan films exhibited superior moisture permeability ($p = 0.006$) and sorption characteristics ($p = 0.001$), indicative of higher hydrophilicity. There were no significant differences in tensile strength ($p = 0.324$) for alginate and chitosan-based formulations. Preliminary testing was conducted using 0.71 µm in diameter microspheres for modeling film dissolution into Lactated Ringer's solution. Experimental release profiles were modeled for each film from which the average release from alginate films ($M_{AGCa} = 81\%$) was estimated to be twice the percentage associated with chitosan films ($M_{CD} = 42\%$). The film comprised of 2.5% (w/v) medium MW chitosan/dextran 70 kDa (5:1) was selected for studying the release of 5-Fluorouracil (5-FU) as a model hydrophobic drug. Diffusion coupled with film disintegration is immediate ($t_B = 0$) in case of encapsulated 5-FU as compared to the control film encapsulating microspheres characterized by $t_B = 70$ min $\pm$ 7 min. This shift in release profile and the ability to modulate the timing of membrane burst can be attributed to the approximate ratio (1: 505) in molecular size between drug and microsphere. This hypothesis has been validated by the film pore size measured to be 430 nm $\pm$ 88 nm using atomic force microscopy.

Keywords: Chitosan, Alginate, Microsphere, 5-FU, Pore size, Atomic force microscopy, Release modeling

Background

5-fluoro-1,2,3,4-tetrahydropyrimidine-2,4-dione known as 5-Fluorouracil (5-FU) is a therapeutic agent used to treat non-melanoma skin, breast, and pancreatic cancer administered orally, intravenously (Adrucil) or topically as a cream (Effudex) (Giannola et al. 2010, 2012; Loven et al. 2002). This agent has a molecular weight of 130 g/mol, is partially soluble in water with a maximum solubility of

1 mg/ml, and has an $LD_{50} = 230$ mg/kg (orally in mice) (Fluorouracil 2012). As shown by the results of a recent *in vitro* study of release of 5-FU from Chitosan-Alginate microcapsules, minimal drug release has been reported in the stomach and small intestine. This is an advantage because 5-FU controlled release is required to avoid the side effects associated with such agents (Shabbear et al. 2012). Woolfson et al. reported that a 5-FU bio-adhesive patch can be used for local delivery to the uterine cervix in the condition of cervical intraepithelial neoplasia (CIN) which is common and potentially malignant, affecting women in

* Correspondence: maryam.mobed-miremadi@sjsu.edu
[1]Department of Biomedical, Chemical and Materials Engineering, San Jose State University, San Jose, CA 95192-0082, USA
Full list of author information is available at the end of the article

a wide age group (Woolfson et al. 1995). There was significant delay in the absorption of 5-FU through patches due to the controlled release of drug from the patches. The half-life was statistically significant (p < 0.01) compared to intravenous route in rabbits and such half-life is observed in humans in the conventional route (Diasio & Harris 1989). The enhancement in drug delivery by the transdermal patch as compared to intravenous delivery was shown by results of a clinical evaluation of 5-FU from transdermal patches on cell-induced tumors in mice indicating a statistical increase in survival time of 7 days (p < 0.01) (Chandrashekar & Prasanth 2008). These patches were based on an ethylcellulose and polyvinylpyrrolidone K-30 (PVP K-30; ratio: 8:2) formulation, di-butylphthalate as a plasticizer, and 2% (v/w) of isopropyl myristate as a permeation enhancer. *In vitro* induction of apoptosis and cell cycle arrest by polyvinylpyrrolidone K-30 has been recently reported (Wang et al. 2003) and isopropyl myristate is classified as a skin irritant (2000).

In parallel it has been extensively documented that the use of biocompatible and biodegradable alginate and chitosan wound dressing films/patches are advantageous in many ways: 1) They kill bacteria and prevent infections caused by systemic invasion (Burkatovskaya et al. 2008); 2) They promote a better healing environment by absorbing the wound exudate (Murakami et al. 2010); 3) They eliminate the need for wiping or washing the wounds in between new dressings (Ayogi et al. 2007); 4) They have good bio-adhesive properties, eliminating the need for surgical adhesives (Khan et al. 2000). The three most common means of film drying mentioned in the above references are air drying, thermostatic, and freeze drying. Meng et al. found that the films dried in open air were drier than expected (Meng et al. 2010). Sezer et al. observed that lyophilizing resulted in thick films with larger pores (Sezer et al. 2007) and the size of the pores in the films was found to increase the concentration of chitosan, thus improving the ability of the wound dressing to absorb the wound fluids. In a more recent study of cross-linked scaffolds for tissue engineering, for the 4–12% (w/v), no change in pore size range as a function of chitosan concentration was observed; but porosity was found to be inversely proportional to the chitosan content (Jana et al. 2012).

In light of the aforementioned findings and the need for a biocompatible transdermal controlled-release mechanism for hydrophobic therapeutic agents, the specific aim of this research is to devise formulations for fabricating chitosan and alginate wound dressings with embedded fluorescent microspheres mimicking a potential drug to be encapsulated in these films. The amounts of polymers, plasticizer, and cross-linking agents used for the film formulations will be varied for this parametric study. The film will be characterized in terms of thickness, elasticity,

tensile strength, sorption, and the kinetics of release will be empirically modeled to calculate the amount of compound released. Lactated Ringer solution will be used as simulated body fluid or wound exudate for wet tests. The ideal film based on the stated criteria will then be chosen to encapsulate 5-FU and studied in terms of drug release profile and pore size characterization.

Materials and methods
Materials
The following chemicals were purchased from Sigma-Aldrich (MO, USA): Medium MW chitosan (44887, 75-85% deacetylated), Medium MW alginic acid (A2033), Dextran 70 kDa (31390), propylene glycol (P4347), lactic acid (L6661), glycerol (G2025), calcium chloride (C5670), polyethylene glycol (81260), sodium citrate dihydrate (W302600), glacial acetic acid (320099), 5-Fluorouracil (858471). Polystyrene microspheres (Catalog No. G700B) internally- dyed with Fluorescent Green (excitation 468 nm /emission 508 nm) were purchased from Duke Scientific Corporation (CA, USA) Lactated Ringer's solution (Catalog No 6E2323) was purchased from Baxter (IL, USA).

Methods
The stoichiometric amounts of polymers, plasticizer, and cross-linking agents used for the film formulations used in this parametric study are outlined in Table 1.

For the chitosan films (CD) and the alginate films (AGCa), the plasticizer (propylene glycol, PG) and cross-linker (calcium chloride, $CaCl_2$) concentrations were varied, respectively. Although the role of cross-linker and plasticizer differ entirely in formulating composite films, the variables were chosen to overcome the documented limitations associated with each biopolymer: addressing stiffness and lack of porosity for chitosan (Jana et al. 2012; Mobed & Chang 1998; Madsen et al. 2011), and regulation of the encapsulated compound release rate for alginate (Shabbear et al. 2012; Simpliciano & Asi 2012). Three films were fabricated for each formulation and measurements were conducted in triplicate unless indicated otherwise.

Film fabrication
All films were fabricated by a casting/solvent evaporation technique.

Chitosan films were fabricated based on the modification of a previously proposed formulation design, specifically with the removal of corn starch and the cytotoxic cross-linking agent glutaraldehyde from the formulation (Wittaya-arrekul & Prasharn 2006). Chitosan, dextran and 5-FU (0.07% w/v) were dispersed in DI water for 1 hr and PG was then added to this solution. This step was followed by the addition of lactic acid at a rate of

Table 1 Custom design formulations for chitosan, alginate and control films

Film name	Polymer1	Polymer2	Plasticizer	Crosslinking Agent	Solvent
Chitosan	Chitosan (g)	Dextran (g)	Propylene Glycol (ml)	Lactic acid (ml)	DI water (ml)
CD 0.0	2.5	0.5	0	3.0	97.0
CD 0.5	2.5	0.5	0.5	3.0	96.5
CD 1.0	2.5	0.5	1.0	3.0	96.0
CD 1.5	2.5	0.5	1.5	3.0	95.5
Alginate	Alginate (g)	N/A	Glycerol (ml)	CaCl$_2$ (g)	DI water (ml)
AGCa 0.04	2.0	-	0.4	0.04	99.6
AGCa 0.08	2.0	-	0.4	0.08	99.6
AGCa 0.12	2.0	-	0.4	0.12	99.6

0.6 ml/min. Chitosan (pK$_a$ = 6.5) is reported to form complexes with negatively-charged moieties such as sodium carboxymethylcellulose, citrates, pectin, acacia, agar, sodium caprylate, stearic acid sodium tri-polyphosphate, lactic acid, malic acid, and alginic acid (Tiwary & Rana 2010; Adusumilli & Bolton 1991; Akbuga & Bergisadi 1996; Suheyla 1997; Dureja et al. 2001; Wang et al. 2001). Although chitosan solubility is a function of molecular weight and degree of deacetylation, the reported solubility threshold for this polyelectrolyte is 4. Lactic acid (pK$_a$ = 3.86) at a concentration 0.025 M acts as a solubilizer for the protonated amine group of the chitosan (Zhao et al. 2009). In this study, a 0.43 M lactic acid solution was used (3% (v/v) and thus cross-linking occurred by means of electrostatic interactions. The homogenized mixture was subsequently stirred to crosslink for 8 hrs. In the case of fluorescent microspheres (1% v/v), the particles were dispensed into the chitosan/dextran solution prior to the cross-linking step and did not undergo the pre-mixing step as is the case for 5-FU.

Alginate films were fabricated based on a modification of a previously proposed methodology (Rhim 2004). The solution was prepared by the gradual addition of alginate powder over a 1 hr period into a stirred solution of DI water, glycerol and 5-FU (0.07% w/v). Subsequently, CaCl$_2$ pellets were added in one shot to cross-link the solution for 8 hrs. In the case of fluorescent microspheres (1% v/v), the particles were dispensed into the alginate/glycerol solution prior to the cross-linking step and did not undergo the pre-mixing step as is the case for 5-FU.

The stirring speed was set to 240 rpm for the dissolution and cross-linking steps for both chitosan and alginate films. This speed was chosen to overcome the gradual thickening of the polymer solutions as a result of cross-linking.

A 25 ml solution of either alginate or chitosan-based formulation was poured into a 6.06-cm diameter glass petri-dish and allowed to degas at 4°C overnight. Subsequently, the solvent (DI water) was allowed to evaporate under ambient conditions (25°C and relative humidity of 55% ± 10%) for 24 hrs. The circular dried patches were cut using scissors to have an area of approximately 10 cm^2 and stored in an air tight container under ambient conditions for 1 week prior to use.

Thickness measurements

Film thickness (*th*) was measured using an EPIPHOT Nikon transmission microscope/camera (Model40) equipped with the NIS-Elements Basic Research Software imaging software. Five points were randomly taken at different locations of the film.

Tensile testing

The modulus of elasticity (*E*) of these wound dressings was estimated using an in-house stress–strain gage consisting of a ScoutPro scale (mg resolution) and an electronic length measurement device, the limits for which are 581.4 g and 25.75 mm, respectively. A 0.5 cm^2 rectangular sample strip was used as test sample. The ends of films were fastened with glue to the unit, and both the scale and length measurement device were zeroed. The knob on the measurement device was then turned to a displacement of 0.5 mm and both the elongation and mass were recorded until the film fractured and the test was completed. Measurements were conducted on films encapsulating microspheres only. Displacements were applied and resultant forces were measured. Stress (σ) is subsequently calculated by dividing force by the cross-sectional area (A) of the film (Eqn. 1) (Callister & Rethwisch 2010). The strain ε is calculated by dividing the film displacement (elongation, ΔL) by the initial length of the stent (L) (Eqn. 2). The elastic modulus (E) is then calculated from Eqn. 3, which is the slope of the linear portion of the stress–strain curve. The tensile strength (*TS*) is also found by the stress–strain curve, which represents the maximum stress the stent can undergo under tensile stretching before the moment of necking (Eqn. 4). Tensile strength was taken as the highest point (F_{max}) of the stress-train curve.

$$\sigma = \frac{F}{A} \tag{1}$$

$$\varepsilon = \frac{\Delta L}{L_0} \tag{2}$$

$$E = \frac{\sigma}{\varepsilon} \tag{3}$$

$$TS = \frac{F_{max}}{A} \tag{4}$$

Sorption ability

Lactated Ringer's solution (pH = 6.5) was used to simulate the behavior of the wound exudate or the area of lesion onto which the film might be applied (Rhim 2004). The biopolymer sorption ability was determined gravimetrically. The weights of strips of completely dried films were determined directly with a digital balance (mg resolution) and immersed into Lactated Ringer's solution for a 24 hr period. The resultant swollen films were gently blotted with filter paper to remove excess surface water and weighed again. The sorption ability of the film is expressed in terms of percentage of weight increased using Eqn. 5.

$$S(\%) = \frac{W_b - W_a}{W_b} \ x \ 100 \tag{5}$$

where

W_b weight of the film before immersing in solution
W_a weight of the film after immersing in solution and blotting

Drug release testing

At the initial stage, 5-FU release profile was simulated by the use of 0.71 μm in diameter microspheres that were mixed into individual formulations and thus embedded in the films to simulate the behavior of a hydrophobic drug. Each film was immersed into a well stirred

beaker simulating a mixing tank containing 100 ml of Ringer's Lactate (pH = 6.5). Under these sink conditions, compound release occurs from all surfaces of the matrix differing from the geometry of the in-use wound dressing, a methodology pursued in previous studies (Gay et al. 1983; Tada et al. 2010). Franz cells or diffusion chambers are used in standard *in vitro* test for diffusion experiments for researching transdermal drug administration (Franz 1978; Smith & Haigh 1992). As reported, this type of testing often yields permeation data that suffer from poor reproducibility compounding the method variation with the sample non-uniformities (Ng et al. 2010; Chilcott et al. 2005). Due to the lack of knowledge on the topography, porosity, surface roughness of the films, and the associated variance, it was decided to proceed with the traditional immersion method (Gay et al. 1983; Tada et al. 2010). The release and film- dissolution profiles were characterized by analyzing the optical density in the supernatant. Samples of 0.5 ml from the supernatant mixed with 0.5 ml of Ringer's Lactate were taken every 5 min and centrifuged at 3,000 g prior to being subjected to spectrophometric analysis using a UV–VIS spectrophotometer (Agilent 8453) at wavelengths of 555 nm for beads and 279 nm for 5-FU, respectively. Throughout the UV–VIS spectrum, the polystyrene beads will scatter light in the Mie domain (van de Hulst 1981). In addition, the size of the beads is comparable to that of bacteria for which optical density characterization due to particle light scattering is conducted in the range of 550–600 nm (Shuler & Kargi 2010) and thus the justification behind the wavelength of 555 nm for microsphere detection. Linear calibration curves for converting optical density to concentration for 5-FU and the beads are presented in Figures 1a and 1b, respectively. In order to create quantitative drug elution responses, the time release plots were compared in terms of three time periods: a) Lag phase defined as the duration of membrane hydration, b) Membrane burst phase defined as the duration of active diffusion of

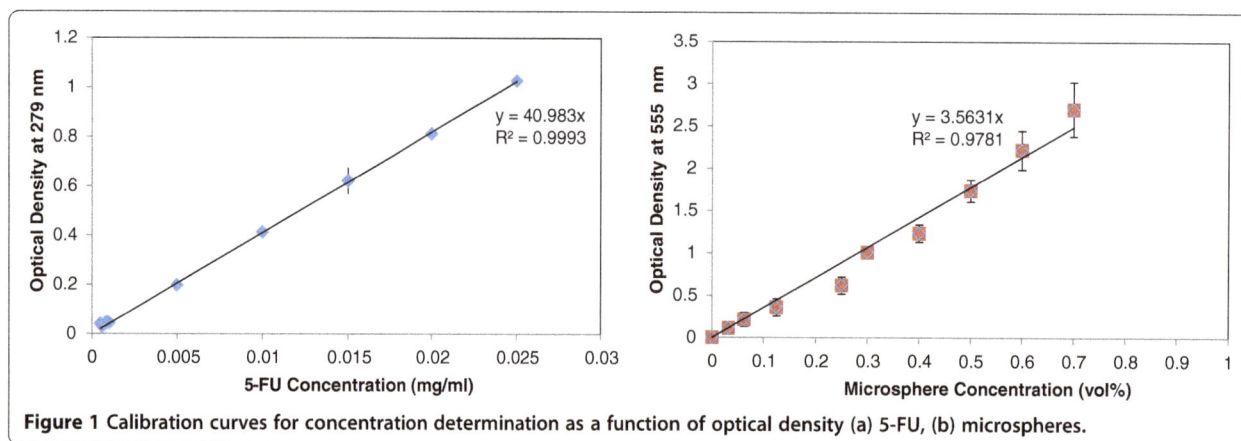

Figure 1 Calibration curves for concentration determination as a function of optical density (a) 5-FU, (b) microspheres.

encapsulated substance from the films, and c) Steady-State Phase defined as the period ($dC/dt = 0$) during which the membrane has entirely disintegrated and the encapsulated substances are free in solution. The burst time (t_B) and release time (t_R) are defined as the duration of lag and burst phases, respectively. The duration of release per formulation as a function of time was also modeled as a function of concentration (C) and the steady state concentration (C_{SS}). For all models residuals (SSE) were minimized using the Solver Tool in Excel 2010.

Pore size measurement

3 cm^2 samples were cut from each film by a razor and glued onto a 15 mm sample disc. This disc was then placed onto the sample holder and held into place by two arms, whereupon the sample holder was placed under the atomic force microscopy (AFM) stage.

Surface imaging was performed on the CD0.5 film formulations using AFM. The characterization was conducted using an Agilent 5500 AFM using non-contact mode PPPHR-NC probes (NanoAndMore, USA). Picoview v1.8 and Gwyddion v2.3 were used as qualitative real-time and quantitative image analysis software, respectively. Experimental parameters were set in Picoview that maximized the trace/retrace profile for optimal imaging. These parameters were: scan speed of 1 line/s, resolution of 256 pixels, sample scale of 20 microns for surface roughness measurements, sample scale of 2 microns for pore size measurements, integral and proportional (I and P) gains of 10, frequency offset at –300 Hz, and stop at percentage at 95%. Four images were obtained at different locations around the sample. These locations were changed through the movement of the stage. The average pore size was obtained by taking the surface profile of the images using Gwyddion across each film. Average surface roughness (R_a) was estimated through a statistical algorithm coded into Gwyddion.

Results

Fabricated circular sample films prepared according to the formulations in Table 1, are depicted in Figure 2.

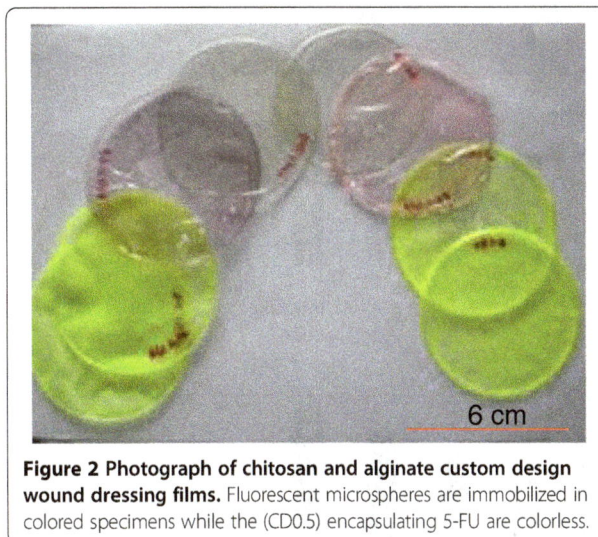

Figure 2 Photograph of chitosan and alginate custom design wound dressing films. Fluorescent microspheres are immobilized in colored specimens while the (CD0.5) encapsulating 5-FU are colorless.

Parametric study

The summary of the parametric study encompassing the variation in film characteristics as a result of formulation is presented in Table 2 and Figures 3a-3e.

A one-tailed student t-test was used to compare film properties as a function of polymer type used (alginate or chitosan) at the 95% confidence interval ($\alpha = 0.05$). Analysis was conducted using the Data Analysis toolbox of Excel 2010.

Alginate films are characterized by significantly higher elastic moduli (p = 0.003, $E_{CD} = 258$ MPa ± 185 MPa .vs. $E_{AGCa} = 1670$ MPa ± 578 MPa) and release windows (p = 0.03, $t_{RCD} = 31$ min ± 5 min .vs. $t_{RAGCa} = 38$ min ± 6 min). A monotonic increase in tensile strength as a function of plasticizer content (PG) for the CD films and cross-linker (CaCl$_2$) content for the AGCa films is observed. Chitosan films exhibit superior sorption characteristics (p = 0.001, $S_{CD} = 1208\% \pm 233\%$ .vs. $S_{AGCa} = 306\% \pm 60.3\%$), with a monotonic increase in sorption capability observed as a function of (PG) content. Chitosan films are also significantly thicker (p = 0.004, $th_{CD} = 1260$ μm ± 255 μm .vs. $th_{AGCa} = 540$ μm ± 169 μm) as well being resistant to burst

Table 2 Average results of the parametric study by wound dressing film

Film name	Thickness (μm)	[E] (kPa)	TS (kPa)	S (%)	t_B (min)	t_R (min)
CD 0.0	1451 ± 74	58 ± 46	18 ± 2	962 ± 24	60 ± 3	30 ± 4
CD 0.5	1431 ± 90	228 ± 40	39 ± 4	1102 ± 40	70 ± 7	50 ± 2
CD 1.0	1258 ± 68	239 ± 30	64 ± 9	1293 ± 62	60 ± 5	25 ± 3
CD 1.5	901 ± 74	507 ± 80	94 ± 11	1474 ± 46	40 ± 6	35 ± 7
AGCa 0.04	592 ± 97	1010 ± 246	27 ± 8	347 ± 24	45 ± 5	55 ± 6
AGCa 0.08	351 ± 143	1915 ± 220	67 ± 7	335 ± 56	30 ± 3	35 ± 1
AGCa 0.12	677 ± 68	2086 ± 400	107 ± 13	237 ± 28	35 ± 1	35 ± 3

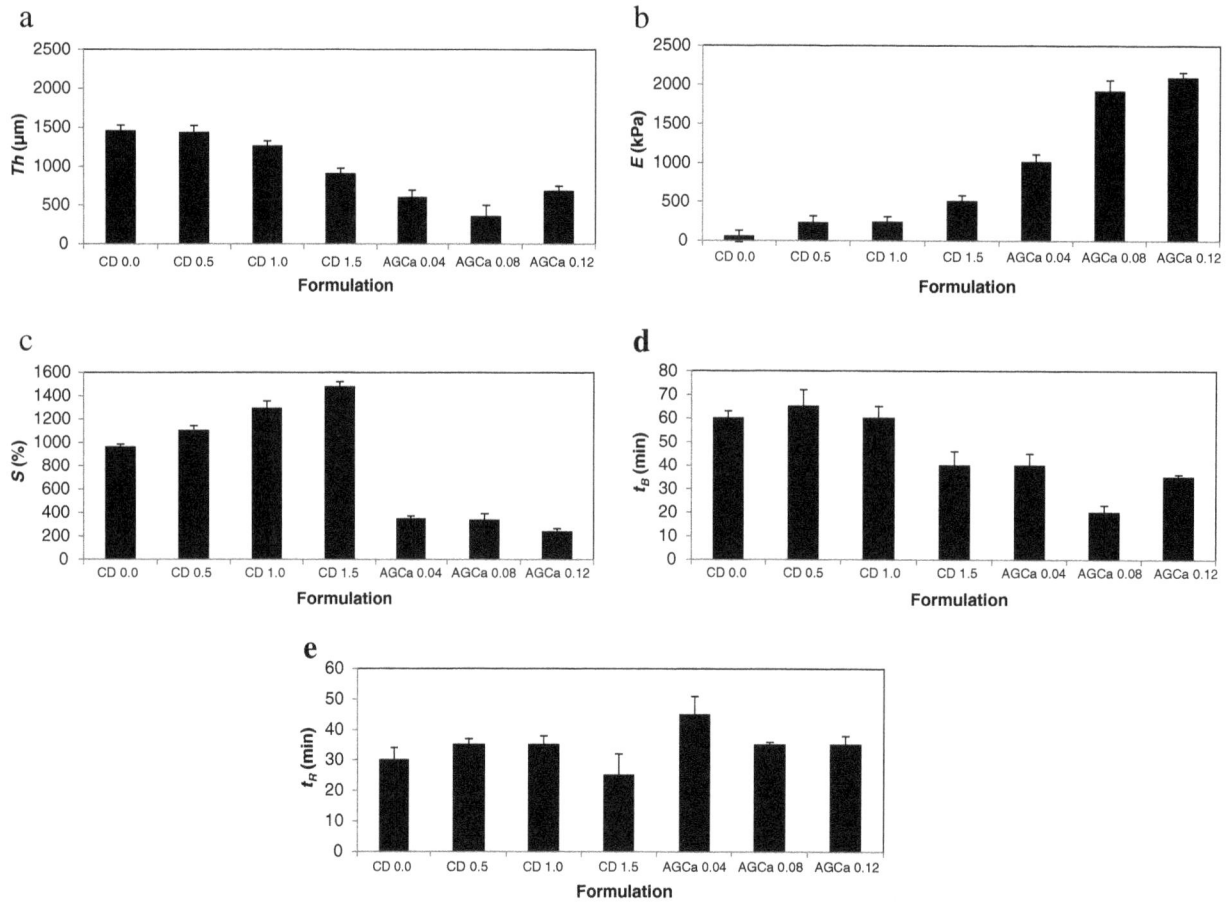

Figure 3 Summary of parametric study for statistically significant film properties by formulation (a) film thickness, (b) elastic Modulus, (c) percent sorption, (d) lag phase duration,(e) duration of release.

(p = 0.027, t_{BCD} = 56 min ± 11 min .vs. t_{BAGCa} = 32 min ± 10 min).

There is no significant difference in tensile strength (p = 0.324, TS_{CD} = 54 kPa ± 33 kPa .vs. TS_{AGCa} = 67 kPa ± 40 kPa) by main polymer type.

Tensile testing

Elastic modulus and tensile strength were determined per formulation from the stress strain curves displayed in Figures 4 and 4b. As mentioned above in average chitosan-based films are significantly more brittle than alginate films. Although both chitosan and alginate films display viscoelastic behaviours, chitosan has linear stress–strain curves whereas alginate has non-linear curves especially after 4% strain. Alginate films fractured at much lower strains (8-10% strain) when compared with chitosan films (18-38% strain).

Microsphere and drug elution profiles

Averaged release profiles by film formulation (N = 3) are plotted in Figures 5a and 5b for chitosan and alginate films. Corresponding empirical piecewise defined mathematical functions modeling the release of encapsulated compounds [C(t)] using Heaviside functions is given by Eqn. (6). Release profiles during burst phase as a function of concentration measurements [R(t)] are presented in Table 3 for each formulation.

$$C(t) = u_{t_B}(t)R(t) - u_{t_R}(t)\left[R(t) - C_{ss}\right] \qquad (6)$$

where

$C(t)$ concentration of encapsulated compound in solution at anytime

$C(t) = 0 \ t < t_B$

$C(t) = R(t) \ t_B \leq t < t_R$

$C(t) = C_{SS} \ t \geq t_R$

$R(t)$ concentration of encapsulated compound in solution during release phase (vol% or mg/ml)

C_{SS} steady state concentration of released compound in solution (vol% or mg/ml)

t_B membrane time to burst (min)

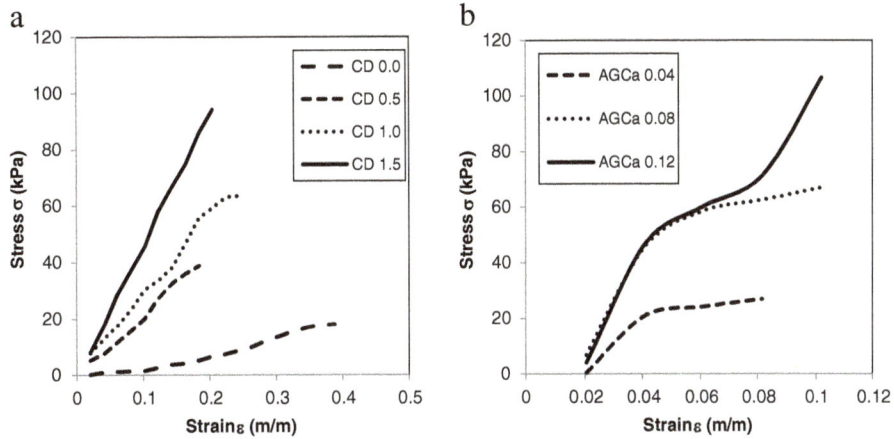

Figure 4 Stress/Strain curves for wound dressings (a) chitosan films, (b) alginate films.

t_R duration of compound release in the supernatant (min)
u_{t_B} Heaviside function related to t_B
u_{t_R} Heaviside function related to t_R

As seen in Figures 5a and 5b, in accordance with the results of the statistical analysis conducted in the parametric study section above by main polymer type, chitosan films are characterized by a longer lag phase while a

longer release times are associated with alginate films. The values of t_B, t_R, C_{ss} used are average values per film and presented in Table 2. The total percentage of microsphere or drug released [M] given by Eqn. 7 can be calculated by film type from the release profile assuming the concentration of compound released in the mixing tank is uniform and the total volume of the Ringer's solution in the tank remains constant (Saterbak et al. 2007).

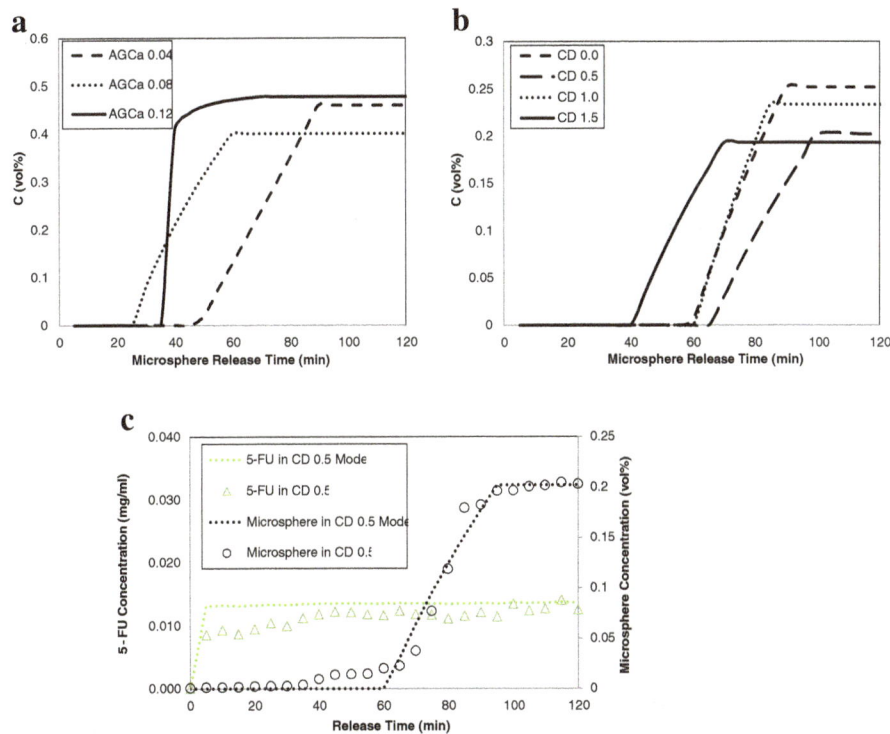

Figure 5 Modeled comparative release profiles for (a) encapsulated microspheres in alginate films, (b) encapsulated microspheres and in chitosan films, (c) release of microspheres and 5-FU from CD0.5 film.

Table 3 Modeling of time release profiles from films (N = 3)

Film	Encapsulated compound	Burst phase model [R(t)]	C_{SS}	SSE	M (%)
CD 0.0	microsphere	$0.606 \ln(t) - 2.4175$	0.251	0.00	49
CD 0.5	microsphere	$0.4817 \ln(t) - 2.016$	0.202	0.00	41
CD 0.5	5-FU	$0.025(t)^{0.2} / (0.35 + (t)^{0.2})$	0.013	0.00	57
CD 1.0	microsphere	$0.667 \ln(t) - 2.774$	0.234	0.00	46
CD 1.5	microsphere	$0.355 \ln(t) - 1.315$	0.193	0.00	34
AGCa 0.04	microsphere	$0.011\, t - 0.530$	0.460	0.03	99
AGCa 0.08	microsphere	$0.515 \ln(t) - 1.157$	0.400	0.00	47
AGCa 0.12	microsphere	$0.5(t-35)^{0.8} / (0.8 + (t-35)^{0.8})$	0.477	0.06	96

$$M = \frac{\frac{1}{V} \int_{t_B}^{t_B + t_R} R'(t)\ dt}{M_0} \ x\ 100 \ \ x\ DF \qquad (7)$$

where

M Percentage of encapsulated volume or mass released
M_0 Initial volume fraction (ml) or amount (mg) per 100 ml
V Ringer's solution volume
DF Dilution factor at sampling

The percentage microsphere release ranges were calculated to be 47-99% and 34%-49% for alginate and chitosan films, respectively.

Since across all formulations, the elasticity of human skin rated at 18.8 MPa (Silver et al. 2007) was not surpassed, elastic modulus was not a criterion for choosing the film formulation for studying the release of the model drug 5-FU. Chitosan, specifically the CD0.5 formulation, was chosen over alginate as the main polymer to encapsulate 5-FU because of its significantly higher sorption ability and superior resistance to membrane burst. Comparative drug elution profiles for the control CD 0.5 film encapsulating microspheres and 5-FU are presented in Figure 5c. Drug release was characterized by the absence of a lag phase ($t_B = 0$) and immediate diffusion. The percentage of 5-FU released from the CD0.5 film is 57% as compared to 41% for the CD0.5 film encapsulating the microspheres.

AFM imaging results

Sample 2D and 3D views for surface roughness and pore size measurements of the CD 0.5 film based on AFM scans, are presented in Figures 6a-6b and Figures 7a-7b, respectively. The film is characterized by an average pore size of 430 nm ± 88 nm and roughness of 1.8 nm. AFM is a technique used for surface characterization and it can only be hypothesized that the recorded surface indentations are pores. However, since 5-FU diffuses through the membrane as proven by the spectrophotometric method and, membrane burst is necessary for the 710 nm microspheres to be released, it is likely that these features are pores.

Discussion

The viscoelastic behaviors of alginate, chitosan, and alginate/chitosan have been characterized experimentally as a function of MW, blend composition, degree of substitution, and extent of cross-linking under compressive and tensile strains has been characterized by multiple sources (Storz et al. 2010; Moresi et al. 2001; Mitchell & Blanshard 1974; Torres et al. 2006; Li et al. 2012; Shon et al. 2007). As reported in the results section, for alginate films beyond 4% strains, non-linear viscoelasticity was observed. For non-blended, non-cross linked alginate gels, strains in the range of 3-10% exhibited linear viscoelastic behavior (Storz et al. 2010). Other studies showed variation of G' (storage modulus) on frequency of oscillation and showed the dependency of viscoelastic behavior on MW and ratio of mannuronic to glucuronic groups (Moresi et al. 2001). On average, the experimental tensile strength is 1000 times lower than values previously reported for identical formulations (Rhim 2004). The reported average elongation at break of 10% coincides with the upper limit reported in this study. The root cause of this discrepancy could reside in the measurement techniques for film thickness. Their results did not indicate that as the concentration of CaCl$_2$ solution increased TS increased, which is not well aligned with the positive correlation observed between the cross-linker concentration and tensile strength in this study. It could be inferred that the increase in TS and the decrease in E by CaCl$_2$ treatment were mainly due to the development of cross-linking between carboxyl group of alginate and Ca^{2+}. Correlations between G' (storage modulus) and G'' (loss modulus) as a function of concentration and oscillatory frequency have been reported on rheological studies conducted on pure chitosan (Torres et al. 2006).

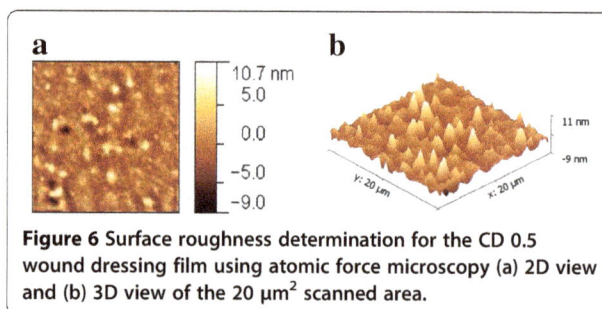

Figure 6 Surface roughness determination for the CD 0.5 wound dressing film using atomic force microscopy (a) 2D view and (b) 3D view of the 20 μm² scanned area.

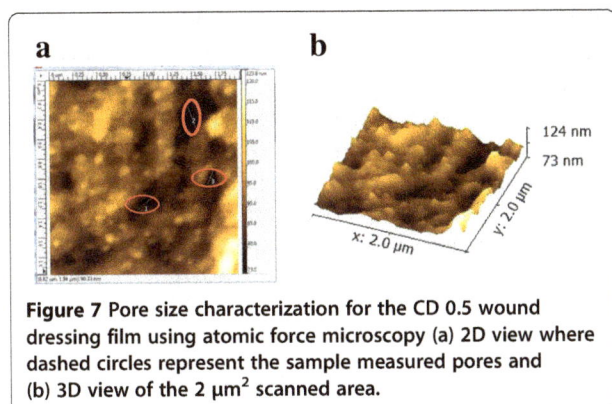

Figure 7 Pore size characterization for the CD 0.5 wound dressing film using atomic force microscopy (a) 2D view where dashed circles represent the sample measured pores and (b) 3D view of the 2 μm^2 scanned area.

G′ is significantly higher than G″ at low frequency oscillations and at high concentrations of chitosan indicating the predominance of elastic behavior. G″ was higher than G′ at lower concentrations indicating the decrease in elastic behavior. Their lower limit of chitosan concentration corresponds to the nominal 0.025 g/mL used in this study. In another study, the tensile strength of chitosan films cross-linked with lactic acid (1.4% w/v in 1% w/v lactic acid), in absence of dextran and PG, was calculated to be 60 MPa, approximately 1000 times higher than the average measured TS for the CD films with an elongation at break of 67% (Khan et al. 2000) as compared to the measured range of 18-38%. Theoretically, in absence of the plasticizer and a lower chitosan concentration, the reported TS should have been lower; however, since the thickness of the films has not been published, it is not possible to narrow down the cause of this discrepancy.

As quantified by tensile testing measurements, alginate films were 10 times stiffer than chitosan films with a much lower fracture strain level. Chitosan is a stiff/rigid polyelectrolyte associated with conferring compressive strength to bio-membranes (Jana et al. 2012; Mobed & Chang 1998). As for the effect of plasticizer content on the reduction in thickness of the chitosan films with increase in PG content, has been reported for other hydrophilic plasticizers such as polyethylene oxide (PEO). This has been explained by the contraction of the three-dimensional film matrices due to strong molecular interactions between chitosan and PEO molecules (Li et al. 2011). In this study chitosan films are significantly thicker than alginate films and CD film thickness decreases with increasing PG content. Cited film thicknesses for chitosan wound dressings range from 0.028 to 0.13 mm (Bhuvaneshwari et al. 2011), approximately 10 times lower than the experimental values. Apart from the difference in measurement methods, it could be hypothesized that the removal of glutaraldehyde as a chemical cross-linker substituted by physical cross-linking in lactic acid, resulted in a less porous matrix, hindering evaporation. Assuming that the porosity of composite films is regulated by the amount of plasticizer (Madsen et al. 2011; Li et al.

2011), it could be inferred that porosity and hence the rate of evaporation increases with increasing PG concentrations, thus resulting in thinner CD films. As indicated by the results of tensile testing, alginate film cross-linking occurred and is dependent on cross-linker concentration. Theoretically, an expected monotonous decrease in film thickness as a function of degree of cross-linking, and thus higher porosity, should have been recorded. It could be inferred that the experimental range for the $CaCl_2$ is this study is not large enough to trigger significant porosity changes.

In this study the chitosan films exhibited significantly higher hydrophilicity and thus a higher affinity for wound exudate simulated by the Lactated Ringer's solution. The results of the sorption ability of the films indicate a growing hydrophilicity with an increase in PG concentration in chitosan films in agreement with previous findings (Ayogi et al. 2007; Wittaya-arrekul & Prasharn 2006) identical to the behavior of PEGylated chitosan derivatives (Bhuvaneshwari et al. 2011). The sorption values of the films are on average 1–5 times higher than the values previously reported (Wittaya-arrekul & Prasharn 2006) due to the higher film thicknesses measured in this study. As the degree of cross-linking increases, alginate water solubility decreases, resulting in lower sorption capability of the alginate films (Rhim 2004).

Atomic force microscopy was used in previous studies looking at the surface roughness and pore size of chitosan and chitosan alginate composite films (Casettaria et al. 2012; Dash et al. 2011; Hu et al. 2007; Karakecili et al. 2007; Zheng et al. 2009; Doulabia et al. 2013). It has been reported that the surface topography of 100% chitosan had a smooth surface with uniformly distributed short spikes, but when an additive was introduced, the surface roughness increased resulting in taller spikes (Doulabia et al. 2013). In these articles the R_a for several formulations/methods of chitosan film fabrication ranges from 0.3- to 4.6 nm (Casettaria et al. 2012; Dash et al. 2011; Hu et al. 2007). The experimental value of 1.8 nm falls within this reported range. The measured pore sizes of 433 nm ± 88 nm are larger than the 2.8 -100 nm previously reported for a scan area of 5 μm^2 as compared to the 2 μm^2 adopted in this study (Hu et al. 2007).

It could be inferred that chitosan rigidity, in addition to thicker-wound dressing walls, are the driving forces behind the statistically significantly higher average membrane burst time as compared to alginate films. Even after membrane burst, the average release from alginate films (M_{AGCa} = 81%) is approximately twice that calculated from chitosan films (M_{CD} = 42%) from which, it could be inferred that alginate films are more porous than chitosan films and hence, the justification of using chitosan for slowing down the release rate of compounds from alginate membranes (Gaserod et al. 1999; Asthana et al. 2012). Theoretically, an increase in $CaCl_2$ for the alginate films in conjunction with an increase in PG content for the chitosan films, should increase the porosity of the films, and hence the diffusion rate (Madsen et al. 2011; Simpliciano & Asi 2012); however, no discernible trends were observed as a result of varying the aforementioned factors. These observations are limited by the unknown distribution of surface-to-through pores for each film obtainable through scanning electron microscopy.

As previously stated, 5-FU release from the CD 0.5 film was characterized by the absence of a lag phase as compared to the microsphere release profile from the same control chitosan-based film. 5-FU has a Stokes radius of 0.372 nm Fournier (2011) as compared to the 710 nm microsphere. Hence, the absence of lag phase for the 5-FU release profiles is attributed attributed to the approximate ratio (1: 505) in molecular size. Given the film pore size of 430 nm ± 88 nm, membrane burst due to osmosis is not necessary for drug diffusion while it is for the larger microspheres. Identical release profiles as in Figure 5c were generated for the co-encapsulation of the drug and the microsphere, demonstrating the immediate diffusion of the drug followed by membrane burst releasing the microspheres (Rook et al. 2012). This two stage release has been recently documented for the release of curcumin co-encapsulated with silica microspheres in chitosan scaffolds (Ahmed et al. 2012) although the mechanism of burst has not been elaborated upon. Revisiting the formulation of the 5-FU film in order to restore the lag phase into the elution profile and thus modulate membrane burst time, the drug should be first encapsulated into microspheres/nanoparticles then immobilized within the wound dressing film, a successful approach adopted to modulate drug release and minimize the membrane burst effects (Tada et al. 2010; Ramadas et al. 2000; Dhoot & Wheatley 2003).

Conclusion and future efforts

Composite films of chitosan and alginate, intended for drug delivery/wound healing applications, were fabricated and characterized. It was revealed by this *in-vitro* evaluation that alginate films are significantly stiffer. Meanwhile chitosan films are thicker, more resistant to osmotic burst and are more hydrophilic as characterized sorption rates. With the successful elimination of glutaraldehyde from the chitosan film formulations, 5-FU was encapsulated as a model drug into a chitosan film (CD 0.5) comprised of 2.5% (w/v) Medium MW chitosan/dextran 70 kDa (5:1) using 0.5 and 3% (v/v) of PG and lactic acid, respectively. The translated microsphere release profiles modeled using Heaviside functions as compared to 5-FU release characterized by the absence of a lag phase, are supported by AFM pore size measurements. Future porosity characterization should encompass scanning electron microscopy (SEM) measurements to determine the nature and directionality of the pores

as well as the distribution of surface-to- through pores. Also, at a known porosity and nominal film thickness, fluorescent spatial and temporal concentration gradients should be measured in order to obtain diffusivity coefficients for optimization of the desired pharmacokinetic flux. *In-vitro* cytotoxicity testing will be added to the protocol to assess the effect of the glutaraldehyde removal from the chitosan formulations.

Abbreviations
5-FU: 5-fluoro-1,2,3,4-tetrahydropyrimidine-2,4-dione; CD: Chitosan-dextran film formulation; AGCa: Alginate film formulation; th: Film thickness (μm); E: Elastic modulus (Pa); TS: Tensile strength (Pa); S: Sorption ability (%); t_B: Membrane time to burst or lag phase duration (min); t_R: Duration of release from membrane or release window5-FU (min); M: Percentage of encapsulated mass or volume released.

Competing interests
The authors declare that they have no competing interests.

Authors' contributions
RKN: has made substantial contributions to 1) conception and design of the wound dressing and bothencapsulants and 2) acquisition of mechanical, sorption moisture permeability and microsphere release data with microspheres and analysis and interpretation of data. SLR: has made substantial contributions in acquisition of mechanical, sorption moisture permeability and microsphere release data with microspheres and analysis and interpretation of data. JJR: has formulated the wound dressings for 5-FU immobilization, replicated the preliminary data with microspheres and acquired and interpreted data for 5-FU release. MK: has been involved in drafting the manuscript and revising it critically for important intellectual content specifically the mechanical properties analysis. MG: has been involved in drafting the manuscript and revising it critically for important intellectual content specifically the Atomic Force Microscopy Analysis.. MMM: has made substantial contributions 1) to conception and design of the wound dressing and both immobilized substances; 2) performed the statistical analysis and modelling of encapsulated substance release profiles 3) drafted the manuscript and revised it critically for general content. All authors read and approved the final manuscript.

Authors' information
RKN: Is a recent graduate of the Biomedical Devices graduate program at SJSU with a Bachelor's Degree in Medical Electronics. SLR: Is a recent graduate of the Biomedical Devices Graduate program at SJSU with a Bachelor's Degree in Chemical Engineering. JJR: Is a graduate student in the Biomedical Engineering program at SJSU with with a Bachelor's Degree in Chemical Engineering. MK: Dr. Keralapura is an Assistant Professor in the department of Electrical Engineering and Director of the Biomedical Systems Laboratory at SJSU. She has extensive experience in fundamental mechanical measurements of tissue and tissue-like media using both rheological techniques with ultrasound strain imaging. She is a highly respected young contributor in the field of Medical Physics. MG: Dr. Goedert is an Adjunct Professor in the department of Biomedical, Chemical and Materials Engineering at SJSU and a French Chemistry Society Prize winner. He has over 20 years of experience in high precision sensor development and 10 years of experience in micro-channel surface characterization. MMM: Dr. Mobed-Miremadi is an instructor and Endowed Chair of Bioengineering at SJSU. She has researched microencapsulation methods extensively and has over 10 years of experience in the biomedical industry specifically in the areas of inkjet bio-printing .bio-reactor design genomics (micro-array fabrication) and medical devices.

Acknowledgements
The authors would like to acknowledge the support from three funding sources namely the C-SUPERB Joint Venture grant ("Bio-Printing of Mammalian Cells"), the CSU Mini-Grant ("Bio-Functionalized Resorbable Drug Eluting Stents") and the SJSU Davidson's College of Engineering Junior Professorship Grant. The authors would also like to thank Mr. William Schlukins and Mr. Alex Yuen for their assistance in the use of AFM.

Mechanical characterization of the films could not have been executed without access to the custom-designed tensile tester granted by Dr. Guna Selvaduray.

Author details
[1]Department of Biomedical, Chemical and Materials Engineering, San Jose State University, San Jose, CA 95192-0082, USA. [2]MSE Biomedical Devices Concentration, San Jose State University, San Jose, CA, USA. [3]MSE Biomedical Engineering, San Jose State University, San Jose, CA, USA. [4]Department of Electrical Engineering, San Jose State University, San Jose, CA, USA.

References
Adusumilli PS, Bolton SM (1991) Evaluation of chitosan citrate complexes as matrices for controlled release formulations using a full 32 factorial design. Drug Dev Ind Pharm 17:1931–1945

Ahmed A, Hearn J, Abdelmagid W, Zhang H (2012) Dual-tuned drug release by nanofibrous scaffolds of chitosan and mesoporous silica microspheres. J Mater Chem 22:25027–25035

Akbuga J, Bergisadi N (1996) 5-Fluorouracil loaded chitosan microspheres: Preparation and release characteristics. J Microencap 13:161–168

Asthana A, Kwang HL, Kim KO, Kim DM, Kim DP (2012) Rapid and cost-effective fabrication of selectively permeable calcium-alginate microfluidic device using "modified" embedded template method. Biomicrofluidics 6(1):012821–0128219

Ayogi S, Onishi H, Machida Y (2007) Novel chitosan wound dressing loaded with minocycline for the treatment of severe burn wounds. Int J Pharm 330(1–2):138–145

Bhuvaneshwari S, Sruthi D, Sivasubramanian V, Niranjana K, Sugunabai J (2011) Development and characterization of chitosan films. IJERA 1(2):292–299

Burkatovskaya M, Tegos GP, Swietlik E, Demidova TN, Castano AP, Hamblin MR (2008) Use of chitosan bandage to prevent fatal infections developing from highly contaminated wounds in mice. Wou Rep Regen 16(3):425–431

Callister W, Rethwisch D (2010) Materials Science and Engineering: An Introduction. John Wiley & Sons, New Jersey

Casettaria L, Vllasaliu D, Castagnino E, Stolnikb S, Howdle S, Illum L (2012) PEGylated chitosan derivatives: Synthesis, characterizations and pharmaceutical applications. Prog Polym Sci 37(5):659–685

Chandrashekar NS, Prasanth VV (2008) Clinical evaluation of 5-fluorouracil form transdermal patches on EAC and DLA cell-induced tumors in mice. APJCP 9:437–440

Chilcott RP, Barai N, Beezer AE, Brain SI, Brown MB, Bunge AL, Burgess SE, Cross S, Dalton CH, Dias M, Farinha A, Finnin BC, Gallagher SJ, Green DM, Gunt H, Gwyther RL, Heard CM, Jarvis CA, Kamiyama F, Kasting GB, Ley EE, Lim ST, McNaughton GS, Morris A, Nazemi MH, Pellett MA, Du Plessis J, Quan YS, Raghavan SL, Roberts M, Romonchuk W, Roper CS, Schenk D, Simonsen L, Simpson A, Traversa BD, Trottet L, Watkinson A, Wilkinson SC, Williams FM, Yamamoto A, Hadgraft J (2005) Inter- and intra-laboratory variation of in vitro diffusion cell measurements: an international multicenter study using quasi-standardized methods and materials. J Pharm Sci 94(3):632–638

Dash M, Chiellini F, Ottenbrite RM, Chiellini E (2011) Chitosan – a versatile semi-synthetic polymer in biomedical applications. Prog Polym Sci 36(8):981–1014

Dhoot NO, Wheatley MA (2003) Microencapsulated liposomes in controlled drug delivery: strategies to modulate drug release and eliminate the burst effect. J Pharmaceut Sci 92(3):679–689

Diasio RB, Harris BE (1989) Clinical pharmacology of 5-fluorouracil. ClinPharmacokinet 16:215–237

Doulabia AZ, Mirzadeh H, Imani M, Samadic N (2013) Chitosan/polyethylene glycol fumarate blend film: Physical and antibacterial properties. Carbohydr Polym 92:48–56

Duke University Medical Center Database (2000) Material safety data sheet. http://www.safety.duke.edu/ohs/MSDS/Absorbase.pdf. Accessed on 27 December 2012

Dureja H, Tiwary AK, Gupta S (2001) Simulation of skin permeability in chitosan membranes. Int J Pharm 213:193–198

Fluorouracil (2012) Drug Bank. http://www.drugbank.ca/drugs/DB00544. Accessed on 27 December 2012

Fournier 2011 Basic transport phenomena in biomedical engineering. Taylor & Francis, New York

Franz T (1978) The finite dose technique as a valid in vitro model for the study of percutaneous absorption. Curr Probl Dermatol 7:58–612

Gaserod O, Sannes A, Skjak-Braek G (1999) Microcapsules of alginate-chitosan. II. A study of capsule stability and permeability. Biomat 20(8):773–783

Gay MH, Williams DL, Kerrigan JH, Creeden DE, Nucefora WA, Orfao SA, Nuwayser ES (1983) Research and Development of Wound Dressing in Maxillofacial Trauma. Defense Technical Information Center, Woburn, MA

Giannola LI, De Caro V, Giandalia G, Siragusa MG, Paderni C, Campisi G, Florena AM (2010) 5-Fluorouracil buccal tablets for locoregional chemotherapy of oral squamous cell carcinoma: formulation, drug release and histological effects on reconstituted human oral epithelium and porcine buccal mucosa. Cur Drug Deli 7(2):109–117

Hu Y, Wu Y, Cai J, Ma Y, WangB XK, He X (2007) Self-assembly and fractal feature of chitosan and its conjugate with metal ions: Cu (II) / Ag (I). Int J Mol Sci 8:1–12

Jana S, Florczyk SJ, Leung M, Zhang M (2012) High-strength pristine porous chitosan scaffolds for tissue engineering. J Mater Chem 22:6291–6299

Karakecili A, Satriano C, Gumusderelioglu M, Marletta G (2007) Surface characteristics of ionically crosslinked chitosan membranes. J Appl Polym Sci 106(6):3884–3888

Khan TA, Peh KK, Ch'ng HS (2000) Mechanical, bio-adhesive strength and biological evaluations of chitosan films for wound dressing. J Pharm PharmaceutSci 3(3):303–311

Li J, Zivanovic S, Davidson PM, Kit K (2011) Production and characterization of thick, thin and ultra-thin chitosan/PEOfilm. Carbohyd Polym 83(2):375–382

Li C, Wang L, Yang Z, Kimm G, Chen H, Ge Z (2012) A viscoelastic chitosan-modified three-dimensional porous poly (L-lactide-co-ε-caprolactone) scafold for cartilage tissue engineering. J Biomater Sci Polym Ed 23(1–4):405–424

Loven K, Stein L, Furst K, Levy S (2002) Evaluation of the efficacy and tolerability of 0.5% fluorouracil cream and 5% fluorouracil cream applied to each side of the face in patients with actinic keratosis. Clin Ther 24(6):990–1000

Madsen B, Joffe R, Peltola H, Nattinen K (2011) Short cellulosic fiber/starch acetate composites micromechanical modeling of Young's modulus. J Compos Mater 45(20):2119–21327

Meng X, Tian F, Yang J, He CN, Xing N, Li F (2010) Chitosan and alginate polyelectrolyte complex membranes and their properties for wound dressing application. J Mater Sci Mater Med 21(5):1751–1759

Mitchell JR, Blanshard JMV (1974) Viscoelastic behaviour of alginate gels. Rheol Acta 13(180–184)

Mobed M, Chang TMS (1998) Comparison of polymerically-stabilized PEG liposomes and physically-adsorbed carboxymethyl and carboxymethyl/ glycolchitin liposomes for biological applications. Biomat 19:1167–1177

Moresi M, Mancini M, Bruno M, Rancini R (2001) Viscoelastic properties of alginate gels by oscillatory dynamic tests. J Texture Stud 32(5–6):375–396

Murakami K, Aoki H, Nakamura S, Nakamura SI, Takikawa M, Hanzawa M, Kishimoto S, Hattori H, Tanaka Y, Kiyosawa T, Sato Y, Ishihara M (2010) Hydrogel blends of chitin/chitosan, fucoidan and alginate as healing-impaired wound dressings. Biomat 31:83–90

Ng SF, Rouse JJ, Sanderson FD, Meidan V, Eccleston GM (2010) Validation of a static Franz diffusion cell system for in vitro permeation studies. AAPS PharmSciTech 11(3):1432–1441

Ramadas M, Paul W, Dileep KJ, Anitha Y, Sharma CP (2000) Lipoinsulin encapsulated alginate-chitosan capsules: intestinal delivery in diabetic rats. J Microencap 17(4):405–411

Rhim JW (2004) Physical and mechanical properties of water resistant sodium alginate films. Swiss Society of Food Science and Technology 37:323–330

Rook J, McNeil M, Mobed-Miremadi M (2012) Characterization of fluorouracil diffusion from chitosan films. In: Abstracts of the 24th Annual California Biotechnology Symposium., Santa Clara, 5–7 January 2012 http://csuperb. org/oars/abstract_view_inline.php?id=222&year=2012&PHPSESSID= 7302ce2bccdcbfefbec2519796af826d. Accessed on 27 December 2012

Saterbak A, McIntire LV, San KY (2007) Bioengineering Fundamentals. Pearson/Prentice Hall, New Jersey

Sezer AD, Hatipoglu F, Cevher E, Ogurtan Z, Bas AL, Ali JA (2007) Chitosan film containing fucoidan as a wound dressing for dermal burn healing: preparation and in vitro/in vivo evaluation. J AAPS PharmSciTech 8(2):E94–E101. doi:10.1208/pt0802039

Shabbear S, Ramanamurthy S, Ramanamurthy KV (2012) Formulation and evaluation of chitosan sodium alginate microcapsules of 5-fluorouracil for colorectal cancer. IJRPC 2(1):7–19

Shon SO, Ji BC, Han YA, Park DJ, Kim IS, Choi JH (2007) Viscoelastic sol–gel state of the chitosan and alginate solution mixture. J Appl Polym Sci 104(3):1408–1414

Shuler ML, Kargi F (2010) Bioprocess Engineering: Basic Concepts. Prentice Hall, New Jersey

Silver FH, Freeman JW, Devore D (2007) Viscoelastic properties of human skin and processed dermis. Skin Res Tech 1:18–23

Simpliciano C, Asi B (2012) Pore size determination and validation using AFM and spectrophotometry. In: Abstracts of the Third Bay Area Biomedical Devices Conference. San Jose State University, 28 March 2012. http://www.engr.sjsu.edu/~bmes/BMDConf2012/abstract4.html. Accessed on 27 December 2012

Smith EW, Haigh JM (1992) In vitro diffusion cell design and validation. Temperature, agitation and membrane effects on betamethasone 17-valerate permeation. Acta Pharm Nord 4(3):171–178

Storz H, Zimmermann U, Zimmerman H, Kuliche WM (2010) Viscoelastic properties of ultra-high viscosity alginates. Rheol Acta 49(2):155–167

Suheyla HK (1997) Chitosan: Properties, preparation and application to micro particulate system. J Microencap 14:689–711

Tada DB, Singh S, Nagesha D, Jost E, Levy CO, Gultepe E, Cormack R, Makrigiorgos M, Sridhar S (2010) Chitosan film containing poly(D, L-lactic-co-glycolic acid) nanoparticles: A platform for localized dual drug release. Pharm Res 27(8):1738–1745

Tiwary AK, Rana V (2010) Cross-linked chitosan films: effect of cross-linking density on swelling parameters. Pak J Pharm Sci 23(4):443–448

Torres MA, Beppu MM, Arruda EJ (2006) Viscous and viscoelastic properties of chitosan solutions and gels. Braz J Food Technol 9(2):101–108

US National Library of Medicine Database (2012), http://dailymed.nlm.nih.gov/dailymed/lookup.cfm?setid=b90e0da7-f702-4f09-9488-74f2bb20e9ac. Accessed on 27 December 2012

van de Hulst HC (1981) Light Scattering by Small Particles. Dover, New York

Wang L, Khor E, Lim LY (2001) Chitosanalginate-CaCl$_2$ system for membrane coat application. J Pharm Sci 90:1134–1142

Wang YB, Lou Y, Luo ZF, Zhang DF, Wang YZ (2003) Induction of apoptosis and cell cycle arrest by polyvinylpyrrolidone K-30 and protective effect of α-tocopherol. Biochem Biophys Res Commun 308(4):878–884

Wittaya-arrekul S, Prasharn C (2006) Development and in vitro evaluation of chitosan-polysaccharides composite wound dressings. Int J Pharm 313:123–128

Woolfson AD, McCafferty DF, McCarron PA, Price JH (1995) Abioadhesive patch cervical drug delivery system for administration of 5-fluorouracil to cervical tissue. J ContRel 35:676–681

Zhao QS, Cheng XJ, Ji QX, Kang CZ, Chen XG (2009) Effect of organic and inorganic acids on chitosan/glycerophosphate thermosensitive hydrogel. J SOL–GEL SCI TECHN 50(1):111–118

Zheng Z, Zhang L, Kong L, Wang A, Gong Y, Zhang X (2009) The behavior of MC3T3-E1 cells on chitosan/poly-L-lysine composite films: Effect of nanotopography, surface chemistry, and wettability. J Biomed Mater Res A 89(2):453–465

3

An integrated experimental and modeling approach to propose biotinylated PLGA microparticles as versatile targeting vehicles for drug delivery

Olivia Donaldson, Zuyi Jacky Huang and Noelle Comolli[*]

Abstract

Polymeric microparticles with covalently attached biotin are proposed as versatile targeting vehicles for drug delivery. The proposed microparticles made of 85/15 poly (lactic-co-glycolic acid) (PLGA) will have biotin available on the outside of the particle for the further attachment with an avidin group. Taking advantage of biotin's high affinity for avidin, and avidin's well-known chemistry, the particle has the potential to be easily coated with a variety of targeting moieties. This paper focuses on the design and resulting effect of adding biotin to PLGA microparticles using an integrated experimental and modeling approach. A fluorescent-tagged avidin (488-streptavidin) was used to confirm the presence and bioavailability of biotin on the outside of the particles. For the purpose of this study, bovine serum albumin (BSA) was used as a model therapeutic drug. Microparticles were created using two different types of polyvinyl alcohol 88 and 98 mol% hydrolyzed, which were then analyzed for their size, morphology, and encapsulation capacity of BSA. Release studies performed in vitro confirmed the slow release of the BSA over a 28-day period. Based on these release profiles, a release kinetics model was used to further quantify the effect of biotinylation of PLGA microparticles on their release characteristics by quantitatively extracting the effective drug diffusivity and drug desorption rate from the release profiles. It was found that the biotinylation of the PLGA microparticles slowed down both the drug desorption and drug diffusion process, which confirmed that biotinylated PLGA microparticles can be used for controlled drug release. The presented technology, as well as the proposed integrated experimental and modeling approach, forms a solid foundation for future studies using a cell-specific ligand that can be attached to avidin and incorporated onto the microparticles for targeted delivery.

Background

Polymeric microparticles have been widely researched for their ability to serve as controlled drug delivery vehicles Brannon-Peppas (1995; Cleland 1997; Shive and Anderson 1997). The main goals of these vehicles are to provide improved drug disposition Putney (1998), protection from metabolic degradation Dziubla et al. (2005), and increased circulation time Putney (1998). The new goal in these designs, however, is to take them a step further and incorporate a method for targeting specific cells Brannon-Peppas and Blanchette (2004; Fung and Saltzman 1997). It is well known that the side effects of

drugs stem from drug interactions in non-targeted cells, such as severe anemia experienced by cancer patients Balkwill (2004; Pegram et al. 1997) and those on antiviral medications (such as HIV and HCV treatments) Sayce et al. (2010). A method to target these cells directly would not only increase the potency of the drugs, but also drastically improve the quality of life for the patients on these treatments.

While biodegradable polymeric microparticles have been investigated for the controlled release of anticancer therapeutics Fung and Saltzman (1997; Datta et al. 2006; Folger et al. 2006), and to a limited extent, for antiviral drugs Datta et al. (2006), the challenge of effective targeting still remains. There are few markers that are specific only to tumor cells, but rather most are simply

* Correspondence: noelle.comolli@villanova.edu
Villanova University, 800 East Lancaster Avenue, Villanova, PA 19085, USA

upregulated and therefore more prevalent Balkwill (2004; Pegram et al. 1997). The challenge of identifying a sole marker to a tumor cell or virus is one that biologist and biochemist are still researching. With an evolving field of possible targets and ligands, the challenge for the engineers then is to create a robust mechanism for incorporation of these new ligands to a polymeric delivery vehicle.

Polymers most commonly used for microparticle drug release include poly(lactic acid) (PLA), poly(glycolic acid) (PGA) and their copolymer poly(lactic-co-glycolic acid) (PLGA) Anderson and Shive (1997; Cao and Shoichet 1998; Panyam et al. 2003). In order to obtain the desired release of the drug for the specific delivery, the amount of glycolic acid can be increased to increase the degradation rate, and therefore speed the release time. The PLGA used was an 85/15 mixture of lactic to glycolic acid. The biodegradation of PLGA occurs through a homogenous hydrolytic chain cleavage mechanism, in which both the surface and the bulk polymer degrade at similar rates Anderson and Shive (1997). The breakdown of PLGA is purely through hydrolysis and does not need the assistance of an enzyme Muthu (2009).

In order to target specific cell lines, a robust targeting strategy is proposed. Taking advantage of the affinity of biotin with avidin, a strong non-covalent bond can easily be created by adding biotin to the surface of the PLGA. Biotin strongly binds to avidin and streptavidin via a combination of van der Waals and hydrophobic interactions. In perfect conditions, a single molecule of avidin would bind to four molecules of biotin. This high affinity makes it possible to have site-specific microparticles that have predetermined antibodies attached to the avidin Datta et al. (2006; Moro et al. 1997). This platform would allow for various avidin-antibody complexes to be connected to the biotin microparticle in order to make a multi-faceted drug delivery system. Since this takes advantage of the same mechanism many biochemical assays use, the chemistry of avidin attachment to an antibody, or antibody fragment, is already well known Moro et al. (1997; Diamandis and Christopoulos 1991; Kocbek et al. 2007). This platform allows both targeted delivery (via the PEG-biotin-avidin), as well as controlled release, and biochemical protection of the drug during the delivery via the PLGA.

Previous research has shown the potential in using polymeric microparticles with a similar linkage but using the reverse order (avidin linked to the polymer) Park et al. (2011). The proposed method is used as a simpler method for attaching the conjugate covalently to the polymer while controlling the length of the 'tethering' arm spacing the conjugate and the polymer. The proposed tethering arm in this case will be a short chain polyethylene glycol (PEG) that can be increased or decreased in length as needed. The attachment of

biotin-PEG to a nanoparticle of PLGA was previously done using a more complex chemistry by Weiss et al., for the proposed use of rapid fluorescent tagging of the nanoparticles Weiss et al. (2007). Although these particles were evaluated for their ability to attach to biotin, the release of drug was not investigated.

In order to better understand the effect that biotin has on the polymeric microparticle, along with the usual in vitro characterization (drug encapsulation, morphology, release rates), release kinetics will further be modeled from experimental data to extract important quantitative information that is essential for the comparative study of the proposed PLGA microparticles. Specifically, drug release from polymeric microparticles undergoes two main phases: (1) the induction (or burst) phase in which an initial burst of protein release is observed due to the desorption of proteins from the surface of mesopores within microparticles and the outer surface of microparticles; (2) the diffusion phase in which the macromolecular drug contained in the occlusions of microparticles diffuse through the pores that are formed during the hydration, degradation and erosion of microparticles. Accordingly, the drug desorption rate determines the dynamics of the initial drug burst, while the diffusion rate determines the subsequent drug release. In this work, a theoretical model of macromolecular drug release presented by Batycky et al. (1997), as shown in Equation (1), is used to quantify drug desorption rate and effective drug diffusivity from drug release profiles. Thus, the effect of polyvinyl alcohol (PVA) surfactants as well as the attachment of biotin to the polymeric microparticles on the drug release process can be quantified.

The mass fraction of released drug, f_{release}, is determined by the following equation:

$$f_{\text{release}} = \varphi_d^{\text{burst}}\left(1 - e^{-k_d t}\right) + \left(1 - \varphi_d^{\text{burst}}\right)\left(1 - \frac{6}{\pi^2}\sum_{j=1}^{\infty}\frac{-e^{-j^2\pi^2\bar{D}_d^*(t-t_d)/r_0^2}}{j^2}\right)$$

(1)

where φ_d^{burst} is the mass fraction of drug involved in the burst phase, k_d is the drug desorption rate constant, $\bar{D}_d^*$ is the effective drug diffusivity, t_d is the drug induction time that allows for the coalescence of micropores and the passage of the macromolecular drug out from the occlusions through the coalescing micropores in microparticles, and r_0 is the initial microparticle radius. The first term of Equation 1 represents the burst phase during which the proteins from the surface of mesopores within microparticles and the outer surface of microparticles are released and during which the micropores within microparticles coalesce for the further release of the encapsulated drug. Following the burst phase is the diffusion phase that is described by the second term of Equation 1 in a Fickian-release manner. The time evolution of released mass, $f_{release}$ predicted from Equation 1, will be compared to the experimental released profile. The values of the parameters that are important for characterizing drug released process, including φ_d^{burst}, k_d, and t_d, are then determined by fitting the model given in Equation 1 to the experimental data. Therefore, Equation 1 is used as a soft-sensor in this work for quantitatively monitoring the drug desorption rate and effective drug diffusivity that cannot be directly determined from the release profiles by eye inspection. These parameters can be used as quantitative criteria for the selection of PLGA microparticles for drug delivery. The primary goal of this paper is to evaluate the effect of the biotinylation of the PLGA microparticles on their morphology and release characteristics.

Materials and methods
Materials
PLGA was purchased from SurModics, located in Birmingham, AL, USA. The EZ-Link®TFPA-PEG₃-Biotin, 488-streptavidin, potassium nitrate, and micro bicinchoninic acid (BCA) protein assay kit were all obtained from Thermo Fisher Scientific (Waltham, MA, USA). A biotin quantification kit was bought from Pierce Biotechnology (Rockford, IL, USA). The ethyl acetate, dichloromethane (DCM), and dimethyl sulfoxide (DMSO) used in the preparation of the microparticle, as well as bovine serum albumin (BSA) and phosphate buffered solution (PBS) were purchased from Sigma-Aldrich (St. Louis, MO, USA). The sodium azide was purchased from Acros Organics (Geel, Belgium). The PVA was brought from Polysciences, Inc. (Warrington, PA, USA).

Biotinylation of PLGA
The PLGA and DMSO were combined in a 10:1 ratio and vortexed until the PLGA dissolved. EZ-Link TFPA-PEG3-Biotin was attached to PLGA in a 20-fold molar excess of biotin (10 mg/mL in DMSO). The amount of biotin was determined using the following equation:

$$V_{biotin} = 1,000 \times \frac{m_{PLGA} MW_{biotin}}{MW_{PLGA} M_{biotin} C_{biotin}} \quad (2)$$

where m_{PLGA} is the mass of PLGA; MW_{biotin} and MW_{PLGA} are the molecular weight of biotin and PLGA, respectively; M_{biotin} is the mole of excess of biotin; C_{biotin} is the concentration of biotin.

The mixture was then photoactivated using UV light for 30 min. After quenching the reaction with approximately 15 mL of deionized (DI) water, the solution was then centrifuged using the Sorvall Legend RT Plus Centrifuge (Thermo Scientific) at 14,000 rpm for approximately 7 h at room temperature. The samples were stored at 4°C until use.

Quantification of biotin
A biotin quantification kit was used to compare the absorbance of a sample to a positive control, biotinylated horseradish peroxidase (HRP). To begin the analysis, a PLGA-biotin pellet was dissolved in ethyl acetate. 4'-hydrocyazobenzene-2-carboxylic acid (HABA)-avidin was then added to both the control and sample. The plate was shaken for approximately 60 s, and the displaced HABA was measured using a BioTek ELx800 UV/Vis microplate reader at a wavelength of 490 nm. The ratio of biotin to PLGA was determined using the recorded absorbance values. All results are presented as the average of triplicate samples with the standard deviation.

PLGA microparticle synthesis
The water-in-oil-in-water method is a common emulsion technique that was performed at room temperature. Briefly, 150 μL of phosphate buffered saline (pH 7.4) with varying amounts of dissolved protein was added to 2 mL of the oil phase (10 mg/mL PLGA in ethyl acetate), and the emulsion was created by adding energy to the solution by homogenizing for 60 s. The primary emulsion was stabilized with the addition of bovine serum albumin (1 mg/mL) to the internal aqueous phase. The primary emulsion was then quickly added to 300 mL of an external aqueous phase (5 wt.% PVA). The emulsion was stabilized through stirring at 500 rpm and the presence of the PVA (either 88 or 98 mol% hydrolyzed). The microparticles hardened while stirring overnight and the ethyl alcohol was evaporated. The microparticles were collected via centrifugation at 14,000 rpm for 90 min. Afterwards, the supernatant was removed and the microparticles were resuspended in DI water. The microparticles were washed two more times and centrifuged at 14,000 rpm at respectively 90 and 30 min. The microparticles were allowed to dry and either used immediately or kept at 4°C until use.

Fluorescent imaging
PLGA-biotin and non-biotinylated PLGA particles were analyzed under fluorescent imaging. Approximately 5 mg of particles were suspended in 1 mL of DI water in an amber microcentrifuge tube. A 5 μL of 488-streptavidin was added to the solution and it was stored in a dark location for at least 90 min. The tube was then centrifuged at 13,300 rpm for 15 min at room temperature. Microparticles were then washed, removing the supernatant, and the pellet was resuspended in a small amount of DI water. The sample was centrifuged at 13,300 rpm for 15 min and the supernatant was removed. The sample was then placed on a glass slide with a cover slip and viewed on a Leica DM 2000 microscope (Leica Microsystems, USA). Images were captured and viewed using a Q imaging Retiga-SRV camera and QCapture Pro 6.0 software (Q Imaging, Surrey, British Columbia, Canada).

Encapsulation efficiency

Protein encapsulation was evaluated by dissolving a known weight of particles (5 mg) in 2 mL DCM. The dissolved particles were mixed with 3 mL of DI water and the solvent-water mixture was shaken overnight at 200 rpm. This provided sufficient time for the protein to be extracted into the water phase. A sample was taken from the water phase and the concentration was found using a BCA protein assay (used per manufacturer's instructions). The BCA assay is a colorimetric assay based on bicinchoninic acid and measures the total protein content in a sample. Negative controls of the particle made with no protein present at all were also performed to ensure that the presence of the degraded lactic acid did not affect the concentration readings.

In vitro release of model drugs

Protein release from the microparticles was evaluated *in vitro* using a known mass of dried microparticles in 30 mL of PBS (with 0.01% NaN3 to prevent bacterial growth). All studies were set so sink conditions would be maintained, specifically, that at no point would the maximum released concentration of protein be greater than 10% the saturation limit for that protein in PBS. Samples of the release medium were removed at designated times using a sample probe with an inline 0.45-μm filter to prevent removal of the microparticles during sampling. Equal volumes of fresh PBS were back-flushed through the filter to ensure a constant volume throughout the study as well as to ensure that any microparticles trapped in the sample probe would be flushed back into the sample container. Samples were kept at -20°C until analysis. Concentrations were found using the BCA protein assay at 490 nm, per manufacturer's instructions.

Particle morphology

Using completely dried microparticles, the size and surface morphology of the particles were observed using the scanning electron microscope. A fraction of the microparticle was taken and placed on a small metal stage, fitted with double-sided carbon tape. The sample was then coated. The sample was placed in a Hitachi S-570 scanning electron microscope (Hitachi America Ltd., Brisbane, CA, USA) for observation under vacuum. The size and distribution of the particles was determined. Using dissolved microparticles, the size of the particle was observed using the Hitachi 7600 transmission electron microscope. Six microliters of the sample was placed on a carbon graph and allowed to dry. The dried sample was sputter coated with a conductive metal and placed in the microscope for examination.

Particle size

Using completely dried microparticles, the size and polydispersity of the particles were observed using the particle size analyzer. A portion of the microparticle was suspended in 4.5 mL of 10 mM of KNO_3. The suspension was approximately 10 mg to 4 mL. The sample was then placed in a Brookhaven 90 plus particle size analyzer (Brookhaven Instruments Corporation, Holtsville, NY, USA) for examination. The KNO_3 was used instead of PBS since the salt solution was necessary to allow for the laser scattering that is needed.

Quantifying drug desorption rate constant and effective drug diffusivity from drug release profiles

While the BSA release percentage can be determined from the release profile directly, the release kinetics parameters such as the effective diffusion rate and the drug desorption cannot be directly determined by eye inspection from experimental data. Thus, release kinetics parameters such as drug desorption rate (k_d) and effective drug diffusivity $\overline{D}_d^*$ in Equation 1 were estimated by fitting $f_{release}$ predicted by Equation 1 to experimental drug release profiles ('*In vitro* release of model drugs' section) via the following procedure:

1. φ_d^{burst} and t_d, which correspond to the mass and time for drug release in the burst phase respectively, were determined from the inflection point of drug release profiles, as the inflection point indicates the switch of drug release from the burst phase to the diffusion phase.
2. k_d and $\overline{D}_d^*$ were determined via a nonlinear least squares approach, which can be represented by Equation 3. This computation was performed by minimizing an objective function consisting of the sum of the squares over N measurements of the differences between the experimental data $f_{release}$ and the model-predicted output $f_{release}$. MATLAB (Mathworks Inc., Natick, MA, USA) routine fmincon was used for solving this parameter estimation problem:

$$\underset{k_d,\overline{D}_d^*}{\text{Min}} \sum_{i=1}^{N} \left[\hat{f}_{release}(i) - f_{release}(i) \right]^2$$

subject to

$$f_{release} = \varphi_d^{burst}\left(1 - e^{-k_d t}\right) + \left(1 - \varphi_d^{burst}\right)$$
$$\left(1 - \frac{6}{\pi^2} \sum_{j=1}^{\infty} \frac{-e^{-j^2 \pi^2 \overline{D}_d^*(t-t_d)/r_0^2}}{j^2}\right).$$

$$(3)$$

Results and discussion

The first step in microparticle production was to synthesize biotinylated PLGA. To do this, TFPA-PEG₃-

Biotin was attached using a UV-initiated reaction. In order to confirm the attachment of the biotin on the polymer, a biotin quantification kit was used. The weight of biotin in relation to polymer was calculated and found to be approximately 0.87 ± 0.37. Once the polymer was confirmed to have biotin, microparticles were made using a water-in-oil-in-water (W/O/W) method. To ensure that the biotin was available to avidin on the outside of the polymer microparticle, a fluorescent assay was performed. Microparticles were combined with streptavidin that was tagged with a green fluorophore (Alexa 488; Life Technologies Corporation, Carlsbad, CA, USA). After incubation, the microparticles were collected, washed, and immediately viewed using a Leica DM 2000 microscope. Presence of green fluorophore (white in Figure 1, Alexa 488 + biotin-PLGA microparticles) around the microparticle indicates that the biotin not only attached to the perimeter, but also was still biologically available for the streptavidin post-microparticle processing. A negative control (Figure 1, Alexa 488 + PLGA microparticles), confirms that the streptavidin is not merely sticking to the polymer surface, but rather to the biotin available on the surface.

Once the microparticle synthesis viability was confirmed, the size and polydispersity were confirmed using a Brookhaven 90 plus particle size analyzer (via dynamic light scattering). Microparticles were made with and without biotin, as well as with and without the model drug (BSA). Microparticles were also made using two different types of PVA (the surfactant for the secondary emulsion step, 88 and 98 mol% hydrolyzed) in an attempt to optimize the microparticle synthesis. Analysis of the samples (Table 1) found that the addition of biotin to the PLGA causes an increase in particle size for both 88 and 98 mol% hydrolyzed PVA. This trend is expected since the presence of the biotin makes the polymer larger as well as more hydrophobic.

The biontinylated PLGA particle containing no BSA created a smaller distribution in the molecular masses of the samples. This could be a result of the biotin itself helping to stabilize the emulsion. Interestingly, the addition of BSA without biotin present increased the particle size; however, the addition of BSA into the biotinylated PLGA particles actually showed a decrease in size. This decrease in size was unexpected but may indicate a specific reaction of the BSA and biotin during the W/O/W emulsion and may not be critical when using the actual cancer therapeutic drug. The particles were created using both 88 and 98 mol% PVA, and it was found that particles generated from the 98 mol% PVA are larger than the particles made from the 88 mol% PVA. The 98 mol% PVA is roughly ten times larger than the 88 mol% PVA particles. The polydispersity of the 98 mol% PVA microparticles was larger than the 88 mol% PVA microparticles as well, indicating that the increase in hydrophilicity in the 98 mol% PVA did not provide an increased stabilizing effect on the emulsion as compared to the 88 mol% PVA.

Microparticles were synthesized using either 88 or 98 mol% PVA as a surfactant as well as with and without the biotin attached to the PLGA. The resulting microparticle size and polydispersity were determined via dynamic light scattering. Results are presented as the average of $n = 4$ samples $\pm$ standard deviation.

In order to evaluate the morphology of the microparticles, both a Hitachi S-4800 scanning electron microscope (SEM) and a Hitachi 7600 transmission electron microscope (TEM) were used to view samples of the biotinylated and plain PLGA microparticles. SEM and TEM images of both 88 and 98 mol% PVA methods are shown in Figure 2. The TEM and SEM images confirm that the addition of the biotin to the PLGA does not change the spherical morphology of the microparticles. The microparticles with and without biotin do not have any morphological differences

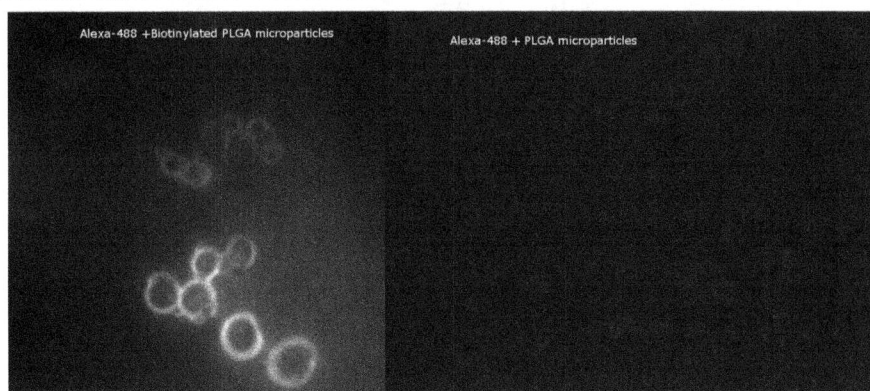

Figure 1 Fluorescent imaging confirms presence of biotin on the outside of microparticles. The presence and bioavailability of biotin on the surface of the PLGA microparticles was confirmed by incubating biotinylated microparticles with streptavidin-488. The particles were washed several times prior to visualization under the fluorescent microscope. Left image, Alexa 488 + biotin-PLGA microparticles. The presence of the 488 fluorophore (white) indicates the biotin was present and capable of attaching the steptavidin. Right image, Alexa 488 + PLGA microparticles.

Table 1 Average size and polydispersity of different microparticle formulations

Type (mol% PVA)	Average size (μm)	Average (PDI) polydispersity index
PLG no BSA (88)	7.81 ± 13.52	0.48 ± 0.55
PLGA BSA (88)	8.06 ± 23.12	0.36 ± 0.037
Biotin BSA (88)	1.83 ± 4.23	0.39 ± 0.18
Biotin no BSA (88)	21.24 ± 15.02	0.021 ± 0.015
PLGA No BSA (98)	21.99 ± 14.91	0.67 ± 0.39
PLGA BSA (98)	30.72 ± 49.99	0.50 ± 0.36
Biotin BSA (98)	16.47 ± 11.85	0.80 ± 0.36
Biotin No BSA (98)	37.97 ± 97.51	0.39 ± 0.52

visible. The particles also have no visible morphological changes (with the exception of size) when changing from 88 to 98 mol% PVA during processing. The ability of the microparticle to keep its spherical shape is an indication that the microparticle morphology is not changed by the addition of the biotin. The agglomeration seen in some of the SEM is believed to be a result of the drying process for imaging and is not expected when the microparticles are in solution.

Once the particle size and morphology was characterized, the mass percentage of BSA encapsulated as well as the encapsulation efficiency of the model drug was determined (Table 2). It was found that for particles made from 88 mol% PVA, the PLGA particles had a slightly higher average encapsulation than the biotinylated particles. On the other hand, when the PVA was changed to 98 mol%, the average encapsulation percentage for the biotinylated particle was higher than the PLGA particle. The 88 mol% PVA, however, had similar encapsulation of BSA between the plain PLGA and biotinylated PLGA microparticles. Although the 98 mol% hydrolyzed PVA may be a better surfactant, there is no clear trend apparent for the effect of biotin on the encapsulation of a drug; further optimization should be done using the actual therapeutic. It was determined that the 98 mol% PVA more effectively stabilized the microparticle during hardening, allowing for the higher encapsulation (shown in Table 2) of BSA. If the particle can be successfully stabilized during the hardening step, it should not swell and allow water in, or drug out. If higher encapsulation corresponds to the larger size, since more internal aqueous phase is retained, the microparticles

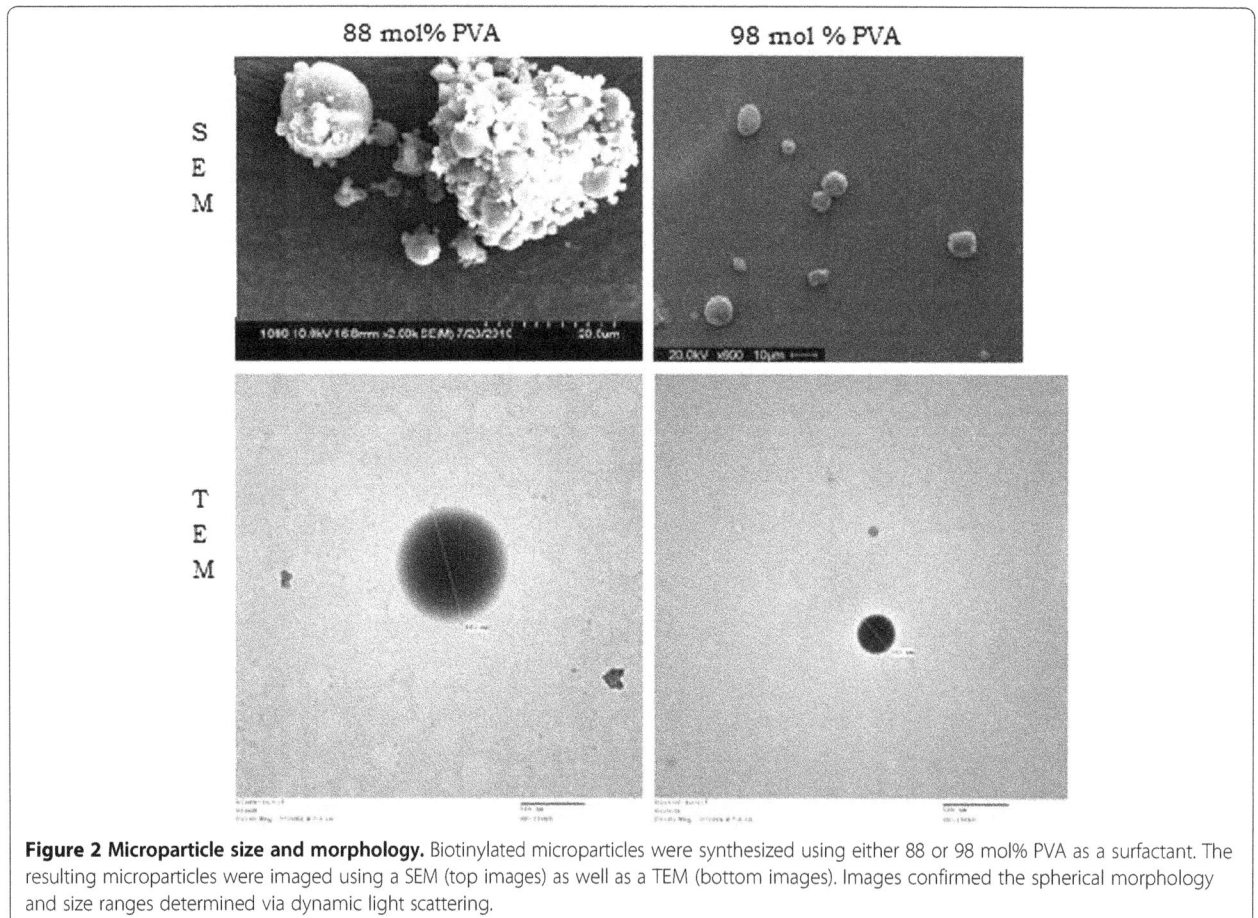

Figure 2 Microparticle size and morphology. Biotinylated microparticles were synthesized using either 88 or 98 mol% PVA as a surfactant. The resulting microparticles were imaged using a SEM (top images) as well as a TEM (bottom images). Images confirmed the spherical morphology and size ranges determined via dynamic light scattering.

Table 2 BSA loading for different microparticle formulations

Type (mol% PVA)	Average mass encapsulation percentage	Average encapsulation efficiency
PLG (88)	4.12	3.21
Biotin (88)	3.30	1.78
PLGA (98)	3.02	4.35
Biotin (98)	12.26	15.57

should be larger. The movement of drug and water in and out during this phase is a common problem during hardening, since the osmotic pressure will readily drive molecules through the oil phase (polymer in solvent) until it fully hardens. This flux creates the largest challenge in designing microparticle with a high drug loading.

Microparticles were synthesized using either 88 or 98 mol% PVA as a surfactant as well as with and without the biotin attached to the PLGA. The resulting microparticle's loading of BSA was quantified dissolving the microparticle and extracting the BSA which was quantified using a micro BCA assay. The BSA loading is presented as both the percentage of the particle mass that is BSA as well as the percentage of the initial BSA that was actually loaded into the microparticle.

Once the amount of drug encapsulated in each type of microparticle was determined, the rate of release of the model drug could be evaluated. *In vitro* release studies were performed under sink condition in PBS at 37°C for all four types of microparticles. Using a micro BCA assay, the mass of BSA released was determined. The cumulative percentage of BSA released over time was calculated and is shown in Figures 3 and 4. It was determined that the release of BSA from the biotinylated and PLGA particle, made from 88 mol% PVA, followed the same trend (Figure 3) over a 28-day period. This indicates that the presence of biotin on the surface of the microparticle does not alter the release characteristics of the microparticles. The same result was found for microparticles made using 98 mol% PVA (Figure 4). For the 28-day period, approximately 80% of the model drug is released for both biotinylated and non-biotinylated microparticles. Comparing Figures 3 and 4, there is no apparent effect of the change in surfactant on release, as expected. The change in the surfactant should mainly change the microparticles stability during formation, leading to potential changes in morphology and encapsulation capacity.

In order to further quantify the effect of the biotinylation of the PLGA microparticles on the drug release characteristics, the effective drug diffusivity and drug desorption rate that directly characterize the drug release process are further determined from the release profiles. Specifically,

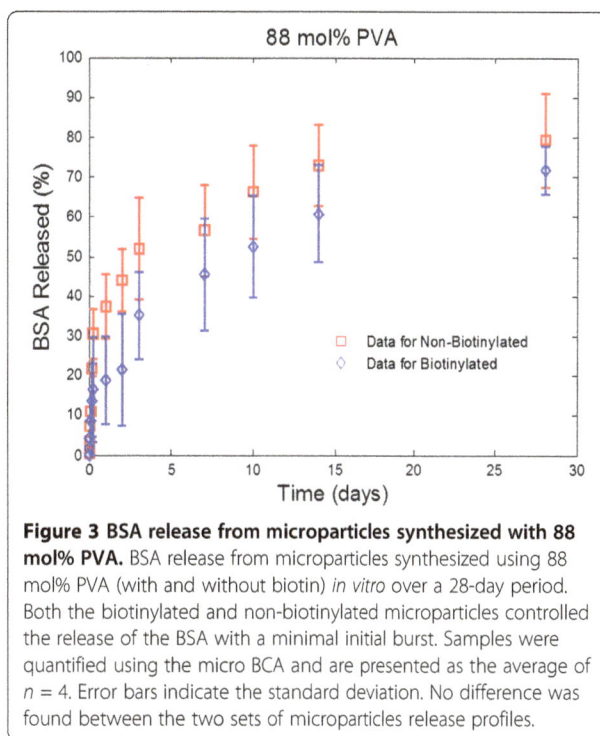

Figure 3 BSA release from microparticles synthesized with 88 mol% PVA. BSA release from microparticles synthesized using 88 mol% PVA (with and without biotin) *in vitro* over a 28-day period. Both the biotinylated and non-biotinylated microparticles controlled the release of the BSA with a minimal initial burst. Samples were quantified using the micro BCA and are presented as the average of $n = 4$. Error bars indicate the standard deviation. No difference was found between the two sets of microparticles release profiles.

the model given by Equation 1 is fitted to the release profiles presented in Figures 3 and 4 via the approach shown in 'Quantifying drug desorption rate constant and effective drug diffusivity from drug release profiles' section. The values of the parameters $\overline{D_d^*}$, k_d, φ_d^{burst} and t_d in

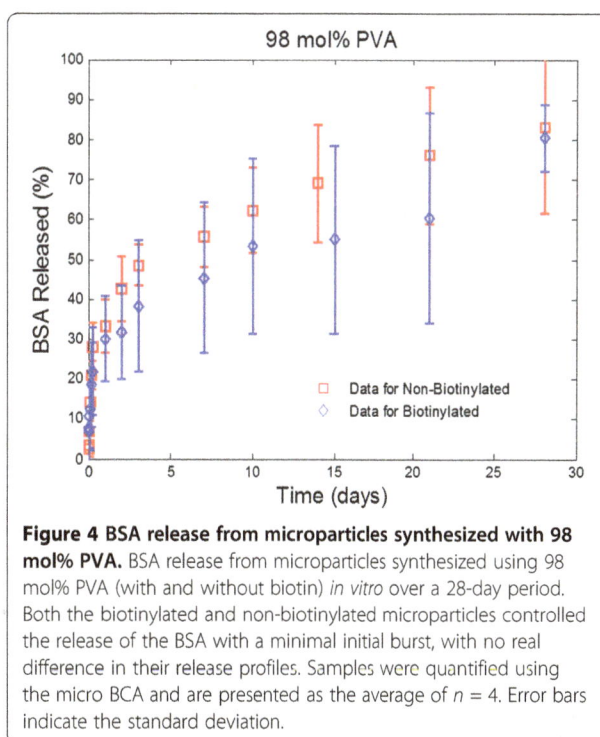

Figure 4 BSA release from microparticles synthesized with 98 mol% PVA. BSA release from microparticles synthesized using 98 mol% PVA (with and without biotin) *in vitro* over a 28-day period. Both the biotinylated and non-biotinylated microparticles controlled the release of the BSA with a minimal initial burst, with no real difference in their release profiles. Samples were quantified using the micro BCA and are presented as the average of $n = 4$. Error bars indicate the standard deviation.

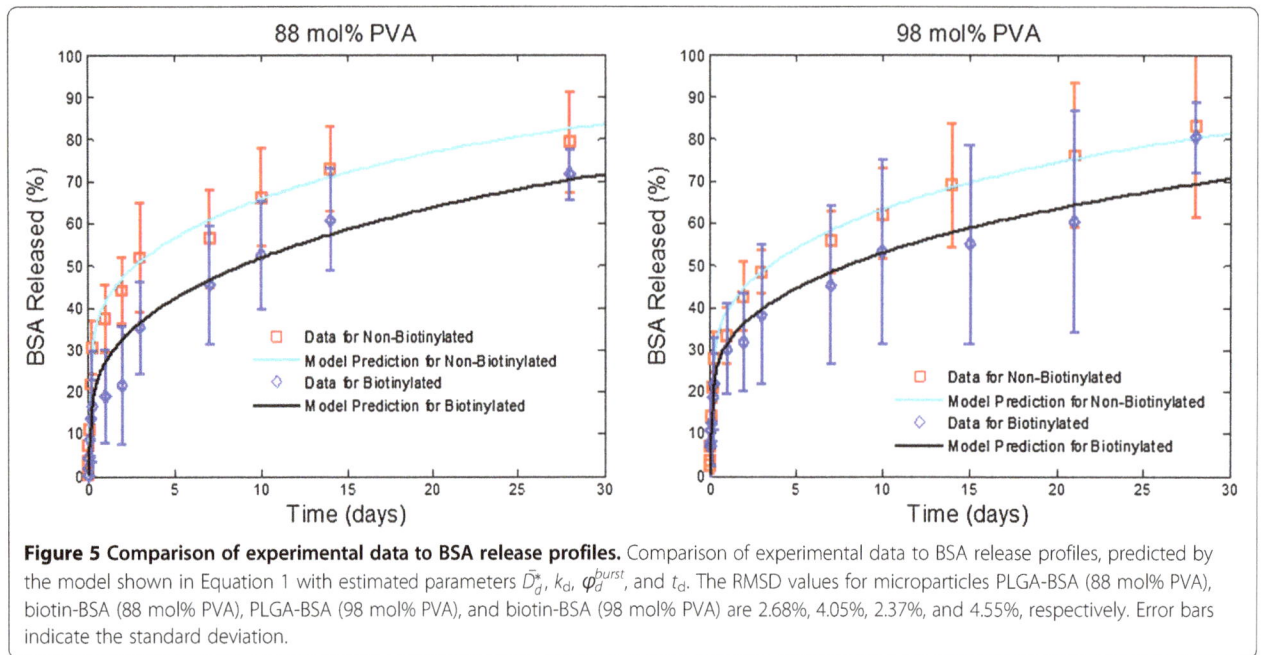

Figure 5 Comparison of experimental data to BSA release profiles. Comparison of experimental data to BSA release profiles, predicted by the model shown in Equation 1 with estimated parameters $\bar{D}_d^*$, k_d, φ_d^{burst}, and t_d. The RMSD values for microparticles PLGA-BSA (88 mol% PVA), biotin-BSA (88 mol% PVA), PLGA-BSA (98 mol% PVA), and biotin-BSA (98 mol% PVA) are 2.68%, 4.05%, 2.37%, and 4.55%, respectively. Error bars indicate the standard deviation.

Equation 1 are estimated. The fitting result is presented in Figure 5, which shows that the drug release profiles predicted by the estimated model fit the experimental data well for all types of microparticles under investigation. In particular, the predicted drug release profiles pass through most error bars shown in the data. The root-mean-square deviation of prediction (RMSD) for microparticles PLGA-BSA (88 mol% PVA), biotin-BSA (88 mol% PVA), PLGA-BSA (98 mol% PVA), and biotin-BSA (98 mol% PVA) is calculated as 2.68%, 4.05%, 2.37%, and 4.55%, respectively. As shown in Figure 5, the model predicts that more BSA is released by the plain PLGA microparticles than the biotinylated PLGA microparticles, and that PLGA microparticles made of a lower percent of PVA release slightly more BSA. This is in a good agreement with the trends shown in the release profiles. A conclusion drawn from these observations is that the deviation of the model prediction from the experimental data is within a reasonably small scale, and that values of parameters $\bar{D}_d^*$, k_d, φ_d^{burst}, and t_d estimated

from release profiles properly characterize the drug release dynamics of all microparticles under investigation.

Table 3 shows the corresponding estimated values of parameters $\bar{D}_d^*$, k_d, φ_d^{burst}, and t_d. It can be seen from Table 3 that $\bar{D}_d^*$ decreases by a factor of 0.65 when microparticles made of 88 mol% PVA were biotinylated. A similar decreasing ratio (i.e., 0.61) is observed in the value of $\bar{D}_d^*$ for biotinylated microparticles that are made of 98 mol% PVA. This means that the attachment of biotin to microparticles slightly slows down the drug diffusion and thus the drug release process. This is expected since the presence of the biotin on the outer surface acts as another layer of diffusion barrier. The value of $\bar{D}_d^*$ decreases by a factor of approximately 0.90 when 98 mol% PVA instead of 88 mol% PVA is used to make microparticles. This implies that increasing the mole percentage of PVA can slightly slow down drug release but to a limited degree. The decrease in drug release at higher mole percentage of PVA may be due to its tendency to remain on the surface of the particles even after hardening. In addition to influencing drug diffusion process, the attachment of biotin and the mole percentage of PVA also affect drug release in the burst phase, in which proteins are desorbed from the outer surface of microparticles. Table 3 shows that the attachment of biotin reduces the value of φ_d^{burst}, the mass fraction of drug that is desorbed during the burst stage. The authors believe this is due to the fact that the attachment of biotin reduces the amount of BSA trapped on the surface during formation due to steric hindrance. Since the attachment of biotin cannot change the desorption pattern of proteins from the

Table 3 Parameter estimation results

Type (mol% PVA)		$\bar{D}_d^*$ (cm^{2}s^{-1})	φ_d^{burst}	K_d (day^{-1})	t_d (day)
88	PLGA BSa	1.5×10^{-13}	0.03	7.60	0.25
	Biotin BSA	9.87×10^{-14}	0.16	8.10	0.25
98	PLGA BSa	1.38×10^{-13}	0.28	8.63	0.25
	Biotin BSA	8.37×10^{-14}	0.22	8.50	0.25

Values of the mass fraction of drug involved in the burst phase (φ_d^{burst}), drug desorption rate constant (k_d), effective drug diffusivity (D_d^*), and drug induction time (t_d) are determined from the drug release profiles.

outer surface, it does not affect drug desorption rate constant (k_d) and drug induction time (t_d). It can also be seen from Table 3 that the mole percentage of PVA has a minor effect on protein desorption during the burst phase. This result, along with the result that the increasing mol% of PVA did not have a drastic effect on the drug loading and/or particle morphology, indicates that the higher mole percentage of PVA is not providing a more stable emulsion during the solvent hardening stage of microparticle synthesis. This also indicates that the increase in PVA mole percentage does not improve the emulsion process, since it cannot guarantee that more of the drug will remain within the microparticle during encapsulation.

Conclusions

Polymeric microparticles created through a water-in-oil-in-water double emulsion effectively demonstrated a controlled release of a model drug. The presence of biotin on the outside of the polymeric microparticles was confirmed using fluorescent imaging. The microparticle synthesis was optimized through the use of PVA consisting of various mole percentages. The effect of the different PVA surfactant on microparticle synthesis determined that 88 and 98 mol% PVA created similar particles that only differed in size and slightly (approximately 10%) in encapsulation. It was determined that increasing the mole percentage of PVA created a more stable emulsion during the hardening phase, allowing for higher encapsulation efficiencies of the model drug (BSA). The release studies found that the attachment of biotin to the PLGA microparticle had only a minor effect on the release trend during the 28-day period. The microparticles still exhibited a controlled release over the 28 days with minimal burst and, therefore, are still believed to be effective as carriers for therapeutic drugs. A release kinetics model was used to further quantify the effective drug diffusivity and drug desorption rate, revealing that the attachment of biotin to microparticles slowed down both drug desorption and drug diffusion processes, while the mole percentage of PVA only has a minor effect on drug release rate.

The presence of the biotin on the microparticle, overall, did not have a negative effect on the microparticles ability to entrap and control the release of the model drug. This indicates that the microparticles can be further investigated for their ability to target using a moiety specific to breast cancer (or other types). This moiety will be attached to avidin and combined with the microparticle using the biotin exposed on the microparticle's surface. Until that specific moiety is identified, the effect of the avidin-targeting moiety on the microparticle cannot be evaluated. The authors believe that this robust linkage system will be valuable compared to other targeting strategies, since the simple chemistry will allow for linkage of a variety of different moieties. Therefore, the same

technology as well as the proposed integrated experimental and modeling approach can be used to target either multiple types of cancer cells, or to include multiple targeting antigens for the same cell on one microparticle.

Competing interests
The authors declare that they have no competing interests.

Authors' contributions
OD carried out the experimentation and drafted the manuscript. ZH completed the modeling and related calculations and helped draft that section of the paper. NC conceived of the study, and participated in its design, coordination and preparation of the manuscript. All authors read and approved the final manuscript.

Acknowledgments
The authors would like to thank the following students for their aid in running the experiments: Colleen Clark, Elizabeth Andrews, Lucille Bell, Erin Wagner, Kaitlin Worden, Will Swalchik, and Sherrie Ann Martin. The authors would also like to thank the Delaware Valley Section of the International Society of Pharmaceutical Engineers for partial funding of this research. NC and ZH gratefully acknowledge the financial support from Villanova University SRF/RSG 2012–2013.

References
Anderson JM, Shive MS (1997) Biodegradation and biocompatability of PLA and PLGA microspheres. Adv Drug Del Rev 28:5–24

Balkwill F (2004) The significance of cancer cell expression of the chemokine receptor CXCR4. Sem Cancer Biol 14(3):171–179

Batycky RP, Hanes J, Langer R, Edwards DA (1997) a theoretical model of erosion and macromolecular drug release from biodegrading microspheres. J Pharm Sci 87(12):1464–1477

Brannon-Peppas L (1995) Recent advances on the use of biodegradable microparticles and nanoparticles in controlled drug delivery. Int J Pharm 116(1):1–9

Brannon-Peppas L, Blanchette JO (2004) Nanoparticle and targeted systems for cancer therapy. Adv Drug Deliv Rev 56(11):1649–1659

Cao X, Shoichet MS (1998) Biodegradation and biocompatibility of PLA and PLGA microspheres. Biomaterials 20:329–339

Cleland JL (1997) Protein delivery from biodegradable microspheres. In: Sanders L (ed) Protein delivery physical systems. Kluwer Academic, Hingham, pp 1–25

Datta S, Ray RD, Nath A, Bhattacharyya D (2006) Recognition based separation of HIV-Tat protein using avidin-biotin interaction in modified microfiltration membranes. J Membr Sci 280:298–310

Diamandis EP, Christopoulos TK (1991) The biotin-(strept)avidin system: principles and applications in biotechnology. Clin Chem 37(5):625–636

Dziubla TD, Karim A, Muzykantov VR (2005) Polymer nanocarriers protecting active enzyme cargo against proteolysis. J Control Release 102(2):427–439

Folger F, Noonpakdee W, Loretz B, Joojuntr S, Salvenmoser W, Thaler M, Bernkop-Schnürch A (2006) Inhibition of malarial topoisomerase II in *Plasmodium falciparum* by antisense nanoparticles. Int J Pharm 319(1–2):139–146

Fung LK, Saltzman WM (1997) Polymeric implants for cancer chemotherapy. Adv Drug Del Rev 26(2–3):209–230

Kocbek P, Obermajer N, Cegnar M, Kos J, Kristl J (2007) Targeting cancer cells using PLGA nanoparticles surface modified with monoclonal antibody.
J Control Release 120(1–2):18–26

Moro M, Pelagi M, Fulci G, Paganelli G, Dellabona P, Casorati G, Siccardi AG, Corti A (1997) Tumor cell targeting with antibody-avidin complexes and biotinylated tumor necrosis factor alpha. Cancer Res 57(10):1922–1928

Muthu M (2009) Nanoparticles based on PLGA and its co-polymer: an overview. Asian J Pharm 3:266–273

Panyam J, Dali M, Sahoo SK, Ma W, Chakravarthi SS, Amidon GL, Levy RJ, Labhasetwar V (2003) Polymer degradation and in vitro release of a

model protein from poly(d, l-lactide-co-glycolide) nano- and microparticles.
J Control Release 92(1–2):173–187

Park J, Mattessich T, Jay SM, Agawu A, Saltzman WM, Fahmy TM (2011) Enhancement of surface ligand display on PLGA nanoparticles with amphiphilic ligand conjugates. J Control Release 156(1):109–115

Pegram MD, Finn RS, Arzoo K, Beryt M, Pietras RJ, Slamon DJ (1997) The effect of HER-2/neu overexpression on chemotherapeutic drug sensitivity in human breast and ovarian cancer cells. Oncogene 15(5):537–547

Putney S (1998) Encapsulation of proteins for improved delivery. Curr Opin Chem Biol 2:548–552

Sayce AC, Miller JL, Zitzmann N (2010) Targeting a host process as an antiviral approach against dengue virus. Trends Microbiol 18(7):323–330

Shive MS, Anderson JMS (1997) Biodegradation and biocompatibility of PLA and PLGA microspheres. Adv Drug Deliv Rev 28(1):5–24

Weiss B, Scheider M, Muys L, Taetz S, Neumann D, Schaefer UF, Lehr CM (2007) Coupling of biotin-(poly(ethylene glycol))amine to poly(d, l-lactide-co-glycolide) nanoparticles for versatile surface modification. Bioconjug Chem 18(4):1087–1094

4

Progress and challenges in biomaterials used for bone tissue engineering: bioactive glasses and elastomeric composites

Qizhi Chen[1*], Chenghao Zhu[1] and George A Thouas[2]

Abstract

Driven by the increasing economic burden associated with bone injury and disease, biomaterial development for bone repair represents the most active research area in the field of tissue engineering. This article provides an update on recent advances in the development of bioactive biomaterials for bone regeneration. Special attention is paid to the recent developments of sintered Na-containing bioactive glasses, borate-based bioactive glasses, those doped with trace elements (such as Cu, Zn, and Sr), and novel elastomeric composites. Although bioactive glasses are not new to bone tissue engineering, their tunable mechanical properties, biodegradation rates, and ability to support bone and vascular tissue regeneration, as well as osteoblast differentiation from stem and progenitor cells, are superior to other bioceramics. Recent progresses on the development of borate bioactive glasses and trace element-doped bioactive glasses expand the repertoire of bioactive glasses. Although boride and other trace elements have beneficial effects on bone remodeling and/or associated angiogenesis, the risk of toxicity at high levels must be highly regarded in the design of new composition of bioactive biomaterials so that the release of these elements must be satisfactorily lower than their biologically safe levels. Elastomeric composites are superior to the more commonly used thermoplastic-matrix composites, owing to the well-defined elastic properties of elastomers which are ideal for the replacement of collagen, a key elastic protein within the bone tissue. Artificial bone matrix made from elastomeric composites can, therefore, offer both sound mechanical integrity and flexibility in the dynamic environment of injured bone.

Keywords: Bioceramic, Elastomer, Composite, Mechanical property, Degradation

Introduction

Tissue engineering is 'the application of principles and methods of engineering and life sciences to obtain a fundamental understanding of structure-function relationships in normal and pathological mammalian tissue, and the development of biological substitutes to restore, maintain, or improve tissue function' (Skalak and Fox 1993). A common approach is to harvest an expansion of living tissue *in vitro* and design of biomaterial scaffolds to provide appropriate structural support to match the tissue of interest. Scaffolds are then loaded with numbers of cells and numbers for implantation, which allows surgeons to manipulate local tissue environments, providing more physiological alternatives to standard approaches in reconstructive surgery (Bell 2000).

There are several requirements of scaffold materials to meet the demands of tissue engineering. Firstly, biocompatibility of the substrate materials is imperative. The material must not elicit an unresolved inflammatory response nor demonstrate immunogenicity or cytotoxicity. As with all materials in contact with the human body, tissue scaffolds must be easily sterilizable to prevent infection (Chaikof et al. 2002). This applies notably for bulk degradable scaffolds, where both the surface and the bulk material must be sterile. In addition, the mechanical properties of the scaffold must be sufficient to prevent structural failure during handling and during the patient's normal activities. A further requirement for a scaffold, particularly in bone engineering, is a controllable interconnected porosity that can direct cells to grow into

* Correspondence: qizhi.chen@monash.edu
[1]Department of Materials Engineering, Monash University, Clayton, Victoria 3800, Australia
Full list of author information is available at the end of the article

a physical structure and to support vascularisation. A typical porosity of 90% as well as a pore diameter of at least 100 μm is known to be compulsory for cell penetration and a proper vascularization of the ingrown tissue (Griffith 2002; Karageorgiou and Kaplan 2005; Levenberg and Langer 2004; Mikos and Temenoff 2000). Other desirable aspect concerns the cost-effectiveness of scaffold processing toward industrial-scale production to reliably generate net-like structures with a nominal range of porosities.

Materials used for bone tissue engineering scaffolds include the following: (1) natural or synthetic polymers such as proteins, thermoplastics, hydrogels, thermoplastic elastomers (Berger et al. 2004; Drotleff et al. 2004; Mano et al. 2004; Tirelli et al. 2002) and chemically cross-linked elastomers (Chen et al. 2008b), (2) bioactive ceramics such as calcium phosphates and bioactive glasses or glass ceramics (Hench 1998; Kim et al. 2004; Levenberg and Langer 2004), (3) composites of polymers and ceramics (Boccaccini et al. 2005; Hedberg et al. 2005; Kim et al. 2004; Niiranen et al. 2004; Yao et al. 2005; Zhang et al. 2004), and (4) metallic materials such as titanium and magnesium alloys (Lefebvre et al. 2008). From the material science point of view, bone is a natural composite of inorganic calcium phosphate apatite and biological polymers including collagens, which are deposited by residence osteocytes. The composite system of polymers and ceramics is apparently a logic choice for bone tissue engineering, as demonstrated by the huge research efforts worldwide using these materials (Boccaccini et al. 2005; Di Silvio and Bonfield 1999; Gittens and Uludag 2001; Hedberg et al. 2005; Jiang et al. 2005; Khan et al. 2004; Kim et al. 2004; Li and Chang 2004; Lu et al. 2005; Luginbuehl et al. 2004; Mano et al. 2004; Maquet et al. 2004; Niiranen et al. 2004; Xu et al. 2004; Yao et al. 2005; Zhang et al. 2004).

The present authors previously reviewed biodegradable thermoplastic polymers and bioactive ceramics, including strategies for fabrication of composite scaffolds with defined microstructure and mechanical properties, and methods of *in vitro* and *in vivo* evaluation (Rezwan et al. 2006). Over the past 10 years, new processes of Na-containing bioactive glasses and new bioactive glass compositions doped with various trace elements have been developed aiming at healthy bone growth and/or vascularization (Rahaman et al. 2011). Meanwhile degradable elastomeric polymers have gained increasing attentions in the field of tissue engineering, mainly because of the inherent structural elasticity of biological tissues. Composite scaffolds made from bioceramics and chemically cross-linked elastomers have proven beneficial in terms of both biocompatibility and their operation over a wide range of elastic moduli (Chen et al. 2010a; Liang et al. 2010). This article aims to provide an update on the progress of biomaterials developed for bone tissue engineering, with a specific focus on bio-active glasses and elastomeric composites that show potentials to advance bone tissue engineering, while the rest of biomaterials in bone tissue engineering are reviewed briefly for a complete overview.

Biodegradable and surface erodible thermoplastic polymers

Based on their mechanical properties, polymeric biomaterials can be classified as elastomers and non-elastomeric thermoplastics. This section will provide a brief review on biodegradable thermoplastics. Comprehensive discussions of these polymers and their physical properties have been provided in great detail elsewhere (Chen and Wu 2005; Gunatillake et al. 2003a; Iroh 1999; Kellomäki et al. 2000; Kumudine and Premachandra 1999; Lu and Mikos 1999; Magill 1999; Middleton and Tipton 2000; Ramakrishna et al. 2004; Rezwan et al. 2006; Seal et al. 2001; Yang et al. 2001).

The most widely utilized biodegradable synthetic polymers for 3D scaffolds in tissue engineering are saturated aliphatic polyesters, typically poly-α-hydroxy esters including poly(lactic acid) (PLA), poly(glycolic acid) (PGA) (Gollwitzer et al. 2005; Seal et al. 2001), poly(ε-caprolactone) (PCL) (Pitt et al. 1981), and their copolymers (Jagur-Grodzinski 1999; Kohn and Langer 1996; Mano et al. 2004; Seal et al. 2001). The chemical properties of these polymers allow hydrolytic degradation through de-esterification. Once degraded, the lactic and glycolic acid monomers are metabolized naturally by tissues. Due to these properties, PLA, PGA, PCL, and their copolymers have successfully been applied in a number of biomedical devices, such as degradable sutures and bone internal fixation devices (Biofix®, Bionx Implants Ltd., Tampere, Finland) which have been approved by the US Food and Drug administration (Mano et al. 2004). However, abrupt release of these acidic degradation products can cause a strong inflammatory response (Bergsma et al. 1993; Martin et al. 1996). In general, their degradation rates decrease in the following order: PGA > PLA > PCL. Their blends have been shown to degrade faster than their pure counterparts (Dunn et al. 2001). Poly lactate-glycolic acid (PLGA) can completely degrade in several months *in vivo*, whereas poly-L-lactate (PLLA) and PCL take 3 to 5 years or more to completely degrade *in vivo* (Rich et al. 2002; Yang et al. 2001).

Of particular significance for applications in tissue engineering is the acidic degradation products of PLA, PGA, PCL, and their copolymers that have been implicated in adverse tissue reactions (Niiranen et al. 2004; Yang et al. 2001). Researchers have incorporated basic compounds to stabilize the pH of the environment surrounding the polymer and to control its degradation, such as bioactive glasses and calcium phosphates (Dunn et al. 2001; Heidemann et al. 2001; Rich et al. 2002). The possibility of counteracting this acidic degradation is another important reason proposed for the use of composites (Boccaccini and Maquet 2003).

Other properties of thermoplastics of special interest include their excellent processability to generate a wide range of degradation rates, mechanical, and chemical properties achieved by the use of various molecular weights and stoichiometric ratios. Scaffolds produced in this can be mechanically strong and matched to specific tissue types, but their compliance is not reversible. Given that elastic stretchability is a major mechanical property of living tissue, including collagens of different bone types, elastomeric polymers that can provide sustainable elasticity and structural integrity are thought to be mechanically more advantageous than thermoplastic (non-elastomeric) polymers. Over the past 10 years, there have been an increasing number of research groups working on the development of biodegradable elastomeric biomaterials for bone tissue engineering applications (Li et al 2012; Kim and Mooney 2000; Niklason et al. 1999; Seliktar et al. 2003; Stegemann and Nerem 2003; Waldman et al. 2004; Wang et al. 2002a).

There is a family of hydrophobic polymers that undergo a heterogeneous hydrolysis process that is predominantly confined to the polymer-water interface. This property is referred to as surface eroding as opposed to bulk-degrading behavior. Three representative surface erodible polymers are poly(anhydrides) (poly (1,3-bis-p-carboxyphenoxypropane anhydride) (Domb and Langer 1999a) and poly (erucic acid dimer anhydride) (Domb and Langer 1999b), poly(ortho esters) (POE) (Andriano et al. 2002; Solheim et al. 2000), and polyphosphazenes (Allcock 2002; Magill 1999; (Laurencin et al. 1993, 1996b). These surface bioeroding polymers have been intensively investigated as drug delivery vehicles. The surface-eroding characteristics offers three key advantages over bulk degradation when used as scaffold materials: (1) retention of mechanical integrity over the degrading lifetime of the device, owing to the maintenance of mass to volume ratio, (2) minimal toxic effects (i.e., local acidity), owing to lower solubility and concentration of degradation products, and (3) significantly enhanced bone ingrowth into the porous scaffolds, owing to the increment in pore size as the erosion proceeds (Shastri et al. 2002).

Biodegradable thermoplastic rubbers

Synthetic elastomers can be divided into two categories: thermoplastic elastomers and cross-linked elastomers, based on the type of 'cross-link' used to join their molecular chains. Unlike cross-linked elastomers, where the cross-link is a covalent bond created during the

vulcanization process, the cross-link in thermoplastic elastomers is a weaker dipole or hydrogen bond, or takes place in one of the phases of the material. Linear thermoplastic elastomers usually consist of two separated microphases: crystalline, hydrogen-bonded hard segments and amorphous soft segments. The crystalline or hard segments function as cross-linkers which provide mechanical strength and stiffness, whereas soft segments provide the flexibility (Hiki et al. 2000).

Poly (ε-caprolactone) copolymers with glycolide or lactide

PCL, PGA, and PLA are rigid and have a poor flexibility. In order to provide better control over the degradation and mechanical properties without sacrificing biocompatibility, PCL-based materials have been copolymerized or blended with other hydroxyacids or polymers to produce elastomeric biomaterials. PCL-based copolymers with glycolide and lactide are elastomeric materials. Poly (lactide-co-caprolactone) (PLACL) synthesized by Cohn and Salomon (2005) demonstrates remarkable mechanical properties, with Young's modulus, UTS, and strain at break being up to 30 MPa, 32 MPa, and 600%, respectively.

The degradation rate of the PCL-based copolymers varies over a wide range by the change in the ratio of monomers. In general, the copolymers degrade faster than each homopolymer alone. PCL-co-GA scaffolds synthesized by Lee et al. (2003), for example, lost 3% of their initial mass after 2-week incubation in PBS and 50% after a 6-week incubation, whereas it takes 6–12 months and 2–3 years for PGA and PCL to degrade, respectively (Cohn and Salomon 2005). PGA-co-CL (PGACL) and PLA-co-CL (PLACL) polymers were initially developed for engineering smooth muscle-containing tissues (e.g., blood vessels and urinary bladder) (Keun Kwon et al. 2005; Lee et al. 2003; Matsumura et al. 2003a, b). Both were soon after investigated for their potential applications in bone tissue engineering (Gupta et al. 2009; Webb et al. 2004; Zilberman et al. 2005).

Polyhydroxyalkanoates

Polyhydroxyalkanoates are aliphatic polyesters as well, but produced by microorganisms under unbalanced growth conditions (Doi et al. 1995; Li et al. 2005). These polyesters are generally biodegradable (via hydrolysis) and thermoprocessable, making them attractive as biomaterials for medical devices and tissue engineering scaffolds (Chen and Wu 2005). Polyhydroxybutyrate has been investigated for the repair of bone, nerves, blood vessels, urinary tissue, and those of the gastrointestinal tract.

Poly 3-hydroxybutarate (P3HB) is rigid and brittle, with a strain at break typically less than 5%. This thermoplastic material can easily be woven or compressed into textiles with a satisfactory flexibility (Chen and Wu 2005). P3HB has been intensively investigated for bone tissue applications and produces a consistently favorable bone tissue adaptation response with no evidence of an undesirable chronic inflammatory response after implantation periods up to 12 months (Duvernoy et al. 1995; Kalangos and Faidutti 1996). Bone is formed close to the material and subsequently becomes highly organized, with up to 80% of the implant surface lying in direct apposition to newly mineralized bone. The materials showed no evidence of extensive structural breakdown *in vivo* during the implantation period of the study (Doyle et al. 1991).

Among the PHAs, poly 4-hydroxybutyrate (Freier 2006; Grabow et al. 2004; Martin and Williams 2003; Martin et al. 1999; Rao et al. 2010) and copolymers of 3-hydroxybutyrate and 4-hydroxybutyrate (Freier 2006; Grabow et al. 2004; Sudesh and doi 2005), including P3HB-co-3HV (3-hydroxyvalerate) (Avella et al. 2000), P3HB-co-3HD (3-hydroxydecanoate) (Avella et al. 2000), and P3HB-co-3HH (3-hydroxyhexanoate), have been demonstrated to have superb elasticity, with an elongation at break of 400 to 1,100%. The major progress for these materials has so far been in cardiovascular tissue engineering (Martin and Williams 2003; Shum-Tim et al. 1999); however, for bone tissue engineering, P3HB-3HH showed improved attachment, proliferation, and differentiation of rabbit bone marrow cells (Wang et al. 2004; Yang et al. 2004) and chondrocytes (Deng et al. 2002, 2003; Zhao et al. 2003a, b; Zheng et al. 2003, 2005) compared to PLLA. Despite the relatively small amount of research on their applications in bone and cartilage engineering, the potential of the above-mentioned soft elastomeric PHAs should not be ignored, and much research is needed to explore their application as bone engineering scaffolds.

Polyurethane

Polyurethanes (PUs) are a large family of polymeric materials with an enormous diversity of chemical compositions, mechanical properties, tissue-specific biocompatibility, and biodegradability (Lamba et al. 1998; Santerre et al. 2005; Zdrahala 1996). PUs are generally synthesized with three components: a diisocyanate, a polyol, and a chain extender (usually a diamine or diol) by step growth polymerization (Ganta et al. 2003; Szycher 1999). The resultant polyurethanes are phase-segregated polymers composed of alternating polydispersed blocks of 'soft' segments (made of macropolyols) and 'hard' segments (made of diisocyanates and chain extenders). Because of the differences in polarity between the hard (polar) and soft (nonpolar) segments, segmented PU elastomers can undergo microphase separation

to form hard and soft domains. The soft domains are rubbery and amorphous at room temperature due to a glass transition temperature of less than 0°C. The hard domains, which result in the induction of hydrogen bonding between urethane and urea groups in the hard segments of adjacent polymer chains, function as physical cross-links that resist flow when stress is applied to the materials (Guelcher 2008). The mechanical properties, as well as the biodegradation rate, can be tuned by modifying the structure of the hard and soft segments and/or changing the relative fractions of the hard and soft segments.

Historically, PUs had been used in permanent medical devices; they were actually subjected to hydrolysis, oxidation, and enzymatic degradation (Jayabalan et al. 2000; Pinchuk 1994). The soft segments generally dominated the degradation characteristics of PUs, and a high content of soft segments tends to increase the degradation rate (Pinchuk 1994). Many attempts were made to resist biodegradation processes (Zdrahala 1996). Converse to this, more recent attempts have been made to enhance the biodegradability of PUs. Over the past two decades, scientists have been utilizing the flexible chemistry of PU materials to design degradable polymers for tissue engineering, including both hard (Saad et al. 1997) and soft types (Alperin et al. 2005; Borkenhagen et al. 1998; Fujimoto et al. 2007; McDevitt et al. 2003). These materials have taken advantage of processes such as hydrolytic mechanisms and have varied molecular structure to control hydrolysis rates.

In contrast to degradation behavior of PLA, PGA, and PLGA, PUs demonstrated no significant pH change in the microenvironment of their degradation products, instead showing a linear degradation rate with no autocatalytic effect (Guan et al. 2005). However, the degradation products of PUs could be toxic when aromatic diisocyanates (e.g. 4,4′-methylenediphenyl diisocyanate and toluene diisocyanate) are used. To address this problem, aliphatic diisocyanates (e.g., lysine diisocyanate (LDI) and 1,4-diisocyanatobutane (BDI)) have been used as the replacements of aromatic diisocyanates (Gunatillake et al. 2003b; Lamba et al. 1998; Pinchuk 1994) in PUs that are designed to be biodegradable.

In general, PUs are recognized to have good blood and tissue compatibility (Fromstein and Woodhouse 2006; Zdrahala and Zdrahala 1999). PUs made with LDI as the diisocyanate demonstrated no significantly detrimental effects on cell viability, growth, and proliferation *in vitro* and *in vivo*. Subcutaneous implantation in rats revealed that LDI-based PUs did not aggravate capsule formation, accumulation of macrophages, or tissue necrosis (Zhang et al. 2002). Excellent reviews on biocompatibility of PUs can be found in a number of books (Fromstein and Woodhouse 2006; Lamba et al. 1998; Zdrahala and

Zdrahala 1999) and a number of topic reviews (Christenson et al. 2007; Griesser 1991; Guelcher 2008; Santerre et al. 2005; Szycher et al. 1996; Zdrahala 1996; Zdrahala and Zdrahala 1999).

Most aliphatic diisocynate-based poly(ester urethane urea)s (PEUU)s have a Young's modulus (at small strains) of several tens of megapascals and an impressively large breaking strain in the range of 100 to 1,000% (Guan and Wagner 2005; Hong et al. 2010). PU rubbers made from PEUU: BDI/PCL, PEUU: BDI/PCL-polycarbonate, and PCUU: BDI/polycarbonate show a super elasticity, with the elongation at break and resilience being 600% to 800% and 99% to 100%, respectively, (Guan and Wagner 2005; Hong et al. 2010).

In addition to their tunable mechanical and biodegradable properties, PU elastomers also have a good processibility. They can be fabricated into highly porous scaffolds by a number of foaming techniques, such as thermally induced phase separation (Guan et al. 2005) salt leaching/freeze-drying (Gogolewski and Gorna 2007; Gogolewski et al. 2006; Spaans et al. 1998a, 1998b), wet spinning (Gisselfalt et al. 2002; Liljensten et al. 2002), and electrospinning (Stankus et al. 2004, 2007). By applying the fabrication techniques mentioned above, different porosities, surface-to-volume ratios, and three-dimensional structures with concomitant changes in mechanical properties can be achieved to suit a wide range of tissue engineering, including bone and soft tissues (Guelcher 2008). Table 1 provides a summary of the applications of PUs in bone tissue engineering.

From the point of view of biodegradation, PHAs and PUs could, in principle, be used in tissue engineering as implants that require a longer retention time or a higher stability in the surrounding environment, but which eventually absorb. This might be useful for tissues with slower healing and remodeling times or with an inability to maintain innate structural integrity (e.g., muscle). Their slow degradation profile (2 to 3 years) has limited their applications in bone tissue engineering, as the healing rate of bone is typically 6 to 12 weeks (Kakar and Einhorn 2008). Hence, suitable elastomeric polymers with faster degradation kinetics that matches the healing profile of bone tissue remain to be explored. For this, recently developed degradable, chemically cross-linked polyester elastomers provide considerable potential (see the 'Biodegradable chemically cross-linked elastomers' section).

Biodegradable chemically cross-linked elastomers
Poly(propylene fumarate)
Poly(propylene fumarate) (PPF) is an unsaturated linear polyester. Like PLA and PGA, the degradation products of PPF (i.e., propylene glycol and fumaric acid) are biocompatible and readily removed from the body. The double bond along the backbone of the polymer permits cross-linking *in situ*, which causes a moldable composite to harden within 10 to 15 min. Mechanical properties and degradation time of the composite may be controlled by varying the PPF molecular weight. Therefore, preservation of the double bonds and control of molecular weight during PPF synthesis are critical issues (Payne and Mikos 2002). PPF has been suggested for use as a scaffold for guided tissue regeneration, often as part of an injectable bone replacement composite (Yaszemski et al. 1995). It also has been used as a substrate for osteoblast cultures (Peter et al. 2000). The development of composite materials combining PPF and inorganic particles, e.g., HA or bioactive glass, has not been investigated to a large extent in comparison with the extensive research efforts dedicated to PLGA- and PLA-based composites.

Poly(polyol sebacate)
Poly(polyol sebacate) (PPS) is a family of cross-linked polyester elastomers, developed for soft tissue engineering (Wang et al. 2002a). Polyol and sebacic acid are both endogenous monomers found in human metabolites (Ellwood 1995; Natah et al. 1997; Sestoft 1985); hence, PPSs generally show little toxicity to host tissues (Chen et al. 2011a; Wang et al. 2003). Poly(glycerol sebacate) (PGS) is the most extensively evaluated member of the PPS family, with most *in vitro* data demonstrating that PGS has a very good biocompatibility (Fidkowski et al. 2005a; Gao et al. 2007; Motlagh et al. 2006; Sundback et al. 2005; Sundback et al. 2004; Wang 2004). Poly(xylitol sebacate) (PXS) has also been developed using xylitol, a well-studied monomer in terms of biocompatibility and pharmacokinetics in humans (Ellwood 1995; Natah et al. 1997; Sestoft 1985; Talke and Maier 1973). As a metabolic intermediate in the mammalian carbohydrate metabolism, xylitol enters the metabolic pathway slowly without causing rapid fluctuations of blood glucose levels (Natah et al. 1997; Winkelhausen and Kuzmanova 1998). Inspired by the good biocompatibility of xylitol, Langer's group was the first to develop PXS (Bruggeman et al. 2008b, 2010). An *in vitro* evaluation of biocompatibility of PXS, poly(sorbitol sebacate) (PSS), and poly(mannitol sebacate) (PMS) polymers showed that they supported primary human foreskin fibroblasts in terms of cellular attachment and proliferation with the exception of PSS and PMS that were synthesised at the ratio of 1:1 (polyol/sebacic acid) (Bruggeman et al. 2008a).

In vivo assessment of PGS was first conducted by subcutaneous implantation of 3-mm-thick material in Sprague–Dawley rats (Wang et al. 2002b, 2003; Wang 2004). This evaluation showed that PGS induced an acute inflammatory response but no chronic inflammation, while PLGA caused both. The PGS implants in rats

Table 1 Bone tissue engineering applications of polyurethanes

Animal models	Polyurethane scaffolds	Major conclusions	Reference
Iliac crest (sheep)	Porous scaffolds synthesized from HMDI, PEO-PPO-PEO, and PCL at various ratios. Pore size, 300 to 2,000 μm; porosity, 85%	At 18 and 25 months, all the defects in the ilium implanted with polyurethane bone substitutes had healed with new bone.	Gogolewski and Gorna (2007), Gogolewski et al. (2006)
		The extent of bone healing depended on the chemical composition of the polymer from which the implant was made.	
		The implants from polymers with the incorporated calcium-complexing additive were the most effective promoters of bone healing, followed by those with vitamin D and polysaccharide-containing polymer.	
		There was no bone healing in the control defects.	
Bone marrow stromal cells	BDI with PCL films	Bone marrow stromal cells were cultured on rigid polymer films under osteogenic conditions for up to 21 days. This study demonstrated the suitability of this family of PEUUs for bone tissue engineering applications.	Kavlock et al. (2007)
Femoral condyle	LTI with PCL-co-PGA-co-PDLLA	Extensive cellular infiltration deep to the implant and new bone formation at 6 weeks	Dumas et al. (2010)
Chondrocytes	Porous scaffolds synthesized from HMDI with PCL and ISO	Although the covalent incorporation of the isoprenoid molecule into the polyurethane chain modified the surface chemistry of the polymer, it did not affect the viability of attached chondrocytes.	Eglin et al. (2010)
		The change of surface characteristics and the more open pore structure of the scaffolds produced from the isoprenoid-modified polyurethane are beneficial for the seeding efficiency and the homogeneity of the tissue-engineered constructs.	

were completely absorbed after 60 days without scarring or permanent damage to tissue structure (Wang 2004). Another *in vivo* investigation via subcutaneously implanted PGS films in the same species has shown that PGS has excellent biocompatibility, inducing only a mild inflammatory response (Pomerantseva et al. 2009). *In vivo* applications of PGS in the nerve (Sundback et al. 2005), vascular (Bettinger et al. 2005b, 2006; Kemppainen and Hollister 2010; Motlagh et al. 2006), and myocardial (Stuckey et al. 2010) tissue engineering consistently show a mild foreign body response in terms of both acute and chronic inflammations. Subcutaneous implantation of PXSs in Lewis rats has shown improved biocompatibility when compared to PLGA implants (Bruggeman et al. 2008b). Up to now, reports on PXS have indicated that these elastomers could be viable candidates as biodegradable medical devices that can offer structural integrity and stability over a clinically required period (Bruggeman et al. 2010). PSS and PMS polymers also exhibit better *in vivo* biocompatibility than PLGA, evidenced by mild acute inflammatory reactions and less fibrous capsules formation during chronic inflammation (Bruggeman et al. 2008a).

PGS was reported to be completely resorbed 60 days after implantation in rats (Wang et al. 2003). This comparatively faster degradation rate of PGS *in vivo* was also reported by Stuckey et al. (2010) who used PGS sheets as

a pericardial heart patch. They found that the PGS patch was completely resorbed after 6 weeks. These examples of *in vivo* degradation indicate that aqueous enzymatic action, combined with dynamic tissue movements and vascular perfusion, might enhance the enzymatic breakdown of ester bonds in PGS and, thus, facilitate the hydrolytic weakening of this material *in vivo*.

Most recently, an *in vitro* enzymatic degradation protocol was reported to be able to simulate and quantitatively capture the features of *in vivo* degradation of PGS-based materials (Liang et al. 2011). In the study, PGS and PGS/Bioglass® composites were subjected to enzymatic degradation in tissue culture medium or a buffer solution at the pH optima in the presence of defined concentrations of an esterase. The *in vitro* enzymatic degradation rates of the PGS-based materials were markedly higher in the tissue culture medium than in the buffered solution at the optimum pH 8. The *in vitro* enzymatic degradation rate of PGS-based biomaterials cross-linked at 125°C for 2 days was approximately 0.5 to 0.8 mm/month in tissue culture medium, which falls within the range of *in vivo* degradation rates (0.2 to 1.5 mm/month) of PGS cross-linked at similar conditions. Enzymatic degradation was also further enhanced in relation to cyclic mechanical deformation.

Briefly, PGS and the related PPS family are rapidly degrading polymers (several weeks) (Chen et al. 2012b; Li et al. 2012; Liang et al. 2011). Up to now, there is only one report on the application of PPS as a scaffolding material for bone tissue engineering (Chen et al. 2010d). Nonetheless, it must be emphasized that *among the above-reviewed degradable polymers, the rapid degradation kinetics of the PPS family best matches the healing profile of bone, which has complete healing rates of 6 to 12 weeks* (Kakar and Einhorn 2008).

Bioactive ceramics

A common feature of bioactive glasses and ceramics is a time-dependent, kinetic modification of the surface that occurs upon implantation. The surface forms a biologically active hydroxycarbonate apatite (HCA) layer, which provides the bonding interface with tissues. The HCA phase that forms on bioactive implants is chemically and structurally equivalent to the mineral phase in bone, providing interfacial bonding (Hench 1991, 1998). The *in vivo* formation of an apatite layer on the surface of a bioactive ceramic can be reproduced in a protein-free and acellular simulated body fluid, which is prepared to have an ionic composition similar to that of the human blood plasma, as described previously (Kokubo et al. 2003). Typical mechanical properties of the bioactive ceramic phases discussed in this article are listed in Table 2.

Dilemmas in developing biomaterials for bone tissue engineering

Since almost two-thirds of the weight of bone is hydroxyapatite $Ca_{10}(PO_4)_6(OH)_2$, it seems logical to use this ceramic as the major component of scaffold materials for bone tissue engineering. Actually, hydroxyapatite and related calcium phosphates (CaP) (e.g., β-tricalcium phosphate) have been intensively investigated (1990; Burg et al. 2000; Hench and Wilson 1999; LeGeros and LeGeros 2002). As expected, calcium phosphates have an excellent biocompatibility due to their close

resemblance to bone mineral chemical and crystal structure (Jarcho 1981; Jarcho et al. 1977). Although they have not shown osteoinductive ability, they certainly possess osteoconductive properties as well as a remarkable ability to bind directly to bone (Denissen et al. 1980; Driskell et al. 1973; Hammerle et al. 1997; Hollinger and Battistone 1986). A large body of *in vivo* and *in vitro* studies have reported that calcium phosphates, no matter in which form (bulk, coating, powder, or porous) or phase (crystalline or amorphous) they are in, consistently support the attachment, differentiation, and proliferation of osteoblasts and mesenchymal cells, with hydroxyapatite being the best one among them (Brown et al. 2001).

Crystalline calcium phosphates have long been known to have very prolonged degradation times *in vivo*, often in the order of years (Rezwan et al. 2006; Vacanti et al. 2000). Nanosized carbonated HA is a stable component of natural bone, though it metabolizes like all tissues. Hence, it would be fundamentally wrong if one expected HA to rapidly degrade in a physiological environment. In fact, clinical investigation has recently demonstrated *that implanted hydroxyapatites and calcium phosphates are virtually inert*, remaining within the body for as long as 6 to 7 years post-implantation (Marcacci et al. 2007). This should make HA less favored as a scaffold material for use in tissue engineering. The degradation rates of amorphous HA and TCP are high, but they are too fragile to build a 3D porous network.

The properties of synthetic calcium phosphates vary significantly with their crystallinity, grain size, porosity, and composition. In general, the mechanical properties of synthetic calcium phosphates decrease significantly with increasing content of amorphous phase, microporosity, and grain size. High crystallinity, low porosity, and small grain size tend to give higher stiffness, higher compressive and tensile strength, and greater fracture toughness (Kokubo 1999a; LeGeros and LeGeros 1999).

Table 2 Mechanical properties of hydroxyapatite, 45 S5 Bioglass®, glass-ceramics, and human cortical bone

Ceramics	Compression strength (MPa)	Tensile strength (MPa)	Elastic modulus (GPa)	Fracture toughness $(MPa\sqrt{m})$	Reference
Hydroxyapatite	>400	approximately 40	approximately 100	approximately 1.0	Hench (1999), LeGeros and LeGeros (1999)
45 S5 Bioglass®	approximately 500	42	35	0.5 to 1	Hench (1999), Hench and Kokubo (1998)
A-W	1,080	215 (bend)	118	2.0	Kokubo (1999b)
Parent glass of A-W	NA	72 (bend)	NA	0.8	Kokubo (1999b)
Bioverit® I	500	140 to 180 (bend)	70 to 90	1.2 to 2.1	Holand and Vogel (1993)
Cortical bone	130 to 180	50 to 151	12 to 18	6 to 8	Keaveny and Hayes (1993), Moore et al. (2001), Nalla et al. (2003), Zioupos and Currey (1998)

NA, not applicable.

It has been reported that the flexural strength and fracture toughness of dense hydroxyapatite are much lower in dry compared to aqueous conditions (de Groot et al. 1990).

Comparing the properties of hydroxyapatite and related calcium phosphates with those of bone (Table 2), it is apparent that the bone has a reasonably good compressive strength, though it is lower than that of hydroxyapatite, and better tensile strength and significantly better fracture toughness than hydroxyapatite. The apatite crystals in the bone tissue make it strong enough to tolerate compressive loading. Combined with macroscale stress fibers, and the typically tubular structure of long bone or mesh-like structure of flatter bone, the high tensile strength and fracture toughness are attributed to flexible collagen fibers. Hence, calcium phosphates alone cannot be used for load-bearing scaffolds in spite of their good biocompatibility and osteoconductivity.

A major challenge in bone tissue engineering is to develop a scaffolding material that is mechanically strong and yet biodegradable. To engineer bone tissue, which is hard and functions to support the body, the scaffold material must be strong and tough. Ideally, the scaffold needs to be degradable, as this biodegradation would avoid the detrimental effects of a persisting foreign substance and allow its gradual replacement with the new bone. Unfortunately, in this context, *mechanical strength and biodegradability counteract each other*. In general, mechanically strong materials (e.g., crystalline hydroxyapatite, Ti alloys, and crystalline polymers) are virtually inert and remain part of the repaired bone, while biodegradable materials (e.g., amorphous hydroxyapatite and glasses) tend to be mechanically fragile. This forms the greatest challenge in the design of bioceramics for bone engineering at load-bearing sites, but there are processing approaches such as sintering of 45 S5 Bioglass® (Chen and Boccaccini 2006a), for example, may offer opportunities to address the above dilemma (see the 'Na-containing silicate bioactive glasses' section).

Na-containing silicate bioactive glasses

The basic constituents of the most bioactive glasses are SiO_2, Na_2O, CaO, and P_2O_5. 45 S5 Bioglass® contains 45% SiO_2, 24.5% Na_2O, 24.4% CaO, and 6% P_2O_5, in weight percent (Hench 1991). In 1969, Hench and co-workers discovered that certain glass compositions had excellent biocompatibility as well as the ability to bond bone (Hench et al. 1971). The bioactivity of this glass system can vary from surface bioactive (i.e., bone bonding) to bulk degradable (i.e., resorbed within 10 to 30 days in tissue) (Hench 1998). Through interfacial and cell-mediated reactions, bioactive glass develops a calcium-deficient, carbonated phosphate surface layer that allows it to chemically bond to host bone (Hench

1997, 1998, 1999; Hench et al. 1971; Hench and Wilson 1993; Pereira et al. 1994; Wilson et al. 1981). It is clearly recognized that for a bond with bone tissue to occur, a layer of biologically active carbonated hydroxyapatite (HCA) must form (Hench and Wilson 1984). This bioactivity is not exclusive to bioactive glasses; hydroxyapatite and related calcium phosphates also show an excellent ability to bond to bone, as discussed further below. The capability of a material to form a secure biological interface with the surrounding tissue is critical in the elimination of scaffold loosening.

An important feature of bioactive glasses for applications in bone tissue engineering is their ability to induce bone tissue growth processes such as enzyme activity (Aksay and Weiner 1998; Lobel and Hench 1996, 1998; Ohgushi et al. 1996), revascularization (Day et al. 2004; Keshaw et al. 2005), osteoblast adhesion and differentiation from mesenchymal stem cells (Gatti et al. 1994; Lu et al. 2005; Roether et al. 2002; Schepers et al. 1991). Another significant finding is that the dissolution products from bioactive glasses, in particular the 45 S5 Bioglass® composition, upregulate osteogenic gene expression and growth factor production (Xynos et al. 2000a). Silicon alone has been found to play a key role in bone mineralization and gene activation, which has led to an increased interest in the substitution of silicon for calcium into synthetic hydroxyapatites. Investigations *in vivo* have shown that bone ingrowth into silicon-substituted HA granules was remarkably greater than that into pure HA (Xynos et al. 2000b).

It has been found that bioactive glass surfaces can release biologically relevant levels of soluble ionic forms of Si, Ca, P, and Na, depending on the processing route and particle size. These released ions induce intracellular and extracellular responses (Xynos et al. 2000a, 2001). For example, a synchronized sequence of genes is activated in the osteoblasts that undergo cell division and synthesize an extracellular matrix, which mineralizes to become bone (Xynos et al. 2000a, 2001). In addition, bioactive glass compositions doped with AgO_2 have been shown to elicit antibacterial properties while maintaining their bioactive function (Bellantone et al. 2002). In recent investigations, 45 S5 Bioglass® has been shown to increase secretion of vascular endothelial growth factor *in vitro* and to enhance vascularization *in vivo*, suggesting that scaffolds containing controlled concentrations of Bioglass® might stimulate neovascularization, which is beneficial to large tissue constructs (Day et al. 2004).

One key reason that makes bioactive glasses a relevant scaffold material is the possibility of controlling a range of chemical properties and, thus, the rate of bioresorption. The structure and chemistry of glasses, in particular sol–gel derived glasses (Pereira et al. 1994); Chen et al. 2010b; Chen and Thouas 2011), can be tailored at a molecular level by modifying the thermal or environmental processing history to vary the composition. It is possible to design glasses with degradation properties specific to a particular application of bone tissue engineering.

It was once reported that crystallization of bioactive glasses, which is necessary to achieve mechanical strength, decreased the level of bioactivity (Filho et al. 1996), even turning a bioactive glass into an inert material (Li et al. 1992). This antagonism between bioactivity and mechanical strength was considered to hamper the application of bioactive glasses. This issue has now been addressed by the discovery that Na-containing glasses (e.g., 45 S5 Bioglass®) can be sintered to form a mechanically strong crystalline phase, which can transform into amorphous calcium phosphate at body temperature and in a biological environment, remaining both bioactive and degradable (Chen and Boccaccini 2006a; Chen and Boccaccini 2006b; Chen et al. 2011c, 2012a; Chen 2011). The loss in mechanical strength due to biodegradation is in the time fashion of tissue engineering, i.e., matching the healing profile of bone. This highly desirable property is a unique feature of this 45 S5 Bioglass®, which has not previously been found in any other material (e.g., hydroxyapatites, Ti-alloys, or polymers).

The above advantages are the reasons why 45 S5 Bioglass® is relatively successfully exploited in clinical treatments of periodontal disease (Perioglas^{TM}) and as a bone filler material (Novabone^{TM}) (Hench 1998). Bioglass® implants have also been used to replace damaged middle ear bones, restoring hearing to patients (Hench 1997). Bioactive glasses have gained new attention recently as promising scaffold materials, either as fillers or coatings of polymer structures, and as porous materials themselves when melt-derived and sol–gel-derived glasses (Boccaccini and Maquet 2003; Boccaccini et al. 2003; Chen and Boccaccini 2006b; Chen et al. 2010c; Chen and Thouas 2011; Jones and Hench 2003a, b; Kaufmann et al. 2000; Laurencin et al. 2002; Livingston et al. 2002; Yuan et al. 2001).

Borate bioactive glasses

While silicate 45 S5 compositions have been widely investigated over the last 50 years, borate- and borosilicate-based compositions have recently been explored (Fu et al. 2012; Rahaman et al. 2011; Yang et al. 2012). Boron is a trace element (see the 'Bioactive glasses doped with trace elements' section). Dietary boron is documented to benefit in bone health (Nielsen 2008; Uysal et al. 2009), as shown by Chapin et al. (1997). In their study, rats developed improved vertebral resistance to crash force after dietary intake of boron (Chapin et al. 1997). Gorustovich et al. (2006, 2008) furthermore found that boron deficiency in mice alters

periodontal alveolar bone remodeling by inhibiting bone formation.

Borate bioactive glasses have been reported to support cell proliferation and differentiation *in vitro* (Fu et al. 2009, 2010a; Marion et al. 2005) and tissue infiltration *in vivo* (Fu et al. 2010b). Boron concentrations in the blood around borate glass pellets implantation in rabbit tibiae were well below the toxic level (Zhang et al. 2010). However, there is a concern associated with the toxicity of boron released into the solution as borate ions, $(BO_3)^{3-}$. It has been reported that some borate glasses exhibited cytotoxicity under static *in vitro* culture conditions (Fu et al. 2010b), although no considerable toxicity was detected under more dynamic culture conditions, suggesting the importance of borate clearance (Fu et al. 2010b).

Borate bioactive glasses have also been reported to degrade faster than their silicate counterparts due to their relative chemical instability (Fu et al. 2009, 2010a, c; Huang et al. 2006a, 2007; Yao et al. 2007). By partially or fully replacing the SiO_2 in silicate glasses with B_2O_3, the complete degradation rate of the glasses can be varied over a wide range, from several days to longer than 2 months (Fu et al. 2009, 2010a, c; Huang et al. 2006a; Yao et al. 2007). Moreover, borate bioactive glasses are more readily converted to an apatite-like composition than the silicate materials (Huang et al. 2006a). The conversion mechanism of bioactive glass to apatite is similar to that of silicate 45 S5 glass, with the formation of a borate-rich layer, similar to the silicate-rich layer of the former (Hench 1998; Huang et al. 2006a, b). The ease of controlling the degradation rate in these borate-based glasses offers new opportunities to regulate the degradation rate of synthetic biomaterials to match injured bone healing rates.

Bioactive glasses doped with trace elements

Bioactive glasses have recently modified by doping with elements such as Cu, Zn, and Sr, which are known to be beneficial for healthy bone growth (Fu et al. 2010a; Hoppe et al. 2011; Wang et al. 2011; Zheng et al. 2012). To understand the biological significances of these types of trace elements in materials, it is useful to consider their abundance in biological tissues. The most abundant compound in the human body is water (65 to 90 wt.%), which contains most of the oxygen and hydrogen (Table 3). Approximately 96% of the weight of the body is comprised of oxygen, carbon, hydrogen, and

Table 3 Elements in the human body (Seeley et al. 2006)

Element	O	C	H	N	Ca	P	K	S	Na	Cl	Mg	Trace element
Wt.%	65.0	18.5	9.5	3.3	1.5	1.0	0.4	0.3	0.2	0.2	0.1	<0.01
At.%	25.5	9.5	63.0	1.4	0.31	0.22	0.06	0.05	0.3	0.03	0.1	<0.01

nitrogen, which are the building blocks of all proteins. The rest (approximately 4%) of the mass of the body exists largely either in the bone and tooth as minerals (Ca, Mg, and P) or in the blood and extracellular fluid as major electrolytes (Na, K, and Cl), referred to here as macroelements (Table 4, reference).

In addition to the macroelements, there are also a large number of elements in lower concentrations (how much...ppm?) for the proper growth, development, and physiology of the body (see the list of known trace elements in the human body (Whitney and Rolfes 2010) below). These elements are referred to as trace elements or micronutrients, and while this list is increasing, it is important to bear in mind that these trace elements are all toxic at high levels. In 1966, for instance, the addition of cobalt compounds to stabilize beer foam in Canada led to cardiomyopathy, which came to be known as *beer drinker's cardiomyopathy* (1967; Barceloux 1999). In brief, the majority of metal elements are needed in the human body as micronutrients (eg., as enzyme cofactors) but are toxic at levels higher than required, partly resulting in excretion or excess storage as deposits. Hence, it is highly important that as a glass degrades *in vivo*, the trace elements in scaffolds must be released at a biologically acceptable rate. In this section, we focus on trace elements doped in bioactive glasses for bone tissue engineering, including strontium, zinc, and copper.

List of known trace elements in the human body, which are all toxic at high levels (Whitney and Rolfes 2010).

- Barium
- Beryllium
- Boron
- Caesium
- Chromium
- Cobalt
- Copper
- Iodine
- Iron
- Lithium
- Molybdenum
- Nickel
- Selenium
- Strontium
- Tungsten
- Zinc

Strontium is chemically closely related to calcium, sharing the same main group with calcium on the periodic table of elements and having a similar atomic radius to the calcium cation (r_{Sr} = 1.16 Å and r_{Ca2+} = 1.0 Å). Because of the above chemical analogy, Sr has long been used as a dope element in the hydroxyapatite products (Chen et al. 2004;

Marie et al. 2001; Wong et al. 2004). *In vivo* investigations have demonstrated that strontium is, in general, a benign element, having pharmacological effects on bone balance in normal bone and in the treatment of osteoporosis (Marie et al. 2001; Marie 2010; Meunier et al. 2002). Moreover, a drug of strontium ranelate has been reported to enhance fracture healing of bone in rats in terms of callus resistance. The group treated with only strontium ranelate showed a significant increase in callus resistance compared to the untreated control group. An added benefit of doping trace elements is the enhanced X-ray imaging contrast.

Zinc is necessary in the function of all cells, binding specific DNA regions to regulate genetic control of cell proliferation (Whitney and Rolfes 2010). Zn is also reported to play a role in bone healing and metabolism (Yamaguchi 1998), with anti-inflammatory roles (Lang et al. 2007). It has been demonstrated that Zn (a) stimulates bone formation *in vitro* by activating protein synthesis in osteoblast cells, (b) increases ATPase activity in bone (Yamaguchi 1998) and inhibits bone resorption of osteoclast cells in mouse marrow cultures (Yamaguchi 1998), and (c) has regulatory effects on bone cells and, thus, on gene expression (Cousins 1998; Kwun et al. 2010). Nonetheless, it has been well documented that an excess of zinc may cause anemia or reduced bone formation (Whitney and Rolfes 2010) as well as systemic cytotoxicity.

Copper is contained in enzymes of the ferroxidase (ceruloplasmin) system which regulates iron transport and facilitates release from storage. A copper deficiency can result in anemia from reduced ferroxidase function. However, excess copper levels cause liver malfunction and are associated with the genetic disorder Wilson's disease. There have been controversial reports on the effects of copper on bone remodelling. On the one hand, Zhang et al. (2003) reported that Cu^{2+} at a concentration of 10^{-6} M inhibits osteoclast activity. Smith et al. (2002) also found that dietary copper deprivation causes a reduction of bone mineral density. On the other hand, Cashman et al. (2001) found that copper supplements over a period of 4 weeks did not affect bone formation or bone resorption, as manifested by biochemical markers. Furthermore, Lai and Yamaguchi (2005) showed that supplementation with copper induced a decrease in bone tissue in rats, showing reduced or absent anabolic effects on bone formation both *in vivo* and *in vitro*.

Perhaps what is positively relevant to bone tissue engineering about copper is that this element has consistently been reported to trigger endothelial cells towards angiogenesis. Finney et al. (2009) found that a significant amount of Cu ions was distributed in human endothelial cells when they were induced to enhance angiogenesis. This phenomenon was believed to indicate the importance of copper ions as angiogenic agent. In another

Table 4 Macroelements and their roles in the human body (Whitney and Rolfes 2010)

Macroelements	Roles
O, C, H, N	In water and the molecular structures of proteins
Ca	Structure of bone and teeth; muscle and nerve activity
P	Structure of bone and teeth; intermediate in REDOX metabolism and production of ATP in energy
Mg	Important in bone structure, muscle contraction, and metabolic processes
Na	Major electrolyte of blood and extracellular fluid; required for the maintenance of pH and osmotic balance; nerve and muscle signaling
K	Major electrolyte of blood and intracellular fluid; required for the maintenance of pH and osmotic balance; nerve and muscle signaling
Cl	Major electrolyte of blood and extracellular and intracellular fluid; required for the maintenance of pH and osmotic balance; nerve and muscle signaling
S	Element of the essential amino acids methionine and cysteine; contained in the vitamins thiamine and biotin. As part of glutathione, it is required for detoxification. Poor growth due to reduced protein synthesis and lower glutathione levels potentially increasing oxidative or xenobiotic damage are consequences of low sulfur and methionine and/or cysteine intake.

work, copper and angiogenesis growth factor FGF-2 were found to have synergistic stimulatory effects on angiogenesis *in vitro* (Gerard et al. 2010). In addition to its function of stimulating proliferation of human endothelial cells (Hu 1998), Cu was shown to promote the differentiation of mesenchymal stem cells towards the osteogenic lineage (Rodriguez et al. 2002).

In summary, although trace elements have beneficial effects on bone remodeling and/or associated angiogenesis, the risk of toxicity at high levels must be highly regarded in the design of composition and degradation rate of bioactive biomaterials so that the release of these elements must be satisfactorily lower than their biologically safe levels.

Biocomposites

The primary disadvantage of bioactive glasses is their mechanical weakness and low fracture toughness (Table 2) due to their amorphous structure. Hence, bioactive glasses alone have limited application in load-bearing situations owing to poor mechanical strength and mismatch with the surrounding bone. However, these materials can be sintered to improve their mechanical properties (Chen et al. 2006a, Chen et al. 2006b), or used in combination with polymers to form composite materials with better bone repair potential (Roether et al. 2002).

Thermoplastic-based composites

From a biological perspective, it is a natural strategy to combine polymers and ceramics to fabricate scaffolds for bone tissue engineering because, structurally, native bone is essentially the combination of a naturally occurring polymer and biological apatite. From the materials science point of view, a single material type does not always provide the necessary mechanical and/or chemical properties desired for this particular application. In these instances, composite materials designed to combine the advantages of both materials may be most appropriate. Polymers and ceramics that degrade *in vivo* should be chosen for designing biocomposites for tissue engineering scaffolds, except for permanent implants. While massive release of acidic degradation from polymers causes inflammatory reactions (Bergsma et al. 1993, 1995; Temenoff et al. 2000), the basic degradation of calcium phosphate or bioactive glasses would buffer these by-products of polymers thereby improving the physiological conditions of tissue environment due to pH control. Mechanically, bioceramics are much stronger than polymers and play a critical role in providing mechanical stability to construct prior to synthesis of new bone matrix by cells. However, as mentioned above, ceramics and glasses are very fragile due to their intrinsic brittleness and flaw sensitivity. To capitalize on their advantages and minimize their shortcomings, ceramic and glass materials can be combined with various polymers to form composite biomaterials for osseous regeneration. Tables 5 and 6 list selected dense and porous ceramic/glass-polymer composites, which have been designed as biomedical devices or scaffold materials for bone tissue engineering, and their mechanical properties.

In general, all of these synthetic composites have good biocompatibility. Kikuchi et al. (1999), for instance, combined TCP with PLA to form a polymer-ceramic composite, which was found to possess the osteoconductivity of β-TCP and the degradability of PLA. The research team led by Laurencin synthesized similar porous scaffolds containing PLGA and HA, which combine the degradability of PLGA with the bioactivity of HA, fostering cell proliferation and differentiation as well as mineral formation (Attawia et al. 1995; Devin et al. 1996; Laurencin et al. 1996a). Other composites of bioactive glass and PLA were observed to form calcium phosphate layers on their surfaces and support rapid and abundant

Table 5 Biocomposites used for bone tissue engineering

Biocomposite		Percentage of ceramic (%)	Compressive (C), tensile (T), flexural (F), and bending (B) strengths (MPa)	Modulus (MPa)	Ultimate strain (%)	Toughness (kJ/m^2)	Reference
Ceramic	Polymer						
HA fiber	PDLLA	2 to 10.5 (vol.)	45 (F)	1.75×10^3 to 2.47×10^3			Deng et al. (2001)
	PLLA	10 to 70 (wt.)	50 to 60 (F)	6.4×10^3 to 12.8×10^3	0.7 to 2.3		Kasuga et al. (2001)
HA	PLGA	40 to 85 (vol.)	22 (F)	1.1×10^3		5.29	Xu et al. (2004), Xu and Simon (2004a, b)
	Chitosan	40 to 85 (vol.)	12 (F)	2.15×10^3		0.092	Xu et al. (2004)
	Chitosan + PLGA	40 to 85 (vol.)	43 (F)	2.6×10^3		9.77	Xu et al. (2004)
	PPhos	85 to 95 (wt.)					Greish et al. (2005)
	Collagen	50 to 72 (wt.)					Rodrigues et al. (2003)
β-TCP	PLLA-co-PEH	75 (wt.)	51 (F)	5.18×10^3			Kikuchi et al. (1999)
	PPF	25 (wt.)	7.5 to 7.7 (C)	191 to 134			Peter et al. (1998)
A/W	PE	10 to 50 (vol.)	18 to 28 (B)	0.9×10^3 to 5.7×10^3			Juhasz et al. (2003a, b), Juhasz et al. (2004)
Ca$_3$(CO$_3$)$_2$	PLLA	30 (wt.)	50	3.5×10^3 to 6×10^3			Kasuga et al. (2003)
Bioglass®	**PGA**	**2 to 1 (wt.)**	**0.5 to 2 (T)**	**0.5 to 2 (T)**	**150 to 600**		Chen et al. (2010a) Chen et al. (2011b), Liang et al. (2010)
Human cortical bone		70 (wt.)	50 to 150 (T) 130 to 180 (C)	12×10^3 to 18×10^3			Keaveny and Hayes (1993), Moore et al. (2001), Nalla et al. (2003), Zioupos and Currey (1998)

Table 6 Properties of porous composites developed for bone tissue engineering

Biocomposite		Percentage of ceramic (wt.%)	Porosity (%)	Pore size (μm)	Strength (MPa)	Modulus (MPa)	Ultimate strain (%)	Reference
Amorphous CaP	PLGA	28 to 75	75	>100		65		Ambrosio et al. (2001), Khan et al. (2004)
β-TCP	Chitosa-gelatin	10 to 70		322 to 355	0.32 to 0.88	3.94 to 10.88		Yin et al. (2003)
HA	PLLA	50	85 to 96	100×300	0.39	10 to 14		Zhang and Ma (1999)
	PLGA	60 to 75	81 to 91	800 to 1800	0.07 to 0.22	2 to 7.5		Guan and Davies (2004)
	PLGA		30 to 40	110 to 150		337 to 1459		Devin et al. (1996)
Bioglass®	PLGA	75	43	89	0.42	51		Laurencin et al. (2002), Lu et al. (2003), Stamboulis et al. (2002)
	PLLA	20 to 50	77 to 80	approximately 100 (macro); approximately 10 (micro)	1.5 to 3.9	137 to 260	1.1 to 13.7	Zhang et al. (2004)
	PLGA	0.1 to 1		50 to 300				Blaker et al. (2004)
	PDLLA	5 to 29	94	approximately 100 (macro); 10 to 50 (micro)	0.07 to 0.08	0.65 to 1.2	7.21 to 13.3	Blaker et al. (2003, 2005), Verrier et al. (2004)
Phosphate glass A/W	PLA-PDLLA	40	93 to 97	98 to 154	0.017 to 0.020	0.075 to 0.12		Navarro, et al. (2004), Li and Chang (2004)
	PDLLA	20 to 40	85.5 to 95.2					
Bioglass	**PGS**	**90**	**>90**	**300 to 500**	**0.4 to 1.0**			Chen et al. (2010d)
Human cancellous bone		70	60 to 90	300 to 400	0.4 to 4.0	100 to 500	1.65 to 2.11	Giesen et al. (2001), Yeni and Fyhrie (2001), Yeni, et al. (2001)

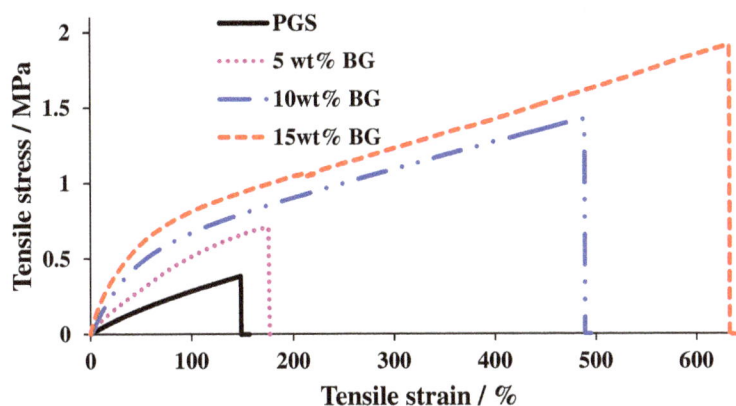

Figure 1 Typical tensile stress–strain curves. Of pure PGS and PGS composites of 5, 10, or 15 wt.% Bioglass®. Note the mechanical strength and strain at rupture increased simultaneously with the addition of Bioglass® filler (Chen et al. 2010a; Liang et al. 2010).

growth of human osteoblasts and osteoblast-like cells when cultured *in vitro* (Blaker et al. 2004; Blaker et al. 2003; Blaker et al. 2005; Boccaccini et al. 2003; Li and Chang 2004; Lu et al. 2003; Maquet et al. 2003, 2004; Navarro et al. 2004; Stamboulis et al. 2002; Verrier et al. 2004; Zhang et al. 2004).

A comparison between dense composites and cortical bone indicates that with thermoplastics, the most promising synthetic composite seems to be HA fiber-reinforced PLA composites (Kasuga et al. 2001), which however exhibit mechanical property values closer to the lower values of the cortical bone. Up to now, the best thermoplastic-based composite scaffolds reported in the literature seem to be those made from combinations of Bioglass® and

PLLA or PDLLA (Blaker et al. 2004; Maquet et al. 2003, 2004; Zhang et al. 2004). These composites have a well-defined porous structure; at the same time, their mechanical properties are close to (but lower than) those of cancellous bone.

Elastomer-based composites

Very recently, our group developed elastomeric composites from PPS and bioceramics (Chen et al. 2010a, 2011b; Liang et al. 2010). There are several advantages of using PPS elastomers over other thermoplastic polymers as a base for a reinforced composite. Firstly, its elastomeric properties make it ideal for a range of tissue repair applications (Bettinger et al. 2005a; Chen et al. 2008a; Fidkowski et al. 2005b; Redenti et al. 2009; Wang et al.

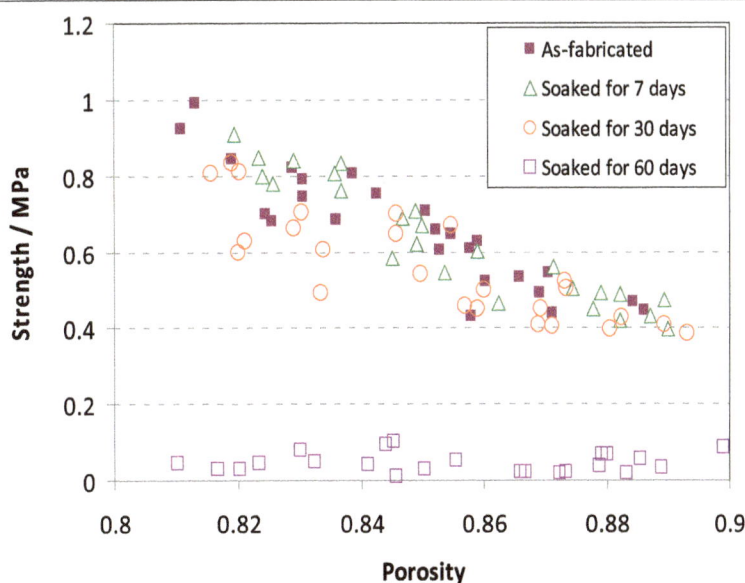

Figure 2 Compressive strength of Bioglass®-PGS scaffolds. During soaking in a tissue culture medium under physiological conditions for up to 2 months (Chen et al. 2010d).

2002b, c). In the case of bone, there is a requirement for some flexibility in the initial phases of bone repair, which involves cartilage deposition before bone formation (Oliveira et al. 2009). Secondly, PPS is acidic and, thus, able to react with alkaline Bioglass® via metallic carboxylation, resulting in a chemical bonding between the inorganic and organic components of the composite (Ma and Wu 2007). Thirdly, the degradation kinetics of PPS are entirely tunable by alternating its cross-link density to such a degree that it can maintain high physical integrity during degradation (Wang et al. 2002b). In addition, the elastic properties (i.e., Young's modulus, elongation at break and resilience) of these composites can be enhanced simultaneously by adding ceramic fillers due to the bound-rubber mechanism (Figure 1) (Chen et al. 2010a, 2011b; Liang et al. 2010). Finally, due to its combination

Table 7 Advantages and disadvantages of synthetic biomaterials used in bone tissue engineering

Biomaterial	Advantages	Disadvantages
Calcium phosphates (e.g. HA, TCP, and biphase CaP)	(1) Excellent biocompatibility	(1) Brittle
	(2) Supporting cell activity	
	(3) Good osteoconductivity	(2) They biodegrade too slowly in the crystalline state and are mechanically too weak in the amorphous state.
Na-containing silicate bioactive glasses	(1) Excellent biocompatibility	(1) Mechanically brittle and weak at the amorphous state
	(2) Supporting cell activity	
	(3) Good osteoconductivity	
	(4) Vasculature	
	(5) Rapid gene expression	
	(6) Tailorable degradation rate	
	(7) Tailorable mechanical strength via sintering, and the issue associated with strength and degradation could be addressed	
Borate bioactive glasses	(1) Tailorable degradation rate	(1) Risk of toxicity due to the release of borate ions
	(2) Tailorable mechanical strength	
Bioactive glass ceramics (e.g., A-W)	(1) Excellent biocompatibility	(1) Brittle
	(2) Supporting cell activity	
	(3) Good osteoconductivity	(2) Slow degradation rate
Bulk biodegradable polymers		
Poly(lactic acid)	(1) Good biocompatibility	(1) Inflammation caused by acid degradation products.
	(2) Biodegradable (with a wide range of degradation rates)	
Poly(glycolic acid)	(3) Bioresorbable	
Poly(lactic-co-glycolic acid)	(4) Good processability	(2) Accelerated degradation rates cause collapse of scaffolds.
Poly(propylene fumarate)	(5) Good ductility	
Poly(polyol sebacate)	(6) Elasticity	
Surface bioerodible polymers		
Poly(ortho esters)	(1) Good biocompatibility	(1) Not completely replaced by new bone tissue
Poly(anhydrides)	(2) Retention of mechanical integrity over the degradative lifetime of the device	
Poly(phosphazene)	(3) Significantly enhanced bone ingrowth into the porous scaffolds, owing to the increment in pore size	
Composites (containing bioactive phases)	(1) Excellent biocompatibility	(1) Still not as good as natural bone matrix
	(2) Supporting cell activity	
	(3) Good osteoconductivity	
	(4) Tailorable degradation rate	(2) Fabrication techniques need to be improved.
	(5) Improved mechanical reliability	

Table 8 List of advantages and disadvantages of biodegradable polymeric biomaterials

Material		Advantages	Disadvantages
Thermoplastic	Non-elastomers	Easy fabrication (by melt or solvent processing)	Rigid
			Lack of flexibility
		Tunable mechanical properties and degradation kinetics	Release of acidic degradation products
			Possibility of foreign body response
Elastomer	Thermoplastic	Easy fabrication	Heterogeneous degradation profile; mechanical failure; much faster than material degradation
		Flexible	
		High elongation	Release of acidic degradation products
		Tunable mechanical properties and degradation kinetics	Possibility of foreign body response
	Cross-linked	Flexible	Relatively difficult processability
		Tightly controlled purity	
		Structure, mechanical properties, and degradation kinetics	Possibility of foreign body response
		Good maintenance of form stability during degradation	Release of acidic degradation products

of satisfactory mechanical strength at the time of implantation and tunable biodegradability postimplantation, sintered 45 S5 Bioglass® ceramics can breakdown and change into nanosized bone minerals under aqueous physiological conditions (Chen and Boccaccini 2006b).

Our group has also has also developed a bone-like composite scaffold from PGS and 45 S5 Bioglass® (Chen et al. 2010d). These reinforced elastomeric scaffolds have similar mechanical properties to that of cancellous bone and exhibit a mechanically steady state over prolonged periods in a physiologic environment (Figure 2). This is very relevant to engineering features in scaffolds to match the lag phase of bone repair (Chen et al. 2010d).

Conclusion

While the ideal tissue-engineered bone substitute should be a material, which is bioresorbable, biocompatible, and supports cell attachment, proliferation, and maturation and which is ultimately resorbed once new bone has formed, allowing this bone to undergo remodelling, this goal has yet to be achieved. To design a scaffold, it is necessary to weigh up the 'pros and cons' of the potential precursor materials, which are summarized in Table 7. Among the bioactive ceramics and glasses listed in Table 7, Na-containing silicon bioactive glasses offer a number of advantages. The role of silicon in biological regulation of osteogenesis and the potential to address the dilemma between mechanical strength and degradation rate make these glasses promising scaffold materials over others, such as HA and related crystalline calcium phosphates. Recent progresses on the development of borate bioactive glasses and trace element-doped bioactive glasses expand the repertoire of bioactive glasses. Although boride and

other trace elements have beneficial effects on bone remodelling and/or associated angiogenesis, the risk of toxicity at high levels must be highly regarded in the design of new composition of bioactive biomaterials so that the release of these elements must be satisfactorily lower than their biologically safe levels.

Between the two main classes of bulk degradable and surface erodible polymer, the bulk degradable type is more promising than the surface-erosive group, considering that being replaced by new bone tissue is one of the most important criteria of an ideal scaffold material. Between thermoplastic and elastomeric polymers, Table 8 provides a comparison of both materials, as discussed earlier. Cross-linked synthetic elastomers (especially polyester elastomers) are the most attractive for use as a substitute of collagen matrix in tissue engineering. This is because, firstly, they are elastic and best match with the elasticity of biological tissue. Secondly, they are able to provide mechanical stability and structural integrity to tissues and organs without causing catastrophic mechanical implant failure, which is an issue remaining with thermoplastic rubbers. Thirdly, polyester elastomers allow closely control of structural and mechanical properties to suit various applications. Lastly and most importantly, polyester elastomers, most of which can safely breakdown to natural metabolic products by simple hydrolysis, have the potential to be tailored in their degradation rates to match healing kinetics of injured bone tissue, which can hardly achieved with current thermoplastics and thermoplastic rubbers. However, establishing the most suitable ceramic or mineral filler material and processing conditions for an elastomer is likely to provide many potential avenues for future research in bone tissue engineering scaffolds.

Competing interest

The authors declare that they have no competing interests.

Authors' contributions

QC initiated and shaped the article. CZ contributed in the 'Polyhydroxyalkanoates' section under the supervision of QC. GAT participated in scientific discussions and revision of the article. All authors read and approved the final manuscript.

Authors' informations

QC received a Ph.D. degree in Biomaterials from Imperial College London. She is currently an academic in the Department of Materials Engineering at Monash University. Previously she was also employed by the National Heart and Lung Institute London and the University of Cambridge. She has produced more than 100 peer-refereed journal articles and book chapters. Her research interests broadly cover polymeric, ceramic, metallic, and composite materials for applications in biomedical engineering. QC acknowledges Australia Research Council Discovery Project grant: Novel artificial bone constructs for rapid vasculature and bone regeneration.

Author details

[1]Department of Materials Engineering, Monash University, Clayton, Victoria 3800, Australia. [2]Department of Zoology, The University of Melbourne, Parkville, Victoria 3010, Australia.

References

(1967) Quebec beer-drinker's cardiomyopathy. JAMA 202:1145

(1990) Handbook of bioactive ceramics. CRC Press, Boca Raton, Florida

Aksay IA, Weiner S (1998) Biomaterials - is this really a field of research? Curr Opin Solid State Mater Sci 3:219–220

Allcock HR (2002) Syntheses of synthetic polymers: polyphosphazenes. In: Atala A, Lanza RP (eds) Methods of tissue engineering. Academic Press, California, pp 597–608

Alperin C, Zandstra PW, Woodhouse KA (2005) Polyurethane films seeded with embryonic stem cell-derived cardiomyocytes for use in cardiac tissue engineering applications. Biomaterials 26:7377–7386. doi:10.1016/j. biomaterials.2005.05.064

Ambrosio AMA, Sahota JS, Khan Y, Laurencin CT (2001) A novel amorphous calcium phosphate polymer ceramic for bone repair: 1. Synthesis and characterization. J Biomed Mater Res 58:295–301

Andriano KP, Gurny R, Heller J (2002) Synthesis of synthetic polymers: poly(ortho esters). In: Atala A, Lanza RP (eds) Methods of tissue engineering. Academic Press, California, pp 619–627

Attawia MA, Herbert KM, Laurencin CT (1995) Osteoblast-like cell adherance and migration through 3-dimensional porous polymer matrices. Biochem Biophys Res Commun 213:639–644

Avella M, Martuscelli E, Raimo M (2000) Review - properties of blends and composites based on poly(3-hydroxy)butyrate (PHB) and poly(3-hydroxybutyrate-hydroxyvalerate) (PHBV) copolymers. J Mater Sci 35:523–545

Barceloux DG (1999) Cobalt. J Toxicol Clin Toxicol 37:201–216

Bell E (2000) Tissue engineering in perspective. In: Lanza RP, Langer R, Vacanti JP (eds) Principles of tissue engineering. Academic Press, California, pp xxxv–xli

Bellantone M, Williams HD, Hench LL (2002) Broad-spectrum bactericidal activity of Ag2O-doped bioactive glass. Antimicrob Agents Chemother 46:1940–1945

Berger J, Reist M, Mayer JM, Felt O, Peppas NA, Gurny R (2004) Structure and interactions in covalently and ionically crosslinked chitosan hydrogels for biomedical applications. Eur J Pharm Biopharm 57:19–34

Bergsma EJ, Rozema FR, Bos RRM, De Bruijn WC (1993) Foreign body reactions to resorbable poly(L-lactide) bone plates and screws used for the fixation of unstable zygomatic fractures. J Oral Maxillofac Surg 51:666–670

Bergsma JE, Debruijn WC, Rozema FR, Bos RRM, Boering G (1995) Late degradation tissue-response to poly(L-lactide) bone plates and screws. Biomaterials 16:25–31

Bettinger CJ, Borenstein JT, Langer RS (2005a) Biodegradable microfluidic scaffolds for vascular tissue engineering. Nanoscale Mater Sci Biol Med 845:25–30

Bettinger CJ, Weinberg EJ, Kulig KM, Vacanti JP, Wang Y, Borenstein JT, Langer R (2005b) Three-dimensional microfluidic tissue-engineering scaffolds using a flexible biodegradable polymer. Adv Mater 18:165–169

Bettinger CJ, Orrick B, Misra A, Langer R, Borenstein JT (2006) Microfabrication of poly (glycerol-sebacate) for contact guidance applications. Biomaterials 27:2558–2565

Blaker JJ, Gough JE, Maquet V, Notingher I, Boccaccini AR (2003) In vitro evaluation of novel bioactive composites based on Bioglass (R)-filled polylactide foams for bone tissue engineering scaffolds. J Biomed Mater Res A 67A:1401–1411

Blaker JJ, Day RM, Maquet V, Boccaccini AR (2004) Novel bioresorbable poly(lactide-co-glycolide) (PLGA) and PLGA/Bioglass (R) composite tubular foam scaffolds for tissue engineering applications. Adv Mater Forum 455–456:415–419

Blaker JJ, Maquet V, Jerome R, Boccaccini AR, Nazhat SN (2005) Mechanical properties of highly porous PDLLA/Bioglass (R) composite foams as scaffolds for bone tissue engineering. Acta Biomater 1:643–652

Boccaccini AR, Maquet V (2003) Bioresorbable and bioactive polymer/Bioglass® composites with tailored pore structure for tissue engineering applications. Compos Sci Technol 63:2417–2429

Boccaccini AR, Notingher I, Maquet V, Jérôme R (2003) Bioresorbable and bioactive composite materials based on polylactide foams filled with and coated by Bioglass® particles for tissue engineering applications. J Mater Sci Mater Med 14:443–450

Boccaccini AR, Blaker JJ, Maquet V, Day RM, Jérôme R (2005) Preparation and characterisation of poly(lactide-co-grycolide) (PLGA) and PLGA/Bioglass® composite tubular foam scaffolds for tissue engineering applications. Mater Sci Eng C 25:23–31

Borkenhagen M, Stoll RC, Neuenschwander P, Suter UW, Aebischer P (1998) In vivo performance of a new biodegradable polyester urethane system used as a nerve guidance channel. Biomaterials 19:2155–2165

Brown S, Clarke I, Williams P (2001) Bioceramics, vol 14. In: Proceedings of the 14th international symposium on ceramics in medicine, Palm Springs, CA, 14–17 November 2001

Bruggeman JP, Bettinger CJ, Nijst CLE, Kohane DS, Langer R (2008a) Biodegradable xylitol-based polymers. Adv Mater 20:1922–1927. doi:10.1002/adma.200702377

Bruggeman JP, de Bruin BJ, Bettinger CJ, Langer R (2008b) Biodegradable poly (polyol sebacate) polymers. Biomaterials 29:4726–4735. doi:10.1016/j. biomaterials.2008.08.037

Bruggeman JP, Bettinger CJ, Langer R (2010) Biodegradable xylitol-based elastomers: in vivo behavior and biocompatibility. J Biomed Mater Res A 95A:92–104. doi:10.1002/jbm.a.32733

Burg KJL, Porter S, Kellam JF (2000) Biomaterial developments for bone tissue engineering. Biomaterials 21:2347–2359. doi:10.1016/s0142-9612(00)00102-2

Cashman KD, Baker A, Ginty F, Flynn A, Strain JJ, Bonham MP, O'Connor JM, Bugel S, Sandstrom B (2001) No effect of copper supplementation on biochemical markers of bone metabolism in healthy young adult females despite apparently improved copper status. Eur J Clin Nutr 55:525–531. doi:10.1038/sj.ejcn.1601177

Chaikof EL, Matthew H, Kohn J, Mikos AG, Prestwich GD, Yip CM (2002) Biomaterials and scaffolds in reparative medicine. Ann N Y Acad Sci 961:96–105

Chapin RE, Ku WW, Kenney MA, McCoy H, Gladen B, Wine RN, Wilson R, Elwell MR (1997) The effects of dietary boron on bone strength in rats. Fundam Appl Toxicol 35:205–215. doi:10.1006/faat.1996.2275

Chen QZ (2011) Foaming technology of tissue engineering scaffolds - a review. J Bubble Sci Tech Eng 3:34–47

Chen QZ, Boccaccini AR (2006a) Coupling mechanical competence and bioresorbability in Bioglass®-derived tissue engineering scaffolds. Adv Eng Mater 8:285–289. doi:10.1002/adem.200500259

Chen QZ, Rezwan K, Armitage D, Nazhat SN, Boccaccini AR (2006) The surface functionalization of 45S5 Bioglass (R)-based glass-ceramic scaffolds and its impact on bioactivity. J Mater Sci-Mater Med. 17(11):979-87

Chen QZ, Thouas GA (2011) Fabrication and characterization of sol–gel derived 45 S5 Bioglass (R)-ceramic scaffolds. Acta Biomater 7:3616–3626. doi:10.1016/ j.actbio.2011.06.005

Chen GQ, Wu Q (2005) The application of polyhydroxyalkanoates as tissue engineering materials. Biomaterials 26:6565–6578

Chen QZ, Wong CT, Lu WW, Cheung KMC, Leong JCY, Luk KDK (2004) Strengthening mechanisms of bone bonding to crystalline hydroxyapatite in vivo. Biomaterials 25:4243–4254. doi:10.1016/j.biomaterials.2003.11.017

Chen QZ, Thompson ID, Boccaccini AR (2006a) 45 S5 Bioglass (R)-derived glass-ceramic scaffolds for bone tissue engineering. Biomaterials 27:2414–2425. doi:10.1016/j.biomaterials.2005.11.025

Chen QZ, Rezwan K, Armitage D, Nazhat SN, Boccaccini AR. The surface functionalization of 45S5 Bioglass (R)-based glass-ceramic scaffolds and its impact on bioactivity. J Mater Sci-Mater Med. 2006;17(11):979-87.

Chen QZ, Bismarck A, Hansen U, Junaid S, Tran MQ, Harding SE, Ali NN, Boccaccini AR (2008a) Characterisation of a soft elastomer poly(glycerol sebacate) designed to match the mechanical properties of myocardial tissue. Biomaterials 29:47–57. doi:10.1016/j.biomaterials.2007.09.010

Chen QZ, Harding SE, Ali NN, Lyon AR, Boccaccini AR (2008b) Biomaterials in cardiac tissue engineering: ten years of research survey. Mater Sci Eng R-Rep 59:1–37. doi:10.1016/j.mser.2007.08.001

Chen QZ, Jin LY, Cook WD, Mohn D, Lagerqvist EL, Elliott DA, Haynes JM, Boyd N, Stark WJ, Pouton CW, Stanley EG, Elefanty AG (2010a) Elastomeric nanocomposites as cell delivery vehicles and cardiac support devices. Soft Matter 6:4715–4726. doi:10.1039/c0sm00213e

Chen QZ, Li YA, Jin LY, Quinn JMW, Komesaroff PA (2010b) A new sol–gel process for producing Na2O-containing bioactive glass ceramics. Acta Biomater 6:4143–4153. doi:10.1016/j.actbio.2010.04.022

Chen QZ, Liang SL, Cook WD (2010c) A new family of elastomeric biocomposites with a potential of wide applications in tissue engineering. In: TERMIS Asia-Pacific Chapter 2010, Sydney, 15–17 September 2010

Chen QZ, Quinn JMW, Thouas GA, Zhou XA, Komesaroff PA (2010d) Bone-like elastomer-toughened scaffolds with degradability kinetics matching healing rates of injured bone. Adv Eng Mater 12:B642–B648. doi:10.1002/adem.201080002

Chen QZ, Liang SL, Wang J, Simon GP (2011a) Manipulation of mechanical compliance of elastomeric PGS by incorporation of halloysite nanotubes for soft tissue engineering applications. J Mech Behav Biomed Mater 4:1805–1818

Chen QZ, Liang SL, Thouas GA (2011b) Synthesis and characterisation of poly (glycerol sebacate)-co-lactic acid as surgical sealants. Soft Matter 7:6484–6492. doi:10.1039/c1sm05350g

Chen QZ, Mohn D, Stark WJ (2011c) Optimization of bioglass (R) scaffold fabrication process. J Am Ceram Soc 94:4184–4190. doi:10.1111/j.1551-2916.2011.04766.x

Chen QZ, Xu JL, Yu LG, Fang XY, Khor KA (2012a) Spark plasma sintering of sol-gel derived 45 S5 Bioglass®-ceramics: mechanical properties and biocompatibility evaluation. Mater Sci Eng C Mater Biol Appl 32:494–502. doi:10.1016/j.msec.2011.11.023

Chen QZ, Yang XY, Li Y, Thouas GA (2012b) A comparative study of in vitro enzymatic degradtion of PXS and PGS. RSC Advances 2:4125–4134. doi:10.1039/C2RA20113E

Christenson EM, Anderson JM, Hittner A (2007) Biodegradation mechanisms of polyurethane elastomers. Corrosion Eng Sci Technol 42:312–323. doi:10.1179/174327807x238909

Cohn D, Salomon AH (2005) Designing biodegradable multiblock PCL/PLA thermoplastic elastomers. Biomaterials 26:2297–2305

Cousins RJ (1998) A role of zinc in the regulation of gene expression. Proc Nutr Soc 57:307–311. doi:10.1079/pns19980045

Day RM, Boccaccini AR, Shurey S, Roether JA, Forbes A, Hench LL, Gabe SM (2004) Assessment of polyglycolic acid mesh and bioactive glass for soft-tissue engineering scaffolds. Biomaterials 25:5857–5866

de Groot K, Lein CPAT, Wolke JGC, de Bliek-Hogervost JMA (1990) Chemistry of calcium phosphate bioceramics. In: Yamamuro T, Hench LL, Wilson J (eds) Handbook of bioactive ceramics. CRC Press, Boca Raton, Florida, pp 3–16

Deng XM, Hao JY, Wang CS (2001) Preparation and mechanical properties of nanocomposites of poly(D, L-lactide) with Ca-deficient hydroxyapatite nanocrystals. Biomaterials 22:2867–2873

Deng Y, Zhao K, Zhang XF, Hu P, Chen GQ (2002) Study on the three-dimensional proliferation of rabbit articular cartilage-derived chondrocytes on polyhydroxyalkanoate scaffolds. Biomaterials 23:4049–4056

Deng Y, Lin XS, Zheng Z, Deng JG, Chen JC, Ma H, Chen GQ (2003) Poly (hydroxybutyrate-co-hydroxyhexanoate) promoted production of extracellular matrix of articular cartilage chondrocytes in vitro. Biomaterials 24:4273–4281. doi:10.1016/s0142-9612(03)00367-3

Denissen HW, Degroot K, Makkes PC, Vandenhooff A, Klopper PJ (1980) Tissue-response to dense apatite implants in rats. J Biomed Mater Res 14:713–721

Devin JE, Attawia MA, Laurencin CT (1996) Three-dimensional degradable porous polymer-ceramic matrices for use in bone repair. J Biomater Sci Polym Ed 7:661–669

Di Silvio L, Bonfield W (1999) Biodegradable drug delivery system for the treatment of bone infection and repair. J Mater Sci Mater Med 10:653–658

Doi Y, Kitamura S, Abe H (1995) Microbial synthesis and characterization of poly(3-hydroxybutyrate-co-3-hydroxyhexanoate). Macromolecules 28:4822–4828

Domb AJ, Langer R (1999a) Poly(1,3-bis-p-carboxyphenoxypropane anhydride). In: Mark JE (ed) Polymer data handbook. Oxford Press, Oxford, pp 303–305

Domb AJ, Langer R (1999b) Poly(erucic acid dimmer anhydride). In: Mark JE (ed) Polymer data handbook. Oxford Press, Oxford, pp 457–459

Doyle C, Tanner ET, Bonfield W (1991) In vitro and in vivo evaluation of polyhydroxybutyrate and of polyhydroxybutyrate reinforced with hydroxyapatite. Biomaterials 12:841–847

Driskell TD, Hassler CR, McCoy LR (1973) The significance of resorbable bioceramics in the repair of bone defects. In: Proc 26th Ann Conf Med Bio 1973, 15:199-206

Drotleff S, Lungwitz U, Breunig M, Dennis A, Blunk T, Tessmar J, Göpferich A (2004) Biomimetic polymers in pharmaceutical and biomedical sciences. Eur J Pharm Biopharm 58:385–407

Dumas JE, Davis T, Holt GE, Yoshii T, Perrien DS, Nyman JS, Boyce T, Guelcher SA (2010) Synthesis, characterization, and remodeling of weight-bearing allograft bone/polyurethane composites in the rabbit. Acta Biomater 6:2394–2406. doi:10.1016/j.actbio.2010.01.030

Dunn AS, Campbell PG, Marra KG (2001) The influence of polymer blend composition on the degradation of polymer/hydroxyapatite biomaterials. J Mater Sci Mater Med 12:673–677

Duvernoy O, Malm T, Ramstrom J, Bowald S (1995) A biodegradable patch used as a pericardial substitute after cardiac-surgery: 6- and 24-month evaluation with CT. Thorac Cardiovasc Surg 43:271–274

Eglin D, Grad S, Gogolewski S, Alini M (2010) Farsenol-modified biodegradable polyurethanes for cartilage tissue engineering. J Biomed Mater Res A 92A:393–408. doi:10.1002/jbm.a.32385

Ellwood KC (1995) Methods available to estimate the energy values of sugar alcohols. Am J Clin Nutr 62:1169S–1174S

Fidkowski C, Kaazempur-Mofrad MR, Borenstein J, Vacanti JP, Langer R, Wang Y (2005a) Endothelialized microvasculature based on a biodegradable elastomer. Tissue Eng 11:302–309

Fidkowski C, Kaazempur-Mofrad MR, Borenstein J, Vacanti JP, Langer R, Wang YD (2005b) Endothelialized microvasculature based on a biodegradable elastomer. Tissue Eng 11:302–309

Filho OP, Latorre GP, Hench LL (1996) Effect of crystallization on apatite-layer formation of bioactive glass 45 S5. J Biomed Mater Res 30:509–514

Finney L, Vogt S, Fukai T, Glesne D (2009) copper and angiogenesis: unravelling a relationship key to cancer progression. Clin Exp Pharmacol Physiol 36:88–94. doi:10.1111/J.1440-1681.2008.04969.x

Freier T (2006) Biopolyesters in tissue engineering applications. In: Meller G, Grassel T (eds) Polymers for regenerative medicine. Advances in Polymer Science. Springer, Heidelberg, pp 1–61

Fromstein JD, Woodhouse KA (2008) Chapter 218 Polyurethane biomaterials. In: Encyclopedia of Biomaterials and Biomedical Engineering. Wnek GE, Bowlin GL (Ed.) 2nd Edition. Informa Healthcare USA Inc., New York. pp 2304-2313

Fu HL, Fu Q, Zhou N, Huang WH, Rahaman MN, Wang DP, Liu X (2009) In vitro evaluation of borate-based bioactive glass scaffolds prepared by a polymer foam replication method. Mater Sci Eng C Mater Biol Appl 29:2275–2281. doi:10.1007/s10856-012-4605-7

Fu H, Rahaman MN, Day DE, Huang W (2012) Long-term conversion of 45 S5 bioactive glass-ceramic microspheres in aqueous phosphate solution. J Mater Sci Mater Med 23:1181–1191. doi:10.1016/J.msec.2009.05.013

Fu Q, Rahaman MN, Bal BS, Bonewald LF, Kuroki K, Brown RF (2010a) Silicate, borosilicate, and borate bioactive glass scaffolds with controllable degradation rate for bone tissue engineering applications. II. In vitro and in vivo biological evaluation. J Biomed Mater Res A 95A:172–179. doi:10.1002/jbm.a.32823

Fu Q, Rahaman MN, Fu H, Liu X (2010b) Silicate, borosilicate, and borate bioactive glass scaffolds with controllable degradation rate for bone tissue engineering applications. I. Preparation and in vitro degradation. J Biomed Mater Res A 95A:164–171. doi:10.1002/jbm.a.32824

Fujimoto KL, Tobita K, Merryman WD, Guan JJ, Momoi N, Stolz DB, Sacks MS, Keller BB, Wagner WR (2007) An elastic, biodegradable cardiac patch induces contractile smooth muscle and improves cardiac remodeling and function in subacute myocardial Infarction. J Am Coll Cardiol 49:2292 2300

Ganta SR, Piesco NP, Long P, Gassner R, Motta LF, Papworth GD, Stolz DB, Watkins SC, Agarwal S (2003) Vascularization and tissue infiltration of a biodegradable polyurethane matrix. J Biomed Mater Res A 64:242–248

Gao J, Ensley AE, Nerem RM, Wang YD (2007) Poly(glycerol sebacate) supports the proliferation and phenotypic protein expression of primary baboon vascular cells. J Biomed Mater Res A 83A:1070–1075. doi:10.1002/jbm.a.31434

Gatti AM, Valdre G, Andersson OH (1994) Analysis of the in vivo reactions of a bioactive glass in soft and hard tissue. Biomaterials 15:208–212

Gerard C, Bordeleau L-J, Barralet J, Doillon CJ (2010) The stimulation of angiogenesis and collagen deposition by copper. Biomaterials 31:824–831. doi:10.1016/j.biomaterials.2009.10.009

Giesen EBW, Ding M, Dalstra M, van Eijden T (2001) Mechanical properties of cancellous bone in the human mandibular condyle are anisotropic. J Biomech 34:799–803

Gisselfalt K, Edberg B, Flodin P (2002) Synthesis and properties of degradable poly(urethane urea)s to be used for ligament reconstructions. Biomacromolecules 3:951–958

Gittens SA, Uludag H (2001) Growth factor delivery for bone tissue engineering. J Drug Target 9:407–429

Gogolewski S, Gorna K, Turner AS (2006) Regeneration of bicortical defects in the iliac crest of estrogen-deficient sheep, using new biodegradable polyurethane bone graft substitutes. J Biomed Mater Res A 77:802–810. doi:10.1002/jbm.a.30669

Gogolewski S, Gorna K (2007) Biodegradable polyurethane cancellous bone graft substitutes in the treatment of iliac crest defects. J Biomed Mater Res A 80:94–101. doi:10.1002/jbm.a.30834

Gollwitzer H, Thomas P, Diehl P, Steinhauser E, Summer B, Barnstorf S, Gerdesmeyer L, Mittelmeier W, Stemberger A (2005) Biomechanical and allergological characteristics of a biodegradable poly(D, L-lactic acid) coating for orthopaedic implants. J Orthop Res 23:802–809

Gorustovich AA, Lopez JMP, Guglielmotti MB, Cabrini RL (2006) Biological performance of boron-modified bioactive glass particles implanted in rat tibia bone marrow. Biomed Mater 1:100–105. doi:10.1088/1748-6041/1/3/002

Gorustovich AA, Steimetz T, Nielsen FH, Guglielmotti MB (2008) A histomorphometric study of alveolar bone modelling and remodelling in mice fed a boron-deficient diet. Arch Oral Biol 53:677–682. doi:10.1016/j.archoralbio.2008.01.011

Grabow N, Schmohl K, Khosravi A, Philipp M, Scharfschwerdt M, Graf B, Stamm C, Haubold A, Schmitz KP, Steinhoff G (2004) Mechanical and structural properties of a novel hybrid heart valve scaffold for tissue engineering. Artif Organs 28:971–979

Greish YE, Bender JD, Lakshmi S, Brown PW, Allcock HR, Laurencin CT (2005) Low temperature formation of hydroxyapatite-poly(alkyl oxybenzoate) phosphazene composites for biomedical applications. Biomaterials 26:1–9

Griesser HJ (1991) degradation of polyurethanes in biomedical applications - a review. Polym Degrad Stab 33:329–354

Griffith LG (2002) Emerging design principles in biomaterials and scaffolds for tissue engineering. Ann N Y Acad Sci 961:83–95

Guan LM, Davies JE (2004) Preparation and characterization of a highly macroporous biodegradable composite tissue engineering scaffold. J Biomed Mater Res A 71A:480–487

Guan JJ, Wagner WR (2005) Synthesis, characterization and cytocompatibility of polyurethaneurea elastomers with designed elastase sensitivity. Biomacromolecules 6:2833–2842. doi:10.1021/bm0503322

Guan J, Fujimoto KL, Sacks MS, Wagner WR (2005) Preparation and characterization of highly porous, biodegradable polyurethane scaffolds for soft tissue applications. Biomaterials 26:3961–3971

Guelcher SA (2008) Biodegradable Polyurethanes: Synthesis and Applications in Regenerative Medicine. Tissue Eng Part B Rev 14:3–17. doi:10.1089/teb.2007.0133

Gunatillake PA, Adhikari R, Gadegaard N (2003a) Biodegradable synthetic polymers for tissue engineering. Eur Cell Mater 5:1–16

Gunatillake PA, Martin DJ, Meijs GF, McCarthy SJ, Adhikari R (2003b) Designing biostable polyurethane elastomers for biomedical implants. ChemInform. doi:10.1002/chin.200341282

Gupta D, Venugopal J, Mitra S, Dev VRG, Ramakrishna S (2009) Nanostructured biocomposite substrates by electrospinning and electrospraying for the mineralization of osteoblasts. Biomaterials 30:2085–2094. doi:10.1016/j.biomaterials.2008.12.079

Hammerle CHF, Olah AJ, Schmid J, Fluckiger L, Gogolewski S, Winkler JR, Lang NP (1997) The biological effect of natural bone mineral on bone neoformation on the rabbit skull. Clin Oral Implants Res 8:198–207

Hedberg EL, Shih CK, Lemoine JJ, Timmer MD, Liebschner MA K, Jansen JA, Mikos AG (2005) In vitro degradation of porous poly(propylene fumarate)/poly(DL-lactic-co- glycolic acid) composite scaffolds. Biomaterials 26:3215–3225

Heidemann W, Jeschkeit S, Ruffieux K, Fischer JH, Wagner M, Krüger G, Wintermantel E, Gerlach KL (2001) Degradation of poly(D, L)actide implants with or without addition of calciumphosphates in vivo. Biomaterials 22:2371–2381

Hench LL (1991) Bioceramics: from concept to clinic. J Am Ceram Soc 74:1487–1510

Hench LL (1997) Sol–gel materials for bioceramic applications. Curr Opin Solid State Mater Sci 2:604–610

Hench LL (1998) Bioceramics. J Am Ceram Soc 81:1705–1728

Hench LL (1999) Bioactive glasses and glasses-ceramics. In: Shackelford JF (ed) Bioceramics -applications of ceramic and glass materials in medicine. Trans Tech Publication, Switzerland, pp 37–64

Hench LL, Kokubo T (1998) Properties of bioactive glasses and glass-ceramics. In: Black J, Hastings G (eds) Handbook of biomaterial properties. Chapman & Hall, London, pp 355–363

Hench LL, Wilson J (1984) Surface-active biomaterials. Science 226:630–636

Hench LL, Wilson J (1993) An introduction to bioceramics. World Scientific, Singapore

Hench LL, Wilson J (1999) An introduction to bioceramics, 2nd edn. Word Scientific, London

Hench LL, Splinter RJ, Allen WC, Greenlee TK (1971) Bonding mechanisms at the interface of ceramic prosthetic materials. J Biomed Mater Res 2:117–141

Hiki S, Miyamoto M, Kimura Y (2000) Synthesis and characterization of hydroxy-terminated [RS]-poly(3-hydroxybutyrate) and its utilization to block copolymerization with -lactide to obtain a biodegradable thermoplastic elastomer. Polymer 41:7369–7379

Holand W, Vogel W (1993) Mechinable and phosphate glass-ceramics. In: Hench LL, Wilson J (eds) An introduction to bioceramics. World Scientific, Singapore, pp 125–137

Hollinger JO, Battistone GC (1986) Biodegradable bone repair materials - synthetic-polymers and ceramics. Clin Orthop Relat Res :290–305

Hong Y, Guan JJ, Fujimoto KL, Hashizume R, Pelinescu AL, Wagner WR (2010) Tailoring the degradation kinetics of poly(ester carbonate urethane)urea thermoplastic elastomers for tissue engineering scaffolds. Biomaterials 31:4249–4258. doi:10.1016/j.biomaterials.2010.02.005

Hoppe A, Guldal NS, Boccaccini AR (2011) A review of the biological response to ionic dissolution products from bioactive glasses and glass-ceramics. Biomaterials 32:2757–2774. doi:10.1016/j.biomaterials.2011.01.004

Hu GF (1998) Copper stimulates proliferation of human endothelial cells under culture. J Cell Biochem 69:326–335. doi:10.1002/(sici)1097-4644(19980601)69:3<326::aid-jcb10>3.0.co;2-a

Huang WH, Day DE, Kittiratanapiboon K, Rahaman MN (2006a) Kinetics and mechanisms of the conversion of silicate (45 S5), borate, and borosilicate glasses to hydroxyapatite in dilute phosphate solutions. J Mater Sci Mater Med 17:583–596

Huang W, Rahaman MN, Day DE, Li Y (2006b) Mechanisms for converting bioactive silicate, borate, and borosilicate glasses to hydroxyapatite in dilute phosphate solution. Phys Chem Glass Eur J Glass Sci Technol Part B 47:647–658

Huang WH, Rahaman MN, Day DE (2007) Conversion of bioactive silicate (45 S5) borate, and borosilicate glasses to hydroxyapatite in dilute phosphate solution. In: Mizuno M (ed) Advances in bioceramics and biocomposites II. Wiley, New Jersey, pp 131–140

Iroh JO (1999) Poly(epsilon-caprolactone). In: Mark JE (ed) Polymer data handbook. Oxford Press, USA, pp 361–362

Jagur-Grodzinski J (1999) Biomedical application of functional polymers. React Funct Polym 39:99–138

Jarcho M, Kay JF, Gumaer KI, Doremus RH, Drobeck HP (1977) Tissue, cellular and subcellular events at a bone-ceramic hydroxylapatite interface. J Bioeng 1:79–92

Jarcho M (1981) Calcium-phosphate ceramics as hard tissue prosthetics. Clin Orthop Relat Res :259–278

Jayabalan M, Lizymol PP, Thomas V (2000) Synthesis of hydrolytically stable low elastic modulus polyurethane-urea for biomedical applications. Polymer Int 49:88–92. doi:10.1002/(sici)1097-0126(200001)49:1<88::aid-pi298>3.0.co;2-7

Jiang G, Evans ME, Jones IA, Rudd CD, Scotchford CA, Walker GS (2005) Preparation of poly(ε-caprolactone)/continuous bioglass fibre composite using monomer transfer moulding for bone implant. Biomaterials 26:2281–2288

Jones JR, Hench LL (2003a) Factors affecting the structure and properties of bioactive foam scaffolds for tissue engineering. J Biomed Mater Res 68 B:36–44

Jones JR, Hench LL (2003b) Regeneration of trabecular bone using porous ceramics. Curr Opin Solid State Mater Sci 7:301–307

Juhasz JA, Best SM, Bonfield W, Kawashita M, Miyata N, Kokubo T, Nakamura T (2003a) Apatite-forming ability of glass-ceramic apatite-wollastonite - polyethylene composites: effect of filler content. J Mater Sci Mater Med 14:489–495

Juhasz JA, Best SM, Kawashita M, Miyata N, Kokubo T, Nakamura T, Bonfield W (2003b) Mechanical properties of glass-ceramic A-W - polyethylene composites: effect of filler content. Bioceramics 15:947–950, TRANS TECH PUBLICATIONS LTD, Zurich-Uetikon

Juhasz JA, Best SM, Brooks R, Kawashita M, Miyata N, Kokubo T, Nakamura T, Bonfield W (2004) Mechanical properties of glass-ceramic A-W-polyethylene composites: Effect of filler content and particle size. Biomaterials 25:949–955

Kakar S, Einhorn TA (2008) Biology and enhancement of skeletal repair. In: Browner BD, Levine AM, Jupiter JB, Trafton PG, Krettek C (eds) Skeletal trauma: basic science, management, and reconstruction. Saunders, Elsevier, Philadelphia, pp 33–50

Kalangos A, Faidutti B (1996) Preliminary clinical results of implantation of biodegradable pericardial substitute in pediatric open heart operations. J Thorac Cardiovasc Surg 112:1401–1402

Karageorgiou V, Kaplan D (2005) Porosity of 3D biomaterial scaffolds and osteogenesis. Biomaterials 26:5474–5491

Kasuga T, Ota Y, Nogami M, Abe Y (2001) Preparation and mechanical properties of polylactic acid composites containing hydroxyapatite fibers. Biomaterials 22:19–23

Kasuga T, Maeda H, Kato K, Nogami M, Hata K, Ueda M (2003) Preparation of poly (lactic acid) composites containing calcium carbonate (vaterite). Biomaterials 24:3247–3253

Kaufmann EABE, Ducheyne P, Shapiro IM (2000) Evaluation of osteoblast response to porous bioactive glass (45 S5) substrates by RT-PCR analysis. Tissue Eng 6:19–28

Kavlock KD, Pechar TW, Hollinger JO, Guelcher SA, Goldstein AS (2007) Synthesis and characterization of segmented poly(esterurethane urea) elastomers for bone tissue engineering. Acta Biomater 3:475–484. doi:10.1016/j.actbio.2007.02.001

Keaveny TM, Hayes WC (1993) Mechanical properties of cortical and trabecular bone. In: Hall BK (ed) Bone. A treatise, volume 7: bone growth. CRC Press, Boca Raton, FL, pp 285–344

Kellomäki M, Heller J, Törmälä P (2000) Processing and properties of two different poly (ortho esters). J Mater Sci Mater Med 11:345–355

Kemppainen JM, Hollister SJ (2010) Tailoring the mechanical properties of 3D-designed poly(glycerol sebacate) scaffolds for cartilage applications. J Biomed Mater Res A 94:9–18

Keshaw H, Forbes A, Day RM (2005) Release of angiogenic growth factors from cells encapsulated in alginate beads with bioactive glass. Biomaterials 26:4171–4179

Keun Kwon I, Kidoaki S, Matsuda T (2005) Electrospun nano- to microfiber fabrics made of biodegradable copolyesters: structural characteristics, mechanical properties and cell adhesion potential. Biomaterials 26:3929–3939

Khan YM, Katti DS, Laurencin CT (2004) Novel polymer-synthesized ceramic composite-based system for bone repair: An in vitro evaluation. J Biomed Mater Res A 69A:728–737

Kikuchi M, Tanaka J, Koyama Y, Takakuda K (1999) Cell culture test of TCP/CPLA composite. J Biomed Mater Res 48:108–110

Kim BS, Mooney DJ (2000) Scaffolds for engineering smooth muscle under cyclic mechanical strain conditions. J Biomech Eng 122:210–215

Kim HW, Knowles JC, Kim HE (2004) Hydroxyapatite/poly(ε-caprolactone) composite coatings on hydroxyapatite porous bone scaffold for drug delivery. Biomaterials 25:1279–1287

Kohn J, Langer R (1996) Bioresorbable and bioerodible materials. In: Ratner BD, Hoffman AS, Schoen FJ, Lemons JE (eds) Biomaterials science: an introduction to materials in medicine. Academic Press, California, pp 64–73

Kokubo T (1999a) Novel biomedical materials based on glasses. In: Shackelford JF (ed) Bioceramics -applications of ceramic and glass materials in medicine. Trans Tech Publications Ltd, Switzerland, pp 65–82

Kokubo T (1999b) A/W glass-ceramic: processing and properties. In: Hench LL, Wilson J (eds) An introduction to bioceramics. Word Scientific, London, pp 75–88

Kokubo T, Kim HM, Kawashita M (2003) Novel bioactive materials with different mechanical properties. Biomaterials 24:2161–2175

Kumudine C, Premachandra JK (1999) Poly(lactic acid). In: Mark JE (ed) Polymer data handbook. Oxford Press, Oxford, pp 70–77

Kwun I-S, Cho Y-E, Lomeda R-AR, Shin H-I, Choi J-Y, Kang Y-H, Beattie JH (2010) Zinc deficiency suppresses matrix mineralization and retards osteogenesis transiently with catch-up possibly through Runx 2 modulation. Bone 46:732–741. doi:10.1016/j.bone.2009.11.003

Lai YL, Yamaguchi M (2005) Effects of copper on bone component in the femoral tissues of rats: anabolic effect of zinc is weakened by copper. Biol Pharm Bull 28:2296–2301. doi:10.1248/bpb.28.2296

Lamba NMK, Woodhouse KA, Cooper SL (1998) Polyurethanes in biomedcial applications. CRC Press, Boca Raton, USA

Lang C, Murgia C, Leong M, Tan L-W, Perozzi G, Knight D, Ruffin R, Zalewski P (2007) Anti-inflammatory effects of zinc and alterations in zinc transporter mRNA in mouse models of allergic inflammation. Am J Physiol Lung Cell Mol Physiol 292:L577–L584. doi:10.1152/ajplung.00280.2006

Laurencin CT, Norman ME, Elgendy HM, Elamin SF, Allcock HR, Pucher SR, Ambrosio AA (1993) Use of polyphosphazenes for skeletal tissue regeneration. J Biomed Mater Res 27:963–973

Laurencin CT, Attawia MA, Elgendy HE, Herbert KM (1996a) Tissue engineered bone-regeneration using degradable polymers: the formation of mineralized matrices. Bone 19:S93–S99

Laurencin CT, ElAmin SF, Ibim SE, Willoughby DA, Attawia M, Allcock HR, Ambrosio AA (1996b) A highly porous 3-dimensional polyphosphazene polymer matrix for skeletal tissue regeneration. J Biomed Mater Res 30:133–138

Laurencin CT, Lu HH, Khan Y (2002) Processing of polymer scaffolds: polymer-ceramic composite foams. In: Atala A, Lanza RP (eds) Methods of tissue engineering. Academic Press, California, pp 705–714

Lee S-H, Kim B-S, Kim SH, Choi SW, Jeong SI, Kwon IK, Kang SW, Nikolovski J, Mooney DJ, Han Y-K, Kim YH (2003) Elastic biodegradable poly(glycolide-co-caprolactone) scaffold for tissue engineering. J Biomed Mater Res A 66A:29–37. doi:10.1002/jbm.a.10497

Lefebvre LP, Banhart J, Dunand DC (2008) Porous metals and metallic foams: current status and recent developments. Adv Eng Mater 10:775–787. doi:10.1002/adem.200800241

LeGeros RZ, LeGeros JP (1999) Dense hydroxyapatite. In: Hench LL, Wilson J (eds) An introduction to bioceramics. Word Scientific, London, pp 139–180

LeGeros RZ, LeGeros JP (2002) Calcium phosphate ceramics: past, present and future. In: Bioceramics 15: proceedings of the 15th international symposium on ceramics in medicine, Sydney, 4–8 December 2002

Levenberg S, Langer R (2004) Advances in tissue engineering. Curr Top Dev Biol 61:113

Li H, Chang J (2004) Preparation and characterization of bioactive and biodegradable Wollastonite/poly(D, L-lactic acid) composite scaffolds. J Mater Sci Mater Med 15:1089–1095

Li P, Yang Q, Zhang F, Kokubo T (1992) The effect of residual glassy phase in a bioactive glass-ceramic on the formation of its surface apatite layer in vitro. J Mater Sci Mater Med 3:452–456

Li H, Du R, Chang J (2005) Fabrication, characterization, and in vitro degradation of composite scaffolds based on PHBV and bioactive glass. J Biomater Appl 20:137–155

Li Y, Thouas GA, Chen QZ (2012) Biodegradable soft elastomers: synthesis/properties of materials and fabrication of scaffolds. RSC Advances. 2:8229–8242 doi:10.1039/c2ra20736b

Liang S-L, Yang X-Y, Fang X-Y, Cook WD, Thouas GA, Chen Q-Z (2011) In vitro enzymatic degradation of poly (glycerol sebacate)-based materials. Biomaterials 32:8486–8496. doi:10.1016/j.biomaterials.2011.07.080

Liang SL, Cook WD, Thouas GA, Chen QZ (2010) The mechanical characteristics and in vitro biocompatibility of poly(glycerol sebacate)-Bioglass (R) elastomeric composites. Biomaterials 31:8516–8529. doi:10.1016/j.biomaterials.2010.07.105

Liljensten E, Gisselfalt K, Edberg B, Bertilsson H, Flodin P, Nilsson A, Lindahl A, Peterson L (2002) Studies of polyurethane urea bands for ACL reconstruction. J Mater Sci Mater Med 13:351–359

Livingston T, Ducheyne P, Garino J (2002) In vivo evaluation of a bioactive scaffold for bone tissue engineering. J Biomed Mater Res 62:1–13

Lobel KD, Hench LL (1996) In-vitro protein interactions with a bioactive gel-glass. Journal of Sol-gel. Sci Technol 7:69–76

Lobel KD, Hench LL (1998) In vitro adsorption and activity of enzymes on reaction layers of bioactive glass substrates. J Biomed Mater Res 39:575–579

Lu HH, El-Amin SF, Scott KD, Laurencin CT (2003) Three-dimensional, bioactive, biodegradable, polymer-bioactive glass composite scaffolds with improved mechanical properties support collagen synthesis and mineralization of human osteoblast-like cells in vitro. J Biomed Mater Res A 64A:465–474

Lu HH, Tang A, Oh SC, Spalazzi JP, Dionisio K (2005) Compositional effects on the formation of a calcium phosphate layer and the response of osteoblast-like cells on polymer-bioactive glass composites. Biomaterials 26:6323–6334

Lu L, Mikos AG (1999) Poly(lactic acid). In: Mark JE (ed) Polymer data handbook. Oxford Press, Oxford, pp 627–633

Luginbuehl V, Meinel L, Merkle HP, Gander B (2004) Localized delivery of growth factors for bone repair. Eur J Pharm Biopharm 58:197–208

Ma LJ, Wu YQ (2007) FTIR spectroscopic study on the interaction between a fluoroionophore and metal ions. Anal Sci 23:799–802

Magill JH (1999) Poly(phosphazenes), bioerodible. In: Mark JE (ed) Polymer data handbook. Oxford Press, Oxford, pp 746–749

Mano J, Sousa RA, Boesel LF, Neves NM, Reis RL (2004) Bioinert, biodegradable and injectable polymeric matrix composites for hard tissue replacement: State of the art and recent developments. Compos Sci Technol 64:789–817

Maquet V, Boccaccini AR, Pravata L, Notingher I, Jerome R (2003) Preparation, characterization, and in vitro degradation of bioresorbable and bioactive composites based on Bioglass (R)-filled polylactide foams. J Biomed Mater Res A 66A:335–346

Maquet V, Boccaccini AR, Pravata L, Notingher I, Jerome R (2004) Porous poly (alpha-hydroxyacid)/Bioglass (R) composite scaffolds for bone tissue engineering. I: preparation and in vitro characterisation. Biomaterials 25:4185–4194

Marcacci M, Kon E, Moukhachev V, Lavroukov A, Kutepov S, Quarto R, Mastrogiacomo M, Cancedda R (2007) Stem cells associated with macroporous bioceramics for long bone repair: 6- to 7-year outcome of a pilot clinical study. Tissue Eng 13:947–955

Marie PJ (2010) The calcium-sensing receptor in bone cells: A potential therapeutic target in osteoporosis. Bone 46:571–576. doi:10.1016/j.bone.2009.07.082

Marie PJ, Ammann P, Boivin G, Rey C (2001) Mechanisms of action and therapeutic potential of strontium in bone. Calcif Tissue Int 69:121–129. doi:10.1007/s002230010055

Marion NW, Liang W, Reilly GC, Day DE, Rahaman MN, Mao JJ (2005) Borate glass supports the in vitro osteogenic differentiation of human mesenchymal stem cells. Mech Adv Mater Struct 12:239–246. doi:10.1080/15376490590928615

Martin C, Winet H, Bao JY (1996) Acidity near eroding polylactide-polyglycolide in vitro and in vivo in rabbit tibial bone chambers. Biomaterials 17:2373–2380

Martin DP, Skraly FA, Williams SF (1999) Polyhydroxy alkanoate compositions having controlled degradation rates., PCTPatent Application No. WO 99/32536

Martin DP, Williams SF (2003) Medical applications of poly-4-hydroxybutyrate: a strong flexible absorbable biomaterial. Biochem Eng J 16:97–105. doi:10.1016/s1369-703x(03)00040-8

Matsumura G, Hibino N, Ikada Y, Kurosawa H, Shin'oka T (2003a) Successful application of tissue engineered vascular autografts: clinical experience. Biomaterials 24:2303–2308

Matsumura G, Miyagawa-Tomita S, Shin'oka T, Ikada Y, Kurosawa H (2003b) First evidence that bone marrow cells contribute to the construction of tissue-engineered vascular autografts in vivo. Circulation 108:1729–1734

McDevitt TC, Woodhouse KA, Hauschka SD, Murry CE, Stayton PS (2003) Spatially organized layers of cardiomyocytes on biodegradable polyurethane films for myocardial repair. J Biomed Mater Res A 66A:586–595. doi:10.1002/jbm.a.10504

Meunier PJ, Slosman DO, Delmas PD, Sebert JL, Brandi ML, Albanese C, Lorenc R, Pors-Nielsen S, de Vernejoul MC, Roces A, Reginster JY (2002) Strontium ranelate: dose-dependent effects in established postmenopausal vertebral osteoporosis - a 2-year randomized placebo controlled trial. J Clin Endocrinol Metab 87:2060–2066. doi:10.1210/jc.87.5.2060

Middleton JC, Tipton AJ (2000) Synthetic biodegradable polymers as orthopedic devices. Biomaterials 21:2335–2346. doi:Cited By (since 1996) 582

Mikos AG, Temenoff JS (2000) Formation of highly porous biodegradable scaffolds for tissue engineering. Electron J Biotechnol 3:114–119

Moore WR, Graves SE, Bain GI (2001) Synthetic bone graft substitutes. Aust N Z J Surg 71:354–361

Motlagh D, Yang J, Lui KY, Webb AR, Ameer GA (2006) Hemocompatibility evaluation of poly(glycerol-sebacate) in vitro for vascular tissue engineering. Biomaterials 27:4315–4324

Nalla RK, Kinney JH, Ritchie RO (2003) Mechanistic fracture criteria for the failure of human cortical bone. Nat Mater 2:164–168

Natah S, Hussien K, Tuominen J, Koivisto V (1997) Metabolic response to lactitol and xylitol in healthy men. Am J Clin Nutr 65:947–950

Navarro M, Ginebra MP, Planell JA, Zeppetelli S, Ambrosio L (2004) Development and cell response of a new biodegradable composite scaffold for guided bone regeneration. J Mater Sci Mater Med 15:419–422

Nielsen FH (2008) Is boron nutritionally relevant? Nutr Rev 66:183–191. doi:10.1111/j.1753-4887.2008.00023.x

Niiranen H, Pyhältö T, Rokkanen P, Kellomäki M, Törmälä P (2004) In vitro and in vivo behavior of self-reinforced bioabsorbable polymer and self-reinforced bioabsorbable polymer/bioactive glass composites. J Biomed Mater Res A 69:699–708

Niklason LE, Gao J, Abbott WM, Hirschi KK, Houser S, Marini R, Langer R (1999) Functional arteries grown in vitro. Science 284:489–493. doi:10.1126/science.284.5413.489

Ohgushi H, Dohi Y, Yoshikawa T, Tamai S, Tabata S, Okunaga K, Shibuya T (1996) Osteogenic differentiation of cultured marrow stromal stem cells on the surface of bioactive glass ceramics. J Biomed Mater Res 32:341–348

Oliveira SM, Mijares DQ, Turner G, Amaral IF, Barbosa MA, Teixeira CC (2009) Engineering endochondral bone: in vivo studies. Tissue Eng Part A 15:635–643. doi:10.1089/ten.tea.2008.0052

Payne RG, Mikos AG (2002) Synthesis of synthetic polymers: poly(propylene fumarate). In: Atala A, Lanza RP (eds) Methods of tissue engineering. Academic Press, California, pp 649–652

Pereira MM, Clark AE, Hench LL (1994) Calcium phosphate formation on sol–gel-derived bioactive glasses in vitro. J Biomed Mater Res 28:693–698

Peter SJ, Miller ST, Zhu G, Yasko AW, Mikos AG (1998) In vivo degradation of a poly(propylene fumarate)/β-tricalcium phosphate injectable composite scaffold. J Biomed Mater Res 41:1–7

Peter SJ, Lu L, Kim DJ, Stamatas GN, Miller MJ, Yaszemski MJ, Mikos AG (2000) Effects of transforming growth factor β1 released from biodegradable polymer microparticles on marrow stromal osteoblasts cultured on poly (propylene fumarate) substrates. J Biomed Mater Res 50:452–462

Pinchuk L (1994) A review of the biostability and carcinogenicity of polyurethanes in medicine and the new generation of 'biostable' polyurethanes. J Biomater Sci Polym Ed 6:225–267

Pitt CG, Gratzl MM, Kimmel GL (1981) Aliphatic polyesters II. The degradation of poly (DL-lactide), poly (ε-caprolactone), and their copolymers in vivo. Biomaterials 2:215–220

Pomerantseva I, Krebs N, Hart A, Neville CM, Huang AY, Sundback CA (2009) Degradation behavior of poly(glycerol sebacate). J Biomed Mater Res A 91A:1038–1047. doi:10.1002/jbm.a.32327

Rahaman MN, Day DE, Bal BS, Fu Q, Jung SB, Bonewald LF, Tomsia AP (2011) Bioactive glass in tissue engineering. Acta Biomater 7:2355–2373. doi:10.1016/j.actbio.2011.03.016

Ramakrishna S, Huang ZM, Kumar GV, Batchelor AW, Mayer J (2004) An introduction to biocomposites. World Scientific, Singapore

Rao U, Kumar R, Balaji S, Sehgal PK (2010) A novel biocompatible poly (3-hydroxy-co-4-hydroxybutyrate) blend as a potential biomaterial for tissue engineering. J Bioactive Compatible Polymers 25:419–436. doi:10.1177/0883911510369037

Redenti S, Neeley WL, Rompani S, Saigal S, Yang J, Klassen H, Langer R, Young MJ (2009) Engineering retinal progenitor cell and scrollable poly(glycerol-sebacate) composites for expansion and subretinal transplantation. Biomaterials 30:3405–3414. doi:10.1016/j.biomaterials.2009.02.046

Rezwan K, Chen QZ, Blaker JJ, Boccaccini AR (2006) Biodegradable and bioactive porous polymer/inorganic composite scaffolds for bone tissue engineering. Biomaterials 27:3413–3431. doi:10.1016/j.biomaterials.2006.01.039

Rich J, Jaakkola T, Tirri T, Närhi T, Yli-Urpo A, Seppälä J (2002) In vitro evaluation of poly(ε-caprolactone-co-DL-lactide)/bioactive glass composites. Biomaterials 23:2143–2150

Rodrigues CVM, Serricella P, Linhares ABR, Guerdes RM, Borojevic R, Rossi MA, Duarte MEL, Farina M (2003) Characterization of a bovine collagen-hydroxyapatite composite scaffold for bone tissue engineering. Biomaterials 24:4987–4997

Rodriguez JP, Rios S, Gonzalez M (2002) Modulation of the proliferation and differentiation of human mesenchymal stem cells by copper. J Cell Biochem 85:92–100. doi:10.1002/jcb.10111

Roether JA, Gough JE, Boccaccini AR, Hench LL, Maquet V, Jerome R (2002) Novel bioresorbable and bioactive composites based on bioactive glass and polylactide foams for bone tissue engineering. J Mater Sci Mater Med 13:1207–1214

Saad B, Hirt TD, Welti M, Uhlschmid GK, Neuenschwander P, Suter UW (1997) Development of degradable polyesterurethanes for medical applications: In vitro and in vivo evaluations. J Biomed Mater Res 36:65–74

Santerre JP, Woodhouse K, Laroche G, Labow RS (2005) Understanding the biodegradation of polyurethanes: From classical implants to tissue engineering materials. Biomaterials 26:7457–7470. doi:0.1016/j.biomaterials.2005.05.079

Schepers E, de Clercq M, Ducheyne P, Kempeneers R (1991) Bioactive glass particulate material as a filler for bone lesions. J Oral Rehabil 18:439–452

Seal BL, Otero TC, Panitch A (2001) Polymeric biomaterials for tissue and organ regeneration. Mater Sci Eng R Rep 34:147–230

Seeley RR, Stephens TD, Rate P (2006) Anatomy and physiology, 8th edn. McGrew Hill, New York

Seliktar D, Nerem RM, Galis ZS (2003) Mechanical strain-stimulated remodeling of tissue-engineered blood vessel constructs. Tissue Eng 9:657–666. doi:10.1089/107632703768247359

Sestoft L (1985) An evaluation of biochemical aspects of intravenous fructose, sorbitol and xylitol administration in man. Acta Anaesthesiol Scand 29:19–29. doi:10.1111/j.1399-6576.1985.tb02336.x

Shastri VP, Zelikin A, Hildgen P (2002) Synthesis of synthetic polymers: poly (anhydrides). In: Atala A, Lanza RP (eds) Methods of tissue engineering. Academic Press, California, pp 609–617

Shum-Tim D, Stock U, Hrkach J, Shinoka T, Lien J, Moses MA, Stamp A, Taylor G, Moran AM, Landis W, Langer R, Vacanti JP, Mayer JE (1999) Tissue engineering of autologous aorta using a new biodegradable polymer. Ann Thorac Surg 68:2298–2304

Skalak R, Fox CF (1993) Tissue engineering. In: Proceedings of a workshop, Granlibakken, Lake Tahoe, California, 26–29 February 1988., pp 26–29

Smith BJ, King JB, Lucas EA, Akhter MP, Arjmandi BH, Stoecker BJ (2002) Skeletal unloading and dietary copper depletion are detrimental to bone quality of mature rats. J Nutr 132:190–196

Solheim E, Sudmann B, Bang G, Sudmann E (2000) Biocompatibility and effect on osteogenesis of poly(ortho ester) compared to poly(DL-lactic acid). J Biomed Mater Res 49:257–263

Spaans CJ, JH Dg, Belgraver VW, Pennings AJ (1998a) A new biomedical polyurethane with a high modulus based on 1,4-butanediisocyanate and ε-caprolactone. J Mater Sci Mater Med 9:675–678. doi:10.1023/a:1008922128455

Spaans CJ, de Groot JH, Dekens FG, Pennings AJ (1998b) High molecular weight polyurethanes and a polyurethane urea based on 1,4-butanediisocyanate. Polymer Bull 41:131–138. doi:10.1007/s002890050343

Stamboulis AG, Boccaccini AR, Hench LL (2002) Novel biodegradable polymer/bioactive glass composites for tissue engineering applications. Adv Eng Mater 4, 105-109+183

Stankus JJ, Guan J, Wagner WR (2004) Fabrication of biodegradable elastomeric scaffolds with sub-micron morphologies. J Biomed Mater Res A 70:603–614

Stankus JJ, Soletti L, Fujimoto K, Hong Y, Vorp DA, Wagner WR (2007) Fabrication of cell microintegrated blood vessel constructs through electrohydrodynamic atomization. Biomaterials 28:2738–2746

Stegemann JP, Nerem RM (2003) Phenotype modulation in vascular tissue engineering using biochemical and mechanical stimulation. Ann Biomed Eng 31:391–402. doi:10.1114/1.1558031

Stuckey DJ, Ishii H, Chen QZ, Boccaccini AR, Hansen U, Carr CA, Roether JA, Jawad H, Tyler DJ, Ali NN, Clarke K, Harding SE (2010) Magnetic resonance imaging evaluation of remodeling by cardiac elastomeric tissue scaffold biomaterials in a rat model of myocardial infarction. Tissue Eng Part A 16:3395–3402. doi:10.1089/ten.tea.2010.0213

Sudesh K, Doi Y (2005) Polyhydroxyalkanoates. In: Bastioli C (ed) Handbook of biodegradable polymers. Rapra Technology Limited, Shawbury, UK, pp 219–256

Sundback CA, Shyu JY, Wu AJ, Sheahan TP, Wang YD, Faquin WC, Langer RS, Vacanti JP, Hadlock TA (2004) In vitro and in vivo biocompatibility analysis of poly (glycerol sebacate) as a potential nerve guide material. Arch Appl Biomater Biomolec Mater 1:37–39

Sundback CA, Shyu JY, Wang YD, Faquin WC, Langer RS, Vacanti JP, Hadlock TA (2005) Biocompatibility analysis of poly(glycerol sebacate) as a nerve guide material. Biomaterials 26:5454–5464. doi:10.1016/j.biomaterials.2005.02.004

Szycher M (1999) Szycher's handbook of polyurethanes. CRC Press, Boca Raton

Szycher M, Reed AM, Soc Plast Engineers INC (1996) Medical-grade polyurethanes: a critical review. In: SPE/ANTEC 1996 Proceedings, vol 1–3, pp 2758–2766

Talke H, Maier KP (1973) Zum Metabolismus von Glukose, Fruktose, Sorbit und Xylit beim Menschen. Transfus Med Hemother 1:49–56

Temenoff JS, Lu L, Mikos AG (2000) Bone tissue engineering using synthetic biodegradable polymer scaffolds. In: Davies JE (ed) Bone engineering. EM Squared, Toronto, pp 455–462

Tirelli N, Lutolf MP, Napoli A, Hubbell JA (2002) Poly(ethylene glycol) block copolymers. J Biotechnol 90:3–15

Uysal T, Ustdal A, Sonmez MF, Ozturk F (2009) Stimulation of bone formation by dietary boron in an orthopedically expanded suture in rabbits. Angle Orthod 79:984–990. doi:10.2319/112708-604.1

Vacanti CA, Bonassar LJ, Vacanti JP (2000) Structure tissue engineering. In: Lanza RP, Langer R, Vacanti JP (eds) Principles of tissue engineering. Academic Press, California, pp 671–682

Verrier S, Blaker JJ, Maquet V, Hench LL, Boccaccini AR (2004) PDLLA/Bioglass (R) composites for soft-tissue and hard-tissue engineering: an in vitro cell biology assessment. Biomaterials 25:3013–3021

Waldman SD, Spiteri CG, Grynpas MD, Pilliar RM, Kandel RA (2004) Long-term intermittent compressive stimulation improves the composition and mechanical properties of tissue-engineered cartilage. Tissue Eng 10:1323–1331. doi:10.1089/ten.2004.10.1323

Wang Y (2004) Biorubber/poly(glycerol sebacate). Informa Healthcare, London, pp 121–128

Wang YD, Ameer GA, Sheppard BJ, Langer R (2002a) A tough biodegradable elastomer. Nat Biotechnol 20:602–606. doi:10.1038/nbt0602-602

Wang YD, Sheppard BJ, Langer R (2002b) Poly(glycerol sebacate) - a novel biodegradable elastomer for tissue engineering. Biol Biomim Mater Properties Funct 724:223–227

Wang XP, Li X, Ito A, Sogo Y (2011) Synthesis and characterization of hierarchically macroporous and mesoporous CaO-MO-SiO(2)-P(2)O(5) (M = Mg, Zn, Sr) bioactive glass scaffolds. Acta Biomater 7:3638–3644. doi:10.1016/j.actbio.2011.06.029

Wang YD, Kim YM, Langer R (2003) In vivo degradation characteristics of poly (glycerol sebacate). J Biomed Mater Res A 66A:192–197

Wang YW, Wu QO, Chen GQ (2004) Attachment, proliferation and differentiation of osteoblasts on random biopolyester poly(3-hydroxybutyrate-co-3-hydroxyhexanoate) scaffolds. Biomaterials 25:669–675. doi:10.1016/s0142-9612(03)00561-1

Webb AR, Yang J, Ameer GA (2004) Biodegradable polyester elastomers in tissue engineering. Expert Opin Biol Ther 4:801–812. doi:10.1517/14712598.4.6.801

Whitney EN, Rolfes SR (2010) Understanding nutrition. Wadsworth Publishing, Belmont

Wilson J, Pigott GH, Schoen FJ, Hench LL (1981) Toxicology and biocompatibility of bioglasses. J Biomed Mater Res 15:805–817

Winkelhausen E, Kuzmanova S (1998) Microbial conversion of -xylose to xylitol. J Ferment Bioeng 86:1–14

Wong CT, Chen QZ, Lu WW, Leong JCY, Chan WK, Cheung KMC, Luk KDK (2004) Ultrastructural study of mineralization of a strontium-containing hydroxyapatite (Sr-HA) cement in vivo. J Biomed Mater Res A 70A:428–435. doi:10.1002/jbm.a.30097

Xu HHK, Quinn JB, Takagi S, Chow LC (2004) Synergistic reinforcement of in situ hardening calcium phosphate composite scaffold for bone tissue engineering. Biomaterials 25:1029–1037

Xu HHK, Simon CG (2004a) Self-hardening calcium phosphate cement-mesh composite: reinforcement, macropores, and cell response. J Biomed Mater Res A 69A:267–278

Xu HHK, Simon CG Jr (2004b) Self-hardening calcium phosphate composite scaffold for bone tissue engineering. J Orthopaedic Res 22:535–543

Xynos ID, Edgar AJ, Buttery LDK, Hench LL, Polak JM (2000a) Ionic products of bioactive glass dissolution increase proliferation of human osteoblasts and induce insulin-like growth factor II mRNA expression and protein synthesis. Biochem Biophys Res Commun 276:461–465

Xynos ID, Hukkanen MVJ, Batten JJ, Buttery LD, Hench LL, Polak JM (2000b) Bioglass ®45 S5 stimulates osteoblast turnover and enhances bone formation in vitro: implications and applications for bone tissue engineering. Calcif Tissue Int 67:321–329

Xynos ID, Edgar AJ, Buttery LDK, Hench LL, Polak JM (2001) Gene-expression profiling of human osteoblasts following treatment with the ionic products of Bioglass® 45 S5 dissolution. J Biomed Mater Res 55:151–157

Yamaguchi M (1998) Role of zinc in bone formation and bone resorption. J Trace Elem Exp Med 11:119–135. doi:10.1002/(sici)1520-670x(1998), 11:2/3<119::aid-jtra5>3.0.co;2-3

Yang S, Leong KF, Du Z, Chua CK (2001) The design of scaffolds for use in tissue engineering. Part I. Traditional factors. Tissue Eng 7:679–689

Yang M, Zhu SS, Chen Y, Chang ZJ, Chen GQ, Gong YD, Zhao NM, Zhang XF (2004) Studies on bone marrow stromal cells affinity of poly (3-hydroxybutyrate-co-3-hydroxyhexanoate). Biomaterials 25:1365–1373. doi:10.1016/j.biomaterials.2003.08.018

Yao J, Radin SS, Leboy P, Ducheyne P (2005) The effect of bioactive glass content on synthesis and bioactivity of composite poly (lactic-co-glycolic acid)/bioactive glass substrate for tissue engineering. Biomaterials 26:1935–1943

Yao A, Wang D, Huang W, Fu Q, Rahaman MN, Day DE (2007) In vitro bioactive characteristics of borate-based glasses with controllable degradation behavior. J Am Ceram Soc 90:303–306. doi:10.1111/j.1551-2916.2006.01358.x

Yang X, Zhang L, Chen X, Sun X, Yang G, Guo X, Yang H, Gao C, Gou Z (2012) Incorporation of B2O3 in CaO-SiO2-P2O5 bioactive glass system for improving strength of low-temperature co-fired porous glass ceramics. J Non-Cryst Solids 358:1171–1179. doi:10.1016/j.jnoncrysol.2012.02.005

Yaszemski MJ, Payne RG, Hayes WC, Langer RS, Aufdemorte TB, Mikos AG (1995) The ingrowth of new bone tissue and initial mechanical properties of a degrading polymeric composite scaffold. Tissue Eng 1:41–52

Yeni YN, Fyhrie DP (2001) Finite element calculated uniaxial apparent stiffness is a consistent predictor of uniaxial apparent strength in human vertebral cancellous bone tested with different boundary conditions. J Biomech 34:1649–1654

Yeni YN, Hou FJ, Vashishth D, Fyhrie DP (2001) Trabecular shear stress in human vertebral cancellous bone: intra- and inter-individual variations. J Biomech 34:1341–1346

Yin YJ, Ye F, Cui JF, Zhang FJ, Li XL, Yao KD (2003) Preparation and characterization of macroporous chitosan-gelatin beta-tricalcium phosphate composite scaffolds for bone tissue engineering. J Biomed Mater Res A 67A:844–855

Yuan H, De Bruijn JD, Zhang X, Van Blitterswijk CA, De Groot K (2001) Bone induction by porous glass ceramic made from Bioglass® (45 S5). J Biomed Mater Res 58:270–276

Zdrahala RJ (1996) Small caliber vascular grafts.2. Polyurethanes revisited. J Biomater Appl 11:37–61

Zdrahala RJ, Zdrahala IJ (1999) Biomedical applications of polyurethanes: a review of past promises, present realities, and a vibrant future. J Biomater Appl 14:67–90

Zhang RY, Ma PX (1999) Poly(alpha-hydroxyl acids) hydroxyapatite porous composites for bone-tissue engineering. I. Preparation and morphology. J Biomed Mater Res 44:446–455

Zhang J-Y, Beckman EJ, Hu J, Yang G-G, Agarwal S, Hollinger JO (2002) Synthesis, biodegradability, and biocompatibility of lysine diisocyanate-glucose polymers. Tissue Eng 8:771–785. doi:10.1089/10763270260424132

Zhang JC, Huang JA, Xu SJ, Wang K, Yu SF (2003) Effects of Cu2+ and pH on osteoclastic bone resorption in vitro. Prog Nat Sci 13:266–2003. doi:10.1080/10020070312331343510

Zhang K, Wang Y, Hillmyer MA, Francis LF (2004) Processing and properties of porous poly(L-lactide)/bioactive glass composites. Biomaterials 25:2489–2500

Zhang X, Jia W, Gua Y, Wei X, Liu X, Wang D, Zhang C, Huang W, Rahaman MN, Day DE, Zhou N (2010) Teicoplanin-loaded borate bioactive glass implants for treating chronic bone infection in a rabbit tibia osteomyelitis model. Biomaterials 31:5865–5874. doi:10.1016/j.biomaterials.2010.04.005

Zhao K, Deng Y, Chen GQ (2003a) Effects of surface morphology on the biocompatibility of polyhydroxyalkanoates. Biochem Eng J 16:115–123. doi:10.1016/s1369-703x(03)00029-9

Zhao K, Deng Y, Chen JC, Chen GQ (2003b) Polyhydroxyalkanoate (PHA) scaffolds with good mechanical properties and biocompatibility. Biomaterials 24:1041–1045

Zheng Z, Deng Y, Lin XS, Zhang LX, Chen GQ (2003) Induced production of rabbit articular cartilage-derived chondrocyte collagen II on polyhydroxyalkanoate blends. J Biomater Sci Polym Ed 14:615–624

Zheng Z, Bei FF, Tian HL, Chen GQ (2005) Effects of crystallization of polyhydroxyalkanoate blend on surface physicochemical properties and interactions with rabbit articular cartilage chondrocytes. Biomaterials 26:3537–3548. doi:10.1016/j.biomaterials.2004.09.041

Zheng K, Yang SB, Wang JJ, Russel C, Liu CS, Liang W (2012) Characteristics and biocompatibility of Na(2)O-K(2)O-CaO-MgO-SrO-B(2)O(3)-P(2)O(5) borophosphate glass fibers. J Non-Cryst Solids 358:387–391. doi:10.1016/j.jnoncrysol.2011.10.004

Zilberman M, Nelson KD, Eberhart RC (2005) Mechanical properties and in vitro degradation of bioresorbable fibers and expandable fiber-based stents. J Biomed Mater Res B Appl Biomater 74B:792–799. doi:10.1002/jbm.b.30319

Zioupos P, Currey JD (1998) Changes in the stiffness, strength, and toughness of human cortical bone with age. Bone 22:57–66

5

Infrared spectroscopic analysis of restorative composite materials' surfaces and their saline extracts

Reem Ajaj[1,2*], Robert Baier[2], Jude Fabiano[3] and Peter Bush[3]

Abstract

This study aims at finding out if multiple attenuated internal reflection-infrared (MAIR-IR) spectroscopic analysis can be used as a tool to differentiate commercial resin composite brands and to find out if different resin composites will have different abilities of leaching materials that are cytotoxic to human gingival fibroblasts (HGFs) Tooth-colored resin fillings have become increasingly popular as restorative materials, which make it important to differentiate the commercial brands for forensic and biological purposes. Fourteen resin composite brands were used in the study. MAIR-IR spectroscopic analysis was used for surface characterization of the organic and inorganic parts of the resin composite samples which were studied as is and after 2 weeks of saline incubation. IR spectroscopy was also done on the saline extracts to find out if different resin composite materials would have different leaching abilities. The saline extracts were also used for the viability testing of HGF cell cultures. One-way analysis of variance test statistics was used to analyze the results. It was found that the resin composite brands have different spectra after saline soaking. It was also found that these resin composite brands possess different leaching abilities with regard to the amount and type of materials and different cytotoxic effects, which were found to be threshold dependent, meaning there is a critical or threshold value of leaching material at or above which the toxic effect will be significant and below which there is no toxic effect. Therefore, IR spectroscopy might be considered as a useful tool for dental resin composite characterization. However, more oral simulating environmental testing methods, different surface characterization methods, and more cell viability testing methods and assays must be considered for more specific results which relate more to the behavior of these dental resin composites in the oral environment.

Keywords: Multiple attenuated internal reflection-infrared, Resin composite, Human gingival fibroblasts, Forensic, Cytotoxic

Introduction

Recent literature has demonstrated how the slightly different inorganic fractions of dental resin composites maybe used for forensic identifications of unknown accident victims, but has not examined the possible additional identifying value from examination of the resinous organic fractions of these same materials. Similarly, certain dental restorative resinous materials have been implicated in providing saline-extractable components that were toxic to

human gingival fibroblast cells (HGFCs), but the identities of these extractable substances have not been revealed by analysis. Recognizing that infrared (IR) spectroscopic analysis of both the resin composites and their saline extracts could provide surface-sensitive information relevant to both the forensic and possible biotoxicity issues previously raised, this investigation set out to determine if IR spectroscopy using the multiple attenuated internal reflection (MAIR)-IR technique could serve these needs. For forensic scientists, it might add to the database collected previously using other methods that led to the use of the portable generator-based X-ray fluorescence (XRF) instrument for nondestructive analysis at crime scenes (Jeffrey et al. 2005) and the Spectral Library for Identification and Classification Explorer (Bush et al. 2008; Ubelaker et al. 2002). For

* Correspondence: raajaj@kau.edu.sa
[1]Section of Biomaterials, Division of Conservative Dental Sciences, School of Dentistry, King Abdulaziz University, Jeddah 22254, Saudi Arabia
[2]Department of Oral Diagnostic Sciences, Division of Biomaterials, State University of New York at Buffalo, 355 Squire Hall, 110 Parker Hall, Buffalo NY 14214, USA
Full list of author information is available at the end of the article

clinicians, it might aid in the appropriate selection for clinical use depending on their cytotoxic behavior.

The main instrumental approach used in our study was MAIR-IR spectrometry for surface compositional analysis of 14 resin composite brands; all of them were included in previous studies of resin composites (Bush et al. 2006, 2007a, 2008; Hermanson et al. 2008). IR spectroscopic analysis was done on the resin composite samples 'as is' and after saline soaking for 2 weeks in an incubator under 37°C to simulate body intra-oral conditions. Saline soaking of the samples was done to evaluate possible surface compositional changes that might occur after these restorations are placed in the patients' mouths. Lee et al. (1995a) reported changes in the infrared spectra of the surfaces of these composites after immersion in 75% ethanol and in artificial saliva (Moi-Stir, Pendopharm, Montreal, Canada). Vankerckhoven et al. (1982) used MAIR-IR spectroscopy to determine the influence of some manipulative factors (polymerization time, temperature, and mechanical treatments such as polishing) on the concentration of unreacted methacrylate groups in scrapings from the surfaces of the resin composites, and all of the tested manipulations caused a decrease in the resin composites' apparent surface double-bond content. A review of the literature did not identify any prior studies that have used MAIR-IR spectroscopy to examine the intact resin composite surface chemistry of as-prepared or saline-extracted resins, as they would appear in the oral cavity.

The second aspect in our study was the IR spectroscopic analysis of the saline extracts of the resin composites. Studying the saline extracts of these composites is significant to know if different brands of resin composites have different leaching abilities with regard to the amount and type of the leached materials and thus have potentially different toxicities to cells in the proximal vicinity of resin composite restorations in the mouth. Evidence of leaching from various fillers has been reported using plasma spectrometry (Soderholm 1983) and atomic absorption spectrophotometry (Soderholm et al. 1984; Soderholm 1990). Leached components from dental composites in oral simulating fluids have also been studied using gas chromatography/mass spectrometry (Lee et al. 1998).

The third aspect of our study was cytotoxicity testing of the saline extracts of the resin composites. This was accomplished by adding the saline extracts to HGFs and using a widely accepted viability and proliferation test method, methylthiol tetrazolium (MTT) assay (Wikipedia, 2012), to obtain the results. The fact that some proportions of residual monomers or short-chain polymers may not react and remain un-bonded after curing of dental composites, in addition to the susceptibility of polymers in dental resin restorations to chemical degradation (Lee et al. 1998), makes it crucial to understand how

these materials might react in the biological environment. Thompson et al. (1982) used ultraviolet spectrophotometry to analyze the un-polymerized materials extracted from cured orthodontic bonding resin in various aqueous solutions and found that orthodontic bonding resins, even when mixed and cured according to the manufacturers' instructions, do leach considerable amounts of un-polymerized components and that precautions should be observed during the polymerization and handling of these materials. High-pressure liquid chromatography was used to analyze different commercial resin composites for the presence of bisphenol-A (BPA) and/or bisphenol-A dimethacrylate (BAD) (estrogen-like components), assuming that these materials could contribute to the overall estrogen load that might result in deleterious side effects, but it was concluded that dental resins in general do not represent a significant source of BPA or BAD exposure (Lewis et al. 1999; Schmalz et al. 1999).

Components eluted from dental resin composites, including diluents (triethylene glycol dimethacrylate (TEGDMA) and decamethacrylate) and some additives (ultraviolet stabilizer TINUVINP), plasticizers (dicyclohexyl phthalate and bis(2-ethylhexyl) phthalate), initiator (triphenyl stibine), coupling agent (γ-methacryloxypropyl trimethoxysilane) and phenyl benzoate, have been shown to make collagen less resistant to trypsin digestion (Lee et al. 1998). Trypsin is an enzyme that acts to degrade protein (proteolytic enzyme or proteinase) (Infoplease, 2012). Collagen is a very important component structure of the bone, teeth, and the gingival and periodontal ligament, all of which can be affected when restorations are placed in contact with or near them. Collagen is produced by fibroblast cells (including HGF). It has also been well established that the resin composite co-monomer TEGDMA causes gene mutation in some cases *in vitro* (Schweikl et al. 2006).

Methods

Fourteen composite samples were collected from commercial sources (Prisma AP.H, SureFil, Quixx, and Esthet.X (Dentsply Caulk, Milford, DE, USA); 4 Seasons, Tetric Evo Ceram, and Heliomolar (Ivoclar Vivadent, Amherst, NY, USA); Filtek Supreme Plus (3M ESPE, St. Paul, MN, USA); Durafill VS and Venus (Heraeus, South Bend, IN, USA); Grandio (VOCO, Cuxhaven, Germany); ICE and Rok (SDI, Bayswater, Australia); and 3D-Direct (Brea, CA, USA). Four samples from each resin composite brand were made, two for use in MAIR-IR spectroscopic analysis and the other two for saline incubation and further analysis of the samples and saline extracts using MAIR-IR spectroscopy (Perkin-Elmer (Waltham, MA, USA) Spectrum 100 FTIR spectrophotometer, with Perkin-Elmer ATR mirror assembly). The samples were made using a mold (ResinKeeper) for composites, manufactured by COSMEDENT® (Manalapan, NJ, USA), and light cured for 40 s using a Spectrum® 800 curing unit (DENTSPLY Caulk) operating at an intensity of approximately 550 mW/cm^2 of halogen light. Two samples from each brand were used for the as-is spectral analysis and another two for the 'saline immersion' and further spectral analysis.

Infrared spectra of the resin samples as is

The MAIR-IR spectroscopic instrument was adjusted during all procedures with the IR spectra wave number ranging from 4,000 to 600 cm^{-1}, transmission in percentage, 10× scan, and 4-cm^{-1} resolution. Two samples were used for each resin composite brand and were clamped to the KRS-5 prism. After sample removal, the readings of the residues were taken (no residues were found).

The other two resin composite samples from each resin composite brand were placed in 45-ml conical tubes and immersed in 10 ml of 0.9% sodium chloride (physiologic saline) solution. They were placed in the incubator (37°C) for 2 weeks and shaken at random times. After the 2-week period, samples were removed from the saline solution using pre-cleaned tweezers and placed on labeled microscopic glass slides under a fume hood until the samples were dry.

Infrared spectra of saline-soaked samples

The same procedures for the as-is samples were applied. Also, the spectra were subtracted from their own baselines (using the spectral subtraction option provided in the instrument software) and converted to absorbance mode then baseline corrected by choosing the 'automatic baseline correction' option in the software. Bands were located, and the heights and bases of the peaks were recorded for calculation of the absorbance of each peak.

Infrared spectra of the saline extracts

For each resin composite material's extract, a standard analytical procedure was applied as follows: 500 μl of the composite saline extract was placed on the germanium prism (does not dissolve in water) using a 100-μl Eppendorf Digital Pipette 4710 (Eppendorf, Hauppauge, NY, USA) and then placed under the fume hood until drying was complete; spectrum of the saline extract was then taken as is, after distilled water leaching, and after distilled water rinsing. The protocol for distilled water leaching was to apply distilled water until it covered the surface of the prism, leaving it for 15 s, and then spilling it, followed by air drying. For distilled water rinsing, distilled water was delivered from a squeeze bottle for 15 s by holding the prism about 20 cm away to produce a shear stress of approximately 1 Pa, and again air drying.

Also, the spectra of the composite saline extracts were converted to absorbance mode, then baseline corrected. Bands were located, and the heights and bases of the peaks were recorded for calculation of the absorbance of each peak.

IR spectroscopy of reference materials

The following materials were collected from commercial sources and are known constituents of the dental resin composite compositions:

- *Ethylene dimethacrylate (EDMA)* cross-linking monomer (Lot no. 283–11, Polyscience, Inc., Rydal, PA, USA)
- *Ethylene glycol dimethacrylate 98% (EGDMA)* (Lot no. 05216CI, Aldrich Chemical Company, Inc., Milwaukee, WI, USA)
- *95% TEGDMA* (Lot# 110 k3657, Sigma®, Seelze, Germany)
- *Bis-A-dimethacrylate* (Lot no. 03924AR, Aldrich Chemical Company, Inc.)
- *(1S)-(+)-Camphorquinone (d-2,3-bornanedione)* (Lot no. 58H3516, Sigma®, Germany)
- *(1R)-(–)-Camphorquinone 99%* (Lot no. 04129TI, Aldrich Chemical Company, Inc.)

Instrument settings were adjusted as described previously. For EDMA, EGDMA and TEGDMA, these monomers were spread over the germanium prisms, and the spectrum for each of them was taken. For the bis-A and camphorquinones, these materials were in powder form and dissolved in acetone to be placed on the germanium prisms. The infrared spectrum of acetone alone, after evaporation, showed no infrared absorption. Acetone was used to dissolve the materials and then placed on the germanium prisms, and the spectra of these materials were taken after thorough drying.

Scanning electron microscopy/energy-dispersive spectroscopy of the saline extracts

An amount of 500 μl of each composite's saline extract was dried on a germanium prism. scanning electron microscopy/energy-dispersive spectroscopy (SEM/EDS) of one specimen (Prisma AP.H) was taken and showed the presence of no elements other than Na, Cl, and Ge. Prisma AP.H was selected randomly, as a typical sample from the larger group. SEM pictures and EDS analysis were taken for three different areas on the germanium prism randomly selected.

Viability testing

Culture medium for the HGFs was prepared using 5 g of minimum essential medium (Alpha medium) from GIBCO™ (Cat. no. 12000–041, Lot no. 397128, Life Technologies, Grand Island, NY, USA), 1.1 g of sodium bicarbonate, 5 ml of L-glutamine 200 mM 100X, 5 ml of antibiotic-antimycotic penicillin-streptomycin, and

50 ml of fetal bovine serum (JM Biosciences, San Diego, CA, USA). Cured resin composite's saline extracts for each brand were filter sterilized using 5-ml syringes (BD Luer-Lok™ Tip, Franklin Lakes, NJ, USA; latex free, sterile) and a 0.45-μm polyvinylidene difluoride filter (Acrodisc LC GELMAN®, Pall, Port, Washington, NY, USA) that fits into the tip of the syringe. The control was pure saline, and the samples were filter sterilized directly before adding them to the cell cultures. Cell cultures were grown to confluence for 10 days (Figure 1).

Cell cultures were replaced into 24-well cell culture plates; each well contained 500 μl of cell culture media. An amount of 50 μl from each extract was filter sterilized and added to the seeded cells (after removal of 50 μl of cell culture media from each well). For the control and each resin composite extract, the experiment was done in triplicate. Forty-eight hours later, 50 μl of the MTT reagent was added to each well. Twenty-four hours later, examination of the cell cultures under a light microscope showed the purple precipitate in all cultures (Figure 2). Cells were transferred to a 96-well microplate with 200 μl of cell culture in each well to enable reading of the formazan titer in the microplate reader machine. The plate was placed in the microplate reader, set at 595 nm wavelength, and readings were taken.

New cell cultures were grown as described above. All steps were repeated the same way, but 100 μl of the composite saline extracts were added to 400 μl of medium in each well. MTT assay was repeated the same way, and photos of the purple formazan precipitate under a light microscope (×40 magnification) were taken (Figure 3). An amount of 200 μl was replaced using the pipette into the 96-well microplates in the same way. Readings were taken using the microplate reader at a 595-nm wavelength. It was noticed that the third well readings of the Durafill, Rok and Venus (corresponding

to the organization numbers 7, 13, and 14 in the microtiter plate, respectively) were not consistent with the readings of the other wells for the same material. So, another 200 μl of the third well of each material was taken after mixing the contents and added to another 96-well microplate tube, and readings were retaken for confirmation.

One-way analysis of variance (ANOVA) statistical comparison was used for both the 50-μl and 100-μl added composite saline extract groups with a significance level of 0.05 for the statistical analysis to compare the MTT precipitate absorbance values of the resin composite's extracts to the controls. Data were transformed because Levene's test for equality of variance values was not fulfilled. Thus, log transformation of variables (log 10) was done, and new variables were computed.

Results and discussion

Infrared spectra of the resin composite samples and their saline extracts were subtracted from their own baselines. For a more accurate evaluation of the intensities of the peaks and fractions of different functional groups, all of the spectra of the saline-soaked samples were subtracted from their own baselines. All of the spectra are baseline corrected by selecting the baseline correction (automatic correction) option. Average readings of the MTT viability testing are presented in Table 1.

Statistical analysis of the 50-μl composite saline extract added to the 450-μl cell culture is presented in Figure 4. One-way ANOVA of MTT precipitate absorbance readings was calculated. The analysis was significant, $F(14,30) = 14.64$, $p < 0.05$. The MTT precipitate value was found to be more with Tetric Evo Ceram (mean difference (M) = −0.65, standard deviation (SD) = 0.02), Filtek Supreme (M = −0.62, SD = 0.01), Quixx (M = −0.67, SD = 0.02), Durafill (M = −0.64, SD = 0.01), ICE (M = −0.64, SD = 0.04), 3D-Direct (M = −0.69, SD = 0.02), Rok (M = −0.70, SD = 0.02), and Venus (M = −0.4, SD = 0.06) as compared to the control (M = −0.84, SD = 0.04). It was noticed that the mean difference values were negative, which means that the above-mentioned resin composite's extracts have higher MTT precipitate absorbance than the control and thus higher metabolic activity (usually taken to equal viability) values.

Statistical analysis of the 100-μl composite saline extract added to the 400-μl cell culture is presented in Figure 5. One-way ANOVA of MTT precipitate absorbance readings was calculated. The analysis was significant, $F(14,30) = 4.75$, $p < 0.05$. The MTT precipitate value was found to be less with Prisma AP.H (M = −0.55, SD = 0.03), 4 Seasons (M = −0.55, SD = 0.09), Tetric Evo Ceram (M = −0.63, SD = 0.03), and Heliomolar (M = −0.58, SD = 0.21) as compared to the control (M = −0.09, SD = 0.04). It was noticed that the mean difference values were positive, which means that the above-mentioned

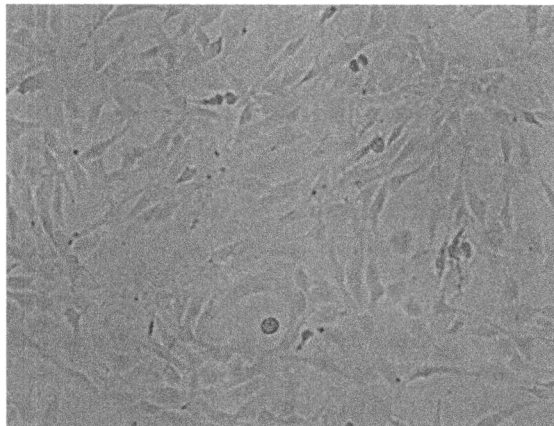

Figure 1 HGF cells after growing to confluence, viewed under a light microscope. ×40 original magnification.

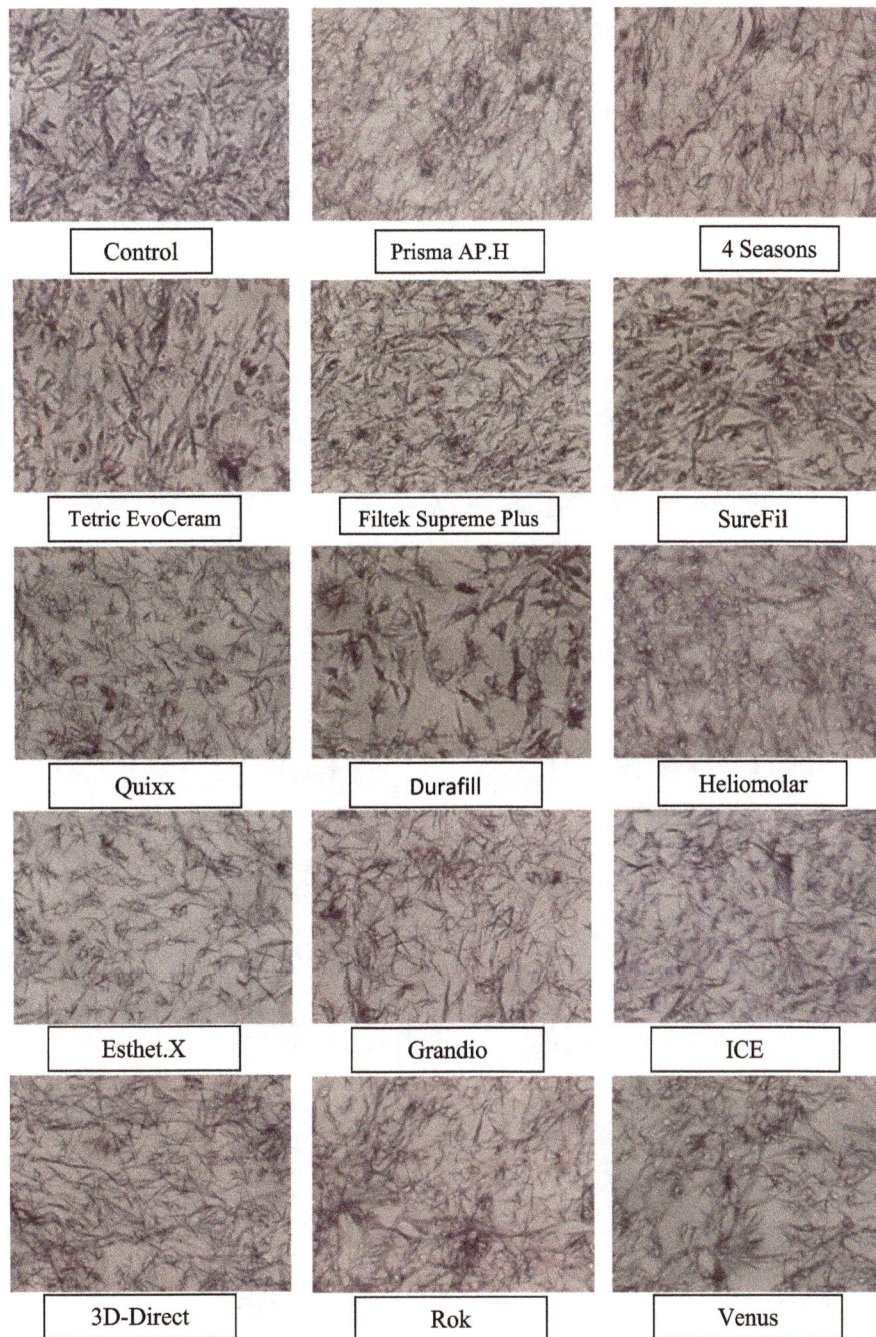

Figure 2 MTT precipitate for the 50-μl added composite saline extracts group under a light microscope. ×40 original magnification.

resin composite's extracts have lower MTT precipitate absorbance than the control and thus statistically higher cytotoxic effects.

Infrared spectroscopic analysis for the transmittance spectra

When quickly viewing the spectra of the resin composites as is, the spectra of all the resin composite brands look almost identical. They have the same general band positions, and the differences between them seem minute or even null. More careful analysis is required in using IR spectroscopy as a tool for differentiating as-prepared resin composite brands.

After analyzing the spectra of the saline-soaked samples, there was a significant change in the intensity of the peaks of all of the resin composite brands. This intensity differs among brands, with the most reduction in intensity shown in Esthet.X and Durafill and minimal reduction shown in

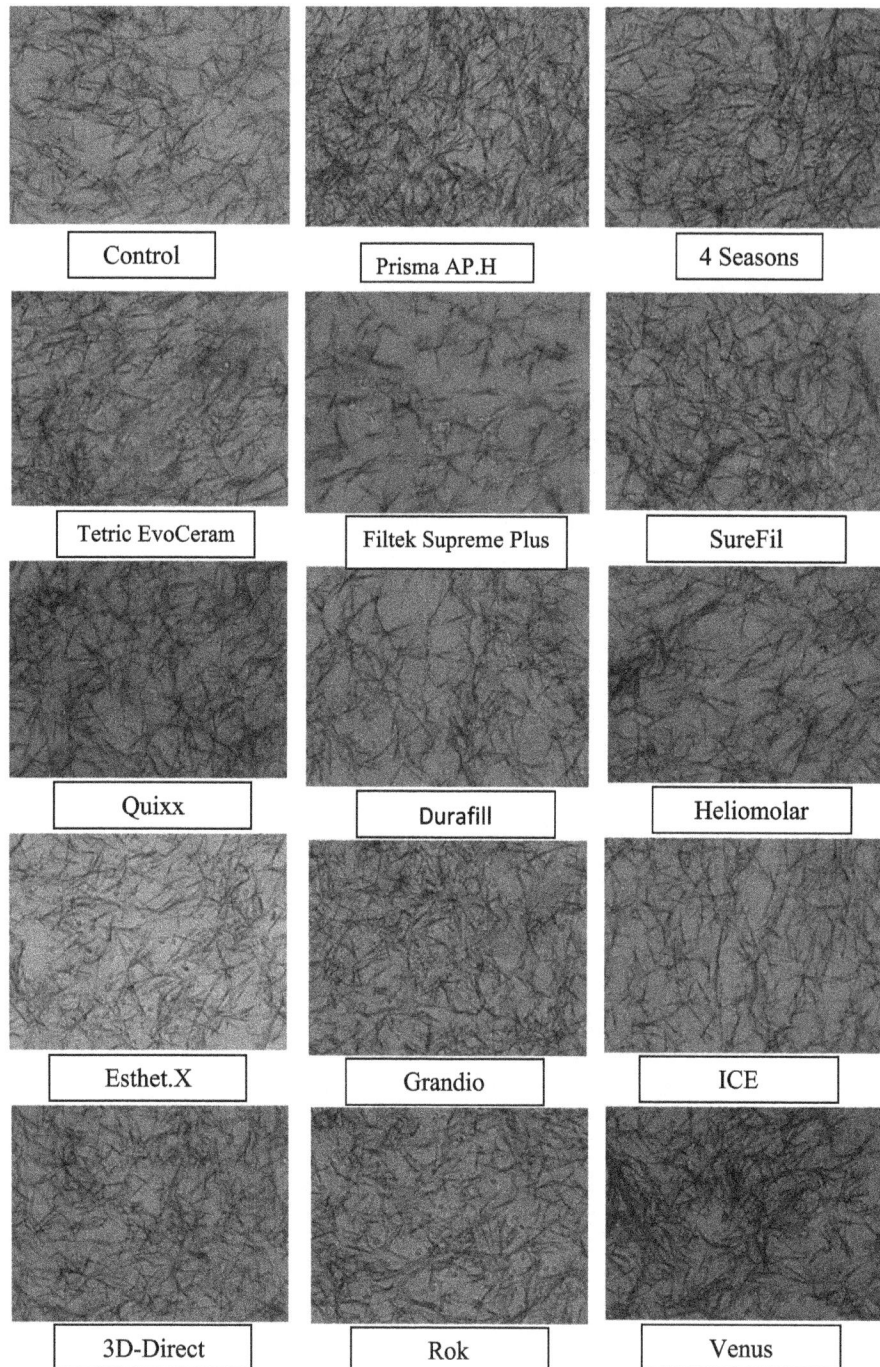

Figure 3 MTT precipitate for the 100-μl added composite saline extracts group under a light microscope. ×40 original magnification.

SureFil and 3D-Direct. The difference in peak intensity reduction among resin composite brands could provide a valuable differentiation tool for forensic purposes in that the saline-soaked samples resemble the resin composite restorations more after placement in the patients' mouths than the as-prepared resins. For that reason, better quantitative analysis of the spectra can be achieved by plotting the spectra in absorbance (Smith 1998).

When analyzing the spectra of the saline extracts, it was noticed that resin composite brands have different leaching abilities as some resin composite brands' saline extracts had more intense peaks than the others. The most intense peaks were shown in Durafill, Esthet.X, and Venus saline extracts, and minimal or even no peaks were shown in Grandio and Heliomolar saline extracts. It was also noticed that after distilled water leaching and

Table 1 Average readings of the MTT viability testing

	50-µl added saline extracts		100-µl added saline extracts	
	Average absorption	Standard deviation ±	Average absorption	Standard deviation ±
	N = 3		N = 3	
Control	0.145	0.010	0.734	0.100
Prisma AP.H	0.165	0.010	0.309	0.070
4 Seasons	0.146	0.001	0.293	0.070
Tetric Evo Ceram	0.228	0.010	0.241	0.020
Filtek Supreme	0.236	0.010	0.394	0.300
SureFil	0.180	0.010	0.655	0.100
Quixx	0.215	0.010	0.375	0.020
Durafill	0.226	0.010	0.771	0.500
Heliomolar	0.174	0.030	0.283	0.100
Esthet.X	0.170	0.010	0.370	0.100
Grandio	0.158	0.020	0.580	0.050
ICE	0.229	0.030	0.500	0.020
3D-Direct	0.204	0.010	0.766	0.100
Rok	0.198	0.010	0.874	0.200
Venus	0.231	0.030	0.500	0.300

distilled water rinsing, all composite saline extract spectra had lost the peaks eventually except in 3D-Direct and ICE saline extracts. For better quantitative analysis of the saline-extracted materials, the spectra of the saline extracts were also plotted in absorbance. As found in previously published analyses of the inorganic elemental compositions of composite resins, there are small but useful discriminating features in their IR spectra characterizing their covalently bound resin and filler components.

IR spectroscopic analysis for the absorbance spectra of the saline-soaked samples

After saline soaking, the IR spectra of the samples showed that all peak positions remained the same, but there was decrease in the intensity of all peaks, which was different among the resin composite brands. The surface characteristics and composition of the saline-soaked samples are believed to be of more interest to study as it resembles the surface of the resin composite restorations after placement in the patients' mouths.

After comparing the shapes of the bands for the absorbance spectra of the saline-soaked samples, it was noticed that 3D-Direct, Rok, ICE, 4 Seasons, Tetric Evo Ceram, Venus, and Grandio have similar band shape in the region between 1,200 and 600 cm^{-1} (silica stretch region). Quixx has a unique band shape in the region of 1,200 to 600 cm^{-1}. Esthet.X and Prisma AP.H have similar bands shape in the region between 1,200 and 600 cm^{-1}. Filtek Supreme, Heliomolar, and Durafill have similar band shape

in the region between 1,200 and 600 cm^{-1}, and they have a unique intense peak at 800 cm^{-1}, yet to be correlated with specific filler components. SureFil has a similar band shape as Filtek Supreme, Heliomolar, and Durafill, but the band at 800 cm^{-1} is less accentuated. From the above qualitative comparison of the bands' shapes among the 14 dental resin composites, it is found that it is possible to categorize resin composite brands according to the shapes of their infrared spectra, at a qualitative 'pattern recognition' level. This finding can help and would add to the database to aid future and forensic discrimination among different dental resin composite brands.

For quantitative comparison, the two major bands (ester band at ≈1,700 cm^{-1} and silicate band at 1,200 to 800 cm^{-1}) were compared in all the absorbance spectra of the saline-soaked samples. Also, comparison of the fraction of the ester band absorbance to the silica band absorbance was made, and it was found that 3D-Direct has the highest ester band absorbance among all other resin composite brands with an absorbance value of ≈0.4, followed by 4 Seasons and Prisma AP.H with a value of ≈0.2. Heliomolar, Rok, SureFil, Grandio, Quixx, Tetric Evo Ceram, and Venus have an ester band absorbance value of ≈0.1. The other resin composite brands have lower ester band absorbance values. It was also found that the highest silica band absorbance was for 3D-Direct too, with an absorbance value of ≈1.0, followed by SureFil and Heliomolar with a silica band absorbance value of ≈0.5. 4 Seasons showed a silica band absorbance value of ≈0.4. Tetric Evo Ceram and Grandio have a silica band absorbance value of ≈0.3, followed by Durafill, Filtek Supreme, Rok, Prisma AP.H, and Venus with a silica band absorbance value of ≈0.2. Esthet.X, ICE, and Quixx were found to have the lowest silica band absorbance among all resin composite brands. These findings show that quantitative difference in band absorbance among the saline-soaked resin composite samples does exist.

The ester/silicate absorbance ratio value represents the fraction of the major organic band to the major inorganic band absorbance. The ester/silicate absorbance values were found to be highest for the resin composite brand Quixx with a value of ≈0.6, followed by Prisma AP.H, 4 Seasons, ICE, and Rok with a value of ≈0.5 and 3D-Direct with a value of ≈0.4. Tetric Evo Ceram, SureFil, Esthet.X, Grandio, and Venus showed a value of ≈0.3. Filtek Supreme, Durafill, and Heliomolar have the lowest fraction of organic ester/inorganic silicate absorbance.

From the above qualitative and quantitative comparisons of the absorbance spectra of the saline-soaked resin composite samples, it was found that the resin composite brands could be categorized into similar or different groups. This can be used as a valuable tool to differentiate resin composite brands for forensic purposes using IR spectroscopic analysis.

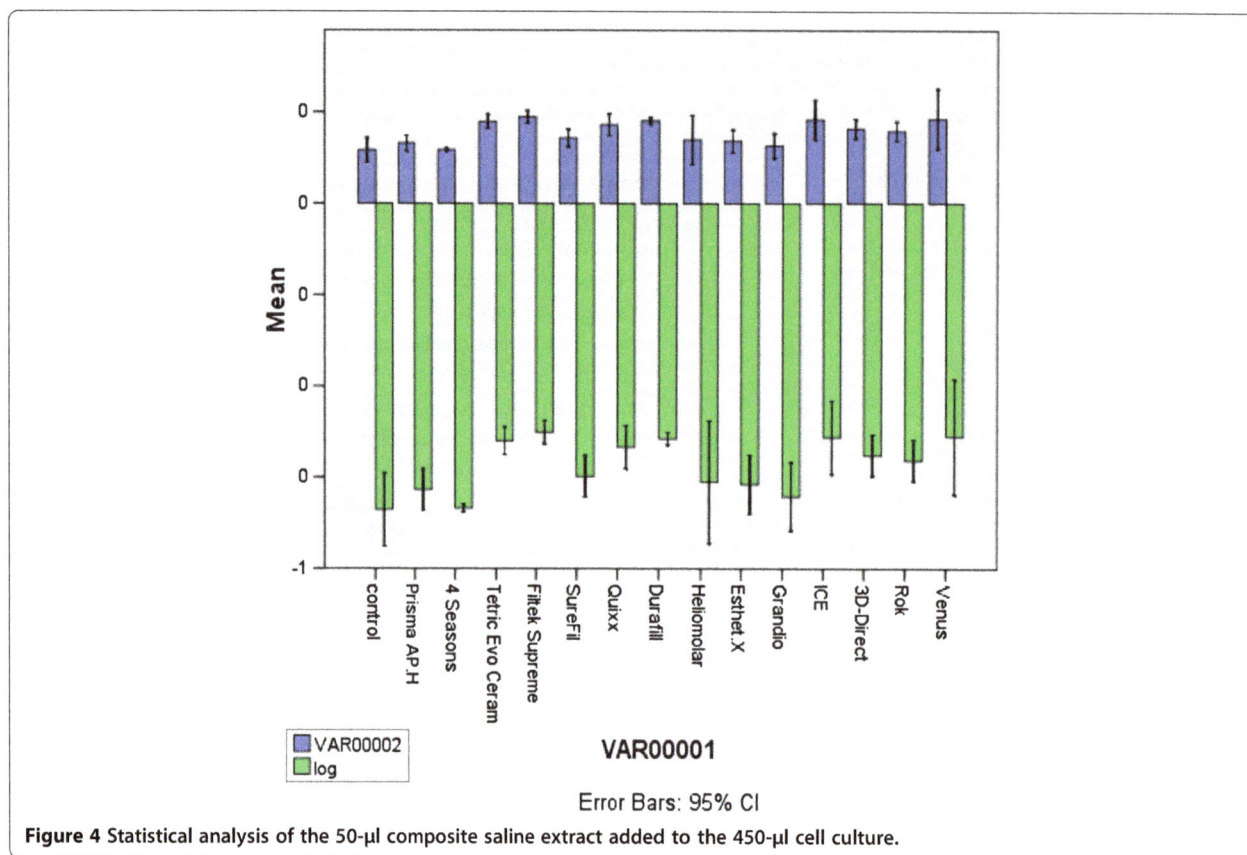

Figure 4 Statistical analysis of the 50-µl composite saline extract added to the 450-µl cell culture.

IR spectra of the pure basic materials

The pure basic materials (urethane dimethacrylate, TEGDMA, bis-GMA, and camphorquinones) are the main materials present in the composition of most of the resin composite brands as supplied by the manufacturers (Air Force Medical Services Public Site, 2012). These mixtures comprise the monomers and photoinitiators. Other materials constituting the composition of the dental resin composite brands are the different fillers. Many studies have been done to study the effects of the monomers in their pure forms on the cellular viability and mutational effects (Schmalz et al. 1999; Janke et al. 2003; Issa et al. 2004; Theilig et al. 2000; Moharamzadeh et al. 2007; Lai et al. 2004).

Upon taking the spectra of different pure basic materials (EDMA, EGDMA, TEGDMA, and bis-A) and photoinitiators (camphorquinones), it was found that the spectra look almost the same as each other. That explains why dental resin composite materials with different combinations of some of these mixtures still look almost the same. It was also noticed that some of these pure material spectra have the same band positions found in the resin composite spectra but with sharper and more intense peaks in the low molecular size pure materials. This could be explained by the fact that dental resin composite surface composition is a polymerized

mixture of materials, so the presence of other bands and convolution of the bands are a logical explanation of the wider and convoluted band spectra. Also, the spectra of the pure materials are missing the wide band at 1,200 to 800 cm^{-1}, which corresponds to the silica stretch found in the dental resin composites. The silica stretch found in the dental resin composites is due to the presence of the inorganic filler particles.

Absorbance spectra of the saline extracts

When the absorbance spectra of the composite saline extracts were evaluated, it was found that different resin composites have different leaching abilities according to the different absorbance bands present in some of the extracts and not present in others. The resin composite brands with intense saline extract absorbance bands are 4 Seasons, Durafill, Prisma AP.H, Quixx, SureFil, and Venus. Although these composites showed the most intense bands, this finding cannot be correlated to the MTT viability findings presented in Table 1 because Durafill, SureFil, and Venus were shown to have minimal or no cytotoxicity to HGF cells when 100 µl of their extracts was added, even though they are having what appeared to be the most leaching materials.

For that reason, quantitative analysis of the absorbance of the major bands was carried out and confirmed the

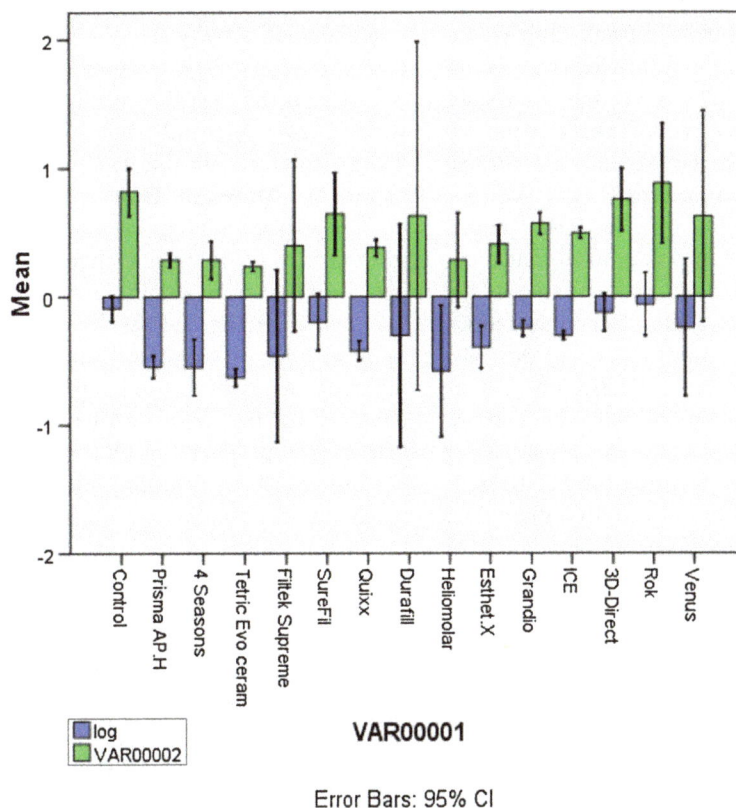

Figure 5 Statistical analysis of the 100-μl composite saline extract added to the 400-μl cell culture.

absence of correlation between the band absorbance and viability findings. The three major peaks found in the saline extract spectra are at ≈1,718 cm^{-1}, two peaks with 1:1 ratio at 1,318 and 1,294 cm^{-1}, and at 1,168 cm^{-1} corresponding to ester bond, aromatic amines, and carboxylic acids/esters, respectively (Smith 1998).

The band at 1,718 cm^{-1} is most intense in Venus followed by 4 Seasons, Durafill, Filtek Supreme, ICE, Prisma AP.H, Quixx, and SureFil. The two peaks at 1,318 and 1,294 cm^{-1} are most intense in Venus followed by 4 Seasons, Durafill, Prisma AP.H, and SureFil. The band at 1,168 cm^{-1} is present in 4 Seasons, Durafill, Prisma AP.H, Quixx, SureFil, and Venus. From the previous findings, it was shown that Venus, 4 Seasons, Durafill, Prisma AP.H, and SureFil have the most leaching abilities. Thus, different composite resin brands have different leaching abilities. Also, it was determined that saline-extractable components of these same resins can have differential effects on the viabilities of HGFCs and that such effects are likely to be concentration dependent.

SEM/EDS of the saline extract

Previous studies were done about leaching of fillers from dental resin composites in distilled water (Soderholm 1983,

1990, 1981). These studies were done using distilled water as the incubation media and concluded that filler particles do leach. None of the resin composites brands used in these studies were the same brands as those used in our study. To investigate whether the filler particles might leach in the saline extract, unfiltered resin composite saline extract dried on germanium prism was analyzed using SEM/EDS and showed only Na, Cl, and Ge (Figures 6 and 7). This might indicate either that no inorganic fillers leach from the resin composite or that the amount of fillers leached is very minute or was skipped during EDS analysis. Also, the leaching of the fillers might be time dependent as the previous studies were done in a 30-day to 6-month period while only 2 weeks of incubation period was used in our study.

MTT viability test

MTT viability assay was done to find out if the composite saline extracts have different cytotoxic effects on HGF cells as it was found that they possess different leaching abilities. One-way ANOVA was used to compare the viability values of the composite saline extracts to the control in each group (the 50-μl added and the 100-μl added composite saline extracts). It is not possible to compare the values between the two groups because each experiment was done in separate cultures on different days. Even though all factors were standardized, different cell lines would have different proliferation rates, and their behavior is not predictable. This explains the difference in the absorbance readings of the controls of both groups. However, comparison between the two groups can be carried out according to how they differ from their own control.

When analyzing the results of the 50-µl added composite saline extract group, it was noticed that there was statistically significant different values between the control and Tetric Evo Ceram, Filtek Supreme, Quixx, Durafill, ICE, 3D-Direct, Rok, and Venus. It was also noticed that these materials had significantly higher MTT precipitate absorbance values than the control. This can be explained either due to the low sensitivity of the MTT test or because of the fact that the MTT test is actually a measure of mitochondrial activity rather than true cell viability, and the addition of a low amount of cytotoxic materials not sufficient to kill the cells will cause

the cells to metabolize these toxins and thus increase the mitochondrial activity. Another possible explanation could be derived from the science of homeopathy. Homeopathy is based on the idea that small doses of a substance that would cause symptoms when administered in large doses will actually activate the defense mechanism against this substance (American Cancer Society, 2012). This can possibly explain why in our studies there was an increase in the cell viability results when low doses of resin composite extracts were administered to HGF cell cultures.

When analyzing the results of the 100-µl composite saline extract, it was found that there were significantly different values of the MTT precipitate readings but in contrast to the 50-µl added resin composite saline extract group. These results indicate lower MTT precipitate absorbance values, which indicate lower viability results. These significantly different values were shown by Prisma AP.H, 4 Seasons, Tetric Evo Ceram, and Heliomolar. When compared to the control from the

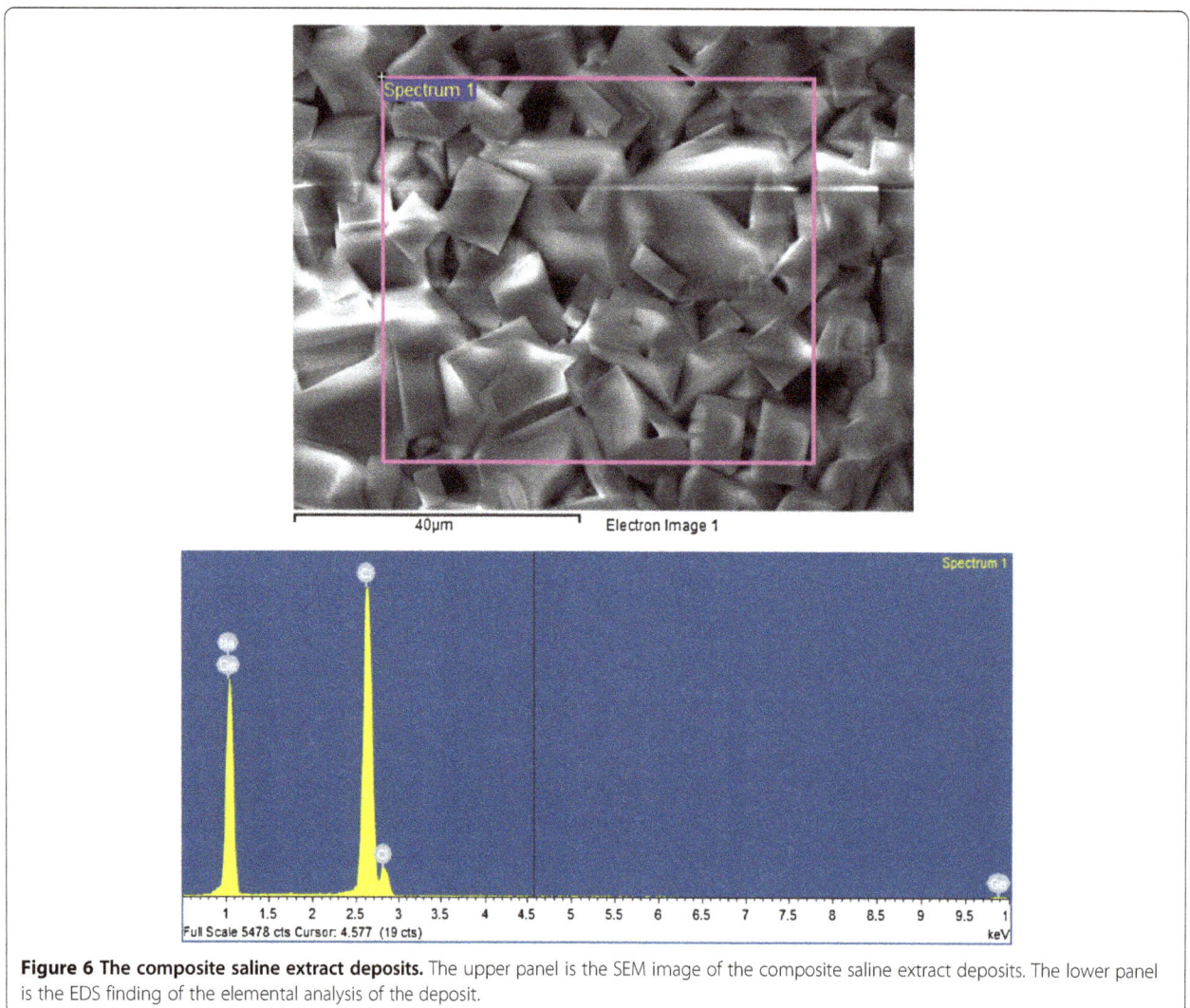

Figure 6 The composite saline extract deposits. The upper panel is the SEM image of the composite saline extract deposits. The lower panel is the EDS finding of the elemental analysis of the deposit.

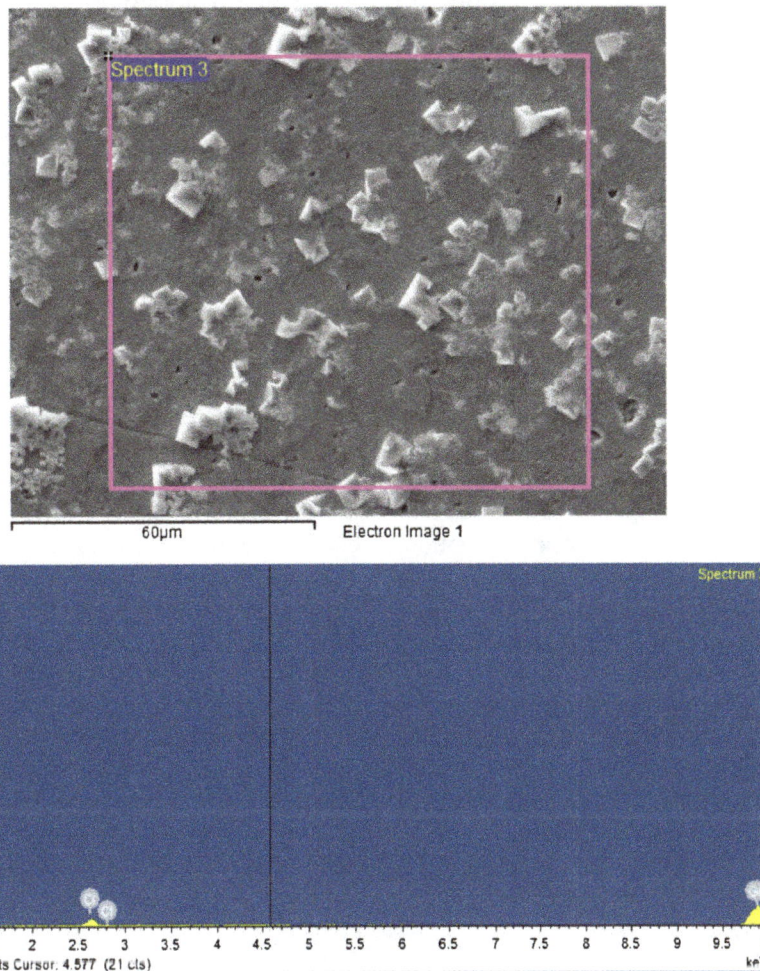

Figure 7 Germanium prism surface with some composite saline extract deposits. The upper panel is the SEM image of the germanium prism surface with some composite saline extract deposits. The lower photo is the EDS finding of the elemental analysis of this surface.

above findings, it was noticed that different resin composite brands do leach materials that possess different cytotoxic effects to HGF cells and that this cytotoxic effect is threshold dependent.

Limitations of this study

Inorganic filler particles in dental composites can leach ions from compounds of silicon, barium, strontium, and sodium (Soderholm 1981, 1983; Soderholm et al. 1984), but it is also likely that those detected elements actually could be present in compounds such as silicates and carbonates that do have IR-detectable covalent bonds. MAIR-IR spectrometry used in our study detects such functional groups and most other covalent bonds (Smith 1998) but will not detect inorganic leaching ions from filler particles, which might also affect cells in the proximal vicinity of the resin composite restorations in the mouth.

Saline at body temperature was used in correspondence to previous laboratory work on other restorative materials (Intermediate Restorative Material (IRM), Geriostore, and Ketac Fil) to study the leachable materials from these dental restoratives (Al-Sabek et al. 2005). Also, saline is harmless to cells and does not give any readings in the MAIR-IR spectrometer (because it contains only ionically bonded salt). This is one of the biggest limitations encountered in our study because the use of saline alone may not absolutely mimic the more complex *in vivo* oral environment in which these resin composite restorations are placed. The oral cavity is subjected to different chemistries frequently during eating of food and drinking of various beverages. Also, food and drinks will subject the oral cavity to major fluctuations in temperatures and degrees of abrasion. All of the changes that occur in the oral cavity can affect the degree and amount of leaching materials from dental resin composite restorations placed in it. Also, the oral environment is subject to the deposition of different amounts of plaque and calculus that consequently may absorb the leaching materials and so affect the duration and frequency of exposure of cells adjacent to the retained debris on these materials (Lee et al. 1995a, 1995b, 1998). Other than foods and drinks, saliva does contain enzymes, and hydrolysis and/or enzyme catalysis can also cause chemical degradation of dental composites (Koin et al. 2008). These factors must be considered although it is difficult to standardize all of these factors as they are not controlled and differ from person to person due to natural differences among people and lifestyles.

Many biological reactions *in vivo* are not immediately cytotoxic and are extended well beyond 24 h. Cytotoxicity assays measure mainly finite effects on cells during the first 12 to 24 h after exposure to toxic substances and are the major category of tests designed for the initial evaluation of materials. Other important processes that should be taken into consideration are inflammation, immune reactions, and mutagenesis for comprehensive testing of the effect of these materials on cells and to more clearly postulate what will happen in the real human model (Hanks et al. 1996).

Depending on only one testing method for ideal surface analysis, characterization, and comparison is not possible. The use of other techniques and adding the results together are very important. Other techniques could be SEM/EDS (Bush et al. 2008; Ubelaker et al. 2002; Hosoda et al. 1990), contact angle goniometry (Galan et al. 2004), XRF (Bush et al. 2007b, 2008), or quantitative light-induced fluorescence (Pretty et al. 2002).

Cells might come into direct contact with these resin restorative materials (e.g., periodontal ligament (PDL) fibroblasts in root-end filling materials, dental pulp fibroblasts in direct pulp capping, gingival fibroblasts in class IV subgingival restorations, and buccal and labial mucosa in bonding resins of orthodontic brackets). Better knowledge of surface characteristics will be crucial for better understanding of how cells in direct contact with these restorations will react. Huang et al. (2002) stated that resinous perforation repair materials inhibit the growth, attachment, and proliferation of human gingival fibroblasts. The study of Al-Sabek et al. (2005) showed preferential attachment of HGFs to the resin ionomer Geriostore when compared with IRM and Ketac Fil but did not explain the reasons for the results. Another study did direct-contact cytotoxicity testing of resin-based restoration materials on HGFs and resulted in finding a time-dependent reduction of their growth with irritation and defective morphology of the fibroblasts in the vicinity of the resin-based materials (Willershausen et al. 1999). Sailynoja et al. (2004) used both direct-contact and extract methods for cytotoxicity testing of UTMA-based hybrid resin and concluded that with increasing incubation temperature to 72°C, cytotoxic effects of the extracts were shown whereas the lower-temperature extracts did not, and that the direct-contact test did not show cytotoxicity. Tuncel et al. (2006) used an agar diffusion method, and cytotoxicity rankings were determined using lysis index scores for cytotoxicity evaluation of three different composites. The study found that the cytotoxicity of the composites increased when fiber reinforced. No chemical analysis of the cytotoxic elements was provided, however.

Another limitation to any *in vitro* model of 'biocompatibility' is the time allotted for incubation of the samples in saline or other media. Although 2 weeks was enough to produce a sufficient amount of leaching materials to be analyzed by IR spectroscopy, it is likely that all commercially available resin-based dental materials will continue to release components that may cause detrimental effects or alter cellular function *in vitro* even after 2 weeks of aging in artificial saliva. Wataha et al. (1999) call attention to the effect of chronic exposure of the cells *in vivo* to these materials with continuous wash out when swallowing versus the one-time subjection of the cells *in vitro* to 2-week accumulated leaching materials.

HGFs were chosen for this study because they are cells in proximal vicinity to dental restorations. PDL fibroblasts would also be affected, could simulate the periapical tissues even better, and are known to be similar to gingival fibroblasts, except that they have a higher production rate of collagen. Also, gingival fibroblasts were chosen due to their easy availability and culturing characteristics (Huang et al. 2002; Hou and Yaeger 1993).

Conclusions

Different resin composite brands have interestingly different surface characteristics after incubation in saline, which were not as readily found in the materials as is. This finding made it possible to categorize the saline-soaked resin composite brands according to their absorbance spectra shapes and values. This might be beneficial addition to the database for forensic discrimination and characterization of different resin composite brands according to a new method, which is IR spectroscopic analysis. It will also fill the gap of studying the organic portion that was not covered by the previous studies, which were concentrating on the inorganic portion.

The fact that the saline-soaked samples were found to have different spectra from the as-is samples and from each other raised the value of studying these resin composite surfaces after incubation in fluids that will more closely simulate the oral environment with fluctuating temperatures and acidity. These fluctuations can be due to different eating and drinking habits. Also, the restorations, when placed in the oral cavity, are subjected to frictional forces and deposition of plaque and calculus that will act on them and change their surface chemistry. All of the above factors should be considered in future studies for better understanding of the surface characteristics of these resin composite brands inside the patients' mouths.

For the biotoxicity aspect, the IR spectra for the saline-soaked samples showed changes in surface characteristics of resin composites. This is of great importance to study as this surface is in contact with oral mucosal cells and was not attended by most of the previous studies. Also, the IR spectroscopy of the saline extract showed that different resin composite brands would have different leaching abilities although these findings are not well correlated to the viability findings. This makes it crucial for future studies to find the correlation between the leached materials and cytotoxicity findings, which was found to be threshold dependent. For the biological effects of these resin composites on the HGF cells, the oral environmental factors mentioned earlier should be considered, and the application of direct viability test and also more than one surface characterization technique are essential for better understanding of the biological effects of resin composite brands to cells in proximal vicinity to them in the oral cavity.

More sensitive and precise viability testing methods in combination with more clinically relevant situations should be the target of future studies. This study focused on differentiating different resin composite brands, which was not the case in previous studies. Previous studies focused on the difference between composites and other restorative dental materials and did not address the wide variety of dental resin composite brands, which were proven by this study to have different surface characteristics and biological behaviors.

Also, the conclusion from these studies is that IR spectrometry (particularly using the very surface-sensitive MAIR technique) can provide valuable reference characteristics for later forensic identification of the distinct resin composites present in unknown trauma victims. MAIR-IR can also identify miniscule amounts of saline-extractable components from resin components that can have differential consequences for the viabilities of neighboring gingival fibroblasts. It should be a new requirement for such analyses that IR spectroscopic identification be attempted given these early successes.

Abbreviations
ANOVA: analysis of variance test statistics; BAD or bis-GMA: bisphenol-A dimethacrylate; BPA: bisphenol-A; EDMA: ethylene dimethacrylate; EDS: energy-dispersive spectroscopy; HGF: human gingival fibroblast; IR: infrared; MAIR-IR: multiple attenuated internal reflection-infrared spectroscopy; MTT: methylthiazol tetrazolium, yellow dye utilized by cells as an *in vitro* test of viability and proliferation; PDL: periodontal ligament; SEM: scanning electron microscopy; TEGDMA: triethylene glycol dimethacrylate; XRF: X-ray fluorescence.

Competing interests
The authors declare that they have no competing interests.

Authors' contributions
RA, RB, and PB came up with the main idea of the research. RA carried out the literature search, data collection and interpretation, and the manuscript preparation. JF contributed to the reading and approval of the manuscript. All authors read and approved the final manuscript.

Acknowledgments
Special thanks to Dr. Rosemary Dziak and Mrs. Nancy Marzec for their guidance and help in the oral biology laboratory section of the research.

Author details
[1]Section of Biomaterials, Division of Conservative Dental Sciences, School of Dentistry, King Abdulaziz University, Jeddah 22254, Saudi Arabia. [2]Department of Oral Diagnostic Sciences, Division of Biomaterials, State University of New York at Buffalo, 355 Squire Hall, 110 Parker Hall, Buffalo NY 14214, USA. [3]Department of Restorative Dentistry, State University of New York at Buffalo, Buffalo NY 14214, USA.

References
USAF Dental Evaluation and Consultation Service, Synopsis of Restorative Resin Composite Systems (Project 05–06) (August/2005). http://www.yumpu.com/en/document/view/4601291/synopsis-of-restorative-resin-composite-systems-air-force-. Accessed April 2013.

Al-Sabek F, Shostad S, Kirkwood KL (2005) Preferential attachment of human gingival fibroblasts to the resin ionomer Geristore. J Endod 31(3):205–208

American Cancer Society (2012) Homeopathy. http://www.cancer.org/Treatment/TreatmentsandSideEffects/ComplementaryandAlternativeMedicine/PharmacologicalandBiologicalTreatment/homeopathy. Accessed 29 Sep 2012

Bush MA, Bush PJ, Miller RG (2006) Detection and classification of composite resins in incinerated teeth for forensic purposes. J Forensic Sci 51(3):636–642

Bush MA, Miller RG, Prutsman-Pfeiffer J, Bush PJ (2007a) Identification through X-ray fluorescence analysis of dental restorative resin materials: a comprehensive study of noncremated, cremated, and processed-cremated individuals. J Forensic Sci 52(1):157–165

Bush MA, Miller RG, Fagen HA, Bush PJ (2007b) The role of dental materials in situations involving high temperatures: a review article in forensic odontology. Minerva Medicolegal 127(2):97–103

Bush MA, Miller RG, Norrlander AL, Bush PJ (2008) Analytical survey of restorative resins by SEM/EDS and XRF: databases for forensic purposes. J Forensic Sci 53(2):419–425

Galan J Jr, Namen FM, Fernando Filho CS (2004) Wettability of some packable resin-based composites. An in vitro study. Eur J Prosthodont Restor Dent 12(3):121–124

Hanks CT, Wataha JC, Sun Z (1996) In vitro models of biocompatibility: a review. Dent Mater 12(3):186–193

Hermanson AS, Bush MA, Miller RG, Bush PJ (2008) Ultraviolet illumination as an adjunctive aid in dental inspection. J Forensic Sci 53(2):408–411

Hosoda H, Yamada T, Inokoshi S (1990) SEM and elemental analysis of composite resins. J Prosthet Dent 64(6):669–676

Hou LT, Yaeger JA (1993) Cloning and characterization of human gingival and periodontal ligament fibroblasts. J Periodontol 64(12):1209–1218

Huang FM, Tai KW, Chou MY, Chang YC (2002) Resinous perforation-repair materials inhibit the growth, attachment, and proliferation of human gingival fibroblasts. J Endod 28(4):291–294

Infoplease (2012) Trypsin. In: The Columbia electronic encyclopedia, 6th edn. Columbia University Press, Available via infoplease. http://www.infoplease.com/ce6/sci/A0849555.html. Accessed 27 Sep 2012

Issa Y, Watts DC, Brunton PA, Waters CM, Duxbury AJ (2004) Resin composite monomers alter MTT and LDH activity of human gingival fibroblasts in vitro. Dent Mater 20(1):12–20

Janke V, von Neuhoff N, Schlegelberger B, Leyhausen G, Geurtsen W (2003) TEGDMA causes apoptosis in primary human gingival fibroblasts. J Dent Res 82(10):814–818

Jeffrey S, Schweitzer JIT, Floyd S, Selavka C, Zeosky G, Gahn N, McClanahan T, Burbine T (2005) Portable generator-based XRF instrument for non-destructive analysis at crime scenes. Nucl Instrum Meth B 241:816–819

Koin PJ, Kilislioglu A, Zhou M, Drummond JL, Hanley L (2008) Analysis of the degradation of a model dental composite. J Dent Res 87(7):661–665

Lai YL, Chen YT, Lee SY, Shieh TM, Hung SL (2004) Cytotoxic effects of dental resin liquids on primary gingival fibroblasts and periodontal ligament cells in vitro. J Oral Rehabil 31(12):1165–1172

Lee SY, Greener EH, Mueller HJ (1995a) Effect of food and oral simulating fluids on structure of adhesive composite systems. J Dent 23(1):27–35

Lee SY, Greener EH, Menis DL (1995b) Detection of leached moieties from dental composites in fluids simulating food and saliva. Dent Mater 11(6):348–353

Lee SY, Huang HM, Lin CY, Shih YH (1998) Leached components from dental composites in oral simulating fluids and the resultant composite strengths. J Oral Rehabil 25(8):575–588

Lewis JB, Rueggeberg FA, Lapp CA, Ergle JW, Schuster GS (1999) Identification and characterization of estrogen-like components in commercial resin-based dental restorative materials. Clin Oral Investig 3(3):107–113

Moharamzadeh K, Van Noort R, Brook IM, Scutt AM (2007) Cytotoxicity of resin monomers on human gingival fibroblasts and HaCaT keratinocytes. Dent Mater 23(1):40–44

Pretty IA, Smith PW, Edgar WM, Higham SM (2002) The use of quantitative light-induced fluorescence (QLF) to identify composite restorations in forensic examinations. J Forensic Sci 47(4):831–836

Sailynoja ES, Shinya A, Koskinen MK, Salonen JI, Masuda T, Matsuda T, Mihara T, Koide N (2004) Heat curing of UTMA-based hybrid resin: effects on the degree of conversion and cytotoxicity. Odontology 92(1):27–35

Schmalz G, Preiss A, Arenholt-Bindslev D (1999) Bisphenol-A content of resin monomers and related degradation products. Clin Oral Investig 3(3):114–119

Schweikl H, Spagnuolo G, Schmalz G (2006) Genetic and cellular toxicology of dental resin monomers. J Dent Res 85(10):870–877

Smith B (1998) Infrared spectral interpretation: a systematic approach. CRC Press, Boca Raton

Soderholm KJ (1981) Degradation of glass filler in experimental composites. J Dent Res 60(11):1867–1875

Soderholm KJ (1983) Leaking of fillers in dental composites. J Dent Res 62(2):126–130

Soderholm KJ, Zigan M, Ragan M, Fischlschweiger W, Bergman M (1984) Hydrolytic degradation of dental composites. J Dent Res 63(10):1248–1254

Soderholm KJ (1990) Filler leachability during water storage of six composite materials. Scand J Dent Res 98(1):82–88

Theilig C, Tegtmeier Y, Leyhausen G, Geurtsen W (2000) Effects of BisGMA and TEGDMA on proliferation, migration, and tenascin expression of human fibroblasts and keratinocytes. J Biomed Mater Res 53(6):632–639

Thompson LR, Miller EG, Bowles WH (1982) Leaching of unpolymerized materials from orthodontic bonding resin. J Dent Res 61(8):989–992

Tuncel A, Ozdemir AK, Sumer Z, Hurmuzlu F, Polat Z (2006) Cytotoxicity evaluation of two different composites with/without fibers and one nanohybrid composite. Dent Mater J 25(2):267–271

Ubelaker DH, Ward DC, Braz VS, Stewart J (2002) The use of SEM/EDS analysis to distinguish dental and osseus tissue from other materials. J Forensic Sci 47(5):940–943

Vankerckhoven H, Lambrechts P, van Beylen M, Davidson CL, Vanherle G (1982) Unreacted methacrylate groups on the surfaces of composite resins. J Dent Res 61(6):791–795

Wataha JC, Rueggeberg FA, Lapp CA, Lewis JB, Lockwood PE, Ergle JW, Mettenburg DJ (1999) In vitro cytotoxicity of resin-containing restorative materials after aging in artificial saliva. Clin Oral Investig 3(3):144–149

Willershausen B, Schafer D, Pistorius A, Schulze R, Mann W (1999) Influence of resin-based restoration materials on cytotoxicity in gingival fibroblasts. Eur J Med Res 4(4):149–155

Wikipedia (2012) MTT assay. http://en.wikipedia.org/wiki/MTT_assay. Accessed 27 Sep 2012

6

Optimization of monomethoxy poly(ethylene glycol) grafting on Langerhans islets capsule using response surface method

Hamideh Aghajani-Lazarjani[1], Ebrahim Vasheghani-Farahani[1*], Sameereh Hashemi-Najafabadi[1], Seyed Abbas Shojaosadati[1], Saleh Zahediasl[2], Taki Tiraihi[3] and Fatemeh Atyabi[4]

Abstract

Langerhans islet transplantation is a much less invasive approach compared with the pancreas transplantation to 'cure' diabetes. However, destruction of transplanted islets by the immune system is an impediment for a successful treatment. Chemical grafting of monomethoxy poly(ethylene glycol) onto pancreatic islet capsule is a novel approach in islet immunoisolation. The aim of this study was to determine an optimized condition for grafting of monomethoxy poly(ethylene glycol) succinimidyl propionate (mPEG-SPA) on islets capsule. Independent variables such as reaction time, the percentage of longer mPEG in the mixture, and polymer concentration were optimized using a three-factor, three-level Box-Behnken statistical design. The dependent variable was IL-2 (interleukin-2) secretion of lymphocytes co-cultured with PEGylated or uncoated control islets for 7 days co-culturing. A mathematical relationship is obtained which explained the main and quadratic effects and the interaction of factors which affected IL-2 secretion. Response surface methodology predicted the optimized values of reaction time, the percentage of longer mPEG in the mixture, and polymer concentration of 60 min to be 63.7% mPEG$_{10}$ and 22 mg/mL, respectively, for the minimization of the secreted IL-2 as response. Islets which were PEGylated at this condition were transplanted to diabetic rats. The modified islets could survive for 24 days without the aid of any immunosuppressive drugs and it is the longest survival date reported so far. However, free islets (unmodified islets as control) are completely destroyed within 7 days. These results strongly suggest that this new protocol provides an effective clinical means of decreasing transplanted islet immunogenicity.

Keywords: Response surface methodology, Diabetes, PEGylation, Pancreatic islets, Transplantation

Introduction

Diabetes mellitus type I, insulin dependent diabetes mellitus (IDDM), is an autoimmune disorder which leads to the destruction of insulin-producing pancreatic islets. Current therapies for IDDM include exogenous insulin therapy and pancreas transplantation. Although daily administration of insulin is the standard protocol, deficient control of blood glucose leads to severe complications, such as heart disease, nephropathy, and hypoglycemia (Hill 2004). Currently, pancreas transplantation is the only available option to 'cure' diabetes, but this procedure requires major surgery and lifelong immunosuppression therapy. Langerhans islet transplantation is another approach to cure diabetes which is much less invasive (Lakey et al. 2006). However, immune destruction of transplanted islets is an impediment for a successful procedure (Devos et al. 2005). There are two major approaches to immunoisolate islets: islet encapsulation and surface modification (Lee and Byun 2010). Conjugation of polyethylene glycol (PEG) to the surface of Langerhans islets, islet PEGylation, is another novel approach for immunoprotection of the islets (Panza et al. 2000; Scott and Chen 2004). This strategy was originally applied for surface modification of red blood cells (RBC) in which PEG covalently attaches to the surface amino acids of the RBC, thereby masking the major and minor blood group antigens from hosting antibodies

* Correspondence: evf@modares.ac.ir
[1]Biotechnology Group, Department of Chemical Engineering, Faculty of Engineering, Tarbiat Modares University, P.O. Box 14115–143, Tehran 1411713116, Iran
Full list of author information is available at the end of the article

(Neu et al. 2003; Sarvi et al. 2006; Hashemi-Najafabadi et al. 2006; Scott and Murad 1998). Based on such encouraging results, PEGylation method was employed for camouflaging pancreatic islets without affecting their viability and functionality. PEGylation also does not increase islets volume (Panza et al. 2000; Lee et al. 2002; Xie et al. 2005). PEGylation reaction happens when the functional group of the polymer, succinimidyl group of mPEG-succinimidyl propionic acid (mPEG-SPA) in this case, conjugates to the amine groups of the collagen matrix of islets, thereby forms a stable amide bond (Jang et al. 2004).

For completing the coverage of the islets surface by PEG without any adverse effects, the reaction factors should be optimized. Contributing factors that affect preceding the reaction are the molecular weight of PEG, polymer concentration, and reaction time (Roberts et al. 2002). Nowadays, many statistical approaches have been known as useful techniques to optimize the process these variables. Box-Behnken or modified central composite design is an independent and nearly rotatable quadratic design, in which the treatment combinations are at the midpoints of the edges of the process space and at the center (Box et al. 1967). Since 1960, Box-Behnken designs have been very popular with experimenters wishing to estimate a second-order model in three or four factors. This popularity is due to these three-level designs' simplicity and high efficiency. Among all the response surface methods, this method requires fewer runs in a three-factor experimental design (Chung et al. 2007).

At present study, three-level, three-factor experiments was designed by Box-Behnken to observe the event of parameters influencing on islets PEGylation. The factors that are considered to have an effect on PEGylation are (a) reaction time, (b) the percentage of longer mPEG in the mixture, and (c) polymer concentration. The dependent variable is the amount of IL-2 (interleukin-2) secreted during co-culturing of islets with lymphocytes (Y).

Methods
Isolation of pancreatic islets and PEGylation
The pancreatic islets were obtained from male Wistar rats (250 to 300 g) as described by Lacy and Kostinovsky (1967). Briefly, 10 mL of Hanks' balanced salt solution (HBSS) containing 2 mg/mL collaganase type V (Sigma Chemical Co., St Louis, MO, USA) was injected into the pancreas. The swollen pancreas was excised and incubated at 37°C for 15 min. The digestion of pancreas stopped by the addition of 20 mL cold HBSS followed by shaking for 1 min to mechanical disruption. Disrupted pancreas was filtrated through 100-μm mesh to remove other tissues. After washing with HBSS

containing fetal bovine serum (FBS), the isolated islets were purified by centrifugation in a discontinuous Histopaque 1.077 (Sigma Chemical Co., St Louis, MO, USA) gradient and then washed with RPMI-1640 culture medium (Sigma Chemical Co., St Louis, MO, USA). Islets were segregated by handpicking and finally cultured overnight in RPMI-1640 medium at 37°C in the humidified atmosphere containing 5% CO_2 to recover.

Activated mPEG (mPEG-SPA) was prepared and characterized as described by Perry and Kwang (2006). For grafting activated mPEG, purified islets were washed twice with HBSS (pH 7.4) followed immediately by suspending in 10 mL HBSS which contained mPEG-SPA. The suspension was incubated at 37°C in humidified air containing 5% CO_2. After incubation, the PEGylated islets were washed twice with RPMI and suspended in culture medium for co-culturing with lymphocytes.

Preparation of splenic lymphocytes
The lymphocyte cells were isolated from the spleen using Ficoll density gradients (Amersham Bioscience, Uppsala, Sweden). The spleen was obtained aseptically from male C57BL/6 mouse. After grinding the spleen in HBSS, the obtained splenocytes were diluted with 3 mL HBSS and layered on 4 mL Ficoll carefully. The cell suspension was centrifuged at $400 \times g$ for 30 min at 4°C, and then the opaque interface containing lymphocytes were transferred into a conical centrifuge tube. After washing with RPMI, the lymphocytes were isolated. The viability of the isolated splenic lymphocytes was determined by staining with trypan blue (Gibco, Paisley, Scotland).

Co-culture of islets with lymphocyte
Thirty islets were co-cultured with the 5×10^5 lymphocytes in each well of 96-well plate in 200 μL of RPMI medium containing 10% FBS in 95% O_2 and 5% CO_2 for 7 days at 37°C. To determine the release of IL-2 from lymphocytes against the islets, 100 μL of the medium was sampled on the fifth day of culturing and frozen at −70°C for subsequent measurement. The same amount of fresh medium was refilled into each well, so the volume of culture medium was maintained at 200 μL/well. Mouse IL-2 was measured by enzyme-linked immunosorbent assay using commercial kit (E-bioscience, San Diego, CA, USA).

Box-Behnken statistical design for optimization
A three-factor, three-level Box-Behnken design was used to optimize and evaluate the main effects, interaction effects, and quadratic effects. This design is suitable for exploring quadratic response surface and constructing second-order polynomial models. This cubic design is given by a set of points at the midpoint of each edge of multidimensional cube and a center point replicate

(Palamakula et al. 2004). The nonlinear computer-generated quadratic model is given in Equation 1:

$$Y = b_0 + b_1A + b_2B + b_3C + b_{12}AB + b_{13}AC \quad (1)$$
$$+ b_{23}BC + b_{11}A^2 + b_{22}B^2 + b_{33}C^2,$$

where Y is the measured response associated with each factor level combination; b_0 is an intercept; b_1 to b_{33} are the regression coefficients; A, B, and C are the independent variables. The dependent and independent variables are shown in Table 1. These high, medium, and low levels are selected from the preliminary experiments.

After generating the polynomial equations relating the dependent and independent variables presented in Table 1, the process was optimized for the desired response. Optimization was performed to obtain the levels A, B and, C, which minimized Y.

Allotransplantation of PEGylated islets

Inbred male Wistar rats were rendered diabetic with a single intraperitoneal injection of 45 mg/kg of streptozotocin (Sigma) freshly dissolved in citrate buffer (pH 4.5) 3 days before transplantation. Only rats with stable non-fasting blood glucose levels of >350 mg/dL

over three continuous measurements were considered diabetic and used for the islet allotransplantation. The blood glucose levels were measured in the tail venous blood using a portable blood glucose meter (Optimum, MediSense, Maiden Head, UK).

For islets transplantation, streptozotocin-induced diabetic recipients were anesthetized by intraperitoneal injection of 90 mg/kg ketamine with 8 mg/kg xylazine. The left kidneys were exposed through a small incision and capsulotomy was performed on the surface of the left kidney, then the unmodified islets or PEGylated islets (1,200 islets/recipient) were injected. After allotransplantation, the glucose levels of non-fasting recipients were measured daily between 10:00 a.m. and 12:00 a.m. Transplantations were considered successful if the blood glucose levels returned to normal level of <120 mg/dL for two consecutive days after islet transplantation, and islet rejection was supposed to have occurred if two consecutive blood glucose levels exceeded 200 mg/dL.

Statistical analysis

Islet survival data are expressed as median ± SD and analyzed using SPSS v.16.0 statistical software. For comparison of the mean values, independent variable t test was used. Significance was determined by one sample t test considering the p value <0.05.

Results and discussion

Box-Behnken statistical design for optimization

The Box-Behnken design for the three factors offers fewer experimental runs as compared with that of central composite models, which require 20 runs (Box et al. 1967). The dependent and independent variables for design-generated experimental runs and the amount of secreted IL-2 for each 18 tests, consisting of 17 runs predicted by software plus a control run, are given in Table 1. Table 2 indicates the analysis of variance of the obtained results. The factors with p value <0.05 are

Table 1 Variables and Box-Behnken design and the obtained and predicted results

Run no.	A	B	C	Experimentally derived IL-2 conc. (pg/mL)	Predicted IL-2 conc. (pg/mL)
	Reaction time (min)	mPEG$_{10}$ in mixture (%)	Polymer conc. (mg/mL)		
0[a]	0	0	0	270.47	101.44
1	30	0	16	160.01	159.07
2	30	50	22	145.32	148.88
3	30	100	16	125.48	126.39
4	30	50	10	130.10	126.57
5	45	100	22	103.45	98.98
6	45	100	10	100.65	103.28
7	45	0	22	126.17	123.55
8	45	0	10	103.49	107.96
9[b]	45	50	16	105.32	101.44
10[b]	45	50	16	115.93	106.19
11[b]	45	50	16	100.21	101.44
12[b]	45	50	16	110.73	106.19
13[b]	45	50	16	98.79	101.44
14	60	0	16	93.72	92.81
15	60	50	10	98.60	95.04
16	60	100	16	95.30	96.24
17	60	50	22	80.47	84.00

[a]Refers that the experiment was done with untreated (free) islets; [b]The center points for calculating experimental error.

Table 2 Analysis of variance table

Factors	p value	F	df	Sum of squares
Reaction time(min)	<0.0001	177.16	1	4647.44
mPEG$_{10}$in mixture (%)	0.0099	16.31	1	427.93
Polymer conc.(mg/mL)	0.1800	2.43	1	63.68
AB	0.0168	12.43	1	325.98
AC	0.0226	10.60	1	278.06
A^2	0.0085	17.61	1	461.92
B^2	0.0741	5.07	1	133.03
Model	0.0010	27.12	9	6402.22
Residual	-	-	5	131.17
Lack of fit	0.2572	3.04	3	107.58

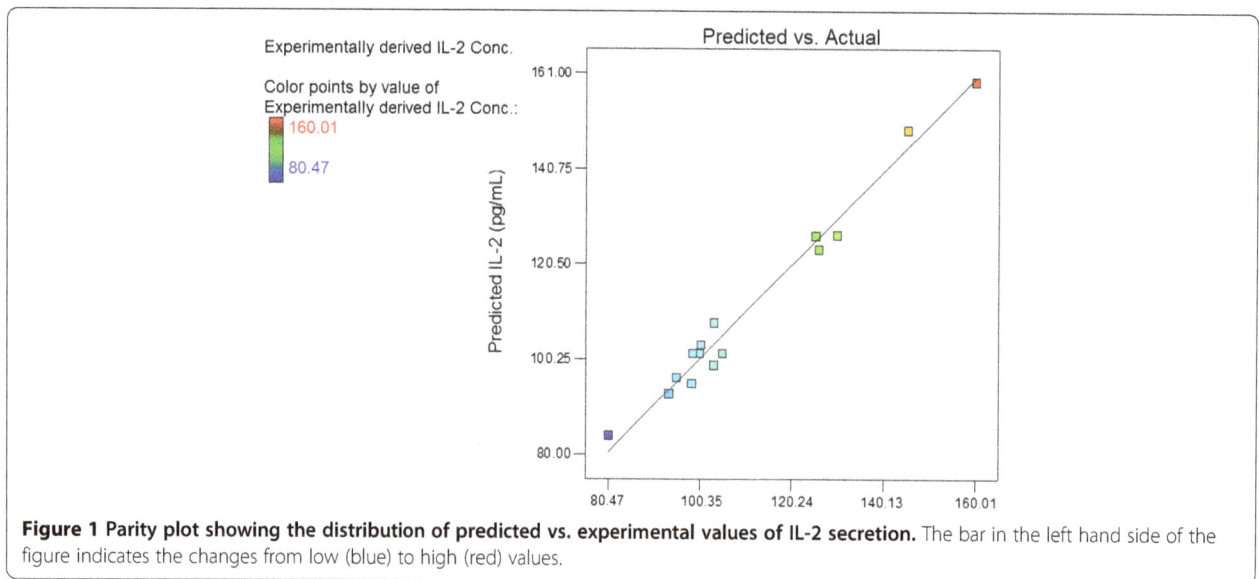

Figure 1 Parity plot showing the distribution of predicted vs. experimental values of IL-2 secretion. The bar in the left hand side of the figure indicates the changes from low (blue) to high (red) values.

significant. From these results, it can be concluded that all experiments resulted in decreasing the IL-2 secretion from lymphocytes which is the result of shielding effect of mPEG-SPA. Jang et al. demonstrated that the lymphocytes co-cultured with PEGylated islets could not affect their viability. Their findings suggested that PEGylation can attenuate the immunogenicity of islets via blocking the recognition of immune cells (Jang et al. 2004). The results in Table 2 indicate that these three factors had a profound effect on the islet masking with mPEG-SPA. At experimental runs of 5, 6, 8, 9, and 13 to 17, the IL-2 secretion decreased 60% more compared to the control run.

The mathematical relationship in the form of polynomial equation for measured response, Y, obtained with statistical package Design Expert (version 7, State Ease Inc., Minneapolis, MN, USA) is in (Equation 2)

$$Y = 101.44 - 24.10A - 7.31B + 2.82C \quad (2)$$
$$+ 9.03AB - 8.34AC + 11.19A^2 + 6B^2$$

This polynomial equation represents the quantitative effect of the process variable (A, B, and C) and their interactions on the response Y. The model F value is 27.12 and implies that the model is significant.

The values of coefficients A, B, and C are related to the effect of these variables on the response. Coefficients with more than 1 factor term and those with higher order terms represent interaction terms and quadratic relationship, respectively. A positive value represents a favorable effect, while the negative value indicates an adverse effect. In this case A, B, AB, and AC have the main effect on the response. The values of A, B, and C were substituted in the equation to obtain the theoretical values of Y. In Figure 1, the parity chart, the experimental response values are plotted versus the predicted

response values. The points are scattered around the diagonal line which indicate the predicted values and the observed values are in good agreement.

Figure 2 is the 2 and 3D views of the main effects of interaction. As it is shown in (a) of Figure 2, IL-2 secretion decreased sharply by increasing time which is completely presented in Figure 3. As shown in (a) of Figure 2, at low reaction times, IL-2 secretion decreased by increasing the percentage of 10 kDa mPEG in the mixture due to the higher shielding effect of long chain polymers on the surface of islets. But at high reaction time, increasing the percentage of 10 kDa mPEG in the mixture of mPEGs (5 and 10 kDa) from 0 to 100 caused an initial decrease in IL-2 secretion with a minimum point occurring at 63.68%. The molecular weight of linear polymers links the chemical and biophysical basis of immunocamouflaging. It is postulated that mPEG with higher molecular weight may be more suitable for surface coating of cells due to its high shielding effect. But Barrou et al. showed that the viscosity of 35 kDa PEG is too high for physiological use. Their results indicated that the optimal chain length at 1.5 mM of PEG is 20 kDa (Neuzillet et al. 2006). The biophysical camouflage of the surface charge is directly proportional to the molecular weight of the grafted polymer. Electrophoretic mobility decreased due to a shift of the shear plane from the surface towards a region of decreased zeta potential. This shift was proportional to the hydrodynamic thickness of the polymer layer and was best achieved with long chain polymers (Lee and Scott 2010). The Flory radii (R_F root mean square of end to end length of the polymer chain and radius of gyration) of the covalently bound polymers can be calculated by this formula: $R_F = aN^{(3/5)}$ ($a = 3.5$ Å, N = number of monomers) (Lee and Scott 2010). According to this formula, the R_F (in nm) values of 5 and

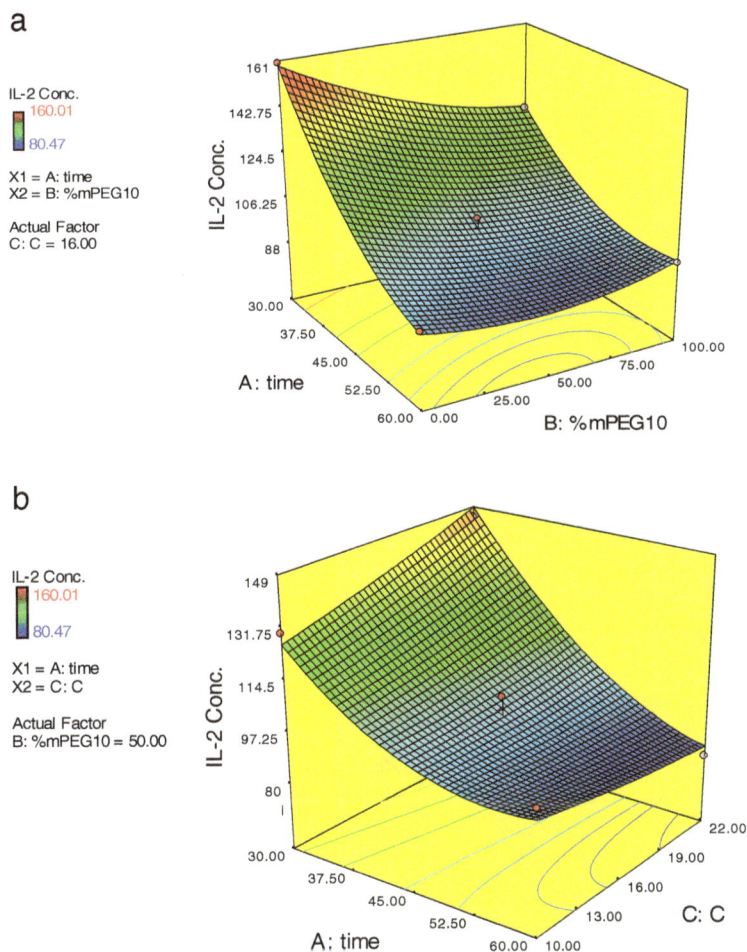

Figure 2 Response surface and contour plots which show the effect of (a) time and mPEG$_{10}$ percent and (b) time and polymer concentration on response Y.

10 kDa polymers are 6 and 9.2 nm, respectively. So longer polymers, because of their higher water absorption, higher volume occupation, and movement in larger space, mask the cell surface charges and antigens better, which causes less stimulation of the immune system.

The steric exclusion effect of the grafted polymer chains primarily prevents protein adsorption (Lee and Scott 2010). This effect maximized when chains are grafted at higher density, i.e., with small separation between chains. But the high-density grafting is difficult to achieve with polymers having a large gyration radius (Lee and Scott 2010). When long and short polymers are attached onto surface together, higher density can be achieved. Therefore, this effect can explain why the minimum point occurred at 63.68%. But it should be mentioned that a 5-kDamPEG derivative will have twice the number of reactive groups as a 10-kDa mPEG derivative at the same mass concentration.

The interaction of reaction time and polymer concentration on the immunological response of PEGylated islets is

shown in Figure 2b. When polymer concentration increased, the IL-2 secretion decreased in slight incline which is presented in the figure. In our previous work, the surface of the Langerhans islets was coated by cyanuric chloride-activated methoxy(polyethylene glycol). The effect of polymer coating, at two different reaction times and polymer concentrations, was investigated. When polymer concentration and reaction time increased to 40 mg/mL and 90 min, respectively; the immunological response to PEGylated islets decreased to 76.5%, compared to untreated islets (Hashemi-Najafabadi et al. 2007). Recently, we reported that by increasing mPEG-succinimidyl carbonate (5 kDa) concentration up to 22 mg/mL, islets were protected from immune cells more efficiently without affecting their viability and functionality (Aghajani-Lazarjani et al. 2010).

According to the correlation given by Equation 2, the minimum IL-2 secretion is 83.5622 pg/mL which is expected to be achieved when reaction time, the percentage of mPEG10 in the mixture, and polymer

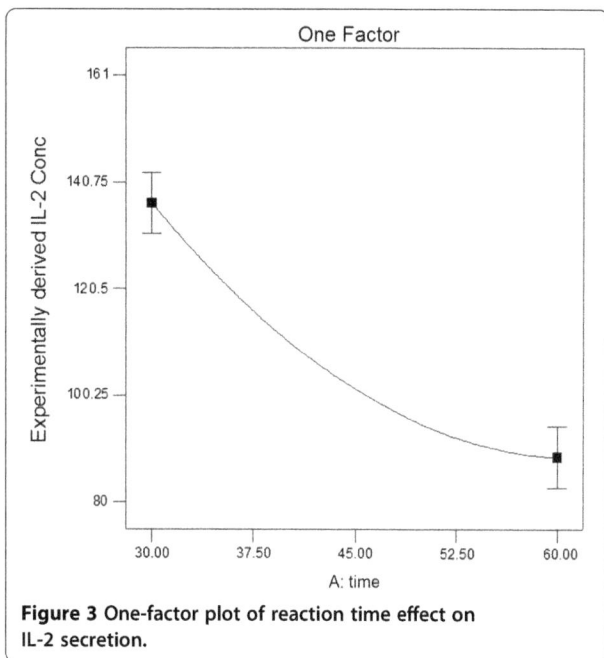

Figure 3 One-factor plot of reaction time effect on IL-2 secretion.

concentration are 60 min, 63.68%, and 22 mg/mL, respectively. The validation experiments were carried out under optimized conditions, and IL-2 secretion was 80.51 ± 6.34 pg/mL. It is in good agreement with the statistically predicted value and confirms the model's authenticity.

Lee et al. (2002) reported that the optimized concentration of mPEG-SPA (5kDa) and reaction time were 0.25% w/v and 1 h, respectively. They used one-factor-at-a-time

method to optimize the reaction condition, so the effect of the factor interactions was not investigated.

Transplantation of islets

To investigate the effect of PEGylation on islet survival, PEGylated islets obtained at the optimum condition, and unmodified islets (control) were transplanted under the capsule of left kidney. Figure 4 shows the changes in non-fasting blood glucose levels of recipients after transplantation. Six hours after transplantation, the first measurement of blood glucose was taken, and sudden decrease was observed in the case of unmodified islets. The same observation was reported by Teramura and Iwata (2009). These results indicate that the islets were damaged and large amounts of insulin released into the blood, so blood glucose decreased transiently.

Unmodified islets survived for 6.3 ± 1.52 days (mean ± SD; $n = 3$). In contrast, normal glycemia was achieved on transplantation of PEGylated islets for 24.0 ± 3.7 days (mean ± SD; $n = 3$).

Lee et al. reported that the PEGylated islets survive for 12 ± 2.6 days. In their research, 12.5 mg/mL mPEG-SPA (5 kDa) was added to the islets solution. The incubation lasted for 1 h at 37°C in a humidified 5% CO_2 atmosphere (Jang et al. 2004), which means that the one-layer PEG is insufficient as immunocamouflage. In another study, they showed triple PEGylations method for islet PEGylation. Although they got long survival time for three recipients, more than 100 days, the median survival time was only 19.0 ± 45.6 days (Lee et al. 2007).

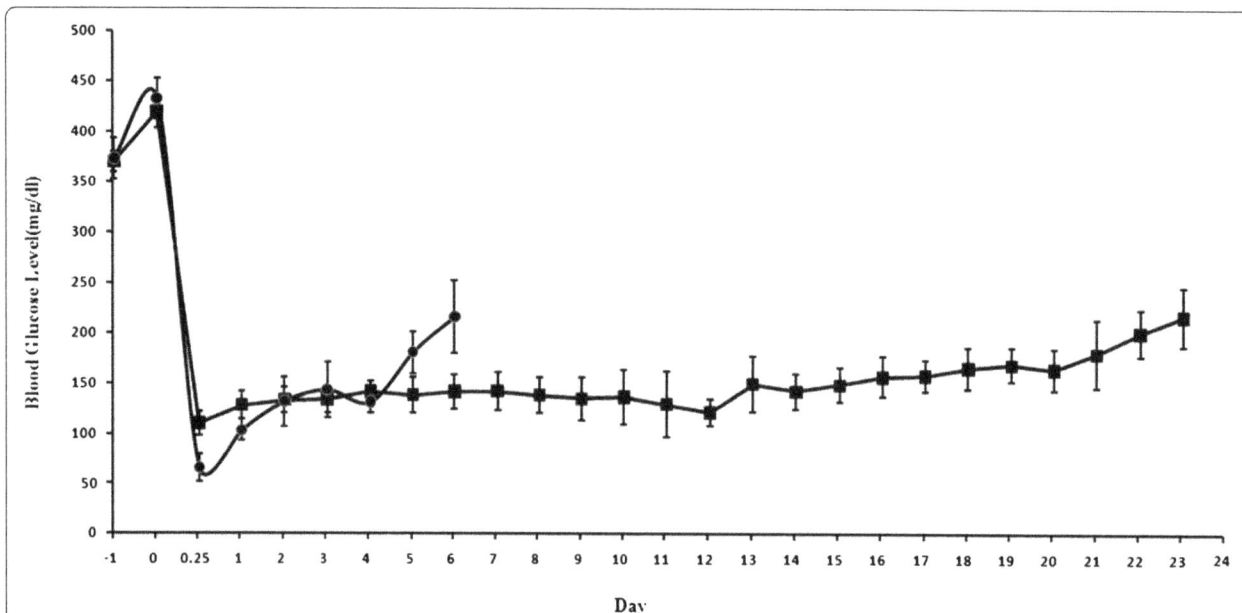

Figure 4 Changes in non-fasting blood glucose levels of streptozotocin-induced diabetic rat. Changes in non-fasting blood glucose levels of streptozotocin-induced diabetic rat after transplantation of unmodified islets as control (black circles) and PEGylated islets (black squres). Graft failure was defined as two consecutive plasma glucose concentrations ≥200 mg/dL.

Conclusions

mPEG-SPA could attach covalently on the surface of Langerhans islets without volume increase and cytotoxicity, but the *in vivo* survival time of PEGylated islets is not long enough. So, recent researches' tendency is to lengthen it. Different approaches have been studied so far such as cyclosporine administration and triple PEGylations method. The aim of this study was to prolong PEGylated islets survival time; therefore, the optimization of PEGylation reaction was studied. The variables of PEGylation reaction were optimized using a response surface method, the Box-Behnken design. The reaction variables such as reaction time, percentage of $mPEG_{10}$ in the mixture, and polymer concentration showed a significant effect on IL-2 secretion from lymphocytes co-cultured with PEGylated islets. The optimum value of these was found to be 60 min, 63.7%, and 22 mg/mL, respectively. The optimized condition was applied for PEGylation, and PEGylated islets were transplanted under the kidney capsule of diabetic rats. The *in vivo* survival time of PEGylated islets increased to 24.0 ± 3.7 days without using immunosuppressive drugs. Thus, single PEGylation at optimum condition intensified islets camouflaging.

Abbreviations

FBS: Fetal bovine serum; HBSS: Hanks' balanced salt solution; IDDM: Insulin dependent diabetes mellitus; IL-2: Interleukin 2; mPEG-SPA: Monomethoxypoly (ethylene glycol) succinimidyl propionate; PEG: Polyethylene glycol; RBC: Red blood cells.

Competing interests

The authors declare that they have no competing interests.

Authors' contributions

HA-L carried out the whole experiments with participation of other authors. EV-F conceived of the study and participated in its design and coordination. SH-N participated in PEGylation of islets. SAS performed design of experiments. SZ participated in isolation of islets. TT participated in transplantation of islets. FA participated in immunoassay experiments. All authors read and approved the final manuscript.

Acknowledgment

This work was supported by grant (no.S88P/3/1518) from the Ministry of Health and Medical Education, Deputy of Research and Food and Drugs Department, Islamic Republic of Iran.

Author details

[1]Biotechnology Group, Department of Chemical Engineering, Faculty of Engineering, Tarbiat Modares University, P.O. Box 14115–143, Tehran 1411713116, Iran. [2]Endocrine Physiology Laboratory, Endocrine Research Centre, Research Institute for Endocrine Sciences, Shahid Beheshti University of Medical Sciences, Tehran 3197619751, Iran. [3]Department of Anatomy, School of Medical Sciences, Tarbiat Modares University, Tehran 1411713116, Iran. [4]Faculty of Pharmacy, Tehran University of Medical Sciences, Tehran 1419733171, Iran.

References

Aghajani-Lazarjani H, Vasheghani-Farahani E, Barani L, Hashemi-Najafabadi S, Shojaosadati SA, Zahediasl S, Tairahi T (2010) Effect of polymer concentration on camouflaging of pancreatic islets with mPEG-succinimidyl carbonate. Artif Cells Blood Substit Biotech Int J 38:250–258

Box GEP, Hunter WG, Hunter JS (1967) Statistics for experimenters: an introduction to design, data analysis, and model building. Wiley, New York

Chung PJ, Goldfrabg HB, Montgomery DG (2007) Optimal three-level designs for response surfaces in spherical experimental regions. J Qual Technol 39:340–354

Devos T, Yan Y, Segers C, Rutgeerts O, Laureys J, Gysemans C, Mathieu C, Waer M (2005) Role of CD4 and CD8 T cells in the rejection of heart or islet xenografts in recipients with xenotolerance in the innate immune compartment. Transplant Proc 37:516–517

Hashemi-Najafabadi S, Vasheghani-Farahani E, Shojaosadati SA, Rasaee MJ, Armstrong JK, Moin M, Pourpak Z (2006) A method to optimize PEG-coating of red blood cells. Bioconj Chem 17:1288–1293

Hashemi-Najafabadi S, Vasheghani-Farahani E, Shojasadati SA, Iwanaga Y (2007) Surface coating of islets of Langerhans by activated polyethylene glycol. Iran J Polym 6:529–538

Hill J (2004) Identifying and managing the complications of diabetes. Nurs Times 100:40–44

Jang JY, Lee DY, Park SJ, Byun Y (2004) Immune reactions of lymphocytes and macrophages against PEG-grafted pancreatic islets. Biomaterials 25:3663–3669

Lacy PE, Kostinovsky M (1967) Method for the isolation of intact islets of Langerhans from the rat pancreas. Diabetes 16:35–39

Lakey JRT, Mirbolooki M, Shapiro AMJ (2006) Current status of clinical islet cell transplantation. In: Hornick PRM (ed) Transplantation immunology: methods and protocols. Humana Press, Totowa, pp 47–103

Lee D, Byun Y (2010) Pancreatic islet PEGylation as an immunological polymeric restraint. Biotech Bioproc Eng 15:76–85

Lee Y, Scott MD (2010) Immunocamouflage: the biophysical basis of immunoprotection by grafted methoxypoly(ethylene glycol) (mPEG). Acta Biomater 6:2631–2641

Lee DY, Yang K, Lee S, Chae SY, Kim KW, Lee MK, Byun Y (2002) Optimization of monomethoxy-polyethylene glycol grafting on the pancreatic islet capsules. J Biomed Mat Res 62:372–377

Lee DY, Park SJ, Lee S, Nam JH, Byun Y (2007) Highly poly(ethylene) glycolylated islets improve long-term islet allograft survival without immunosuppressive medication. Tissue Eng 13:2133–2141

Neu B, Jonathan KA, Timothy CF, Herbert JM (2003) Surface characterization of poly(ethylene glycol) coated human red blood cells by particle electrophoresis. Biorheology 40:477–487

Neuzillet Y, Giraud S, Lagorce L, Eugene M, Debre P, Richard F, Barrou B (2006) Effects of the molecular weight of peg molecules (8, 20 and 35 KDA) on cell function and allograft survival prolongation in pancreatic islets transplantation. Trans Proc 38:2354–2355

Palamakula A, Nutan MTH, Khan MA (2004) Response surface methodology for optimization and characterization of limonene-based coenzyme Q10 self-nanoemulsified capsule dosage form. AAPS PharmSciTech 5:1–8

Panza JL, Wagner RW, Rilo HLR, Harsha Rao R, Beckman EJ, Russell AJ (2000) Treatment of rat pancreatic islets with reactive PEG. Biomaterials 21:1155–1164

Perry R, Kwang N (2006) Monofunctional polyethylene glycol aldehydes. USA Patent7041855

Roberts MJ, Bentley MD, Harris JM (2002) Chemistry for peptide and protein PEGylation. Adv Drug Del Rev 54:459–476

Sarvi F, Vasheghani-Farahani E, Shojaosadati SA, Hashemi-Najafabadi S, Moin M, Pourpak Z (2006) Surface treatment of red blood cells with monomethoxypoly(ethylene glycol) activated by succinimidyl carbonate. Iran J Polym 15:525–534

Scott MD, Chen AM (2004) Beyond the red cell: pegylation of other blood cells and tissues. Transfus Clin Biol 11:40–46

Scott MD, Murad KL (1998) Cellular camouflage: fooling the immune system with polymers. Curr Pharm Res 4:423–438

Teramura Y, Iwata H (2009) Surface modification of islets with PEG-lipid for improvement of graft survival in intraportal transplantation. Transplantation 88:624–630

Xie D, Smyth CA, Eckstein C, Bilbao G, Mays J, Eckhoff DE, Contreras JL (2005) Cytoprotection of PEG-modified adult porcine pancreatic islets for improved xenotransplantation. Biomaterials 26:403–412

7

Enzyme-modified indium tin oxide microelectrode array-based electrochemical uric acid biosensor

Nidhi Puri, Vikash Sharma, Vinod K Tanwar, Nahar Singh, Ashok M Biradar and Rajesh[*]

Abstract

We fabricated a miniaturized electrochemical uric acid biosensor with a 3-aminopropyltriethoxysilane (APTES)-modified indium tin oxide (ITO) microelectrode array (μEA). The ITO-μEA on a glass plate was immobilized with the enzyme uricase, through a cross-linker, bis[sulfosuccinimidyl]suberate (BS^3). The enzyme-immobilized electrode (uricase/BS^3/APTES/ITO-μEA/glass) was characterized by atomic force microscopy and electrochemical techniques. The cyclic voltammetry and impedance studies show an effective binding of uricase at the μEA surface. The amperometric response of the modified electrode was measured towards uric acid concentration in aqueous solution (pH 7.4), under microfluidic channel made of polydimethylsiloxane. The μEA biosensor shows a linear response over a concentration range of 0.058 to 0.71 mM with a sensitivity of 46.26 μA mM^{-1} cm^{-2}. A response time of 40 s reaching a 95% steady-state current value was obtained. The biosensor retains about 85% of enzyme activity for about 6 weeks. The biosensor using μEA instead of a large single band of electrode allows the entire core of the channel to be probed though keeping an improved sensitivity with a small volume of sample and reagents.

Keywords: Uric acid, Self-assembled monolayer, Microfluidic, PDMS, Amperometric sensor

Background

Lab-on-a-chip has become a very popular concept since its inception about a decade ago. It possesses many remarkable features which include the ability to fully integrate all preparation, detection, and analytical processes into a single chip no bigger than the size of a microscopic slide, high throughput, short analysis time, small volume, and high sensitivity (Lee and Lee 2004). When compared to their macrocounterparts, scaling down of electrochemical systems via a microelectrode array (μEA) is anticipated to make the sample size as well as concentration smaller and the electron transfer faster (Wang et al. 2009). Measurement of the concentration of species relies on the chemical phenomena that involve charge transfer (like redox reactions) from or to the electrode. The amount of charge transferred is a direct signal of the concentration of the species, and this can be measured as the charge itself (coulometry), as a current (amperometry; Wang et al. 2009;

Quintino et al. 2005), or as a voltage (potentiometry; Wilson and Dewald 2001; Muñoz and Palmero 2005).

A variety of methods for fabrication of microfluidic devices with subsequent bonding to form channels include photolithography-lift-off (Schoning et al. 2005), soft lithography (Kwakye and Baeumner 2003), and laser ablation (Eswara and Dutt 1974). Whitesides et al. (2001) has reviewed the relevance of soft lithography for the fabrication of microchannels which is faster, cheaper, and more suitable for most biological applications. In device fabrication, the surface alteration and fluid management within the chip are mainly controlled by the surface property of the material, while the detection modem is governed by its optical property (Henaresa et al. 2008).

Uric acid (UA), a final outcome of purine metabolism in biological systems, is an important biological molecule present in body fluids, such as blood and urine. Its level can be used as a marker for the detection of disorders associated with purine metabolism and can reveal the status of immunity (Eswara and Dutt 1974). Its normal

* Correspondence: rajesh_csir@yahoo.com
Polymer and Soft Material Section, CSIR-National Physical Laboratory, Dr. K.S. Krishnan Road, New Delhi 110012, India

level in serum is between 0.13 and 0.46 mM (2.18 to 7.7 mg dL^{-1}; Raj and Ohsaka 2003) and in urinary excretion is between 1.49 and 4.46 mM (25 to 74 mg dL^{-1}; Matos et al. 2000a, b). The presence of abnormal UA levels leads to gout, chronic renal disease, some organic acidemias, leukemia, pneumonia, and Lesch-Nyhan syndrome (Burtic and Ashwood 1994).

In this study, we describe an indium tin oxide microelectrode array (ITO-μEA) printed over a glass plate for the quantitative detection of uric acid in aqueous solution integrated into a polydimethylsiloxane (PDMS)-made microfluidic channel. The surface of ITO-μEA was modified with a self-assembled monolayer (SAM) of 3-aminopropyltriethoxysilane (APTES), which was immobilized with the enzyme uricase through a cross-linker, bis[sulfosuccinimidyl]suberate (BS^3), by forming a strong amide bonding at both ends with free available amino groups of APTES and uricase. The modified electrode (uricase/BS^3/APTES/ITO-μEA/glass) was characterized by atomic force microscopy (AFM), cyclic voltammetry (CV), and electrochemical impedance spectroscopy (EIS) in the presence of $[Fe(CN)_6]^{3-}$ as a redox probe.

Results and discussion

Figure 1 depicts a scheme for the fabrication of uricase/ BS^3/APTES/ITO-μEA/glass, wherein the free NH_2 groups present at the surface of the APTES/ITO-μEA/glass electrode have been utilized for the covalent immobilization of uricase through a cross-linking reagent, BS^3. Figure 2 shows the experimental arrangement of μEA for the detection of uric acid.

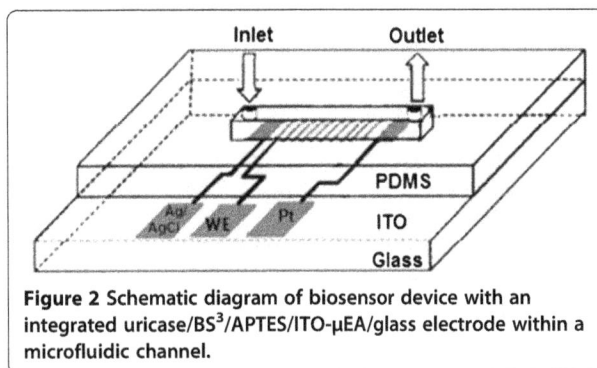

Figure 2 Schematic diagram of biosensor device with an integrated uricase/BS^3/APTES/ITO-μEA/glass electrode within a microfluidic channel.

Surface morphology

The surface characterization of the biosensor was carried out by taking the AFM images of the modified electrode before and after the enzyme immobilization. The AFM image of the APTES/ITO-μEA/glass (Figure 3a) shows densely arrayed APTES molecules with an average image height of 8.6 nm, which is more than the average thickness of APTES monolayer of 3 to 5 nm, indicating a strong polarity in the amino end group of APTES molecule resulting in increased inclined angle of the APTES molecule chain. This makes the APTES molecule chain more disordered and get piled up easily (Wang et al. 2005). The AFM image of uricase-immobilized APTES/ ITO-μEA/glass (Figure 3b) exhibits a regular island-like structure. Since the lateral size of the visible image depends upon the convolution effect that arises between the sample and the AFM tip, an increase of about 8 nm in height in the AFM image was found with the uricase-modified APTES/ITO-μEA/glass surface. This increase

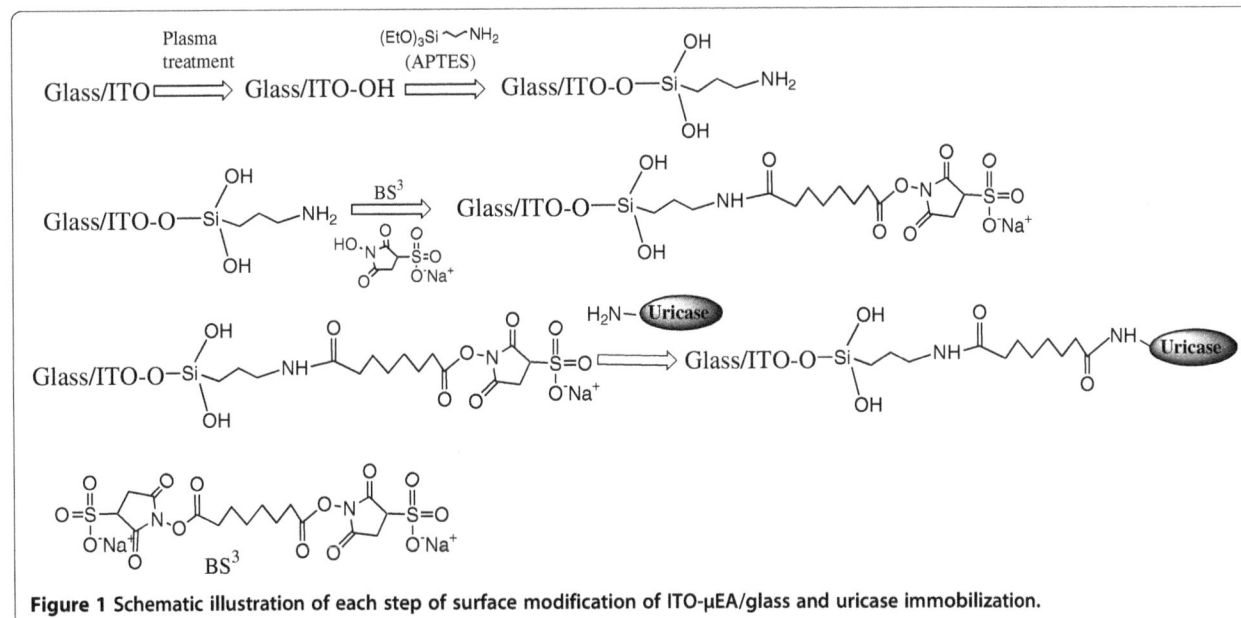

Figure 1 Schematic illustration of each step of surface modification of ITO-μEA/glass and uricase immobilization.

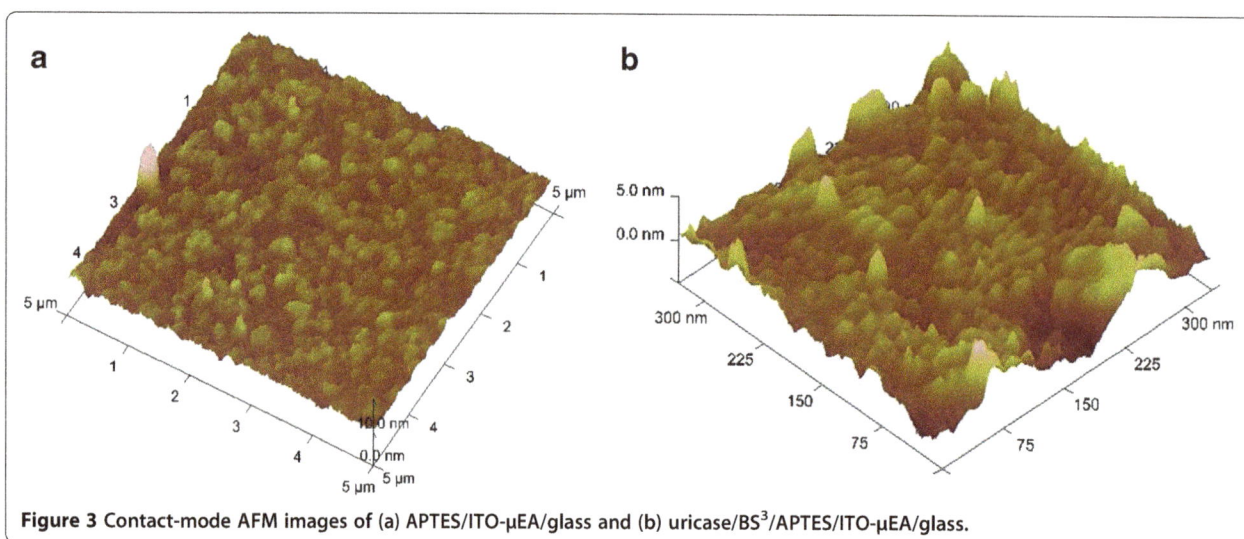

Figure 3 Contact-mode AFM images of (a) APTES/ITO-µEA/glass and (b) uricase/BS³/APTES/ITO-µEA/glass.

in the height of the image is in accordance with the size of the uricase molecule (Akgöl et al. 2008).

Electrochemical characterization of uricase/BS³/APTES/ITO-µEA/glass electrode

The uricase/BS³/APTES/ITO-µEA/glass was characterized by cyclic voltammetry and electrochemical impedance spectroscopy. All electrochemical measurements were performed in phosphate-buffered saline (PBS) solution (pH 7.4) containing 0.1 M KCl and 2 mM $[Fe(CN)_6]^{3-}$ under a PDMS-made microchannel. Figure 4 shows cyclic voltammograms of the modified electrode before and after the enzyme immobilization. The inset shows a magnified view of CVs that correspond to BS³/APTES/ITO-µEA/glass and uricase/BS³/APTES/ITO-µEA/glass electrode.

In all CV experiments, the third cycle was considered as a stable one since no significant changes were observed in the subsequent cycles. The bare ITO-µEA/glass shows a quasi-reversible cyclic voltammogram with a peak-to-peak separation between the oxidation and reduction potential (ΔE_p) of 166.51 mV. Upon modification with SAM of APTES, it shows a more reversible signal with decreased ΔE_p of 144.86 mV between the oxidation and reduction peaks and increased oxidation and reduction current of the redox probe. This is attributed to an increased interfacial concentration of the anionic probe $[Fe(CN)_6]^{3-}$ due to its strong affinity towards the polycationic (NH_2) layer (Zhao et al. 2005). A further modification with a cross-linker, bis[sulfosuccinimidyl]suberate, results in a significant increase in ΔE_p of 152.50 mV with a reduction in the redox current due to a repulsive interaction of polyanions (SO_3-) with the anionic probe $[Fe(CN)_6]^{3-}$, at the surface interface, confirming the formation of a cross-linker, BS³, layer over the surface of APTES/ITO-µEA/glass. This trend of CV curve with an increased ΔE_p of 195.32 mV

and a decreased redox current was further observed after the immobilization of enzyme molecules at the surface of the modified electrode (BS³/APTES/ITO-µEA/glass). This indicates an efficient covalent bonding of insulating enzyme molecules to APTES/ITO-µEA/glass through the cross-linker BS³, which perturbs the interfacial electron transfer considerably, at the electrode surface.

The electrochemical behavior of the modified electrode was characterized by EIS, using an AC signal of 5-mV

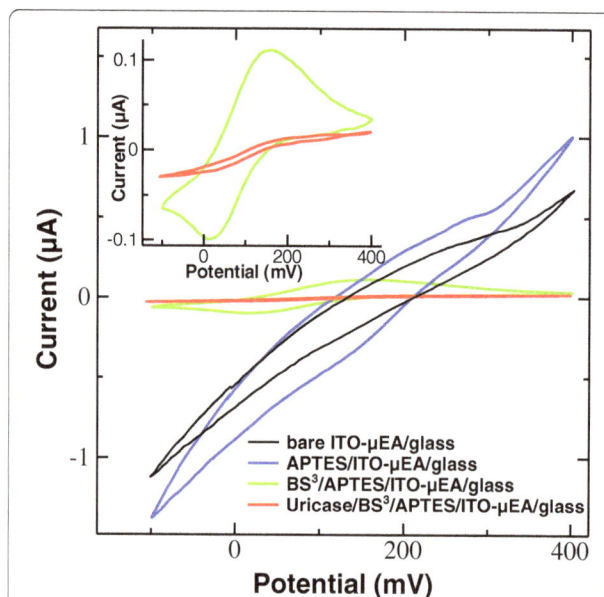

Figure 4 Cyclic voltammograms of bare ITO-µEA/glass, APTES/ITO-µEA/glass, BS³/APTES/ITO-µEA/glass, and uricase/BS³/APTES/ITO-µEA/glass. In 0.1 M KCl solution containing 2 mM $[Fe(CN)_6]^{3-}$. The scan rate is 25 mV s^{-1}. The third cycle voltammogram is shown. The inset shows enlarged CVs of BS³/APTES/ITO-µEA/glass and uricase/BS³/APTES/ITO-µEA/glass.

amplitude, at a formal potential of the redox couple, at a frequency range of 1 to 100,000 Hz. The Nyquist plot, given in Figure 5, shows the impedance spectra taken in each step of the surface modification. The Nyquist plot was fitted using Randles equivalent circuit, as shown in the inset of Figure 5, and a computer software evaluated EIS parameters. The impedance spectrum showed a semi-circle region over a high-frequency range and a linear part at low frequencies, the radius of which corresponds to the charge transfer resistance (R_{et}). The impedance result is modeled by an electronic equivalent circuit, shown in the inset, for the solution resistance (Rs), the Warburg impedance (Zw), which resulted from the diffusion of ions in a bulk electrolyte, the double layer capacitance (Cdl), and the R_{et} for the electrochemical reaction (Randles 1947). Figure 5 shows the electrochemical impedance spectra of the bare and the modified ITO-μEA/glass before and after the immobilization of the enzyme uricase, and the corresponding electron transfer resistance values are listed in Table 1. The bare ITO-μEA/glass shows an R_{et} value of 2.11 KΩ cm^2 which decreased to 1.10 KΩ cm^2 upon treatment with APTES, indicating an easy electronic transport at the electrode surface interface. However, further treatment with a cross-linker, BS^3, and on subsequent immobilization with the enzyme uricase result in increased R_{et} values of 2.90 and 5.17 KΩ cm^2, respectively. The results obtained in EIS are in well agreement with the trend observed in CV measurements, which further confirms the formation of the uricase/BS^3/APTES/ITO-μEA/glass.

Amperometric response of uricase/BS^3/APTES/ITO-μEA/glass

Chronoamperometric response study was carried out with uricase/BS^3/APTES/ITO-μEA/glass as the working

Table 1 ΔE_p and R_{et} of biosensor before and after ITO-μEA surface modifications and enzyme immobilization

	ΔE_p (mV)	R_{et} (KΩ)	R_{et} (KΩ cm^2)[a]
Bare/ITO-μEA/glass	166.51	145.82	2.11
APTES/ITO-μEA/glass	144.86	76.23	1.10
BS^3/APTES/ITO-μEA/glass	152.50	200.13	2.90
Uricase/BS^3/APTES/ITO-μEA/glass	195.32	356.79	5.17

[a]Surface area of ITO-μEA is 1.45 mm^2; ΔE_p, change in peak-to-peak separation in redox potential; R_{et}, charge transfer resistance.

electrode at a bias voltage of 0.26 V vs. Ag/AgCl in PBS (pH 7.4) containing 2 mM [Fe(CN)$_6$]$^{3-}$ as the redox mediator. Figure 6 shows a chronoamperometric response of the uricase/BS^3/APTES/ITO-μEA/glass electrode as a function of uric acid concentration in PBS, in the presence of a redox mediator. An increasing order of amperometric response was observed with increasing uric acid concentration, and 95% steady-state current response to uric acid was obtained in about 40 s.

Figure 7 shows the steady-state current dependence calibration curve to uric acid concentration. The amperometric response of the device to uric acid concentration was found to be linear in the range of 0.058 to 0.71 mM with a correlation coefficient of 0.992 (n = 5). The lowest detection limit of the electrode was 0.0084 mM at a signal-to-noise ratio of 3. The sensitivity of the enzyme electrode was calculated from the slope (m) of the linearity curve, and it was found to be 46.26 μA mM^{-1} cm^{-2}. The stability of the biosensor was studied, under the storage condition of 4°C to 5°C, by

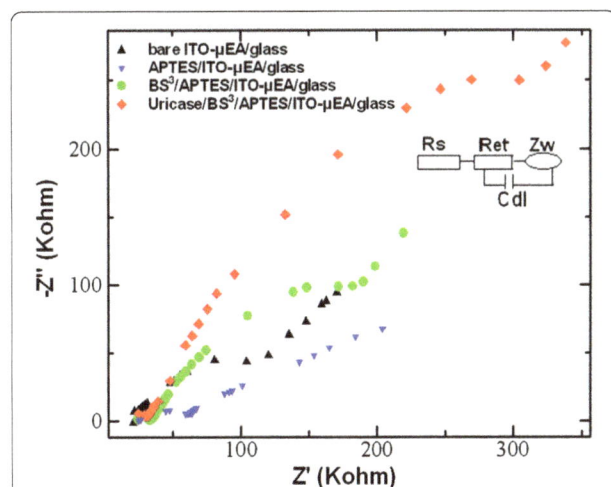

Figure 5 Nyquist plot obtained for bare ITO-μEA/glass, APTES/ITO-μEA/glass, BS^3/APTES/ITO-μEA/glass, and uricase/ BS^3/APTES/ITO-μEA/glass. In 0.1 M KCl solution containing 2 mM [Fe(CN)$_6$]$^{3-}$.

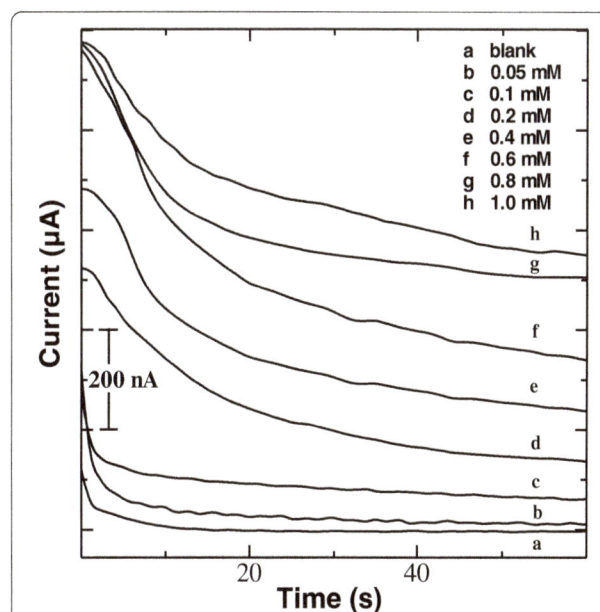

Figure 6 Chronoamperometric response curve of uricase/BS^3/ APTES/ITO-μEA/glass with different concentrations of uric acid.

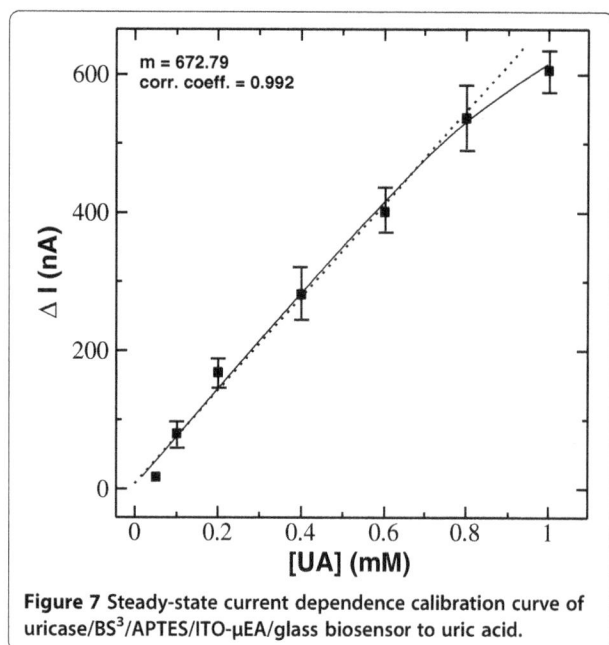

m = 672.79
corr. coeff. = 0.992

Figure 7 Steady-state current dependence calibration curve of uricase/BS^3/APTES/ITO-µEA/glass biosensor to uric acid.

be due to the limited stability of the enzyme over the silane matrix over a period of time.

Conclusions

A biosensor was fabricated by immobilizing the enzyme uricase on SAM of APTES via the cross-linker BS^3 on an ITO-µEA/glass plate. The uricase/BS^3/APTES/ITO-µEA/ glass electrode was characterized by electrochemical techniques and AFM, whereas the amperometric response was studied as a function of uric acid concentration. The biosensor with uricase/BS^3/APTES/ITO-µEA/glass electrode showed a linear range of 0.058 to 0.71 mM with a lower detection limit of 8.4 µM. The response time was found to be 40 s reaching a 95% steady-state current value. The efficient bonding of the enzyme on the electrode surface exhibits an improved sensitivity of 46.26 µA mM^{-1} cm^{-2}. The microfluidic channel provided the controlled volume of the sample to be tested in close proximity to the uricase/BS^3/APTES/ ITO-µEA/glass electrode for a fast electrochemical reaction, wherein the µEA helped in moving the electron through interconnected microelectrode bands having a comparatively reduced area to a single-band -like structure. The easy method of fabrication and the small sample volume requirement together with the high sensitivity towards uric acid measurement make this biosensor advantageous over the recently reported uric acid biosensors (Table 2). This system may

continuously monitoring the current response for 0.6 mM uric acid, at an interval of 1 week. The biosensor retained its current response to uric acid with a slow decrement of about 12% over a period of 6 weeks. The biosensor shows a sharp decrement up to about 40% to its initial activity on the basis of amperometric current response that might

Table 2 Characteristics of some recently reported amperometric uric acid biosensors

Matrix	Response time (s)	Stability	Linear range	Sensitivity	Bias potential (V)	Reference
SAM of heteroaromatic thiol/Au	-	1 day	1 to 300 µM	0.0149 ± 0.05 µA mM^{-1}	~0.4	Raj and Ohsaka (2003)
o-Aminophenol-aniline copolymer	-	~50 days	0.0001 to 0.5 mmol dm^{-3}	~2.65 µA mM^{-1}	0.4	Pan et al. (2006)
Ir-C	41	-	0.1 to 0.8 mM	16.60 µA mM^{-1}	0.25	Luo et al. (2006)
Polyaniline	-	~60 days	0.0036 to 1.0 mmol dm^{-3}	~1 µA mM^{-1}	0.4	Kan et al. (2004)
Polyaniline-PPy	70	4 weeks	2.5 × 10^{-6} to 8.5 ×10^{-5} M	1.12 µA mM^{-1}	0.4	Arslan (2008)
Poly(allylamine)/poly (vinyl sulfate)			10^{-6} to 10^{-3} M	<5 µA mM^{-1}	0.6	Hoshi et al. (2003)
Polyaniline			0.001 to 1.0 mM dm^{-3}	<7 µA mM^{-1}	0.4	Jiang et al. (2007)
ZnS quantum dots		20 days	5.0 × 10^{-6} to 2.0 × 10^{-3} mol L^{-1}	2.2 µA mM^{-1}	0.45	Zhang et al. (2006)
Carbon paste	70	>100 days	Up to 100 µmol dm^{-3}		0.34	Dutra et al. (2005)
Polypropylene			4.82 to 10.94 mg dL^{-1}	0.0029 µA mM^{-1}		Chen et al. (2005)
SAM of PET and DTB on Au	80 to 100		5 to 150 µM	3.4 ± 0.08 nA cm^{-2} µM^{-1}	−0.1	Behera and Raj (2007)
ZnO nanorods			5.0 × 10^{-6} to 1.0 × 10^{-3} mol L^{-1}			Zhang et al. (2004)
GNP-based uric acid biosensor	25		0.07 to 0.63 mM	19.27 µA mM^{-1}		Ahuja et al. (2011)
APTES-based ITO-µEA/glass electrode	40	6 weeks	0.058 to 0.71 mM	46.26 µA mM^{-1} cm^{-2}	0.26	Present work

be optimized further for selectivity with different interferants using a real blood sample or serum for other species of biomedical importance.

Methods
Materials

The enzyme uricase (EC 1.7.3.3, 9 units mg^{-1} from *Bacillus fastidiosus*) was procured from Sigma-Aldrich Corp. (St. Louis, MO, USA). APTES was purchased from Merck Chemicals (Darmstadt, Germany). BS^3 was obtained from Pierce Biotechnology (Rockford, IL, USA). Uric acid with 99% purity was purchased from CDH (New Delhi, India). Positive photoresist 1300–31 was purchased from Shipley (Marlborough, MA, USA), and negative photoresist SU-8-2025 is from MicroChem Corp. (Newton, MA, USA). Sylgard 184 PDMS was purchased from Dow Corning (Midland, MI, USA). Other chemicals were of analytical grade and used without further purification.

Apparatus

AFM images were obtained on a VEECO/diCP2 scanning probe microscope (Plainview, NY, USA). CV and EIS measurements were done on a PGSTAT302N AUTOLAB instrument from Eco Chemie (Utrecht, The Netherlands). Impedance measurements were performed in the presence of a redox probe, $[Fe(CN)_6]^{3-}$, at scanning frequencies from 1 to 100,000 Hz. All measurements were carried out on a three-electrode system with uricase/BS^3/APTES/ITO-μEA/glass as the working electrode, Ag/AgCl as the reference electrode, and platinum as the counter electrode.

Fabrication of uricase/BS^3/APTES/ITO-μEA/glass-embedded microfluidic device

The ITO-coated glass plates (2×3 cm^2) with a typical resistance of approximately 40 Ω were cleaned by ultrasonic cleaning, in chronological order, in soapy water (Extran, Merck Millipore, Billerica, MA, USA), acetone, ethanol, isopropyl alcohol, and double-distilled (DD) water for 10 min each, and drying in vacuum. These cleaned ITO glass plates were then exposed to oxygen plasma in a plasma chamber for 5 min. SAM of APTES was prepared over an ITO glass plate by immersing it in 2% ethanolic solution of APTES for 1.5 h, under ambient conditions. The glass plate was then rinsed with ethanol in order to remove the majority of non-bonded APTES from the surface of the substrate and dried it under N$_2$.

A three-electrode system with a pattern of μEA (1.45 mm^2) as the working electrode and Ag/AgCl and Pt as the reference and counter electrodes, respectively, was printed over a glass plate. The μEA pattern was transferred over an APTES-coated ITO-glass by photolithography using Shipley positive photoresist 1300–31,

followed by etching of remaining exposed ITO coating by treating with a suspension of zinc dust and dilute HCl. The ITO-μEA is composed of 42 interconnected microbands having a band width of 65 μm and a band length of 225 μm with an interspacing bandgap of 65 μm, prepared according to a procedure reported earlier (Chen and White 2011). The Ag pattern was deposited over the glass plate by e-beam evaporation technique, followed by treatment with 1 mM solution of FeCl$_3$ for 10 s, and washed with DD water to obtain an Ag/AgCl reference electrode (Polk et al. 2006). The APTES/ITO-μEA/glass electrode was then modified with SAM of BS^3 by treating it with 5 mM BS^3 solution prepared in sodium acetate buffer (pH 5.0) for 1.0 h, washed with DD water, and dried under N$_2$. The BS^3-treated APTES/ITO-μEA/glass electrode (BS^3/APTES/ITO-μEA/glass) was immersed in PBS solution containing approximately 3 U of uricase (pH 7.4) for a period of 1.5 h. The enzyme electrode so formed was washed thrice with PBS (pH 7.4) to remove the excess unbound enzyme and was finally dried under N$_2$ at room temperature and stored at 4°C.

A PDMS microfluidic channel was prepared to deliver a controlled sample volume over a glass plate comprising a three-electrode system with a pattern of a uricase/BS^3/APTES/ITO-μEA/glass working electrode, Ag/AgCl reference electrode, and Pt counter electrode for response measurement. A master was created with Shipley negative photoresist SU-8-2025 on a smooth glass plate by spin coating with a speed of 1,000 rpm and exposing through a photomask under highly intense UV light for 20 s. The resulting master structure was used as a mold to create PDMS blocks equipped with microfluidic channels of 75 μm in height and 0.5 mm × 16 mm area with sample inlet and outlet ports ($D = 1$ mm) punched at the two ends of the microchannel for fluid flow.

Competing interests
The authors declare that they have no competing interests.

Authors' contributions
NP carried out the EIS studies. VS contributed in the soft lithography for making required microfluidic channels using PDMS. VKT carried out the fabrication of ITO-μEA on a glass substrate. NS participated in the AFM characterization of the biosensor. AMB participated in the scientific and technical discussions. R contributed in the implementation and conclusion of this conceptual work. All authors read and approved the final manuscript.

Acknowledgements
We are grateful to Prof. R.C. Budhani, Director, National Physical Laboratory, New Delhi, India, for providing the facilities. One of the authors, Nidhi Puri, is thankful to CSIR, India, for the financial assistance.

References
Ahuja T, Tanwar VK, Mishra SK, Kumar D, Biradar AM, Rajesh (2011) Immobilization of uricase enzyme on self-assembled gold nanoparticles for application in uric acid biosensor. J Nanosci Nanotech 11:4692–4701. doi:10.1166/jnn.2011.4158

Akgöl S, Öztürk N, Karagözler AA, Uygun DA, Uygun M, Denizli A (2008) A new metal-chelated beads for reversible use in uricase adsorption. J Mol Cat B: Enzym 51:36–41. doi:10.1016/j.molcatb.2007.10.005

Arslan F (2008) An amperometric biosensor for uric acid determination prepared from uricase immobilized in polyaniline-polypyrrole film. Sensors 8:5492–5500. doi:10.3390/s8095492

Behera S, Raj CR (2007) Mercaptoethylpyrazine promoted electrochemistry of redox protein and amperometric biosensing of uric acid. Biosens Bioelectron 23:556–561. doi:10.1016/j.bios.2007.06.012

Burtic CA, Ashwood ER (1994) Teitz textbook of clinical chemistry, 2nd edn. Saunders, Philadelphia

Chen IJ, White IM (2011) High-sensitivity electrochemical enzyme-linked assay on a microfluidic interdigitated microelectrode. Biosens Bioelectron 26:4375–4381. doi:10.1016/j.bios.2011.04.044

Chen J-C, Chung H-H, Hsu C-T, Tsai D-M, Kumar AS, Zen J-M (2005) A disposable single-use electrochemical sensor for the detection of uric acid in human whole blood. Sens Actu B: Chem 110:364–369. doi:10.1016/j.snb.2005.02.026

Dutra RF, Moreira KA, Oliveira MIP, Araujo AN, Montenegro MCBS, Filho JLL, Silva VL (2005) An inexpensive biosensor for uric acid determination in human serum by flow-injection analysis. Electroanalysis 17:701–705. doi:10.1002/elan.200403142

Eswara Dutt VVS, Mottola HA (1974) Determination of uric acid at the microgram level by a kinetic procedure based on a pseudo-induction period. Anal Chem 46:1777–1781. doi:10.1021/ac60348a041

Henaresa TG, Mizutania F, Hisamotob H (2008) Current development in microfluidic immunosensing chip. Anal Chim Acta 611:17–30. doi:10.1016/j.aca.2008.01.064

Hoshi T, Saiki H, Anzai J (2003) Amperometric uric acid sensors based on polyelectrolyte multilayer films. Talanta 61:363–368. doi:10.1016/S0039-9140(03)00303-5

Jiang Y, Wang A, Kan J (2007) Selective uricase biosensor based on polyaniline synthesized in ionic liquid. Sens Actu B: Chem 124:529–534. doi:10.1016/j.snb.2007.01.016

Kan J, Pan X, Chen C (2004) Polyaniline–uricase biosensor prepared with template process. Biosens Bioelectron 19:1635–1640. doi:10.1016/j.bios.2003.12.032

Kwakye S, Baeumner AJ (2003) A microfluidic biosensor based on nucleic acid sequence recognition. Anal Bioanal Chem 376:1062–1068. doi:10.1007/s00216-003-2063-2

Lee SJ, Lee SY (2004) Micro total analysis system (μ-TAS) in biotechnology. Appl Microbiol Biotechnol 64:289–299. doi:10.1007/s00253-003-1515-0

Luo YC, Do JS, Liu CC (2006) An amperometric uric acid biosensor based on modified Ir–C electrode. Biosens Bioelectron 22:482–488. doi:10.1016/j.bios.2006.07.013

Matos RC, Angnes L, Araujo MC, Saldanha TC (2000a) Modified microelectrodes and multivariate calibration for flow injection amperometric simultaneous determination of ascorbic acid, dopamine, epinephrine and dipyrone. Analyst 125:2011–2015. doi:10.1039/B004805O

Matos RC, Augelli MA, Lago CL, Angnes L (2000b) Flow injection analysis-amperometric determination of ascorbic and uric acids in urine using arrays of gold microelectrodes modified by electrodeposition of palladium. Anal Chim Acta 404:151–157. doi:10.1016/S0003-2670(99)00674-1

Muñoz E, Palmero S (2005) Analysis and speciation of arsenic by stripping potentiometry: a review. Talanta 65:613–620. doi:10.1016/j.talanta.2004.07.034

Pan X, Zhou S, Chen C, Kan J (2006) Preparation and properties of an uricase biosensor based on copolymer of o-aminophenol-aniline. Sens Actu B: Chem 113:329–334. doi:10.1016/j.snb.2005.03.086

Polk BJ, Stelzenmuller A, Mijares G, MacCrehan W, Gaitan M (2006) Ag/AgCl microelectrodes with improved stability for microfluidics. Sens and Actu B: Chem 114:239–247. doi:10.1016/j.snb.2005.03.121

Quintino MDSM, Winnischofer H, Nakamura M, Araki K, Toma HE, Angnes L (2005) Amperometric sensor for glucose based on electrochemically polymerized tetraruthenated nickel-porphyrin. Anal Chim Acta 539:215–222. doi:10.1016/j.aca.2005.02.057

Raj CR, Ohsaka T (2003) Voltammetric detection of uric acid in the presence of ascorbic acid at a gold electrode modified with a self-assembled monolayer of heteroaromatic thiol. J Electroanal Chem 540:69–77. doi:10.1016/S0022-0728(02)01285-8

Randles JBB (1947) Kinetics of rapid electrode reactions. Discuss Faraday Soc 1:11. doi:10.1039/DF9470100011

Schoning MJ, Jacobs M, Muck A, Knobbe DT, Wang J, Chatrathi M, Spillmann S (2005) Amperometric PDMS/glass capillary electrophoresis-based biosensor microchip for catechol and dopamine detection. Sens Actu B: Chem 108:688–694. doi:10.1016/j.snb.2004.11.032

Wang YP, Yuan K, Li QL, Wang LP, Gu SJ, Pei XW (2005) Preparation and characterization of poly(n-isopropylacrylamide) films on a modified glass surface via surface initiated redox polymerization. Mat Lett 59:1736–1740. doi:10.1016/j.matlet.2005.01.048

Wang J, Tian B, Chatrathi MP, Escarpa A, Pumera M (2009) Effects of heterogeneous electron-transfer rate on the resolution of electrophoretic separations based on microfluidics with end-column electrochemical detection. Electrophoresis 30:3334–3338. doi:10.1002/elps.200800845

Whitesides GM, Ostuni E, Takayama S, Jiang X, Ingber DE (2001) Soft lithography in biology and biochemistry. Annu Rev Biomed Eng 3:335–373. doi:10.1146/annurev.bioeng.3.1.335

Wilson MM, Dewald HD (2001) Stripping potentiometry of indium in aqueous chloride solutions. Microchem J 69:13–19. doi:10.1016/S0026-265X(00)00184-3

Zhang F, Wang X, Ai S, Sun Z, Wan Q, Zhu Z, Xian Y, Jin L, Yamamoto K (2004) Immobilization of uricase on ZnO nanorods for a reagentless uric acid biosensor. Anal Chim Acta 519:155–160. doi:10.1016/j.aca.2004.05.070

Zhang F, Li C, Li X, Wang X, Wan Q, Xian Y, Jin L, Yamamoto K (2006) ZnS quantum dots derived a reagentless uric acid biosensor. Talanta 68:1353–1358. doi:10.1016/j.talanta.2005.07.051

Zhao J, Bradbury CR, Huclova S, Potapova I, Carrara M, Fermin DJ (2005) Nanoparticle-mediated electron transfer across ultrathin self-assembled films. J Phy Chem B 109:22985–22994. doi:10.1021/jp054127s

8

Structure and biocompatibility of poly(vinyl alcohol)-based and agarose-based monolithic composites with embedded divinylbenzene-styrene polymeric particles

Lydia G Berezhna[1*], Alexander E Ivanov[1], André Leistner[2], Anke Lehmann[2], Maria Viloria-Cols[1] and Hans Jungvid[1]

Abstract

Macroporous monolithic composites with embedded divinylbenzene-styrene (DVB-ST) polymeric particles were prepared by cryogelation techniques using poly(vinyl alcohol) or agarose solutions. Scanning electron microscopy images showed multiple interconnected pores with an average diameter in the range of 4 to 180 μm and quite homogeneous distribution of DVB-ST particles in the composites. Biocompatibility of the composites was assessed by estimation of the C5a fragment of complement in the blood serum and concentration of fibrinogen in the blood plasma which contacted the composites. A time-dependent generation of C5a fragment indicated weak activation of the complement system. At the same time, the difference in fibrinogen concentration, one of the most important proteins in the coagulation system of the blood, between the pristine blood plasma and the plasma, circulated through the monolithic columns, was insignificant.

Keywords: Cryogels, Poly(vinyl alcohol), Agarose, Divinylbenzene-styrene particles, C5a fragment of complement, Fibrinogen

Background

Macroporous monolithic materials produced from hydrophilic polymers by cryogelation techniques find more and more applications in biomedical science and technology. The areas for material applications are separation and purification of proteins (Dainiak et al. 2004), bioaffinity screening techniques (Hanora et al. 2005), synthesis of adsorbents for water purification (Le Noir et al. 2007), or development of tissue-engineering scaffolds (Bolgen et al. 2007) including those used for wound healing (Dainiak et al. 2010). Cryogelation is a process of gel formation, which takes place in a semi-frozen state. Under such condition, the gel forms in narrow unfrozen zones with high concentrations of reagents. At the same time, the crystals of frozen aqueous medium act as porogens and form large interconnected pores appearing after defrosting (Kirsebom et al. 2009). The cryogelation techniques allow preparation of elastic mechanically stable monolithic matrices easily permeable to aqueous solutions of proteins and suspensions of cells. The monolithic gels exhibit multiple interconnected pores of 1 to 100 μm in diameter, which can be controlled by changing the conditions of synthesis (Kirsebom et al. 2009; Plieva et al. 2007). Various types of monolithic porous polymer structures, MPPS™, produced by cryogelation of hydrophilic polymers were developed by Protista (www.protista.se). An attractive feature of the cryogelation technique is the possibility to embed suspended dispersed materials such as polymeric microparticles into the monoliths, which allows for increased absorption capacity and targeting the adsorbents against selected compounds (Ivanov et al. 2012; Koc et al. 2011; Özgür et al. 2011). Recently, polystyrene microparticles were shown to be effective adsorbents for liver toxins such as bilirubin, bile acid, and aromatic amino acids (Weber et al. 2008). In terms of this concept, the composite cryogels with embedded adsorptive microparticles can be considered as promising materials for extracorporeal blood purification and removal of liver toxins, pro-inflammatory factors, xenobiotics, or

* Correspondence: blg_lida@yahoo.com

[1]Protista Biotechnology AB, Bjuv SE-26722, Sweden

Full list of author information is available at the end of the article

even cancer cells from the human blood. Apart from extracorporeal blood purification, the removal of toxic metabolites from human or animal plasma is challenging in view of production of cell growth media. In particular, removal of toxic metabolites related to liver failure or pro-inflammatory cytokine tumor necrosis factor is a need for the studies performed with cultured hepatocytes (Saich et al. 2007; Jones and Czaja 1998). The above production processes and research applications are required for adsorbent devices with low resistance to the flow of plasma combined with effective adsorption of various toxins. The permeable monolithic composites suggest an adequate type of the adsorbent.

However, materials that are supposed to be in contact with human organisms or blood plasma must be biocompatible, in particular, concerning their interaction with clotting and complement systems of the blood (Kirkpatrick et al. 1998). Both agarose and poly(vinyl alcohol) (PVA) were earlier used to synthesize biomaterials (Varoni et al. 2012; Paradossi et al. 2003) Correspondingly, the aim of the present study was to assess the possibility of biomedical applications of the composite cryogels based on PVA or agarose with embedded divinylbenzene-styrene (DVB-ST) polymeric microparticles. To achieve this, the estimation of the C5a component of complement and fibrinogen in the human blood serum or plasma during their contact with the above composite materials has been performed. In view of the different chemical structures of the chosen polymers and, consequently, different methods of cryogelation, the difference in their microstructure and interaction with embedded polymeric microparticles can be expected. Thus, another aim of the study was the characterization of the microstructure of monolithic composites.

Results and discussion
Microstructure of the composite and non-composite monoliths
Micrograph comparison of particle-free agarose and PVA slices, observed by scanning electron microscopy (SEM) at low magnification (Figure 1), has shown that both types of the cryogel monoliths had quite similar structures with evenly distributed interconnected pores, with an average pore size within the range 4 to 100 μm. However, at higher magnification, the difference between the structure of agarose and PVA gels could be observed. The PVA cryogel has porous thick walls of about 10 to 30 μm (Figure 1d) in contrast with the smooth thin (approximately 2 μm) walls of the agarose gel. No visible pores in the walls of the agarose matrix could be seen on the SEM micrographs (Figure 1b). In theory, the different porosities of the particle-free monoliths gels may result in different adsorption characteristics due to an increase in the total binding surface area of the PVA monoliths with porous walls.

Both PVA- and agarose-based composites that consist of DVB-ST polymeric particles connected by sheets and bridges of the cryogelated polymers and exhibited large cavities (up to 180 μm in diameter) are detectable throughout the materials, whereas the mean pore size was 30 μm (Figure 2). Some pores of the larger sized particles observed in the

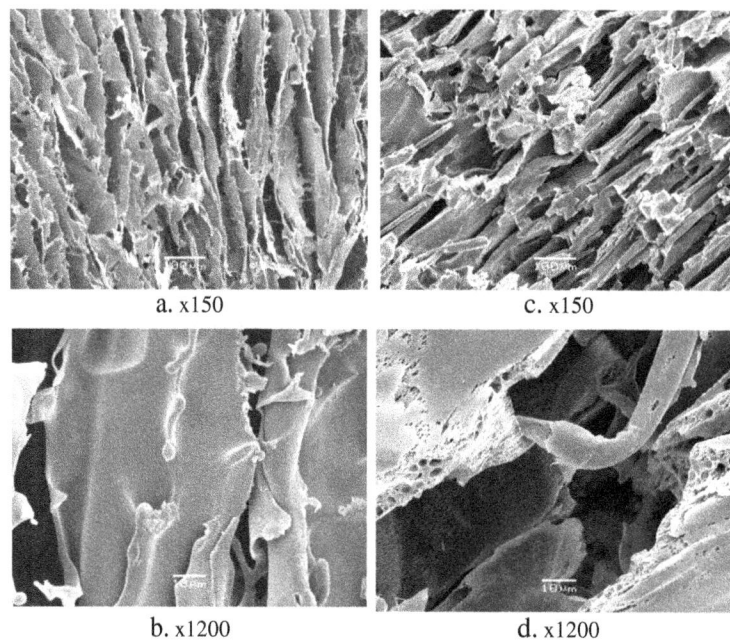

a. x150   c. x150

b. x1200   d. x1200

Figure 1 SEM micrographs of slices of the particle-free monolithic cryogels. They are prepared from (**a**, **b**) 3% agarose solution and (**c**, **d**) 5% PVA solution.

a. x35   b. x35

Figure 2 SEM micrographs of slices of cryogel monoliths. At (**a**) 3% agarose and (**b**) 5% PVA with embedded DVB-ST polymeric particles.

composites probably originated from critical point drying, which resulted in 96% shrinkage of the particle-free agarose and PVA, but not of the composites. Resistance of the composites to shrinkage could be explained by the higher mechanical stability provided by the particles.

Important features of the cryogel composites were high density and homogeneous distribution of the embedded particles in the polymer matrices. There was no big difference between the content of DVB-ST beads in the slices obtained from the top (Figure 3a,c) and the bottom (Figure 3b,d) of the agarose- and PVA-based composites. It is worth noting that some parts of the DVB-ST polymeric particles were tightly embedded or even wrapped in the cryogel matrix, whereas some other parts remained free (Figure 3).

The presence of relatively narrow (4 μm in diameter) pores in cryogel monoliths, as well as the high packing density of DVB-ST polymeric particles, could be a restricting factor for the passage of whole blood through the composites. Some types of leukocytes with sizes of 10 to 21 μm might be stuck in the narrow pores. However, this does not preclude an application possibility of these materials for purification of blood plasma, which could be easily percolated through the monoliths by peristaltic pumping. According to the obtained data, the throughput capacity of the cryogel composites under a pressure of 1 kPa was 0.5 mL/min.

Adsorption of fibrinogen

For biocompatibility of new biomaterials, determination of fibrinogen as well as some fragments of complement called 'split products' in the contacting blood is a mandatory test according to the ISO regulatory documents (ENSAI 2009). Fibrinogen is one of the most important proteins in the coagulation process. Absorption of the fibrinogen homodimer on the surface of any material in contact with blood can mediate the formation of platelet aggregates, further leading to thrombosis (Vogler and Siedlecki 2009; Beugeling 1979). Concentration of the fibrinogen in blood plasma samples before

and after their passage through the monoliths, both noncomposite and composite, is illustrated in Figure 4.

Judged by the obtained data, there was no statistically significant depletion of fibrinogen from the blood plasma which contacted the composites in comparison with the plasma passed through the pumping system without monoliths. The low adsorption of fibrinogen on the monoliths may be explained by two independent reasons. The first one is that this protein is known to be rapidly (within 1 to 2 min) adsorbed by a number of artificial materials, which then gradually release the protein in the plasma (Brash et al. 1988). This phenomenon is due to the displacement of fibrinogen by the other plasma proteins, and it is called the Vroman effect (Brash et al. 1988; Vroman et al. 1980). The equilibrium amount of fibrinogen adsorbed from the blood plasma on nonporous polystyrene was found to be as low as 250 ng/cm^2 (Tsai et al. 1999). This could be a reason why the fibrinogen concentration in the samples of plasma percolated through the monoliths for 30 min was almost the same as that found in the native plasma. The second reason is that the embedded DVB-ST microparticles might have too little fraction of pores accessible for the adsorption of fibrinogen, a relatively large protein with a molecular weight of 340 kDa, and a hydrodynamic radius of 12.7 nm determined at physiological conditions (Wasilewska et al. 2009). The latter hypothesis can be checked by the following calculation. The outer specific surface area of 75-μm DVB-ST beads of known bulk density (0.28 g/cm^3) can be estimated as *ca.* 0.14 m^2/g. Note that the total specific surface area of the beads (S_{BET} = 641 m^2/g) is 4,500 times larger due to the inner porosity of the beads (Table 1). Thus, the DVB-ST beads contained in a composite cryogel monolith (500 mg, see the 'Methods' section) might have a maximum outer surface area of 0.07 m^2 available for this protein adsorption. One can evaluate the corresponding equilibrium amount of fibrinogen adsorbed on the outer surface as *ca.* 180 μg/monolithic gel. Under the conditions of the experiment and at ±0.3 mg/mL standard deviation of fibrinogen concentration measurements (see Figure 4), the mentioned adsorbed amount would be within the limits of the experimental error (±1 mg fibrinogen/gel). Since there

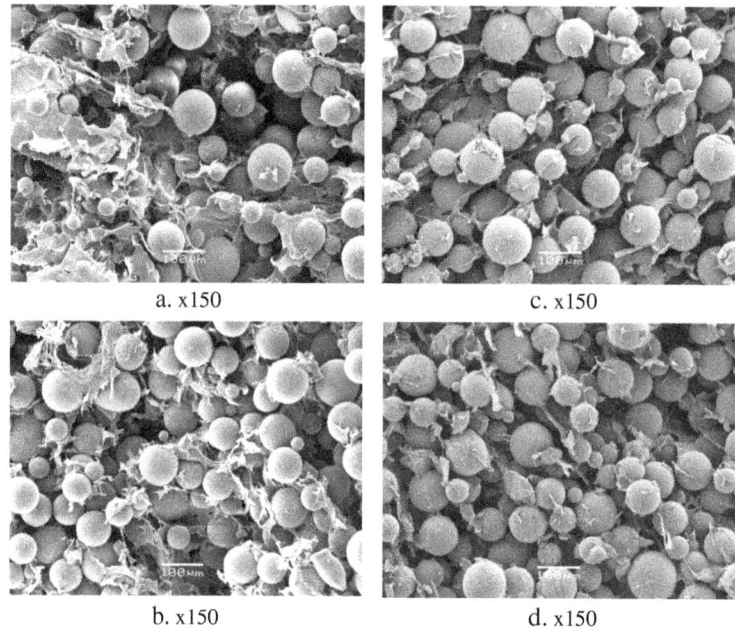

Figure 3 SEM micrographs of slices of 3% agarose and 5% PVA cryogel monoliths. (**a, b**) 3% agarose and (**c, d**) 5% PVA cryogel monoliths with embedded DVB-ST polymeric particles.

were no statistically confident differences between the fibrinogen concentrations in the native and the percolated plasma samples (Figure 4), neither the pores of the DVB-ST beads nor the cryogelated polymer structures (PVA or agarose) (see Table 1) could provide essentially more binding sites for fibrinogen than the outer surface of the beads. The composite monoliths adsorbed much lower amounts of fibrinogen than one could expect from DVB-ST polymeric particles with the above specific surface area and a pronounced fraction of macropores (V_{macro} = 0.22 cm^3/g,

Table 1). This could be a consequence of low accessibility of pores or, possibly, shielding of the bead surface by microscopic fragments of cryogelated PVA or agarose. Fibrinogen adsorption is known to mediate adhesion of platelets, the phenomenon involved in the formation of thrombus (Pulanic and Rudan 2005). The insignificant fibrinogen adsorption on the composite cryogel monoliths can indicate low thrombogenicity of the materials, though this requires for a more detailed investigation of the blood coagulation system brought into contact with the composites.

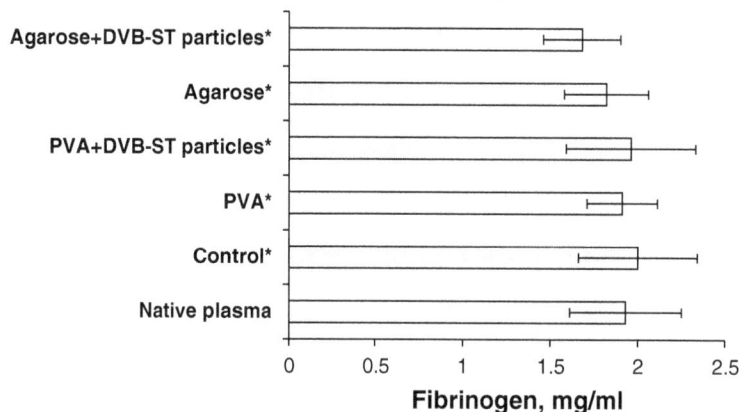

Figure 4 Fibrinogen concentration in blood plasma (M ± SD; *n* = 3). It is percolated through the different types of cryogel monoliths: particle-free 5% PVA, particle-free 3% agarose, PVA, and agarose composites, with embedded (DVB-ST) polymeric particles. The control is plasma percolated through the pumping system without monoliths. Asterisk signifies that $p \geq 0.05$ in comparison with the native plasma.

Table 1 Characteristics of DVB-ST particles

Characteristic	Value
Particle diameter (d_p)	75 μm
Specific surface area (S_{BET})	641 m^2/g
Volume of micropores (V_{micro})	0.27 cm^3/g
Volume of mesopores (V_{meso})	0.88 cm^3/g
Volume of macropores (V_{macro})	0.22 cm^3/g
Average pore diameter (d_{pore})	7.7 nm

Activation of complement system

Determination of C5a is one of the methods for the estimation of complement system activation that resulted from the contact of blood with artificial materials (ENSAI 2009). C5a anaphylatoxin, formed as a part of the sequential cascade of complement activation pathways, is known as a potent pro-inflammatory agent, which takes part in leucocyte activation, cytokine production, release of histamine etc. (Sarma and Ward 2010). In the present study, the time-dependent generation of C5a in the human blood serum during its contact with particle-free or composite monoliths has been examined. The concentration of C5a in the serum incubated with slices of particle-free PVA and agarose cryogels increased with time and, at the final point (30 min), was 1.5 to 2 times higher than the initial value (Figure 5a,b). These data demonstrated that both macroporous gels of agarose and PVA activated the complement system in the human serum, which is in agreement with earlier reported data (Lyle et al. 2010; Black and Sefton 2000; Arima et al. 2009) where the complement activation through the alternative pathway by such polymers as agarose (Lyle et al. 2010) and PVA (Black and Sefton 2000) hydrogels has been shown. In those experiments, the concentrations of the Bb fragment and the soluble form of the terminal membrane attack complex SC5b-9 were measured during the contact of hydrogels with blood serum (Black and Sefton 2000). It is relevant to note that hydrophobic polymers such as polyethylene or polydimethylsiloxane brought into contact with serum caused a much weaker activation of the complement compared with that in hydrophilic polymers such as PVA or cellulose. Heavily hydroxylated chains of these polymers are believed to trigger the alternative pathway. In our experiments, complement activation due to contact of serum with the composite monoliths seemed to be relatively weak because the positive control serum containing zymosan A exhibited a much higher 11-fold increase in the concentration of C5a, already after 10 min of incubation (Figure 5c).

There was no statistically valid difference between the time dependent generation of C5a fragment in the serum contacting the particle-free cryogels and the composites (Figure 5a,b). A slight decrease in C5a concentration, observed for both composites after 30 min of the contact,

can be explained by adsorption of the C5a fragment on the embedded DVB-ST polymeric particles. It seems like the observed activation of the complement was mainly due to the interaction of its components with the hydrophilic PVA and agarose.

Conclusions

Macroporous monolithic composites based on cryogelated PVA and agarose with embedded DVB-ST microparticles were prepared. The composites were densely loaded with homogeneously distributed microparticles and exhibited multiple interconnected pores with the size of 4 to 180 μm. The porous structure allowed for a facile percolation of blood plasma through the monoliths. The percolation did not influence fibrinogen concentration in the plasma, which indicated the absence of any significant effect of the materials on the blood coagulation system. The contact of both particle-free and composite monoliths with blood serum caused a slight increase in the C5a concentration in the serum, which was mainly due to the interaction of complement components with cryogelated PVA and agarose. Embedment of DVB-ST polymeric microparticles into the cryogel monoliths may be advised for the further investigation of these composites as a promising substance for the synthesis of blood compatible materials.

Methods

Materials

Poly(vinyl alcohol) with an average molecular weight of 1.3×10^5 g/mol (Mowiol 18–88) was purchased from Clariant GmbH (Frankfurt-on-Main, Germany). Agarose (electrophoretic grade) was from Fluka (Buchs, Belgium). DVB-ST polymeric particles (d_p = 75 μm) were produced by Polymerics GmbH (Berlin, Germany); their characteristics are listed in Table 1. Glutaraldehyde (GA) (50% aqueous solution), zymosan A from *Saccharomyces cerevisiae*, and other chemicals were obtained from Sigma-Aldrich (St. Louis, MO, USA) or Merck (Darmstadt, Germany). The commercially available ELISA kits from GENTAUR (Kampenhout, Belgium) and CUSABIO (Wuhan, China) were used for immunoassays of fibrinogen and C5a fragment of complement in the blood plasma and serum, respectively.

Methods

Preparation of cryogel monoliths

Cryogel-based monoliths were prepared in plastic syringes according to the earlier reported technique (Plieva et al. 2006). In brief, PVA-based monoliths (3.2 cm in length and 0.9 cm in diameter) were prepared from pre-cooled 5% PVA aqueous solution adjusted to pH 1.0 to 1.2 with 1 M HCl by cryogelation in an air bath cryostat (Arctest, Finland) at −12°C overnight. GA at a concentration of 1%

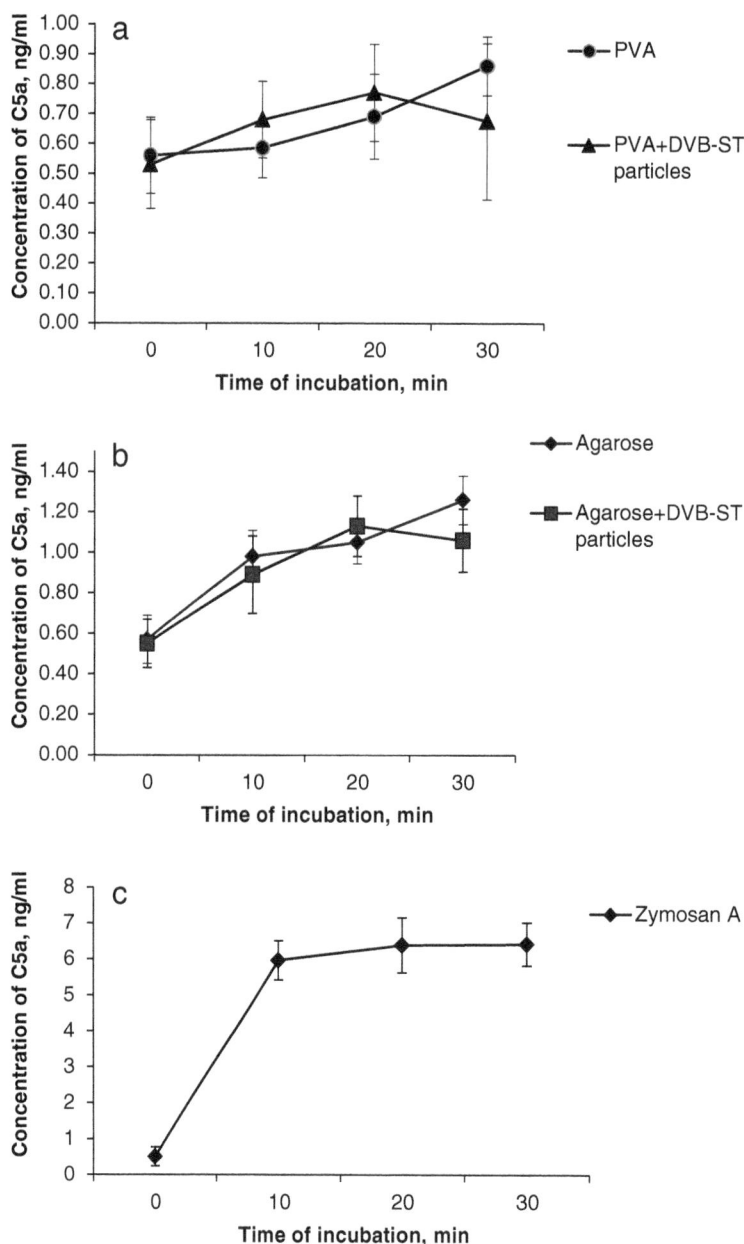

Figure 5 Generation of C5a fragment in the human serum (M ± SD; $n = 3$). This took place after incubation with four different types of cryogels: (**a**) slices of particle-free 5% PVA cryogel and PVA composite embedded with DVB-ST polymeric particles, (**b**) slices of particle-free 3% agarose and agarose composite with embedded the DVB-ST polymeric particles, and (**c**) positive control (zymosan A, 4 mg/mL in serum). Asterisk signifies that $p \leq 0.05$ in comparison with the positive control.

was used as a cross-linker. Decimolar aqueous solution of $NaBH_4$ was circulated through the columns by peristaltic pumping for the blocking of residual aldehyde groups. The monoliths treated in that way were washed with distilled water to neutralize pH, placed into 30% v/v aqueous ethanol solution, and stored in a fridge at +4°C.

Agarose-based cryogel monoliths were prepared from 3% agarose solution, adjusted to pH 12 with 5 M NaOH.

The alkaline agarose solution pre-warmed in a water bath at 60°C was placed into a liquid bath cryostat (Lauda, Germany) at −32°C for 30 min for fast freezing and then kept in the air bath cryostat at −12°C overnight. After defrosting, all monoliths were washed with distilled water to neutralize pH, placed into 30% ethanol solution, and stored in a fridge at 4°C. Before the contact with blood serum or blood plasma, the cryogel monoliths were washed with distilled water and PBS (pH 7.4).

In order to synthesize the composite monoliths, DVB-ST polymeric particles were washed with 2-propanol, distilled water, suspended in 5% w/v PVA or 3% w/v agarose solutions, and rinsed with the same solutions. The suspensions were adjusted to 250 mg/mL concentration of DVB-ST particles in the polymer solutions, which were put through cryogelation as described above.

Scanning electron microscopy examination

The monoliths have been examined under a scanning electron microscope (JEOL JSM-5600LV, JEOL Ltd., Akishima, Tokyo, Japan). In order to prepare the samples for SEM, the top and bottom of each cryogel monolith were cut off to get 4-mm-thick slices, which were further incubated in 2.5% GA solution in 0.12 M sodium phosphate buffer (pH 7.2) overnight. The samples were dehydrated by soaking in aqueous ethanol of increasing concentration (from 0% to 99.5% v/v), critical point dried, and coated with gold/palladium (40:60).

Preparation of blood plasma and serum

Blood samples were taken from two voluntary healthy donors into special plastic tubes to obtain the blood serum or plasma (1 mL of 0.1 M sodium citrate to 9 mL of blood). Tubes with the blood without anticoagulant were immediately placed in a water bath at 37°C for clot formation. Afterwards, all test tubes were centrifuged for 10 min at 5,000 rpm; serum and/or plasma were aspirated, pooled into the plastic Falcon tubes, and placed in a fridge at 4°C.

Percolation of blood plasma through the monoliths and fibrinogen assay

Fibrinogen concentration in blood plasma was measured before and after its circulation through the cryogel monoliths. Plasma (3.6 mL) was circulated through each of the tested cryogel monoliths (0.9 cm × 3.2 cm) by peristaltic pumping (WATSON MARLOW 101 U, Sweden) at 0.3 mL/min flow rate for 30 min. Instruction from the commercially available column for extracorporeal blood purification TORAYMYXIN PMX-20R (Toray Industries, Tokyo, Japan) has been used for calculations of monolith parameters, time of circulation and volume of plasma, passing through (Toray Industries 2007). Control plasma was passed through the empty syringe. Having been passed through monoliths, the plasma samples were stored at −20°C and assayed after defrosting according to the manual for the ELISA kit.

Contact of blood serum with monoliths and C5a assay

Monoliths sliced into 3-mm-thick disks were incubated in the blood serum (1 mL) for 0, 10, 20, and 30 min (37°C, water bath). Having been incubated with monoliths, serum was frozen, stored at −20°C, and assayed after

defrosting according to the manual of the ELISA kit. Zymosan A prepared as described earlier (Craddock et al. 1977) was used at a concentration of 4 mg/mL in the serum as a positive control for the complement system activation. According to the data presented in the literature, the concentration of zymosan required for complement activation *in vitro* is in the range of 1.0 to 5.0 mg/mL (Craddock et al. 1977; Yuanyuan et al. 2001; Fehra and Jacob 1977).

Statistical analysis

All calculations and construction of calibration curves were performed using the Origin (Microcal Software Inc., Northampton, MA, USA) and Excel (Microsoft Corporation, Redmond, WA, USA) programs. Significant difference between each two different groups of data was examined using the Student's t test.

Competing interests

The authors declare that they have no competing interests.

Authors' contributions

LGB developed the concept of the study; made the choice of conditions for blood and plasma experiments and performed them; performed cryogel synthesis, SEM studies, and interpretation of SEM results; and wrote the draft version of the manuscript. AEI participated in cryogel synthesis and the choice of conditions for blood and plasma experiments, wrote some parts of the draft version, such as discussion of fibrinogen adsorption and complement activation by alternative pathway. AL (ALeis) participated in the synthesis and characterization of DVB-ST polymeric particles and participated in the writing the draft version, especially regarding the terms used for composite and noncomposite monoliths. AL (AnLehm) synthesized and characterized DVB-ST polymeric particles. MVC participated in blood and plasma experiments, SEM studies, and interpretation of SEM results. HJ participated in the writing of the draft version and defined the sets of experimental data to be included into the manuscript. All authors read and approved the final manuscript.

Acknowledgements

This study was funded by the 'Monolithic Adsorbent Columns for Extracorporeal Medical Devices and Bioseparations' project (MONACO-EXTRA, FP7 reference number 218242) within the Marie Curie Industry-Academia Partnerships and Pathways programmer of the European Commission.

Author details

[1]Protista Biotechnology AB, Bjuv SE-26722, Sweden. [2]Polymerics GmbH, Berlin D-12681, Germany.

References

Arima Y, Kawagoe M, Toda M, Iwata H (2009) Complement activation by polymers carrying hydroxyl groups. ACS Appl Mater Interfaces 1:2400–2407

Beugeling T (1979) The interaction of polymer surfaces with blood. Journal of Polymer Science: Polymer Symposium 66:419–428

Black JP, Sefton MV (2000) Complement activation by PVA as measured by ELIFA (enzyme-linked immunoflow assay) for SC5b-9. Biomaterials 21:2287–2294

Bolgen N, Plieva F, Galaev I, Mattiasson B, Piskin E (2007) Cryogelation for preparation of novel biodegradable tissue-engineering scaffolds. J Biomater Sci Polymer Edn 18:1165–1179

Brash JL, Scott CF, Hove P, Wojciechowski P, Colman RW (1988) Mechanism of transient adsorption from plasma to solid surfaces: role of the contact and fibrinolytic systems. Blood 71:932–939

Craddock PR, Agustin J, Dalmasso P, Brigham KL, Jacob HS (1977) Pulmonary vascular leukostasis resulting from complement activation by dialyzer cellophane membranes. J Clin Invest 59:879–888

Dainiak MB, Kumar A, Plieva FM, Galaev IY, Mattiasson B (2004) Integrated isolation of antibody fragments from microbial cell culture fluids using supermacroporous cryogels. J Chromatogr 1045(A):93–98

Dainiak MB, Allan IU, Savina IN, Cornelio L, James ES, James SL, Mikhalovsky SV, Jungvid H, Galaev IY (2010) Gelatin–fibrinogen cryogel dermal matrices for wound repair: preparation, optimisation and in vitro study. Biomaterials 31:67–76

ENSAI (2009) IS EN ISO 10993–4: biological evaluation of medical devices - part 4: Selection of tests for interactions with blood. National Standards Authority of Ireland, Dublin

Fehra J, Jacob HS (1977) In vitro granulocyte adherence and in vivo margination: two associated complement-dependent functions. J Exp Med 146:641–652

Hanora A, Bernaudat F, Plieva FM, Dainiak MB, Bulow L, Galaev IY, Mattiasson B (2005) Screening of peptide affinity tags using immobilised metal affinity chromatography in 96-well plate format. J Chromatogr 1087(A):38–44

Ivanov AE, Kozynchenko OP, Mikhalovska LI, Tennison SR, Jungvid H, Gun'ko VM, Mikhalovsky SV (2012) Activated carbons and carbon-containing poly(vinyl alcohol) cryogels: characterization, protein adsorption and possibility of myoglobin clearance. Phys Chem Chem Phys 14:16267–16278

Jones BE, Czaja MJ (1998) Intracellular signaling in response to toxic liver injury. Am J Physiol 275(G):874–878

Kirkpatrick CJ, Bittinger F, Wagner M, Kohler H, van Kooten TG, Klein CL, Otto M (1998) Current trends in biocompatibility testing. Journal of Engineering in Medicine 212:75–84

Kirsebom H, Rata G, Topgaard D, Mattiasson B, Galaev IY (2009) Mechanism of cryopolymerization: diffusion-controlled polymerization in a nonfrozen microphase. An NMR Study Macromolecules 42:5208–5214

Koc I, Baydemir G, Bayram E, Yavuz H, Denizli A (2011) Selective removal of 17β-estradiol with molecularly imprinted particle-embedded cryogel systems. J Hazard Mater 192:1819–1826

Le Noir M, Plieva F, Hey T, Guieysse B, Mattiasson B (2007) Macroporous molecularly imprinted polymer/cryogel composite systems for the removal of endocrine disrupting trace contaminants. J Chromatogr 1154(A):158–164

Lyle DB, Bushar GS, Langone JJ (2010) Screening biomaterials for functional complement activation in serum. J Biomed Mater Res 92(A):205–213

Özgür E, Bereli N, Türkmen D, Ünal S, Denizli A (2011) PHEMA cryogel for in-vitro removal of anti-dsDNA antibodies from SLE plasma. Mater Sci Eng C 31:915–920

Paradossi G, Cavalieri F, Chiessi E (2003) Poly(vinyl alcohol) as versatile biomaterial for potential biomedical application. Journal of Material Science: Materials in Medicine 14:687–691

Plieva FM, Karlsson M, Aguilar MR, Gomez D, Mikhalovsky S, Galaev IY, Mattiasson B (2006) Pore structure of macroporous monolithic cryogels prepared from poly(vinyl alcohol) J Applied Polymer Sci 100:1057–1066

Plieva FM, Galaev IY, Mattiasson B (2007) Macroporous gels prepared at subzero temperatures as novel materials for chromatography of particulate-containing fluids and cell culture applications. J Sep Sci 30:1657–1671

Pulanic D, Rudan I (2005) The past decade: fibrinogen. Coll Antropol 29:341–349

Saich R, Selden C, Rees M, Hodgson H (2007) Characterization of pro-apoptotic effect of liver failure plasma on primary human hepatocytes and its modulation by molecular adsorbent recirculation system therapy. Artif Organs 31:732–742

Sarma JV, Ward PA (2010) The complement system. Cell Tissue Res 343:227–235

Toray Industries (2007) TORAYMYXIN PMX-20R: extracorporeal removal of endotoxin in septic shock. Brochure of Toray Industries, Tokyo, Japan, http://www.gdmedical.ch/PDFs/TORAYMYXIN_depliant_rev3%20LIGHT.pdf

Tsai WB, Grunkemeier JM, Horbett TA (1999) Human plasma fibrinogen adsorption and platelet adhesion to polystyrene. J Biomed Mater Res 44:130–139

Varoni E, Tschon M, Palazzo B, Nitti P, Martini L, Rimondini L (2012) Agarose gel as biomaterial or scaffold for implantation surgery: characterization, histological and histomorphometric study on soft tissue response. Connect Tissue Res 53(6):548–554

Vogler EA, Siedlecki CA (2009) Contact activation of blood plasma coagulation. Biomaterials 30:1857–1869

Vroman L, Adams AL, Fischer GC, Munoz PC (1980) Interaction of high molecular weight kininogen, factor XII and fibrinogen in plasma at interfaces. Blood 55:156–159

Wasilewska M, Adamczyk Z, Jachimska B (2009) Structure of fibrinogen in electrolyte solutions derived from dynamic light scattering (DLS) and viscosity measurements. Langmuir 25:3698–3704

Weber V, Linsberger I, Hauner M, Leistner A, Leistner A, Falkenhagen D (2008) Neutral styrene divinylbenzene copolymers for adsorption of toxins in liver failure. Biomacromolecules 9:1322–1328

Yuanyuan X, Minghe M, Ippolito GC, Schroeder HW, Carroll MC, Volanakis JE (2001) Complement activation in factor D-deficient mice. PNAS 98(25):14577–14582

The effect of polymer and CaCl$_2$ concentrations on the sulfasalazine release from alginate-N,O-carboxymethyl chitosan beads

Moslem Tavakol, Ebrahim Vasheghani-Farahani[*] and Sameereh Hashemi-Najafabadi

Abstract

In this study, pH-sensitive blended polymeric beads were prepared by ionic gelation of mixed alginate and N,O-carboxymethyl chitosan (NOCC) solutions in aqueous media containing calcium chloride. To prepare drug-loaded beads, sulfasalazine (SA) as a model drug was added to the initial aqueous polymer solution. These beads were characterized and evaluated in vitro as potential carriers for colon-specific drug delivery. A 3^2 full factorial experimental design was employed to evaluate the effect of polymer and CaCl$_2$ concentrations on swelling and drug release behavior of the beads in simulated gastrointestinal tract fluid. It was found that the rate of swelling and drug release decreased significantly with increasing polymer and CaCl$_2$ concentrations, but polymer concentration was more effective than CaCl$_2$ concentration. The beads prepared using 4.5% polymer concentration and 4% CaCl$_2$ concentration retained approximately 60% of the loaded drug before approaching the simulated colonic fluid. Based on the results, the alginate-NOCC beads prepared with high polymer concentration could be potentially suitable polymeric carriers for colon-specific delivery of SA.

Keywords: Alginate, N,O-carboxymethyl chitosan, Ionic gelation, Blending, Sulfasalazine, Experimental design, Colon-specific drug delivery

Introduction

The use of polysaccharides in the formulation of colon-specific drug delivery carriers has gained increasing interest lately (Bajpai and Sonkusley 2002; Mahkam 2010; Mladenovska et al. 2007; Prabhu et al. 2008; Saboktakin et al. 2011; Tavakol et al. 2009). Micro- and nanoparticles prepared from some polysaccharides are attractive carriers for colon-specific drug delivery due to their favorite properties such as pH-sensitive swelling behavior, stability in the upper portion of the gastrointestinal tract, and suitable degradability by specific colonic enzymes (Assaad et al. 2011; Kim et al. 2012; Liu et al. 2007a; Sinha and Kumria 2001; Tavakol et al. 2009; Vandamme et al. 2002).

N,O-carboxymethyl chitosan (NOCC) is a chitosan derivative bearing a carboxymethyl substituent at some of the amino and primary hydroxyl sites of the glucosamine units of the chitosan structure. Biodegradability, biocompatibility, excellent water solubility, gel formation ability,

and amphoteric polyelectrolyte characteristics make this material suitable for biomedical applications (Dolatabadi-Farahani et al. 2006; Fan et al. 2006; Lin et al. 2005; Tavakol et al. 2009; Upadhyaya et al. 2013; Zhang et al. 2004). Physically cross-linked carboxymethyl chitosan beads can be prepared by the dropping of aqueous low molecular weight (MW) carboxymethyl chitosan solution into CaCl$_2$ solution (Liu et al. 2007b).

Alginate is a polyanionic copolymer of mannuronic and guluronic acid residues. Physically cross-linked Ca-alginate microparticles have been extensively studied as a potential carrier for oral drug delivery (Bajpai and Sharma 2004; Murata et al. 1993; Pasparakis and Bouropoulos 2006; Zhu et al. 2011). This system has major limitations such as rapid drug release caused by physical instability and high solubility of Ca-alginate beads in neutral and weak alkali media (George and Abraham 2006; Ma et al. 2010; Tavakol et al. 2009; Xing et al. 2003). To overcome these limitations, various approaches have been examined for the preparation of modified beads by blending and/or coating through polyelectrolyte complexation with polymers such as chitosan

* Correspondence: evf@modares.ac.ir
Biotechnology Group, Faculty of Chemical Engineering, Tarbiat Modares University, P.O. Box 14115–143, Tehran, Iran

and chitosan derivatives (Chen et al. 2004; El-Sherbiny 2010; El-Sherbiny et al. 2010; Gong et al. 2011; Jayant et al. 2009; Lin et al. 2005; Meng et al. 2011; Mladenovska et al. 2007; Pasparakis and Bouropoulos 2006; Tavakol et al. 2009; Vandenberg et al. 2001; Zhu et al. 2011).

Lin et al. (2005) prepared a complex of alginate blended with NOCC by ionic gelation in Ca^{2+} solution. These beads demonstrated excellent pH sensitivity and could be a suitable polymeric carrier for site-specific bioactive protein drug delivery in the intestine. They used one-factor-at-a-time method to investigate the effect of polymer concentration and alginate/NOCC ratio on the properties of the beads, which are not useful in investigating interactions between factors. El-Sherbiny et al. (2010) prepared a new pH-sensitive hydrogel containing calcium-cross-linked blend of alginate and methacrylic (or acrylic) acid-grafted carboxymethyl chitosan. These beads showed high swelling degree and drug release percentage in simulated gastric fluid. To overcome these shortcomings, the beads were coated with poly(ethylene glycol)-g-chitosan copolymer (El-Sherbiny 2010). This modification resulted in minimizing the swelling degree and loss of protein drug in the gastric fluid and preferably releasing the drug mostly in the intestine (El-Sherbiny 2010).

In our recent study, blended polymeric beads of alginate and NOCC were prepared and then coated by chitosan (Tavakol et al. 2009). The effect of coating as well as drying procedure on the properties of the beads, prepared at constant polymer and $CaCl_2$ concentrations, were evaluated. It was found that the rate of swelling and drug release decreased for air-dried and coated beads in comparison with freeze-dried and uncoated ones, respectively (Tavakol et al. 2009).

In the present study, a 3^2 full factorial design was performed to investigate the effect of polymer and $CaCl_2$ concentrations, their interaction on the morphology and swelling characteristics of alginate-NOCC beads, as well as sulfasalazine (SA) release from these carriers in simulated gastrointestinal fluid.

Methods
Materials
Chitosan (MW approximately 2×10^5) with an 85% degree of deacetylation was provided from Sigma-Aldrich Corporation (St. Louis, MO, USA). Sodium alginate was obtained from BDH Laboratory (London, England, UK). Calcium chloride, monochloroacetic acid and isopropyl alcohol were purchased from Merck (Darmstadt, Germany). Sulfasalazine was obtained from Zhejiang Jiuzhou Pharmaceutical Co. Ltd (Zhejiang, China). NOCC was synthesized according to the literature (Chen et al. 2004) and characterized by the method described by Sugimoto et al. (1998). All the other used chemicals, solvents, and reagents were of analytical grade.

Preparation of beads
Firstly, aqueous alginate and NOCC solutions, with concentrations of 1.5%, 3%, and 4.5% (w/v), were prepared separately. Next, equal volumes of these solutions were mixed to form a homogenous blend solution which was maintained for 5 h for the complete removal of bubbles. The final pH of the solution was found to be approximately 7.5 ± 0.1. Five milliliters of these solutions was dropped into a 30-ml gently stirred $CaCl_2$ solution with distinct concentrations of 1%, 2.5%, and 4% (w/v) through a syringe needle (0.4 mm in diameter) at a dropping rate of 1.0 ml/min. The distance of the needle tip from the gelling solution surface was 10 cm. The prepared beads were allowed to harden in the calcium chloride solution for 30 min. These beads were filtered, washed with distilled water three times, and dried at 40°C for 24 h or freeze-dried. The freeze-dried beads were obtained through rapid freezing at –80°C, followed by drying in a freeze drier (Zirbus, Denmark).

To prepare drug-loaded beads, SA with a final concentration of 1% (w/v) was added to the initial aqueous alginate solution with continuous stirring, and the pH of the solution was adjusted to 7.5 by adding 2 M NaOH. This solution was used for the preparation of SA-loaded beads by the same procedure described for the preparation of unloaded counterparts.

Characterization of beads
The shape and surface characteristics of the beads were investigated by optical microscopy. The diameter of the beads was determined using an optical microscope and digital micrometer, and the average values were taken for at least 25 beads.

Drug content and encapsulation efficiency determination
Encapsulation efficiency (wt.%) was calculated from the difference between the amount of SA dissolved in aqueous polymer solution and that of SA released in gelation medium divided by the amount of SA dissolved in aqueous polymer solution. For this purpose, the concentration of SA in gelation and washing solution was determined spectrophotometrically at 359 nm. Drug content (wt.%) was determined as the ratio of encapsulated SA weight to the total weight of the dried beads. This was accomplished by immersion of drug-loaded beads in sodium phosphate buffer at pH 7.4. The total released drug after 24 h was determined spectrophotometrically and was considered as encapsulated SA.

Swelling studies
The swelling characteristics of beads were determined by immersing them in dry state into conical flask containing 40 ml of release medium that were incubated at 37°C under shaking at 150 rpm. At first, the dry beads

Table 1 Full factorial experimental design levels of polymer and CaCl$_2$ concentrations

Experimental run	1	2	3	4	5	6	7	8	9
Polymer concentration (g/100 ml)	1.5	1.5	1.5	3.0	3.0	3.0	4.5	4.5	4.5
CaCl$_2$ concentration (g/100 ml)	1.0	2.5	4.0	1.0	2.5	4.0	1.0	2.5	4.0

were swollen in 0.1-M HCl solution at pH 1.2 (simulated gastric fluid) for 2 h. Afterwards, the beads were transferred to a sodium phosphate buffer solution at pH 6.8 (simulated small intestinal fluid) and kept for 3 h. Subsequently, they were transferred to a sodium phosphate buffer solution at pH 7.4 (simulated colonic fluid) until complete dissolution was obtained. At specific time intervals, samples were taken out from the swelling medium and blotted with a piece of paper towel to absorb excess water on the surface. The degree of swelling, $\Phi(\tau)$, at each time was calculated using the following expression:

$$\Phi(\tau) = (\Psi_\tau - \Psi_0)/\Psi_0 \qquad (1)$$

where Ψ_τ and Ψ_0 are the sample weights at time τ and in the dry state, respectively. Each experiment was repeated three times.

Drug release studies

The SA release from drug-loaded beads was studied using the same conditions as described in the swelling studies. At predetermined time intervals, 2 ml of samples were withdrawn from the dissolution medium and immediately replaced by the same volume of fresh medium. The amount of SA released from the beads was determined spectrophotometrically (UV–vis Varian Cary 50, Varian, Inc., Palo Alto, CA, USA) at 359 nm using previously calibrated standard curves at different pH values. To determine the release in 0.1-M HCl solution, the pH of the release medium was adjusted to 7.4 by adding NaOH, and the concentration of SA was determined from the calibration curve at this pH. Each experiment was repeated three times.

Experimental design and statistical analysis

A full factorial design with two parameters at three levels, as shown in Table 1, was applied. The experiments were carried out in random order to avoid any systematic error in the experimental data. Each experiment was repeated three times. Statistical software, Design Expert 7 (Stat-Ease, Inc., Minneapolis, MN, USA) and Minitab 14 (Minitab, State College, PA, USA), were used to analyze the experimental data.

Results and discussion

Characterization of beads

In our previous study (Dolatabadi-Farahani et al. 2006), the synthesized NOCC was analyzed by proton nuclear magnetic resonance spectroscopy based on a method described in the literature (Sugimoto et al. 1998). The degree of substitution of the carboxymethyl groups on the

Figure 1 The photographs of wet and dried alginate-NOCC beads taken under an optical microscope. (**a**) Freeze-dried SA-loaded bead, (**b**) unloaded wet bead, (**c**) SA-loaded wet bead, and air dried SA-loaded beads prepared at different polymer concentrations (*w/v*): (**d**) 1.5%, (**e**) 3%, and (**f**) 4.5%.

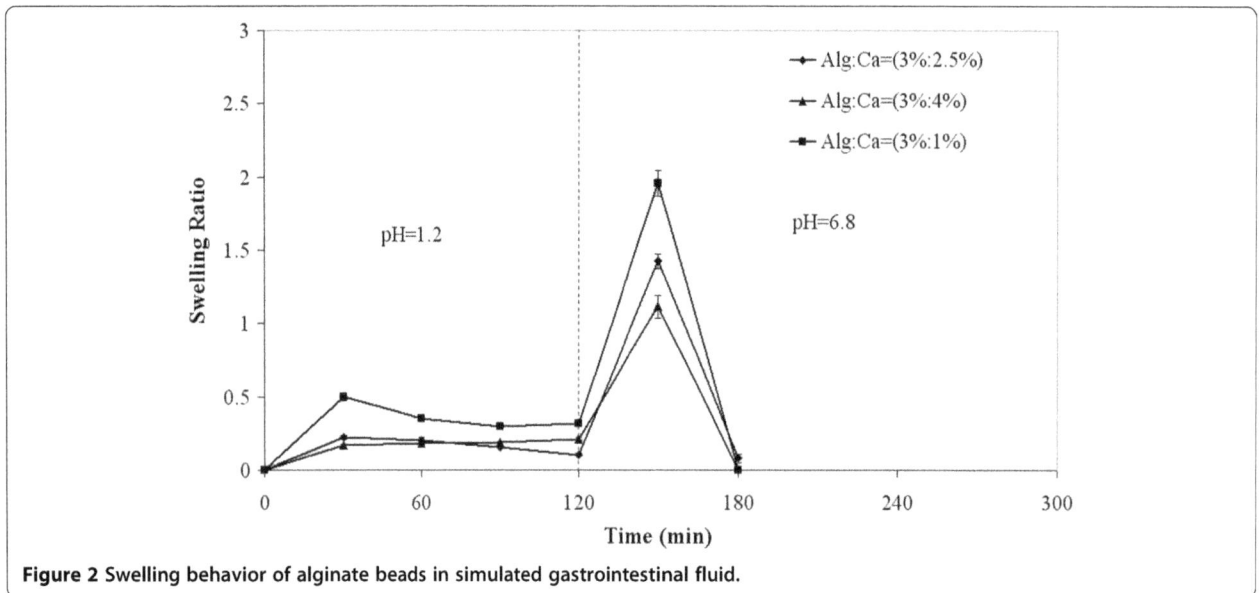

Figure 2 Swelling behavior of alginate beads in simulated gastrointestinal fluid.

amino and primary hydroxyl sites was approximately 20.3% and 19.2%, respectively.

As expected, NOCC or alginate beads formed upon dropwise addition of aqueous NOCC or alginate solution into $CaCl_2$ solution, due to ionic cross-linking between the carboxylate ions ($-COO^-$) on NOCC or alginate, established by Ca^{2+}. Thus, after the dropping of mixed alginate-NOCC solution into calcium chloride solution, alginate entangled through the NOCC network and vice versa, resulting in the formation of interpenetrating polymeric network. Zhang et al. (2004) showed that the blend membranes of carboxymethyl chitosan-alginate are miscible in the ratio from 1:1 to 1:5 and exhibited good

mechanical properties due to strong electrostatic force and hydrogen bonding between different groups of two polymers.

The photographs of wet, freeze-dried, and air-dried beads, taken under an optical microscope, are shown in Figure 1. The diameters of the wet beads were 1.20 ± 0.10 mm independent of calcium chloride and polymer concentrations. After drying, the bead diameters slightly decreased from 0.70 ± 0.55 to 0.45 ± 0.60 mm with decreasing polymer concentration, but the effect of calcium chloride concentration was not significant. The wet beads were spherical in shape with a smooth surface. In the case of the beads prepared with 1.5% (w/v) polymer

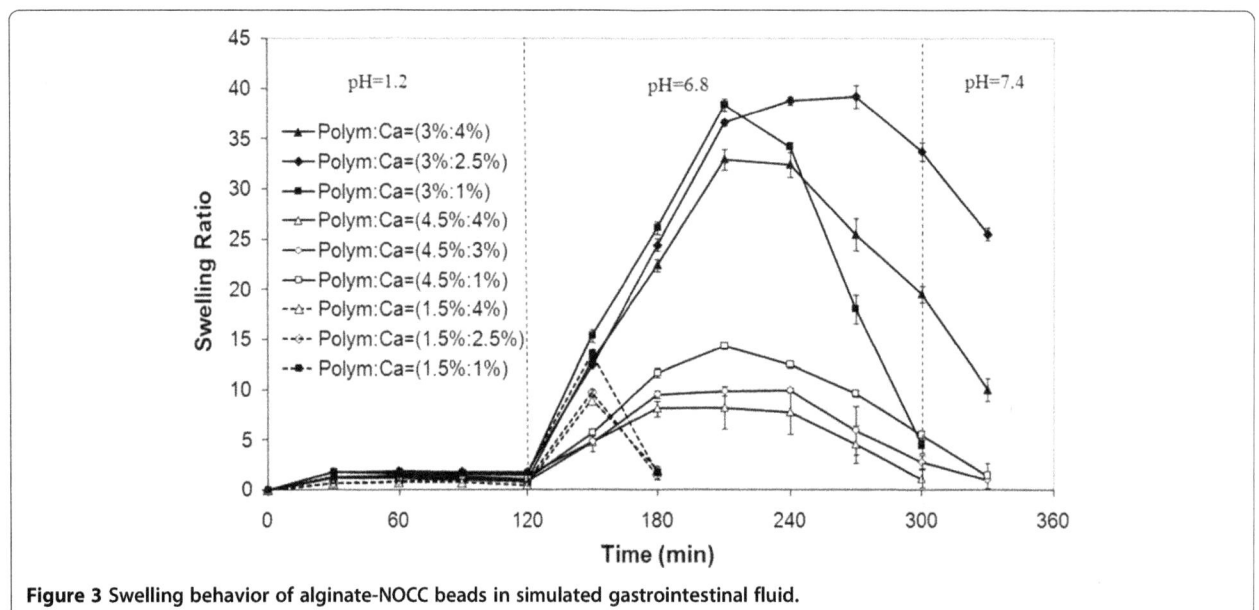

Figure 3 Swelling behavior of alginate-NOCC beads in simulated gastrointestinal fluid.

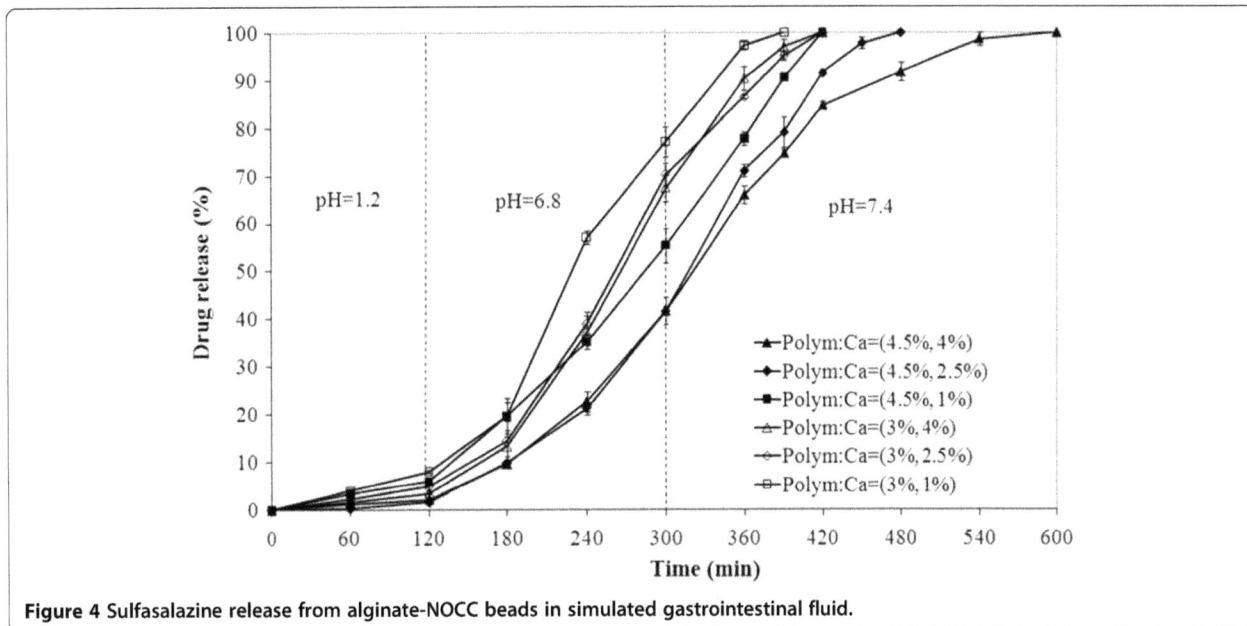

Figure 4 Sulfasalazine release from alginate-NOCC beads in simulated gastrointestinal fluid.

concentration, the spherical shape of the beads changed to an irregular shape with a collapsed center and some cracks on the surface (Figure 1c). The beads prepared with higher polymer concentration (3.0% and 4.5% (w/v)) remained almost spherical with a rather rough surface and compact structure.

Swelling studies

The swelling behavior of alginate and alginate-NOCC beads in simulated gastrointestinal fluid is shown in Figures 2 and 3, respectively. The swelling behavior of NOCC beads could not be studied due to the formation of very mechanically weak beads that lost their shape and were destroyed in the washing or drying steps.

At pH 1.2, the swelling of alginate and alginate-NOCC beads was hindered due to the formation of strong hydrogen bonds between -COOH and -OH groups of

both polymer polar chains (Tavakol et al. 2009) and increased electrostatic attraction between protonated amine groups of NOCC and carboxyl groups of alginate in alginate-NOCC beads (Zhang et al. 2004).

At pH 6.8, alginate and alginate-NOCC beads began to swell noticeably due to the swelling force that resulted from the presence of counterions which neutralized the ionized carboxylic groups on alginate and NOCC and electrostatic repulsion between the ionized carboxylic groups (Lin et al. 2005; Tavakol et al. 2009). This phenomenon can also be related to ion-exchange between the Ca^{2+} ions in the hydrogel network and Na^+ ions in the phosphate buffer solution (Bajpai and Tankhiwale 2006). Finally, the beads start to disintegrate, owing to the highly hydrated structure and almost complete removal of calcium ions (Bajpai and Tankhiwale 2006). The appearance of turbidity and observation of precipitate in the swelling

Table 2 Analysis of variance of responses

Source	SS	DF	MS	F value	p value	
Model	9,493.77	4	2,373.44	144.36	<0.0001	Significant
A (polymer concentration)	8,665.86	1	8,665.86	527.10	<0.0001	
B (CaCl$_2$ concentration)	672.71	1	672.71	40.92	<0.0001	
A^2	72.27	1	72.27	4.40	0.0477	
B^2	82.93	1	82.93	5.04	0.0351	
Residual	361.69	22	16.44			
Lack of fit	43.67	4	10.92	0.62	0.6554	Not significant
Pure error	318.02	18	17.67			
Total	9,855.46	26				

This analysis of responses is in terms of total drug that remained in the beads (before exposure to the simulated colonic tract fluid) SS, sum of squares; DF, degree of freedom; MS, mean square; F value, factor effect value; p value, probability value.

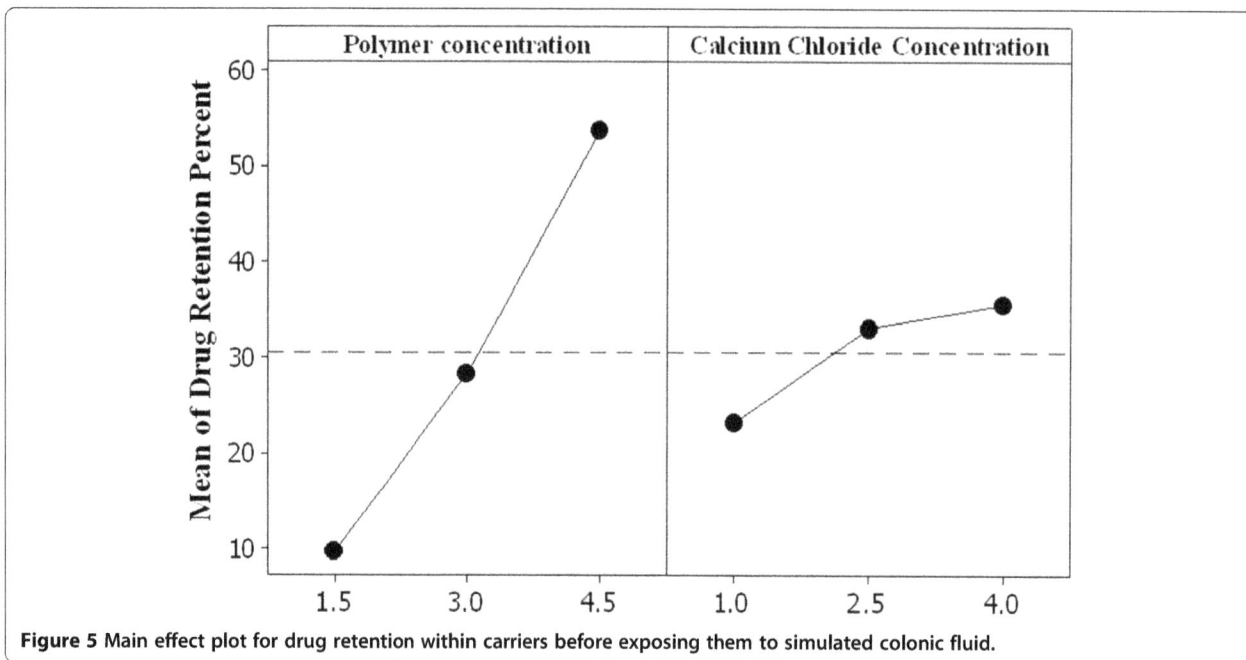

Figure 5 Main effect plot for drug retention within carriers before exposing them to simulated colonic fluid.

medium, especially in the case of the beads prepared at higher calcium chloride concentration, are also indicative of the ion-exchange process.

At neutral and basic media, the swelling degree of alginate-NOCC beads, prepared at constant polymer concentration, decreased with increasing $CaCl_2$ concentration ($p < 0.05$), due to increased cross-linking density of the network. At these media, the swelling and disintegration rate decreased significantly with increasing polymer

concentration ($p < 0.05$). This may be related to increased elastic force that resulted from (1) hydrogen bonding between amine and hydroxyl groups of NOCC and alginate, (2) electrostatic attraction between ionized amine and carboxyl groups of NOCC and alginate, and (3) physical cross-linking as a result of polymer chain entanglements.

Comparing Figures 2 and 3, the disintegration rate of alginate-NOCC beads in PBS was significantly lower than that of alginate beads. This can be related to the

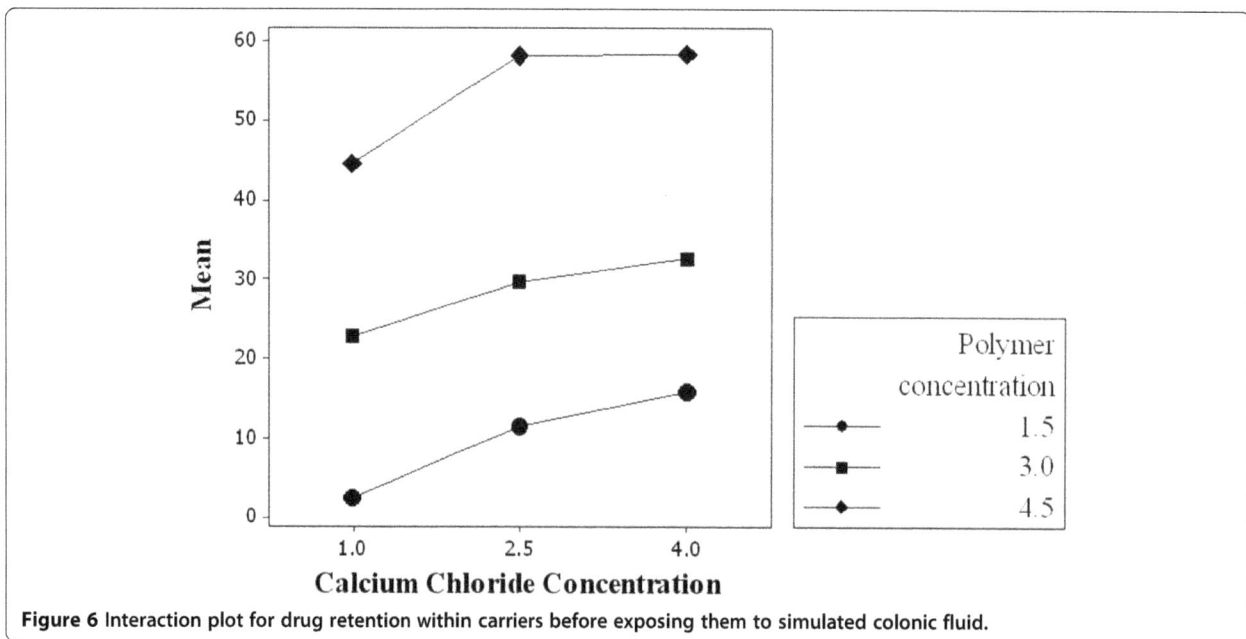

Figure 6 Interaction plot for drug retention within carriers before exposing them to simulated colonic fluid.

presence of strong hydrogen bonds between the hydroxyl and amine groups of alginate and the NOCC and electrostatic attraction between the ionized amine and carboxyl groups of these polymers, which resist to disintegration of network.

Drug release studies

SA release from alginate-NOCC beads in simulated gastrointestinal fluid as a function of polymer and $CaCl_2$ concentrations is shown in Figure 4. Drug release profiles indicate that the SA release from beads at pH 1.2 is relatively slow. This is due to the limited swelling degree of hydrogel network and solubility of SA at this pH. Subsequently, the SA release rate at pH 6.8 and 7.4 increased significantly ($p < 0.05$) in accordance with the swelling behavior of beads (Figure 3) and solubility of SA. The SA release rate decreased with increasing polymer and $CaCl_2$ concentrations in accordance with the swelling behavior of the beads.

Analysis of variance, performed by Minitab 14 software, for the effect of polymer and $CaCl_2$ concentrations on the total drug that remained in the beads before exposing them to the simulated colonic tract fluid (during the initial 5 h of release time) is given in Table 2. This analysis shows that the effects of both factors and their quadratic terms were significant, without having a significant interaction effect.

As shown in the main effect plot presented in Figure 5, the effect of polymer concentration on the retention of drug within the beads was higher than that of $CaCl_2$ concentration. According to the interaction plot of these factors presented in Figure 6, the effect of $CaCl_2$ concentration increasing from 2.5% to 4% became smaller at higher polymer concentrations. This can be related to the formation of densely cross-linked polymeric layer on the surface of droplets which resists to Ca^{2+} diffusion into the beads' core, leading to the formation of beads with unreacted or partially reacted core with smaller resistance to swelling and drug release.

As shown in Figure 4, the alginate-NOCC beads prepared at the polymer concentration of 4.5% and $CaCl_2$ concentration of 4% or 2.5% retained approximately 60% of loaded drug before exposure to the simulated colonic fluid. This is a promising property for the application of optimized alginate-N,O-carboxymethyl chitosan gel beads as colon-specific delivery system.

The swelling and drug release characteristics of the alginate-NOCC beads can be tuned by the modulation of polymer and $CaCl_2$ concentrations. Therefore, the beads can be further evaluated for the release of drugs in the different segments of the gastrointestinal tract. This strategy is more convenient than the coating procedure used in a previous study (Tavakol et al. 2009).

Conclusions

Calcium cross-linked alginate-NOCC beads prepared in the present study demonstrated distinct pH-sensitive swelling and drug release behavior. Sulfasalazine release rate was slow in acidic medium but increased at pH 6.8 and pH 7.4, which is in accordance with the swelling rate of the beads and SA solubility. The rate of SA release decreased with increasing polymer and $CaCl_2$ concentrations, but polymer concentration was more effective. No burst effect was observed for SA release from these pH-sensitive carriers. It was previously shown that the drug release behavior of SA-loaded alginate-NOCC beads was improved by chitosan coating (Tavakol et al. 2009). Based on these results, a suitable polymeric carrier for colon-specific delivery of SA can be developed by either the increasing concentration of alginate-NOCC blend solution or chitosan coating.

Competing interest
The authors declare that they have no competing interests.

Authors' contributions
MT carried out the whole experiment. EV-F directed the study, and SH-N worked on the design of experiments. All authors read and approved the final manuscript.

Authors' information
MT is a PhD student. EVF is a professor and supervisor. SHN is an assistant professor and co-supervisor.

Acknowledgments
The authors acknowledge the SpringerOpen production team for the language editing of this manuscript.

References
Assaad E, Wang YJ, Zhu XX, Mateescua MA (2011) Polyelectrolyte complex of carboxymethyl starch and chitosan as drug carrier for oral administration. Carbohydr Polym 84:1399–1407

Bajpai SK, Sonkusley SJ (2002) Hydrogels for colon-specific oral drug delivery: an *in vitro* drug release study (II). Iran Polym J 11:187–196

Bajpai SK, Sharma S (2004) Investigation of swelling/degradation behaviour of alginate beads crosslinked with Ca^{2+} and Ba^{2+} ions. React Funct Polym 59:129–140

Bajpai SK, Tankhiwale R (2006) Investigation of water uptake behavior and stability of calcium alginate/chitosan bi-polymeric beads: part A. React Funct Polym 66:645–658

Chen SC, Wu YC, Mi FL, Lin YH, Yu LC, Sung HW (2004) A novel pH-sensitive hydrogel composed of N, O-carboxymethyl chitosan and alginate cross-linked by genipin for protein drug delivery. J Control Release 96:285–300

Dolatabadi-Farahani T, Vasheghani-Farahani E, Mirzadeh H (2006) Swelling behaviour of alginate-N, O-carboxymethyl chitosan gel beads coated by chitosan. Iran Polym J 15:405–415

El-Sherbiny IM (2010) Enhanced pH-responsive carrier system based on alginate and chemically modified carboxymethyl chitosan for oral delivery of protein drugs: preparation and *in-vitro* assessment. Carbohydr Polym 80:1125–1136

El-Sherbiny IM, Abdel-Bary EM, Harding DRK (2010) Preparation and *in vitro* evaluation of new pH-sensitive hydrogel beads for oral delivery of protein drugs. J Appl Polym Sci 115:2828–2837

Fan L, Du Y, Zhang B, Yang J, Zhou J, Kennedy JF (2006) Preparation and properties of alginate/carboxymethyl chitosan blend fibers. Carbohydr Polym 65:447–452

George M, Abraham TE (2006) Polyionic hydrocolloids for the intestinal delivery of protein drugs: alginate and chitosan—a review. J Control Release 114:1–14

Gong R, Li C, Zhu S, Zhang Y, Du Y, Jiang J (2011) A novel pH-sensitive hydrogel based on dual cross-linked alginate/N-α-glutaric acid chitosan for oral delivery of protein. Carbohydr Polym 85:869–874

Jayant RD, McShane MJ, Srivastava R (2009) Polyelectrolyte-coated alginate microspheres as drug delivery carriers for dexamethasone release. Drug Deliv 16:331–340

Kim MS, Park SJ, Gu BK, Kim C-H (2012) Ionically crosslinked alginate-carboxymethyl cellulose beads for the delivery of protein therapeutics. Appl Surf Sci 262:28–33

Lin YH, Linang HF, Chung CK, Chen MC, Sung HW (2005) Physically crosslinked alginate/N, O-carboxymethyl chitosan hydrogels with calcium for oral delivery of protein drugs. Biomaterials 26:2105–2113

Liu M, Fan J, Wang K, He Z (2007a) Synthesis, characterization, and evaluation of phosphated cross-linked Konjac glucomannan hydrogels for colon-targeted drug delivery. Drug Deliv 14:397–402

Liu Z, Jiao Y, Zhang Z (2007b) Calcium-carboxymethyl chitosan hydrogel beads for protein drug delivery system. J Appl Polym Sci 103:3164–3168

Ma L, Liu M, Liu H, Chen J, Gao C, Cui D (2010) Dual crosslinked pH- and temperature-sensitive hydrogel beads for intestine-targeted controlled release. Polym Adv Tech 21:348–355

Mahkam M (2010) Novel pH-sensitive hydrogels for colon-specific drug delivery. Drug Deliv 17:158–163

Meng X, Li P, Wei Q, Zhang H-X (2011) pH Sensitive alginate-chitosan hydrogel beads for carvedilol delivery. Pharm Dev Technol 16:22–28

Mladenovska K, Raicki RS, Janevik EI, Ristoski T, Pavlova MJ, Kavrakovski Z, Dodov MG, Goracinova K (2007) Colon-specific delivery of 5-aminosalicylic acid from chitosan-calcium-alginate microparticles. Int J Pharm 342:124–136

Murata Y, Nakada K, Miyamoto E, Kawashima S, Seo SH (1993) Influence of erosion of calcium-induced alginate gel matrix on the release of Brilliant Blue. J Control Release 23:21–26

Pasparakis G, Bouropoulos N (2006) Swelling studies and in vitro release of verapamil from calcium alginate and calcium alginate–chitosan beads. Int J Pharm 323:34–42

Prabhu S, Kanthamneni N, Ma C (2008) Novel combinations of rate-controlling polymers for the release of leuprolide acetate in the colon. Drug Deliv 15:119–125

Saboktakin MR, Tabatabaie RM, Maharramov A, Ramazanovb MA (2011) Synthesis and in vitro evaluation of carboxymethyl starch–chitosan nanoparticles as drug delivery system to the colon. Int J Biol Macromol 48:381–385

Sinha VR, Kumria R (2001) Polysaccharides in colon-specific drug delivery. Int J Pharm 224:19–38

Sugimoto M, Morimoto M, Sashiwa H, Saimoto H, Shigemasa Y (1998) Preparation and characterization of water-soluble chitin and chitosan derivatives. Carbohydr Polym 36:49–59

Tavakol M, Vasheghani-Farahani E, Dolatabadi-Farahani T, Hashemi-Najafabadi S (2009) Sulfasalazine release from alginate-N, O-carboxymethyl chitosan gel beads coated by chitosan. Carbohydr Polym 77:326–330

Upadhyaya L, Singh J, Agarwal V, Tewaria RP (2013) Biomedical applications of carboxymethyl chitosans. Carbohydr Polym 91:452–466

Vandamme TF, Lenourry A, Charrueau C, Chaumeil JC (2002) The use of polysaccharides to target drugs to the colon. Carbohydr Polym 48:219–231

Vandenberg GW, Drolet C, Scott SL, Jdl N (2001) Factors affecting protein release from alginate–chitosan coacervate microcapsules during production and gastric/intestinal simulation. J Control Release 77:297–307

Xing L, Dawei C, Liping X, Rongqing Z (2003) Oral colon-specific drug delivery for bee venom peptide: development of a coated calcium alginate gel beads-entrapped liposome. J Control Release 93:293–300

Zhang L, Guo J, Peng X, Jin Y (2004) Preparation and release behavior of carboxymethylated chitosan/alginate microspheres encapsulating bovine serum albumin. J Appl Polym Sci 92:878–882

Zhu AM, Chen JH, Liu QL, Jiang YL (2011) Controlled release of berberine hydrochloride from alginate microspheres embedded within carboxymethyl chitosan hydrogels. J Appl Polym Sci 120:2374–2380

10

Current progress on bio-based polymers and their future trends

Ramesh P Babu[1,2*], Kevin O'Connor[3] and Ramakrishna Seeram[4,5,6]

Abstract

This article reviews the recent trends, developments, and future applications of bio-based polymers produced from renewable resources. Bio-based polymers are attracting increased attention due to environmental concerns and the realization that global petroleum resources are finite. Bio-based polymers not only replace existing polymers in a number of applications but also provide new combinations of properties for new applications. A range of bio-based polymers are presented in this review, focusing on general methods of production, properties, and commercial applications. The review examines the technological and future challenges discussed in bringing these materials to a wide range of applications, together with potential solutions, as well as discusses the major industry players who are bringing these materials to the market.

Keywords: Bio-based polymers, Renewable resources, Biotechnologies, Sustainable materials

Review

Introduction

Bio-based polymers are materials which are produced from renewable resources. The terms bio-based polymers and biodegradable polymers are used extensively in the literature, but there is a key difference between the two types of polymers. Biodegradable polymers are defined as materials whose physical and chemical properties undergo deterioration and completely degrade when exposed to microorganisms, carbon dioxide (aerobic) processes, methane (anaerobic processes), and water (aerobic and anaerobic processes). Bio-based polymers can be biodegradable (e.g., polylactic acid) or nondegradable (e.g., biopolyhethylene). Similarly, while many bio-based polymers are biodegradable (e.g., starch and polyhydroxyalkanoates), not all biodegradable polymers are bio-based (e.g., polycaprolactone).

Bio-based polymers still hold a tiny fraction of the total global plastic market. Currently, biopolymers share less than 1% of the total market. At the current growth rate, it is expected that biopolymers will account for just over 1% of polymers by 2015 (Doug 2010).

The worldwide interest in bio-based polymers has accelerated in recent years due to the desire and need to find non-fossil fuel-based polymers. As indicated by ISI Web of Sciences and Thomas Innovations, there is a tremendous increase in the number of publication citations on bio-based polymers and applications in recent years, as shown in Figure 1 (Chen and Martin 2012).

Bio-based polymers offer important contributions by reducing the dependence on fossil fuels and through the related positive environmental impacts such as reduced carbon dioxide emissions. The legislative landscape is also changing where bio-based products are being favored through initiatives such as the *Lead Market Initiative* (European Union) and *BioPreferred* (USA). As a result, there is a worldwide demand for replacing petroleum-derived raw materials with renewable resource-based raw materials for the production of polymers.

The first generation of bio-based polymers focused on deriving polymers from agricultural feedstocks such as corn, potatoes, and other carbohydrate feedstocks. However, the focus has shifted in recent years due to a desire to move away from food-based resources and significant breakthroughs in biotechnology. Bio-based polymers similar to conventional polymers are produced by bacterial fermentation processes by synthesizing the building blocks (monomers) from renewable resources, including lignocellulosic biomass (starch and cellulose), fatty acids, and organic waste. Natural bio-based polymers are the other class of bio-based polymers which

* Correspondence: babup@tcd.ie
[1]Centre for Research Adoptive Nanostructures and Nano Devices, Trinity College, Dublin 2, Ireland
[2]School of Physics, Trinity College Dublin, Dublin 2, Ireland
Full list of author information is available at the end of the article

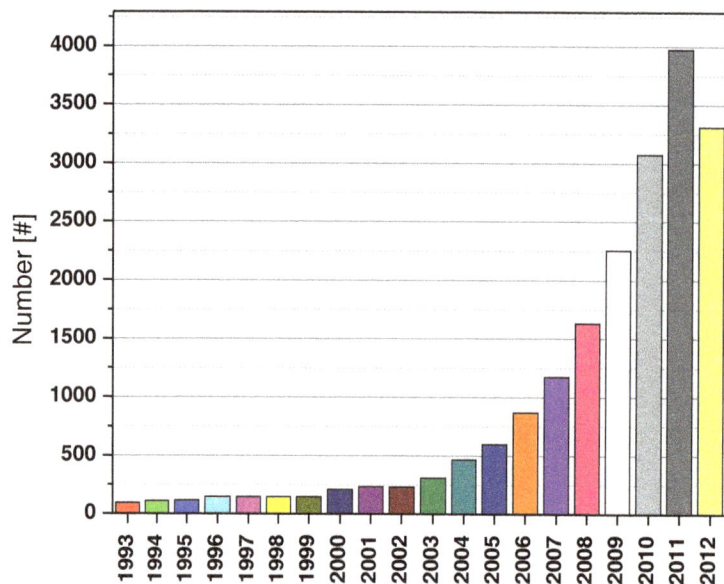

a: (Source: ISI web of sciences)

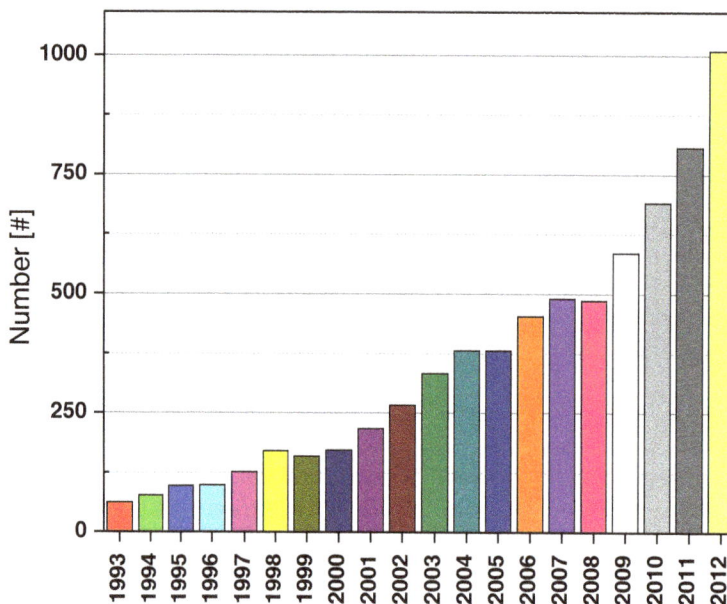

b: (Source: Thomas Innovations)

Figure 1 Citation trends of (a) publications and (b) patents on bio-based polymers in recent years.

are found naturally, such as proteins, nucleic acids, and polysaccharides (collagen, chitosan, etc.). These bio-based polymers have shown enormous growth in recent years in terms of technological developments and their commercial applications. There are three principal ways to produce bio-based polymers using renewable resources:

(1) Using natural bio-based polymers with partial modification to meet the requirements (e.g., starch)

(2) Producing bio-based monomers by fermentation/ conventional chemistry followed by polymerization (e.g., polylactic acid, polybutylene succinate, and polyethylene)

(3) Producing bio-based polymers directly by bacteria (e.g., polyhydroxyalkanoates).

In this paper, an overview of bio-based polymers made from renewable resources and natural polymers derived from plant and animal origins is presented. The review will focus on the preparation, properties, applications, and future trends for bio-based polymers. This paper discusses the use of renewable resources such as ligno-cellulosic biomass to create monomers and polymers that can replace petroleum-based polymers, such as polyester, polylactic acids, and other natural bio-based polymers, which are presented in Figure 2.

Polylactic acid

Polylactic acid (PLA) has been known since 1845 but not commercialized until early 1990. PLA belongs to the family of aliphatic polyesters with the basic constitutional unit lactic acid. The monomer lactic acid is the hydroxyl carboxylic acid which can be obtained via bacterial fermentation from corn (starch) or sugars obtained from renewable resources. Although other renewable resources can be used, corn has the advantage of providing a high-quality feedstock for fermentation which results in a high-purity lactic acid, which is required for an efficient synthetic process. L-lactic acid or D-lactic acid is obtained depending on the microbial strain used during the fermentation process.

PLA can be synthesized from lactic acid by direct polycondensation reaction or ring-opening polymerization of lactide monomer. However, it is difficult to obtain high molecular weight PLA via polycondensation reaction because of water formation during the reaction. Nature Works LLC (previously Cargill Dow LLC) has developed a low-cost continuous process for the production of PLA (Erwin et al. 2007). In this process, low molecular weight pre-polymer lactide dimers are formed during a condensation process. In the second step, the pre-polymers are converted into high molecular weight PLA via ring-opening polymerization with selected catalysts. Depending on the ratio and stereochemical nature of the monomer (L or D), various types of PLA and PLA copolymers can be obtained. The final properties of PLA produced are highly dependent on the ratio of the D and L forms of the lactic acid which are listed in Table 1 for various blend ratios (Garlotta 2001).

PLA is a commercially interesting polymer as it shares some similarities with hydrocarbon polymers such as polyethylene terephthalate (PET). It has many unique characteristics, including good transparency, glossy appearance, high rigidity, and ability to tolerate various types of processing conditions.

PLA is a thermoplastic polymer which has the potential to replace traditional polymers such as PET, PS, and PC for packaging to electronic and automotive applications (Majid et al. 2010). While PLA has similar mechanical properties to traditional polymers, the thermal properties are not attractive due to low Tg of 60°C. This problem can be overcome by changing the stereochemistry of the polymer and blending with other polymers and processing aids to improve the mechanical properties, e.g., varying the ratio of L and D isomer ratio strongly influences the crystallinity of the final polymer. However, much more work is required to improve the properties of PLA to suit various applications.

Currently, Nature Works LLC, USA, is the major supplier of PLA sold under the brand name Ingeo, with a production capacity of 100,000 ton/year. There are other manufactures of PLA based in the USA, Europe, China, and Japan developing various grades of PLA suitable for different industrial sectors such as automobile, electronics, medical devices, and commodity applications, which are mentioned in Table 2) (Doug 2010; Ravenstijn 2010).

PLA is widely used in many day-to-day applications. It has been mainly used in food packing (including food trays, tableware such as plates and cutlery, water bottles, candy wraps, cups, etc.). Although PLA has one of the highest heat resistances and mechanical strengths of all bio-based polymers, it is still not suitable for use in electronic devices and other engineering applications. NEC Corporation (Japan) recently produced a PLA with carbon and kenaf fibers with improved thermal and flame retardancy properties. Fujitsu (Japan) developed a polycarbonate blend with PLA to make computer housings. In recent years, PLA has been employed as a membrane material for use in automotive and chemical industry.

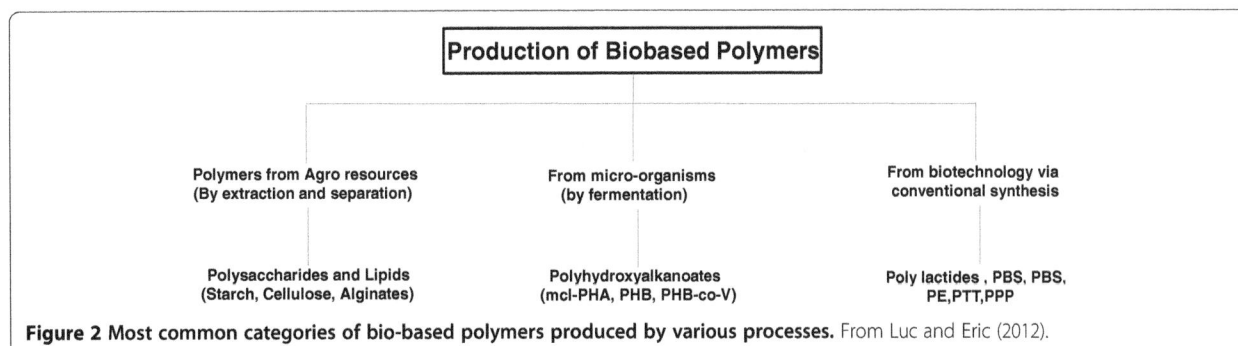

Figure 2 Most common categories of bio-based polymers produced by various processes. From Luc and Eric (2012).

Table 1 Variation in glass transition and melting temperature of PLA with various ratios of L-monomer composition

Copolymer ratio	Glass transition (Tg), °C	Melting temperature (Tm), °C
100:0 (L/DL)-PLA	63	178
95:5 (L/DL)-PLA	59	164
90:10 (L/DL)-PLA	56	150
85:15 (L/DL)-PLA	56	140
80:20 (L/DL)-PLA	56	125

The ease of melt processing has led to the production of PLA fibers, which are increasingly accepted in a wide variety of textiles from dresses to sportswear, furnishing to drapes, and soft nonwoven baby wipes to tough landscape textiles. These textiles can outperform traditional textiles made from synthetic counterparts. Bioresorbable scaffolds produced with PLA and various PLA blends are used in implants for growing living cells. The US Food and Drug Administration (FDA) has approved the use of PLA for certain human clinical applications (Dorozhkin 2009; Garlotta 2001). In addition, PLA-based materials have been used for bone support splints. Applications of PLA-based polymers in various fields are listed in Table 3.

Polyhydroxyalkanoates

Polyhydroxyalkanoates (PHAs) are a family of polyesters produced by bacterial fermentation with the potential to replace conventional hydrocarbon-based polymers. PHAs occur naturally in a variety of organisms, but microorganisms can be employed to tailor their production in cells. Polyhydroxybutyrate (PHB), the simplest PHA, was discovered in 1926 by Maurice Lemoigne as a constituent of the bacterium *Bacillus megaterium* (Lemoigne 1923).

PHA can be produced by varieties of bacteria using several renewable waste feedstocks. A generic process to produce PHA by bacterial fermentation involves fermentation, isolation, and purification from fermentation broth. A large fermentation vessel is filled with mineral medium and inoculated with a seed culture that contains bacteria. The feedstocks include cellulosics, vegetable oils, organic waste, municipal solid waste, and fatty acids depending on the specific PHA required. The carbon source is fed into the vessel until it is consumed and cell growth and PHA accumulation is complete. In general, a minimum of 48 h is required for fermentation time. To isolate and purify PHA, cells are concentrated, dried, and extracted with solvents such as acetone or chloroform. The residual cell debris is removed from the solvent containing dissolved PHA by solid-liquid separation process. The PHA is then precipitated by the addition of an alcohol (e.g., methanol) and recovered by a precipitation process (Kathiraser et al. 2007).

More than 150 PHA monomers have been identified as the constituents of PHAs (Steinbüchel and Valentin 1995). Such diversity allows the production of bio-based polymers with a wide range of properties, tailored for specific applications. Poly-3-hydroxybutyrate was the first bacterial PHA identified. It has received the greatest attention in terms of pathway characterization and industrial-scale production. It possesses similar thermal and mechanical properties to those of polystyrene and polypropylene (Savenkova et al. 2000). However, due to its slow crystallization, narrow processing temperature range, and tendency to 'creep', it is not attractive for many applications, requiring development in order to overcome these shortcomings (Reis et al. 2008). Several companies have developed PHA copolymers with typically 80% to 95% (*R*)-3-hydroxybutyric acid monomer and 5% to 20% of a second monomer in order to improve the properties of PHAs. Some specific examples of PHAs include the following:

Table 2 Global suppliers of PLA

Company	Location	Brand name	Production/planned capacity (kton/year)
Nature Works	USA	Ingeo	140 (by 2013)
Futerro	Belgium	Futerro	1.5 (by 2010)
Tate & Lyle	Netherlands	Hycail	0.2 (by 2012)
Purac	Netherlands	Purasorb	0.05
Hiusan Biosciences	China	Hisun	5
Jiangsu Jiulding	China		5
Teijin	Japan	Biofront	1
Toyobo	Japan	Vylocol	0.2
Synbra	Netherlands	Biofoam	50

Table 3 Application of PLA and their blends in various fields

Polymer	Applications	Reference
PLGA/PGA	Ovine pulmonary valve replacement	Williams et al. 1999; Sodian et al. 1999, 2000; Cheng et al. 2009
PLA/chitosan PLA/PLGA/ chitosan PLA	Drug carrier/drug release	Jeevitha and Kanchana 2013; Jayanth and Vinod 2012; Nagarwal et al. 2010; Chandy et al. 2000; Valantin et al. 2003
PLGA and copolymers	Degradable sutures	Rajev 2000
PLA/HA composites	Porous scaffolds for cellular applications	Jung-Ju et al. 2012
PLA-CaP and PLGA-CaP	Bone fixation devices, plates, pins, screws, and wires, orthopedic applications	Huan et al. 2012
PDLLA	Coatings on metal implants	Schmidmaier et al. 2001
PLA/PLGA	Use in cell-based gene therapy for cardiovascular diseases, muscle tissues, bone and cartilage regeneration, and other treatments of cardiovascular and neurological conditions	Coutu et al. 2009; Kellomaki et al. 2000; Papenburg et al. 2009
PLA and PLA blends	Packaging films, commodity containers, electrical appliances, mobile phone housings, floor mats, automotive spare parts	Rafael et al. 2010
PLA	Textile applications	Gupta et al. 2007; Avinc and Akbar 2009

PLGA, polylactic acid-co-glycolic acid; CaP, calcium phosphates; HA, hydroxyapatite.

- Poly(3HB): Poly(3-hydroxybutyrate)
- Poly(3HB-co-3HV): Poly(3-hydroxybutyrate-co-3-hydroxyvalerate), PHBV
- Poly(3-HB-co-4HB): Poly(3-hydroxybutyrate-co-4-hydroxybutyrate)
- Poly(3HB-co-3HH): Poly(3-hydroxyoctanoate-co-hydroxyhexanoate)
- Poly(3HO-co-3HH): Poly(3-hydroxyoctanoate-co-hydroxyhexanoate)
- Poly (4-HB): Poly(4-hydroxybutyrate).

The copolymer poly(3HB-co-3HV) has a much lower crystallinity, decreased stiffness and brittleness, and increased tensile strength and toughness compared to poly (3HB) while remaining biodegradable. It also has a higher melt viscosity, which is a desirable property for extrusion and blow molding (Hanggi 1995).

The first commercial plant for PHBV was built in the USA in a joint venture between Metabolix and Archer Daniels Midland. However, the joint venture between these two companies ended in 2012. Currently, Tianan Biologic Material Co. in China is the largest producer of PHB and PHB copolymers. Tianan's PHBV contains about 5% valerate which improves the flexibility of the polymer. Tainjin Green Biosciences, China, invested along with DSM to build a production plant with 10-kton/year capacity to produce PHAs for packing and biomedical applications (DSM press release 2008). The current global manufacturers of PHB-based polymers are listed in Table 4 (Doug 2010; Ravenstijn 2010).

PHA polymers are thermoplastic, and their thermal and mechanical properties depend on their composition.

The Tg of the polymers varies from –40°C to 5°C, and the melting temperatures range from 50°C to 180°C, depending on their chemical composition (McChalicher and Srienc 2007). PHB is similar in its material properties to polypropylene, with a good resistance to moisture and aroma barrier properties. Polyhydroxybutyric acid synthesized from pure PHB is relatively brittle and stiff. PHB copolymers, which may include other fatty acids such as beta-hydroxyvaleric acid, may be elastic (McChalicher and Srienc 2007).

PHAs can be processed in existing polymer-processing equipment and can be converted into injection-molded components: film and sheet, fibers, laminates, and coated articles; nonwoven fabrics, synthetic paper products, disposable items, feminine hygiene products, adhesives, waxes, paints, binders, and foams. Metabolix has received FDA clearance for use of PHAs in food contact applications. These materials are suitable for a wide range of food packing applications including caps and closures, disposable items such as forks, spoons, knives, tubs, trays, and hot cup lids, and products such as housewares, cosmetics, and medical packaging (Philip et al. 2007).

PHA and its copolymers are widely used as biomedical implant materials. Various applications of PHA and their polymer blends are listed in Table 5. These include sutures, suture fasteners, meniscus repair devices, rivets, bone plates, surgical mesh, repair patches, cardiovascular patches, tissue repair patches, and stem cell growth. Changing the PHA composition allows the manufacturer to tune the properties such as biocompatibility and polymer degradation time within desirable time frames

Table 4 Global suppliers of various types of PHAs

Company	Location	Brand name	Production/planned capacity (kton/year)
Bio-on	Italy	Minerv	10
Kaneka	Singapore		10 (by 2013)
Meredian	USA		13.5
Metabolix	USA	Mirel	50
Mitsubishi Gas Chemicals	Japan	Biogreen	0.05
PHB Industrial S/A	Brazil	Biocycle	0.05
Shenzen O'Bioer	China		
TEPHA	USA	ThephaFLEX/ThephELAST	
Tianan Biological Materials	China	Enmat	2
Tianjin Green Biosciences	China	Green Bio	10
Tianjin Northern Food	China		
Yikeman Shandong	China		3

under specific conditions. PHAs can also be used in drug delivery due to their biocompatibility and controlled degradability. Only a few examples of PHAs have been evaluated for this type of applications, and it remains an important area for exploitation (Tang et al. 2008).

Polybutylene succinate

Polybutylene succinate (PBS) is an aliphatic polyester with similar properties to those of PET. PBS is produced by condensation of succinic acid and 1,4-butanediol. PBS can be produced by either monomers derived from petroleum-based systems or the bacterial fermentation route. There are several processes for producing succinic acid from fossil fuels. Among them, electrochemical synthesis is a common process with high yield and low cost. However, the fermentation production of succinic acid has numerous advantages compared to the chemical process. Fermentation process uses renewable resources and consumes less energy compared to chemical process. Several companies (solely or in partnership) are now scaling bio-succinate production processes which have traditionally suffered from poor productivity and high downstream processing costs. Mitsubishi Chemical (Japan) has developed biomass-derived succinic

acid in collaboration with Ajinomoto to commercialize bio-based PBS. DSM and Roquette are developing a commercially feasible fermentation process for the production of succinic acid 1,4-butanediol and subsequent production of PBS. Myriant and Bioamber have developed a fermentation technology to produce monomers. There are several companies around the world developing technologies for the production of PBS, as listed in Table 6, including North America and China (Doug 2010; Ravenstijn 2010).

Conventional processes for the production of 1,4-butanediol use fossil fuel feedstocks such as acetylene and formaldehyde. The bio-based process involves the use of glucose from renewable resources to produce succinic acid followed by a chemical reduction to produce butanediol. PBS is produced by transesterification, direct polymerization, and condensation polymerization reactions. PBS copolymers can be produced by adding a third monomer such as sebacic acid, adipic acid, and succinic acid which is also produced by renewable resources (Bechthold et al. 2008).

PBS is a semicrystalline polyester with a melting point higher than that of PLA. Its mechanical and thermal properties depend on the crystal structure and the degree of crystallinity (Nicolas et al. 2011). PBS displays similar

Table 5 Application of PHAs and their blends in various fields

PHA polymer type	Applications	Reference
P(3HB), P(3HB-co-3HHX) and blends	Scaffolds, nerve regeneration, soft tissue, artificial esophagus, drug delivery, skin regeneration, food additive	Yang et al. 2002; Chen and Qiong 2005; Bayram and Denbas 2008; Tang et al. 2008; Clarinval and Halleux 2005
mcl-PHA/scl-PHA	Cardiac tissue engineering, drug delivery, cosmetics, drug molecules	Sodian et al. 2000; Wang et al. 2003; de Roo et al. 2002; Zhao et al. 2003; Ruth et al. 2007
P(4HB) and P(3HO)	Heart valve scaffolds, food additive	Clarinval and Halleux 2005; Valappil et al. 2006
P(3HB-co-4HB), P(3HB-co-3HV)	Drug delivery, scaffolds, artificial heart values, patches to repair gastrointestinal tracts, sutures	Türesin et al. 2001; Williams et al. 1999; Chen et al. 2008; Freier et al. 2002; Kunze et al. 2006; Volova et al. 2003
PHB, Mirel P103	Commodity applications, shampoo and cosmetic bottles, cups and food containers	Philip et al. 2007; Amass et al. 1998; Walle et al. 2001

Table 6 Global producers of PBS

Company	Location	Brand name/polymer type	Production/planned capacity (kton/year)
BASF	Germany	PBS	
Dupont de Nemours	USA	PBST	
Hexing Chemical	China	PBS	3
Ube	Japan	NA	NA
IPC-CAS	China	PBS, PBSA	5
IRE Chemical	Korea	Enpol, PBS, PBSA	3.5
Kingfa	China	PBSA	1
Mitsubishi Gas Chemical	Japan	PBS, PES, PBSLa	3
Showa	Japan	Bionelle PBS, PBSA, PBS	3
SK Chemicals	Korea	Skygreen	NA
DSM	Netherlands	NA	NA

NA, not available; PBSA, poly(butylene succinate adipate).

crystallization behavior and mechanical properties to those of polyolefin such as polyethylene. It has a good tensile and impact strength with moderate rigidity and hardness. The Tg is approximately −32°C, and the melting temperature is approximately 115°C. In comparison with PLA, PBS is tougher in nature but with a lower rigidity and Young's modulus. By changing the monomer composition, mechanical properties can be tuned to suit the required application (Liu et al. 2009a, b).

PBS and their blends have found commercial applications in agriculture, fishery, forestry, construction, and other industrial fields which are listed in Table 7. For example, PBS has been employed as mulch film, packaging, and flushable hygiene products and also used as a non-migrant plasticizer for polyvinyl chloride (PVC). In addition, it is used in foaming and food packaging application. The relatively poor mechanical flexibility of PBS limits the applications of 100% PBS-based products. However, this can be overcome by blending PBS with PLA or starch to improve the mechanical properties significantly, providing properties similar to that of polyolefin (Eslmai and Kamal 2013; Zhao et al. 2010).

Bio-polyethylene

Polyethylene (PE) is an important engineering polymer traditionally produced from fossil resources. PE is produced by polymerization of ethylene under pressure, temperature, in the presence of a catalyst. Traditionally, ethylene is produced through steam cracking of naphtha or heavy oils or ethanol dehydration. With increases in oil prices, microbial PE or green PE is now being manufactured from dehydration of ethanol produced by microbial fermentation. The concept of producing PE from bioethanol is not a particularly new one. In the 1980s, Braskem made bio-PE and bio-PVC from bioethanol. However, low oil prices and the limitations of the biotechnology processes made the technology unattractive at that time (de Guzman 2010).

Currently, bio-PE produced on an industrial scale from bioethanol is derived from sugarcane. Bioethanol is also derived from biorenewable feedstocks, including sugar beet, starch crops such as maize, wood, wheat, corn, and other plant wastes through microbial strain and biological fermentation process. In a typical process, extracted sugarcane juice with high sucrose content is anaerobically fermented to produce ethanol. At the end of the fermentation process, ethanol is distilled in order to remove water and to yield azeotropic mixture of hydrous ethanol. Ethanol is then dehydrated at high temperatures over a solid catalyst to produce ethylene and, subsequently, polyethylene (Guangwen et al. 2007; Luiz et al. 2010).

Table 7 Applications of PBS and their blends

Polymer type	Applications	Reference
PBS/PLA blend	Packaging films, dishware, fibers, medical materials	Weraporn et al. 2011; Liu et al. 2009 a, b; Bhatia et al. 2007; Lee and Wang 2006
PBS and blends	Drug encapsulation systems	Cornelia et al. 2011
PBS/starch	Barrier films	Jian-Bing et al. 2011
PBS and copolymers	Industrial applications	Jun and Bao-Hua 2010 a, b
PBS ionomers	Orthopedic applications	Jung et al. 2009

Bio-based polyethylene has exactly the same chemical, physical, and mechanical properties as petrochemical polyethylene. Braskem (Brazil) is the largest producer of bio-PE with 52% market share, and this is the first certified bio-PE in the world. Similarly, Braskem is developing other bio-based polymers such as bio-polyvinyl chloride, bio-polypropylene, and their copolymers with similar industrial technologies. The current Braskem bio-based PE grades are mainly targeted towards food packing, cosmetics, personal care, automotive parts, and toys. Dow Chemical (USA) in cooperation with Crystalsev is the second largest producer of bio-PE with 12% market share. Solvay (Belgium), another producer of bio-PE, has 10% share in the current market. However, Solvay is a leader in the production of bio-PVC with similar industrial technologies. China Petrochemical Corporation also plans to set up production facilities in China to produce bio-PE from bioethanol (Haung et al. 2008).

Bio-PE can replace all the applications of current fossil-based PE. It is widely used in engineering, agriculture, packaging, and many day-to-day commodity applications because of its low price and good performance. Table 8 shows applications of bio-PE in different fields where it can replace conventional PE.

Bio-based natural polymers

This group consists of naturally occurring polymers such as cellulose, starch, chitin, and various polysaccharides and proteins. These materials and their derivatives offer a wide range of properties and applications. In this section, some of the natural bio-based polymers and their applications in various fields are discussed.

Starch

Starch is a unique bio-based polymer because it occurs in nature as discrete granules. Starch is the end product of photosynthesis in plants - a natural carbohydrate-based polymer that is abundantly available in nature from various sources including wheat, rice, corn, and potato. Essentially, starch consists of the linear polysaccharide amylose and the highly branched polysaccharide amylopectin. In particular, thermoplastic starch is of growing interest within the industry. The thermal and mechanical properties of starch can vary greatly and depend upon such factors as the amount of plasticizer present. The T_g varies between $-50°C$ and $110°C$, and the modulus is similar to polyolefins (Jane 1995). Several challenges exist in producing

commercially viable starch plastics. Starch's molecular structure is complex and partly nonlinear, leading to issues with ductility. Starch and starch thermoplastics suffer from the phenomenon of retrogradation - a natural increase in crystallinity over time, leading to increased brittleness. Plasticizers need to be found to create starch plastics with mechanical properties comparable to polyolefin-derived packaging. Plasticized starch blends and composites and/or chemical modifications may overcome these issues, creating biodegradable polymers with sufficient mechanical strength, flexibility, and water barrier properties for commercial packaging and consumer products (Maurizio et al. 2005).

Novamont is one of the leading companies in processing starch-based products (Li et al. 2009). The company produces various types of starch-based products using proprietary blend formulations. There are other companies around the world producing starch-based products in a similar scale for various applications, which are listed in Table 9 (Doug 2010; Ravenstijn 2010).

Applications of thermoplastic starch polymers include films, such as for shopping, bread, and fishing bait bags, overwraps, flushable sanitary product, packing materials, and special mulch films. Potential future applications could include foam loose-fill packaging and injection-molded products such as 'take-away' food containers. Starch and modified starches have a broad range of applications both in the food and non-food sectors. In Europe in 2002, the total consumption of starch and starch derivatives was approximately 7.9 million tons, of which 54% was used for food applications and 46% in non-food applications (Frost & Sullivan report 2009).

The largest users of starch in the European Union (30%) are the paper, cardboard, and corrugating industries (Frost & Sullivan report 2009). Other important fields of starch application are textiles, cosmetics, pharmaceuticals, construction, and paints, which are listed in Table 10. In the medium and long term, starch will play an increasing role in the field of 'renewable raw materials' for the production of biodegradable plastics, packaging material, and molded products.

Cellulose

Cellulose is the predominant constituent in cell walls of all plants. Cellulose is a complex polysaccharide with crystalline morphology. Cellulose differs from starch where glucose units are linked by β-1,4-glycosidic bonds,

Table 8 Application of bio-PE polymer and their blends

Polymer type	Applications	Reference
Bio-PE	Plastics bags, milk and water bottles, food packaging films, toys	Vona et al. 1965; Aamer et al. 2008
Bio-PE and blends	Agricultural mulch films	Kasirajan and Ngouajio 2012

Table 9 Global suppliers of starch-based products

Company	Location	Brand name	Production/planned capacity (kton/year)
Novamont	Italy	Mater-Bi	120
Japan Corn Starch	Japan	Ever Corn	NA
Biotec	Germany	Bioplast	NA
Rodenberg	Netherlands	Solanyl	50
BIOP	Germany	Biopar	5
Plantic	Australia	Plantic	7.5
Wuhan Huali Environment Protection Sci. & Tech	China	PSM	15
Biograde	China	Cardia	3
PSM	USA	Plaststarch	NA
Livan	Canada	Livan	10

whereas the bonds in starch are predominantly α-1,4 linkages. The most important raw material sources for the production of cellulosic plastics are cotton fibers and wood. Plant fiber is dissolved in alkali and carbon disulfide to create viscose, which is then reconverted to cellulose in cellophane form following a sulfuric acid and sodium sulfate bath. There are currently two processes used to separate cellulose from the other wood constituents (Yan et al. 2009). These methods, sulfite and prehydrolysis kraft pulping, use high pressure and chemicals to separate cellulose from lignin and hemicellulose, attaining greater than 97% cellulose purity. The main derivatives of cellulose for industrial purposes are cellulose acetate, cellulose esters (molding, extrusion, and films), and regenerated cellulose for fibers.

Cellulose is a hard polymer and has a high tensile strength of 62 to 500 MPa and elongation of 4% (Bisanda and Ansell 1992; Eichhorn et al. 2001). In order to overcome the inherent processing problems of cellulose, it is necessary to modify, plasticize, and blend with other polymers. The mechanical and thermal properties vary from blend to blend depending on the composition. The T_g of cellulosic derivatives ranged between 53°C and 180°C (Picker and Hoag 2002).

Eastman Chemical is a major producer of cellulosic polymers. FKuR launched a biopolymer business in the year 2000 and has a capacity of 2,800 metric ton/year of various cellulosic compounds for different applications (Doug 2010). The major producers of cellulose-based compounds are listed in Table 11 (Doug 2010; Ravenstijn 2010).

There are three main groups of cellulosic polymers that are produced by chemical modification of cellulose for various applications. Cellulose esters, namely cellulose nitrate and cellulose acetate, are mainly developed for film and fiber applications. Cellulose ethers, such as carboxymethyl cellulose and hydroxyethyl cellulose, are widely used in construction, food, personal care, pharmaceuticals, paint, and other pharmaceutical applications (Kamel et al. 2008). Finally, regenerated cellulose is the largest bio-based polymer produced globally for fiber and film applications. Regenerated cellulose fibers are used in textiles, hygienic disposables, and home furnishing fabrics because of its thermal stability and modulus (Kevin et al. 2001).

Chemically pure cellulose can be produced using a certain type of bacteria. Bacterial cellulose is characterized by its purity and high strength. It can be used to produce articles with relatively high strength. Currently, applications for bacterial cellulose outside food and biomedical fields are rather limited because of its high price. The other applications include acoustic diaphragms, mining, paints, oil gas recovery, and adhesives.

Table 10 Application of starch and their blends in various fields

Polymer type	Applications	Reference
Starch	Orthopedic implant devices as bone fillers	Ashammakhi and Rokkanen 1997
Starch/ethylene vinyl alcohol/HA starch/polycaprolactone blends	Bone replacement/fixation implants, orthopedic applications	Mainil et al. 1997; Mendes et al. 2001; Marques and Reis 2005
Starch/cellulose acetate blends with methylmethacrylate and acrylic acid	Bone cements	Espigares et al. 2002
Modified starch	Food applications	Jaspreet et al. 2007; Fuentes et al. 2010
Starch derivatives	Drug delivery	Asha and Martins 2012
Thermoplastic starch	Packaging, containers, mulch films, textile sizing agents, adhesives	Zhao et al. 2008; Maurizio et al. 2005; Ozdemir and Floros 2004; Dave et al. 1999; Guo et al. 2005; Kumbar et al. 2001; Li et al. 2011

Table 11 Global suppliers of cellulosic products

Company	Location	Brand name
Innovia films	UK	Nature Flex
Eastman Chemical	USA	Tenite
FKuR	Germany	Biograde
Sateri	China	Sateri

However, the low yields and high costs of bacterial cellulose represent barriers to large-scale industrial applications (Prashant et al. 2009). Table 12 summarizes the applications of cellulose and their compounds in different fields.

Chitin and chitosan

Chitin and chitosan are the most abundant natural amino polysaccharide and valuable bio-based natural polymers derived from shells of prawns and crabs. Currently, chitin and chitosan are produced commercially by chemical extraction process from crab, shrimp, and prawn wastes (Roberts 1997). The chemical extraction of chitin is quite an aggressive process based on demineralization by acid and deproteination by the action of alkali followed by deacetylated into chitosan (Roberts 1997). Chitin can also be produced by using enzyme hydrolysis or fermentation process, but these processes are not economically feasible on an industrial scale (Win and Stevens 2001). Currently, there are few industrial-scale plants of chitin and chitosan worldwide located in the USA, Canada, Scandinavia, and Asia (Ravi Kumar 2000).

Chitosan displays interesting characteristics including biodegradability, biocompatibility, chemical inertness, high mechanical strength, good film-forming properties, and low cost (Marguerite 2006; Virginia et al. 2011; Liu et al. 2012). Chitosan is being used in a vast array of widely varying products and applications ranging from pharmaceutical and cosmetic products to water treatment and plant protection. For each application, different properties of chitosan are required, which changes with the degree of acetylation and molecular weight. Chitosan is compatible with many biologically active components incorporated in

cosmetic product composition (Ravi Kumar 2000). Due to its low toxicity, biocompatibility, and bioactivity, chitosan has become a very attractive material in such diverse applications as biomaterials in medical devices and as a pharmaceutical ingredient (Bae and Moo-Moo 2010; Ramya et al. 2012). Chitosan has application in shampoos, rinses, and permanent hair-coloring agents. Chitosan and its derivatives also have applications in the skin care industry. Chitosan can function as a moisturizer for the skin, and because of its lower costs, it might compete with hyaluronic acid in this application (Bansal et al. 2011; Valerie and Vinod 1998; Hafdani and Sadeghinia 2011).

Pullulan

Pullulan is a linear water-soluble polysaccharide mainly consisting of maltotriose units connected by α-1,6 glycosidic units. Pullulan was first reported by Bauer (1938) and is obtained from the fermentation broth of *Aureobasidium pullulans*. Pullulan is produced by a simple fermentation process using a number of feedstocks containing simple sugars (Bernier 1958; Catley 1971; Sena et al. 2006). Pullulan can be chemically modified to produce a polymer that is either less soluble or completely insoluble in water. The unique properties of this polysaccharide are due to its characteristic glycosidic linking. Pullulan is easily chemically modified to reduce the water solubility or to develop pH sensitivity, by introducing functional reactive groups, etc. Due to its high water solubility and low viscosity, pullulan has numerous commercial applications including use as a food additive, a flocculant, a blood plasma substitute, an adhesive, and a film (Zajic and LeDuy 1973; Singh et al. 2008; Cheng et al. 2011). Pullulan can be formed into molding articles which can resemble conventional polymers such as polystyrene in their transparency, strength, and toughness (Leathers 2003).

Pullulan is extensively used in the food industry. It is a slow-digesting macromolecule which is tasteless as well as odorless, hence its application as a low-calorie food additive providing bulk and texture. Pullulan possesses oxygen barrier property and good moisture retention,

Table 12 Application of cellulose and their compounds in various fields

Polymer type	Applications	Reference
Cellulose esters	Membranes for separation	Kumano and Fujiwara 2008
Carboxylated methyl cellulose	Drug formulations, as binder for drugs, film-coating agent for drugs, ointment base	Chambin et al. 2004; Obae and Imada 1999; Westermark et al. 1999; Hirosawa et al. 2000
Cellulose acetate fibers	Wound dressings	Orawan et al. 2008; Abdelrahman and Newton 2011
Hydroxyethyl cellulose	Spray for clothes polluted with pollen	Hori et al. 2005
Modified celluloses, cellulose whiskers, microfibrous cellulose	Barrier films, water preservation in food packing	Amit and Ragauskas 2009
Cellulose nanofibers	Textile applications	Zeeshan et al. 2013
Cellulose particles	Chromatographic applications, chiral separations	Levison 1993; Arshady 1991a, b

and also, it inhibits fungal growth. These properties make it an excellent material for food preservation, and it is used extensively in the food industry (Conca and Yang 1993). In recent years, pullulan has also been studied for biomedical applications in various aspects, including targeted drug and gene delivery, tissue engineering, wound healing, and even in diagnostic imaging medium (Rekha and Chrndra 2007). Other emerging markets for pullulan include oral care products (Barkalow et al. 2002) and formulations of capsules for dietary supplements and pharmaceuticals (Leathers 2003), leading to increased demand for this unique biopolymer.

Collagen and gelatin

Collagen is the major insoluble fibrous protein in the extracellular matrix and in connective tissue. In fact, it is the single most abundant protein in the animal kingdom. There are at least 27 types of collagens, and the structures all serve the same purpose: to help tissues withstand stretching. The most abundant sources of collagen are pig skin, bovine hide, and pork and cattle bones. However, the industrial use of collagen is obtained from nonmammalian species (Gomez-Guille et al. 2011). Gelatin is obtained through the hydrolysis of collagen. The degree of conversion of collagen into gelatin depends on the pretreatment, function of temperature, pH, and extraction time (Johnston-Banks 1990).

Collagen is one of the most useful biomaterials due to its biocompatibility, biodegradability, and weak antigenicity (Maeda et al. 1999). The main application of collagen films in ophthalmology is as drug delivery systems for slow release of incorporated drugs (Rubin et al. 1973). It was also used for tissue engineering including skin replacement, bone substitutes, and artificial blood vessels and valves (Lee et al. 2001).

The classical food, photographic, cosmetic, and pharmaceutical applications of gelatin is based mainly on its gel-forming properties. Recently in the food industry, an increasing number of new applications have been found for gelatin in products in line with the growing trend to replace synthetic agents with more natural ones (Gomez-Guille et al. 2011). These include emulsifiers, foaming agents, colloid stabilizers, biodegradable film-forming materials, and microencapsulating agents.

Alginates

Alginate is a linear polysaccharide that is abundant in nature as it is synthesized by brown seaweeds and by soil bacteria (Draget et al. 1997). Sodium alginate is the most commonly used alginate form in the industry since it is the first by-product of algal purification (Draget 2000). Sodium alginate consists of α-l-guluronic acid residues (G blocks) and β-d-mannuronic acid residues (M blocks), as well as segments of alternating guluronic and mannuronic acids.

Although alginates are a heterogeneous family of polymers with varying content of G and M blocks depending on the source of extraction, alginates with high G content have far more industrial importance (Siddhesh and Edgar 2012). The acid or alkali treatment processes used to make sodium alginate from brown seaweeds are relatively simple. The difficulties in processing arise mainly from the separation of sodium alginate from slimy residues (Black and Woodward 1954). It is estimated that the annual production of alginates is approximately 38,000 tons worldwide (Helgerud et al. 2009).

Alginates have various industrial uses as viscosifiers, stabilizers, and gel-forming, film-forming, or water-binding agents (Helga and Svein 1998). These applications range from textile printing and manufacturing of ceramics to production of welding rods and water treatment (Teli and Chiplunkar 1986; Qin et al. 2007; Xie et al. 2001). The polymer is soluble in cold water and forms thermostable gels. These properties are utilized in the food industry in products such as custard creams and restructured food. The polymer is also used as a stabilizer and thickener in a variety of beverages, ice creams, emulsions, and sauces (Iain et al. 2009).

Alginates are widely used as a gelling agent in pharmaceutical and food applications. Studies into their positive effects on human health have broadened recently with the recognition that they have a number of potentially beneficial physiological effects in the gastrointestinal tract (Peter et al. 2011; Mandel et al. 2000). Alginate-containing wound dressings are commonly used, especially in making hydrophilic gels over wounds which can produce comfortable, localized hydrophilic environments in healing wounds (Onsoyen 1996). Alginates are used in controlled drug delivery, where the rate of drug release depends on the type and molecular weight of alginates used (Alexnader et al. 2006; Goh et al. 2012). Additionally, dental impressions made with alginates are easy to handle for both dentist and patient as they fast set at room temperature and are cost-effective (Onsoyen 1996). Recent studies show that alginates can be effective in treating obesity, and currently, various functional alginates are being evaluated in human clinical trials (Georg et al. 2012).

Current status and future trends

The use of bio-based feedstocks in the chemical sector is not a novel concept. They have been industrially feasible on a large scale for more than a decade. However, the price of oil was so cost-effective, and the development of oil-based products created so many opportunities that bio-based products were not prioritized at the time. Several factors, such as the limitations and uncertainty in supplies of fossil fuels, environmental considerations, and technological developments, accelerated the advancement of bio-based polymers and products. It took more than a century to evolve the fossil fuel-based chemical industry; however, the bio-based polymer industry is already catching up with fossil fuel-based chemical industry, which has augmented in the last 20 years. Thanks to advancements in white biotechnology, the production of bio-based polymers and other chemicals from renewable resources has become a reality. The first-generation technologies mainly focused on food resources such as corn, starch, rice, etc. to produce bio-based polymers. As the food-versus-fuel debate ascended, the focus of technologies diverted to cellulose-based feedstocks, focusing on waste from wood and paper, food industries, and even stems and leaves and solid municipal waste streams. More and more of these technologies are already in the pipeline to align with the abovementioned waste streams; however, it may take another 20 years to develop the full spectrum of chemicals based on these technologies (Michael et al. 2011).

Challenges that need to be addressed in the coming years include management of raw materials, performance of bio-based materials, and their cost for production. Economy of scale will be one of the main challenges for production of bio-based monomers and bio-based polymers from renewable sources. Building large-scale plants can be difficult due to the lack of experience in new technologies and estimation of supply/demand balance. In order to make these technologies economically viable, it is very important to develop (1) logistics for biomass feedstocks, (2) new manufacturing routes by replacing existing methods with high yields, (3) new microbial strains/enzymes, and (4) efficient downstream processing methods for recovery of bio-based products.

The current bio-based industry focus is mainly on making bio-versions of existing monomers and polymers. Performance of these products is well known, and it is relatively easy to replace the existing product with similar performance of bio-versions. All the polymers mentioned above often display similar properties of current fossil-based polymers. Recently, many efforts are seen towards introducing new bio-based polymers with higher performance and value. For example, Nature Works LLC has introduced new grades of PLA with higher thermal and mechanical properties. New PLA-tri block copolymers have been reported to behave like thermoplastic elastomer. Many developments are currently underway to develop various polyamides, polyesters, polyhydroxyaloknates, etc. with a high differentiation in their final properties for use in automotive, electronics, and biomedical applications.

The disadvantage of some of the new bio-based polymers is that they cannot be processed in all current processing equipment. There is vast knowledge on additive-based chemistry developed for improving the performance and processing of fossil fuel-based polymers, and this knowledge can be used to develop new additive chemistry to improve the performance and properties of bio-based polymers (Ray and Bousmina 2005). For bio-based polymers like PLA and PHA, additives are being developed to improve their performance, by blending with other polymers or making new copolymers. However, the additive market for bio-based polymers is still very small, which makes it difficult to justify major development efforts according to some key additive supplier companies.

The use of nanoparticles as additives to enhance polymer performance has long been established for petroleum-based polymers. Various nano-reinforcements currently being developed include carbon nanotubes, graphene, nanoclays, 2-D layered materials, and cellulose nanowhiskers. Combining these nanofillers with bio-based polymers could enhance a large number of physical properties, including barrier, flame resistance, thermal stability, solvent uptake, and rate of biodegradability, relative to unmodified polymer resin. These improvements are generally attained at low filler content, and this nano-reinforcement is a very attractive route to generate new functional biomaterials for various applications.

Even though new bio-based polymers are produced on an industrial scale, there are still several factors which need to be determined for the long-term viability of bio-based polymers. It is expected that there will be feedstock competition as global demand for food and energy increases over time. Currently, renewable feedstocks used for manufacturing bio-based monomers and polymers often compete with requirements for food-based products. The expansion of first-generation bio-based fuel production will place unsustainable demands on biomass resources and is as much a threat to the sustainability of biochemical and biopolymer production as it is to food production (Michael et al. 2011). Indeed the European commission has altered its targets downwards for first-generation biofuels since October 2012, indicating its preference for non-food sources of sugar for biofuel production (EurActiv.com 2012). Several initiatives are underway to use cellulose-based feedstocks for the production of usable sugars for biofuels, biochemicals, and biopolymers (Jong et al. 2010).

Conclusions

Bio-based polymers are closer to the reality of replacing conventional polymers than ever before. Nowadays, bio-based polymers are commonly found in many applications from commodity to hi-tech applications due to advancement in biotechnologies and public awareness. However, despite these advancements, there are still some drawbacks which prevent the wider commercialization of bio-based polymers in many applications. This is mainly due to performance and price when compared with their conventional counterparts, which remains a significant challenge for bio-based polymers.

Competing interests

The authors declare that they have no competing interests.

Authors' contributions

RPB contributed in writing the whole manuscript. KOC contributed in providing the information on applications and policy information of bio-based polymers. SR contributed in providing the outline for the manuscript. All authors read and approved the final manuscript.

Acknowledgments

RPB would like to acknowledge the financial support from the Environmental Protection Agency, Ireland, under grant no. 2008-ET-LS-1-S2.

Author details

[1]Centre for Research Adoptive Nanostructures and Nano Devices, Trinity College, Dublin 2, Ireland. [2]School of Physics, Trinity College Dublin, Dublin 2, Ireland. [3]School of Biomolecular and Biomedical Sciences, Centre for Synthesis and Chemical Biology, UCD Conway Institute, and Earth Institute, University College Dublin, Belfield, Dublin 4, Ireland. [4]NUSNNI, National University of Singapore, 2 Engineering Drive 3, Singapore 117581, Singapore. [5]Institute of Materials Research and Engineering, Singapore 117602, Singapore. [6]Jinan University, Guangzhou, China.

References

Aamer AS, Fariha H, Abdul H, Safia A (2008) Biological degradation of plastics: a comprehensive review. Biotechnol Adv 26:246–265

Abdelrahman T, Newton H (2011) Wound dressings: principles and practice. Surgery 29:491–495

Alexnader DA, Kong HJ, Mooney DJ (2006) Alginate hydrogels as biomaterials. Macromolecular Biosciences 6:623–633

Amass W, Amass A, Tighe B (1998) A review of biodegradable polymers: uses, current developments in the synthesis and characterization of biodegradable polyesters, blends of biodegradable polymers and recent advances in biodegradation studies. Polymer International 47:89–144

Amit S, Ragauskas AJ (2009) Water transmission barrier properties of biodegradable films based on cellulosic whiskers and xylan. Carbohydr Polym 78(2):357–360

Arshady R (1991a) Beaded polymer supports and gels: 2. Physicochemical criteria and functionalization. J Chromatogr 586:199–219

Arshady R (1991b) Beaded polymer supports and gels: 1. Manufacturing techniques. J Chromatogr 586:181–197

Asha R, Martins E (2012) Recent applications of starch derivatives in nanodrug delivery. Carbohydr Polym 87(2):987–994

Ashammakhi N, Rokkanen P (1997) Absorbable polyglycolide devices in trauma and bone surgery. Biomaterials 18(1):3–9

Avinc A, Akbar K (2009) Overview of poly (lactic acid) fibres. Part I: production, properties, performance, environmental impact, and end-use applications of poly (lactic acid) fibres. Fiber Chemistry 41(6):391–401

Bae KP, Moo-Moo K (2010) Applications of chitin and its derivatives in biological medicine. Int J Mol Sci 11:5152–5164

Bansal V, Pramod KS, Nitin S, Omprakask P, Malviya R (2011) Applications of chitosan and chitosan derivatives for drug delivery. Adva Biol Res 5:28–37

Barkalow DG, Chapedelaine AH, Dzija MJ (2002) Improved pullulan free edible film compositions and methods of making same. PCT International Application WO 02/43657, US 01/43397, 21 Nov

Bauer R (1938) Physiology of Dematium pullulans de Bary. Zentralbl Bacteriol Parasitenkd Infektionskr Hyg Abt2 98:133–167

Bayram C, Denbas EB (2008) Preparation and characterization of triamcinolone acetonide-loaded poly(3-hydroxybutyrate-co-3-hydroxyhexanoate) (PHBHx) microspheres. J Bioactive and Compatible Polymer 23:334–347

Bechthold I, Bretz K, Kabasci S, Kopitzky R, Springer A (2008) Succinic acid: a new platform chemical from biobased polymers from renewable resources. Chemical Engg Technol 31:647–654

Bernier B (1958) The production of polysaccharides by fungi active in the decomposition of wood and forest litter. Can J Microbiol 4:195–204

Bhatia A, Gupta RK, Bhattacharaya SN, Choi HJ (2007) Compatibility of biodegradable PLA and PBS blends for packaging applications. Korea Aust Rheol J 19:125–131

Bisanda ETN, Ansell MP (1992) Properties of sisal-CNSL composites. J Mater Sci 27:1690–1700

Black WAP, Woodward FN (1954) Alginates from common British brown marine algae. In Natural plant hydrocolloids. Adv Chem Ser Am Chem Soc 11:83–91

Catley BJ (1971) Utilization of carbon sources by Pullularia pullulans for the elaboration of extracellular polysaccharides. Appl Microbiol 22:641–649

Chambin DC, Debray C, Rochat-Gonthier MH, Le MM, Pourcelot M (2004) Effects of different cellulose derivatives on drug release mechanism studied at a pre-formulation stage. J Controll Release 95(1):101–108

Chandy T, Das GS, Rao GH (2000) 5-Fluorouracil-loaded chitosan coated polylactic acid microspheres as biodegradable drug carriers for cerebral tumours. J Microencapsul 5:625–631

Chen GQ, Qiong W (2005) The application of polyhydroxyalkanoates as tissue engineering materials. Biomaterials 26:6565–6578

Chen GQ, Martin KP (2012) Plastics derived from biological sources: present and future: a technical and environmental review. Chem Rev 112:2082–2099

Chen QZ, Harding SE, Ali NN, Lyon AR, Boccaccini AR (2008) Biomaterials in cardiac tissue engineering: ten years of research survey. Materials Sci Eng: Reports 59:1–37

Cheng KC, Demirci A, Catchmark JM (2011) Pullulan: biosynthesis, production, and applications. Appl Microbiol Biotechnol 92:29–44

Cheng Y, Deng S, Chen P, Ruan R (2009) Polylactic acid (PLA) synthesis and modifications: a review. Front Chem China 4:259–264

Clarinval AM, Halleux J (2005) Classification of biodegradable polymers. In: Smith R (ed) Biodegradable polymers for industrial applications. Woodhead, Cambridge

Conca KR, Yang TCS (1993) Edible food barrier coatings. In: Ching C, Kaplan DL, Thomas EL (ed) Biodegradable polymers and packaging. Technomic, Lancaster, pp 357–369

Cornelia TB, Erkan TB, Elisabete DP, Rui LR, Nuno MN (2011) Performance of biodegradable microcapsules of poly(butylene succinate), poly(butylene succinate-co-adipate) and poly(butylene terephthalate-co-adipate) as drug encapsulation systems. Colloids Surf B Biointerfaces 84:498–507

Coutu DL, Yousefi AM, Galipeau J (2009) Three-dimensional porous scaffolds at the crossroads of tissue engineering and cell-based gene therapy. J Cell Biochem 108:537–546

Dave AM, Mehta MH, Aminabhavi TM, Kulkarni AR, Soppimath KS (1999) A review on controlled release of nitrogen fertilizers through polymeric membrane devices. Polymer- Plastics Technol Eng 38:675–711

de Roo G, Kellerhals MB, Ren Q, Witholt B, Kessler B (2002) Production of chiral R-3-hydroxyalkanoic acids and R-3-hydroxyalkanoic acid methylesters via hydrolytic degradation of polyhydroxyalkanoate synthesized by pseudomonads. Biotechnol Bioeng 77:717–722

de Guzman D (2010) Bioplastic development increases with new applications. http://www.icis.com/Articles/2010/10/25/9402443/bioplastic-development-increases-with-new-applications.html. Accessed October 2010

Dorozhkin SV (2009) Calcium orthophosphate-based biocomposites and hybrid biomaterials. J Mater Sci 44:2343–2387

Doug S (2010) Bioplastics: technologies and global markets. BCC research reports PLS050A. http://www.bccresearch.com/report/bioplastics-technologies-markets-pls050a.html

Draget KI (2000) Alginates. In: Philips O, Williams A (ed) Handbook of hydrocolloids. Woodhead, Philadelphia, p 379

Draget KI, Skjåk-Braek G, Smidsrød O (1997) Alginate based new materials. Int J Biol Macromol 21:47–55

DSM press release (2008) DSM invests in development of bio-based materials. http://www.observatorioplastico.com/detalle_noticia.php?no_id=73274&seccion=mercado&id_categoria=80002. Accessed March 2008

Eichhorn SJ, Baillie CA, Zafeiropouls N, Mwaikambo LY, Ansell MP, Dufresne A, Entwistle KM, Herrera-Franco PJ, Escamilla GC, Groom L, Hughes M, Hill C, Rials IG, Wild PM (2001) Review: current international research into cellulosic fibres and composites. J Material Sc 36:2107–2131

Erwin TH, David AG, Jeffrey JK, Robert JW, Ryan PO (2007) The eco-profiles for current and near-future NatureWorks® polylactide (PLA) production. Industrial Biotechnology 3:58–81

Eslmai H, Kamal RM (2013) Elongational rheology of biodegradable poly(lactic acid)/poly[(butylene succinate)-co-adipate] binary blends and poly(lactic acid)/poly[(butylene succinate)-co-adipate]/clay ternary nanocomposites. J Appl Polym Sci 127:2290–2306

Espigares I, Elvira C, Mano JF, Vlazquez B, Roman JS, Reis RL (2002) New partially degradable and bioactive acrylic bone cements based on starch blends and ceramic fillers. Biomaterials 23(8):1883–1895

EurActiv.com (2012) EU calls time on first-generation biofuels. http://www.euractiv.com/climate-environment/eu-signals-generation-biofuels-news-515496. Accessed Oct 2012

Frost & Sullivan report (2009) Global bio-based plastic market, M4AI-39. Chapter 5. http://www.frost.com/sublib/display-report.do?ctxixpLink=FcmCtx9&searchQuery=Global+bio-based+plastic+market%2C+2009&bdata=aHR0cDovL3d3dy5mcm9zdC5jb20vc3JjaaC9jYXRhbG9nLXNIYXJjaC5kbz9wdWJsaaWNhNhdGlvbllYJzPTlwMDkmcXVlcnlIUZXh0PUdsb2JhbCtiaW8tYmFzZWQrcGxhc3RpYyttYXJrZXQrMjAwOSZmaWx0ZXJJT2g2g2ZmFsc2UmcGFnZVNpemU9MUMGFnZVNpemU9MTJAfkBTZWFyY2ggUmVzdWx0c0B%2BQDEzNjMxMDQzNjA1NjA1NjA1NjA1NTJAfkBTZWFyY2ggUmVzdWx0c0B%3D&ctxixpLabel=FcmCtx10&id=M4A1-01-00-00-00. Accessed 22 Dec 2009

Freier T, Kunze C, Nischan C (2002) In vitro and in vivo degradation studies for development of a biodegradable patch based on poly(3-hydroxybutyrate). Biomaterials 23:2649–2657

Fuentes Z, Riquelme MJN, Sánchez-Zapata E, Pérez JAÁ (2010) Resistant starch as functional ingredient: a review. Food Res Int 43:931–942

Garlotta D (2001) A literature review of poly (lactic acid). J Polyms and the Envir 9(2):63–84

Georg JM, Kristensen M, Astrup A (2012) Effect of alginate supplementation on weight loss in obese subjects completing a 12-week energy restricted diet: a randomized controlled trail. Am J Clin Nutr 96:5–13

Goh GH, Heng PWS, Chan LW (2012) Alginates as a useful natural polymer for microencapsulation and therapeutic applications. Carbohydr Polym 88:1–12

Gomez-Guille MC, Gimenez B, Lopez CME, Montero MP (2011) Functional bioactive properties of collagen and gelatin from alternative sources: a review. Food Hydrocolloids 25:1813–1827

Guangwen C, Shulian L, Fengjun J, Quan Y (2007) Catalytic dehydration of bioethanol to ethylene over TiO_2/γ-Al_2O_3 catalyst in microchannel reactors. Catal today 125:111–119

Guo M, Liu M, Zhan F, Wu L (2005) Preparation and properties of a slow-release membrane-encapsulated urea fertilizer with superabsorbent and moisture preservation. Ind Eng Chem Res 44:4206–4211

Gupta B, Revagade N, Hilborn J (2007) Poly(lactic acid) fiber: an overview. Prog Polym Sci 34:455–482

Hafdani FN, Sadeghinia N (2011) A review on applications of chitosan as a natural antimicrobial. World Academy of Sci Engg Technol 50:252–256

Hanggi JU (1995) Requirements on bacterial polyesters as future substitute for conventional plastics for consumer goods. FEMS Microbioly Rev 16:213–220

Haung YM, Li H, Huang XJ, Hu YC, Hu Y (2008) Advances of bio-ethylene. Chin J Bioprocess Eng 6:1–6

Helga E, Svein V (1998) Biosynthesis and applications of alginates. Polym Degradation and Stability 59:85–91

Helgerud T, Gaserød O, Fjreide T, Andresen PO, Larsen CK (2009) Alginates. In: Imeson A (ed) Food stabilisers, thickeners and gelling agents. Wiley Blackwell, Oxford, pp 50–72

Hirosawa E, Danjo K, Sunada H (2000) Influence of granulating method on physical and mechanical properties, compression behavior, and compactibility of lactose and microcrystalline cellulose granules. Drug Dev Ind Pharm 26:583–593

Hori K, Nojiri H, Nonomura M, Okuda F, Yanagida H, et al. (2005) Allergen inactivator. US Patent 197319, 20 Nov 2005

Huan Z, Joseph GL, Sarit BB (2012) Fabrication aspects of PLA-CaP/PLGA-CaP composites for orthopedic applications: a review. Acta Biomater 8(6):1999–2016

Iain AB, Seal CJ, Wilcox M, Dettmar PW, Pearson PJ (2009) Applications of alginates in food. In: Brend HAR (ed) Alginates: biology and applications. Microbiology monographs 13. Springer, Hiedelberg, pp 211–228

Jane J (1995) Starch properties, modifications and applications. J Macromolecular Sci 32:751–757

Jaspreet S, Lovedeep K, McCarthy OJ (2007) Factors influencing the physico-chemical, morphological, thermal and rheological properties of some chemically modified starches for food applications—a review. Food Hydrocolloids 21:1–22

Jayanth P, Vinod L (2012) Biodegradable nanoparticles for drug and gene delivery to cells and tissue. Adv Drug Deliv Rev 64:61–71

Jeevitha D, Kanchana A (2013) Chitosan/PLA nanoparticles as a novel carrier for the delivery of anthraquinone: synthesis, characterization and in vitro cytotoxicity evaluation. Colloids Surf B Biointerfaces 101(1):126–134

Jian-Bing Z, Ling J, Yi-Dong L, Madhusudhan S, Tao L, Yu-Zhong W (2011) Bio-based blends of starch and poly(butylene succinate) with improved miscibility, mechanical properties, and reduced water absorption. Carbohydr Polym 83:762–768

Johnston-Banks FA (1990) Gelatin. In: Harris P (ed) Food gels. Elsevier, London, pp 233–289

Jong ED, Higson A, Walsh P, Maria W (2010) Bio-based chemicals: value added products from biorefineries. IEA Bioenergy Task 42 Biorefinery:1–34. http://www.iea-bioenergy.task42-biorefineries.com/publications/reports/?eID=dam_frontend_push&docID=2051. Accessed 15 Feb 2012

Jun X, Bao-Hua G (2010a) Microbial succinic acid, its polymer poly(butylene succinate), and applications. Microbiology Monographs 14:347–388

Jun X, Bao-Hua G (2010b) Poly(butylene succinate) and its copolymers: research, development and industrialization. Biotechnol J 5:1149–1163

Jung S, Lim E, Jong HK (2009) New application of poly(butylene succinate) (PBS) based ionomer as biopolymer: a role of ion group for hydroxyapatite (HAp) crystal formation. J Mater Sci 44:6398–6403

Jung-Ju K, Guang-Zhen J, Hye-Sun Y, Seong-Jun C, Hae-Won K, Ivan BW (2012) Providing osteogenesis conditions to mesenchymal stem cells using bioactive nanocomposite bone scaffolds. Mater Sci Eng C 32:2545–2551

Kamel S, Ali N, Jahangir K, Shah SM, El-Gendy (2008) Pharmaceutical significance of cellulose: a review. Express polymer Letters 2:758–778

Kasirajan S, Ngouajio M (2012) Polyethylene and biodegradable mulches for agricultural applications: a review. Agronomy Sustainable Dev 32(2):501–529

Kathiraser Y, Aroua MK, Ramachandran KB, Tan IKP (2007) Chemical characterization of medium-chain-length polyhydroxyalkanoates (PHAs) recovered by enzymatic treatment and ultrafiltration. J Chem Tech Biotech 82:847–855

Kellomaki M, Niiranen H, Puumanen K, Ashammakhi N, Waris T, Tormala P (2000) Bioabsorbable scaffolds for guided bone regeneration and generation. Biomaterials 21:2495–2505

Kevin JE, Charles MB, John DS, Paul AR, Brian DS, Michael CS, Debra T (2001) Advances in cellulose eater performance and applications. Progress in Polymer Sci 26:1605–1688

Kumano A, Fujiwara N (2008) Cellulose triacetate membranes for reverse osmosis. In: Normam AGF, Li N, Winston Ho WS, Matsuura T (ed) Advanced membrane technology and application. Wiley, New Jersey, pp 21–43

Kumbar SG, Kulkarni AR, Dave AM, Aminabha TM (2001) Encapsulation efficiency and release kinetics of solid and liquid pesticides through urea formaldehyde cross-linked starch, guar gum, and starch + guar gum matrices. J Appli Polym Sci 82:2863–2866

Kunze C, Edgar Bernd H, Androsch R (2006) In vitro and in vivo studies on blends of isotactic and atactic poly (3-hydroxybutyrate) for development of a dura substitute material. Biomaterials 27:192–201

Leathers TD (2003) Biotechnological production and applications of pullulan. Appl Microbiol Biotechnol 62:468–473

Lee SH, Wang S (2006) Biodegradable polymers/bamboo fiber composite with bio-based coupling agent. Compos Part A37:80–91

Lee HC, Anuj S, Lee Y (2001) Biomedical applications of collagen. International J of Pharmaceutics 221:1–22

Lemoigne M (1923) Production d'acide β-oxybutyrique par certaines bact'eries du groupe du Bacillus subtilis. CR. Hebd. Seances Acad. Sci 176:1761

Levison PR (1993) Cellulosics as ion-exchange materials. In: Kennedy JF, Phillips GO, Williams PA (ed) Cellulosics: materials for selective separations and other technologies. Ellis Horwood, Chichester, pp 25–36

Li G, Yong H, Chen C (2011) Discussion on application prospect of starch-based adhesives on architectural gel materials. Adv Materials Res 250:800–803

Li S, Juliane H, Martin KP (2009) Product overview and market projection of emerging biobased products. PRo-BIP 1:1–245

Liu L, Yu J, Cheng L, Qu W (2009a) Mechanical properties of poly(butylene succinate) (PBS) biocomposites reinforced with surface modified jute fibre. Composites Part A: Appl Sci Manufacturing 40:669–674

Liu LF, Yu JY, Cheng LD, Yang XJ (2009b) Biodegradability of PBS composite reinforced with jute. Polym Degrade Stab 94:90–94

Liu M, Zhang Y, Wu C, Xiong S, Zhou C (2012) Chitosan/halloysite nanotubes bionanocomposites: structure, mechanical properties and biocompatibility. Int J Biological Macromol 51:566–575

Luc A, Eric P (2012) Biodegradable polymers. In: Environmental silicate nano-biocomposites. Green energy and technology. Springer, Hiedelberg, pp 13–39

Luiz A, De Castro R, Morschbacker (2010) A method for the production of one or more olefins, an olefin, and a polymer. US 2010/0069691A1, 18 Mar 2010

Maeda M, Tani S, Sano A, Fujioka K (1999) Microstructure and release characteristics of the minipellet, a collagen based drug delivery system for controlled release of protein drugs. J Controlled Rel 62:313–324

Mainil V, Rahn B, Gogolewski S (1997) Long-term in vivo degradation and bone reaction to various polylactides: 1. One-year results. Biomaterials 18:257–266

McChalicher CW, Srienc F (2007) Investigating the structure–property relationship of bacterial PHA block copolymers. J Biotechnology 132:296–302

Ravi Kumar MNV (2000) A review of chitin and chitosan applications. Reactive Functional Polym 46:1–27

Majid J, Elmira AT, Muhammad I, Muriel J, St'ephane D (2010) Poly-lactic acid: production, applications, nanocomposites, and release studies. Comprehensive Rev Food Sci Safety 9(5):552–571

Mandel KG, Daggy BP, Brodie DA, Jacoby HI (2000) Review article: alginate-raft formulations in the treatment of heartburn and acid reflux. Aliment Pharmacol Ther 14:669–690

Marguerite R (2006) Chitin and chitosan: properties and applications. Progress in Polym Sci 31:603–632

Marques AP, Reis RL (2005) Hydroxyapatite reinforcement of different starch-based polymers affect osteoblast-like cells adhesion/spreading and proliferation. Mater Sci Engg 25(2):215–229

Maurizio A, Jan JDV, Maria EE, Sabine F, Paolo V, Maria GV (2005) Biodegradable starch/clay nanocomposite films for food packaging applications. Food Chem 93(3):467–474

Mendes RL, Reis YP, Bovell AM, Cunha CA, Blitterswijk V, de Bruijn JD (2001) Biocompatibility testing of novel starch-based materials with potential application in orthopedic surgery: a preliminary study. Biomaterials 22:2057–2064

Michael C, Dirk C, Harald K, Jan R, Joachim V (2011) Policy paper on bio-based economy in the EU: level playing field for bio-based chemistry and materials.. www.bio-based.eu/policy/en. Accessed December 2012

Nagarwal RC, Singh PN, Kant S, Maiti P, Pandit JK (2010) Chitosan coated PLA nanoparticles for ophthalmic delivery: characterization, in-vitro and in-vivo study in rabbit eye. J Biomed Nanotechnol 6:648–656

Nicolas J, Floriane F, Francoise F, Alan R, Jean PP, Patrick F, Rene SL (2011) Synthesis and properties of poly(butylene succinate): efficiency of different transesterfication catalysts. J Polym Sci Part A: Polym Chem 49:5301–5312

Obae HI, Imada K (1999) Morphological effect of microcrystalline cellulose particles on tablet tensile strength. Int J Pharm 182:155–164

Onsoyen E (1996) Commercial applications of alginates. Carbohydrates in Europe 14:26–31

Orawan S, Uracha R, Pitt S (2008) Electrospun cellulose acetate fiber mats containing asiaticoside or Centella asiatica crude extract and the release characteristics of asiaticoside. Polymer 49(19):4239–4247

Ozdemir M, Floros JD (2004) Active food packaging technologies. Crit Rev Food Sci Nutr 44:185–193

Papenburg BJ, Liu J, Higuera G, Barradas AMC, Boer J, Blitterswijk VCA, Wessling M, Stamatialis D (2009) Development and analysis of multi-layer scaffolds for tissue engineering. Biomaterials 30:6228–6239

Peter WD, Vicki S, Richardson JC (2011) The key role alginates play in health. Food Hydrocolloids 25:263–266

Philip S, Keshavarz T, Roy I (2007) Polyhydroxyalkanoates: biodegradable polymers with a range of applications. J Chemical Tech Biotech 2(3):233–247

Picker KM, Hoag SW (2002) Characterization of the thermal properties of microcrystalline cellulose by modulated temperature differential scanning calorimetry. J Pharmaceutical Sci 91:342–349

Prashant RC, Ishwar BB, Shrikant AS, Rekha SS (2009) Microbial cellulose: fermentive production and applications. Food Technol Biotechnol 47:107–124

Qin Y, Cai L, Feng D, Shi B, Liu J, Zhang W, Shen Y (2007) Combined use of chitosan and alginate in the treatment of waste water. J Appl Polym Sci 104:3181–3587

Rafael A, Loong TL, Susan EM, Selke HT (2010) Poly(lactic acid): synthesis, structures, properties, processing and applications. Chapter 28:457–467

Rajev AJ (2000) The manufacturing techniques of various drug loaded biodegradable poly(lactide-co-glycolide) (PLGA) devices. Biomaterials 21:2475–2490

Ramya R, Venkatesan, Jayachanndran Kim S, Sudha PN (2012) Biomedical applications of chitosan: an overview. J Biomaterial Tissue Engg 2:100–111

Ravenstijn JTJ (2010) The state-of-the art on bioplastics: products, markets, trends and technologies. Polymedia, Lüdenscheid

Ray SS, Bousmina M (2005) Biodegradable polymers and their layered silicate nanocomposites: in greening the 21st century materials world. Progress Material Sci 50:962–1079

Reis KC, Pereira J, Smith AC, Carvalho CWP, Wellner N, Yakimets I (2008) Characterization of polyhydroxybutyrate-hydroxyvalerate (PHB-HV)/maize starch blend films. J Food Engg 89:361–369

Rekha MR, Chrndra PS (2007) Pullulan as a promising biomaterial for biomedical applications: a perspective. Trends in Biomaterials and Artificial Organs 20:21–45

Roberts GAF (1997) Chitosan production routes and their role in determining the structure and properties of the product. In: Domard M, Roberts AF, Vårum KM (ed) Advances in Chitin Science, vol. 2, National Taiwan Ocean University, Taiwan. Jacques Andre, Lyon, pp 22–31. 1998

Rubin AL, Stenzel KH, Miyata T, White MJ, Dune M (1973) Collagen as a vehicle for drug delivery: preliminary report. J of Clinical Pharmacology 13:309–312

Ruth KG, Hartmann R, Egli T, Zinn M, Ren Q (2007) Efficient production of (R)-3-hydroxycarboxylic acids by biotechnological conversion of polyhydroxyalkanoates and their purification. Biomacromolecules 8:279–286

Savenkova L, Gercberga Z, Nikolaeva V, Dzene A, Bibers I, Kalina M (2000) Mechanical properties and biodegradation characteristics of PHB-based films. Process Biochem 35:537–579

Schmidmaier G, Wildemann A, Stemberger A, Has MR (2001) Biodegradable poly (D, L-lactide) coating of implants for continuous release of growth factors. J Biomed Mater Res 58:449–455

Sena RF, Costelli MC, Gibson LH, Coughlin RW (2006) Enhanced production of pullulan by two strains of A. pullulans with different concentrations of soybean oil in sucrose solution in batch fermentations. Brazilian J Chem Eng 2:507–515

Siddhesh NP, Edgar KJ (2012) Alginate derivatization: a review chemistry, properties and applications. Biomaterials 33:3279–3305

Singh RS, Saini GK, Kennedy JF (2008) Pullulan: microbial sources, production and applications. Carbohydr Polym 73:515–531

Sodian R, Hoerstrup SP, Sperling JS, Daebritz S, Martin DP, Moran AM, Kim BS, Schoen FJ, Vacanti JP, Mayer JE (2000) Early in vivo experience with tissue engineered trileaflet heart valves. Circulation 102:22–29

Sodian R, Sperling JS, Martin DP (1999) Tissue engineering of a trileaflet heart valve-early in vitro experiences with a combined polymer. Tissue Engg 5:489–494

Steinbüchel A, Valentin HE (1995) Diversity of bacterial polyhydroxyalkanoic acids. FEMS Microbiol Lett 128:219–228

Tang H, Ishii D, Mahara A, Murakami S, Yamaoka T, Sudesh K, Samian R, Fujita M, Maeda M, Iwata T (2008) Scaffolds from electrospun polyhydroxyalkanoate copolymers: fabrication, characterization, bio absorption and tissue response. Biomaterials 29:1307–1317

Teli MD, Chiplunkar V (1986) Role of thickeners in final performance of reactive prints. Textile Dyer Printer 19:13–19

Türesin F, Gürsel I, Hasirci V (2001) Biodegradable polyhydroxyalkanoate implants for osteomyelitis therapy: in vitro antibiotic release. J Biomaterials Sci 12 (2):195–207. Polymer Edition

Valantin MA, Aubron-Olivier C, Ghosn J, Laglenne E, Pauchard M, Schoen H (2003) Polylactic acid implants to correct facial lipoatrophy in HIV-infected patients: results of the open-label study. Vega Aids 17:2471–2477

Valappil S, Misra S, Boccaccini A, Roy I (2006) Biomedical applications of polyhydroxyalkanoates, an overview of animal testing an in vivo responses. Expert Rev Med Devices 3:853–868

Valerie D, Vinod DV (1998) Pharmaceutical applications of chitosan. Pharmaceutical Sci Technol Today 1:246–253

Virginia E, Marie G, Eric P, Luc A (2011) Structure and properties of glycerol plasticized chitosan obtained by mechanical kneading. Carbohydrate Polym 83:947–952

Volova T, Shishatskaya E, Sevastianov V, Efremov S, Mogilnaya O (2003) Results of biomedical investigations of PHB and PHB/PHV fibers. Biochem Eng J 16:125–133

Vona IA, Costanza JR, Cantor HA, Robert WJ (1965) Manufacture of plastics, vol 1. Wiley, New York, pp 141–142

Walle GAM, de Koning GJM, Weusthuis RA, Eggink G (2001) Properties, modifications and applications of biopolyesters. Adv Biochem Eng Biotechnol 71:264–291

Wang Z, Itoh Y, Hosaka Y, Kobayashi I, Nakano Y, Maeda I, Umeda F, Yamakawa J, Kawase M, Yagi K (2003) Novel transdermal drug delivery system with polyhydroxyalkanoate and starburst polyamidoamine dendrimer. J Biosci and Bioengg 95(5):541–543

Weraporn PA, Sorapong P, Narongchai OC, Ubon I, Puritud J, Sommai PA (2011) Preparation of polymer blends between poly (L-lactic acid), poly (butylene succinate-co-adipate) and poly (butylene adipate-co-terephthalate) for blown film industrial application. Energy Procedia. 9:581–588

Westermark S, Juppo AM, Kervinen L, Yliruusi J (1999) Microcrystalline cellulose and its microstructure in pharmaceutical processing. Eur J Pharm Biopharm 48:199–206

Williams SF, Martin DP, Horowitz DM, Peoples OP (1999) PHA applications: addressing the price performance issue I. Tissue engineering. Int J Biol Macromolecules 25:111–121

Win NN, Stevens WF (2001) Shrimp chitin as substrate for fungal chitin deacetylase. Appl Microbiol Biotechnol 57:334–341

Xie ZP, Huang Y, Chen YL, Jia Y (2001) A new gel casting of ceramics by reaction of sodium alginate and calcium iodate at increased temperature. J Mat Sci Lett 20:1255–1257

Yan YF, Krishnaiah D, Rajin M, Bono A (2009) Cellulose extraction from palm kernel cake using liquid phase oxidation. J Engg Sci Tech 4:57–68

Yang F, Li X, Li G, Zhao N, Zhang X (2002) Study on chitosan and PHBHHx used as nerve regeneration conduit material. J Biomedical Engg 19:25–29

Zajic JE, LeDuy A (1973) Flocculant and chemical properties of a polysaccharide from Pullularia pullulans. Appl Microbiol 25:628–635

Zeeshan K, Gopiraman M, Yuichi H, Kai W, Kim IS (2013) Cationic-cellulose nanofibers: preparation and dyeability with anionic reactive dyes for apparel application. Carbohydrate 91:434–443

Zhao RX, Torley P, Halley PJ (2008) Emerging biodegradable materials: starch- and protein-based bio-nanocomposites. J Mat Sci 43:3058–3071

Zhao K, Tian G, Zheng Z, Chen JC, Chen GQ (2003) Production of D-(−)-3-hydroxyalkanoic acid by recombinant *Escherichia coli*. FEMS Microbiol Lett 218:59–64

Zhao P, Liu W, Wu Q, Ren J (2010) Preparation, mechanical, and thermal properties of biodegradable polyesters/poly(lactic acid) blends. J of Nanomaterials 2010:1–8

11

Evaluation of polyphenylene ether ether sulfone/nanohydroxyapatite nanofiber composite as a biomaterial for hard tissue replacement

Manickam Ashokkumar and Dharmalingam Sangeetha[*]

Abstract

The present work is aimed at investigating the mechanical and *in vitro* biological properties of polyphenylene ether ether sulfone (PPEES)/nanohydroxyapatite (nHA) composite fibers. Electrospinning was used to prepare nanofiber composite mats of PPEES/nHA with different weight percentages of the inorganic filler, nHA. The fabricated composites were characterized using Fourier transform infrared spectroscopy (FTIR)-attenuated total reflectance spectroscopy (ATR) and scanning electron microscopy (SEM)-energy dispersive X-ray spectroscopy (EDX) techniques. The mechanical properties of the composite were studied with a tensile tester. The FTIR-ATR spectrum depicted the functional group as well as the interaction between the PPEES and nHA composite materials; in addition, the elemental groups were identified with EDX analysis. The morphology of the nanofiber composite was studied by SEM. Tensile strength analysis of the PPEES/nHA composite revealed the elastic nature of the nanofiber composite reinforced with nHA and suggested significant mechanical strength of the composite. The biomineralization studies performed using simulated body fluid with increased incubation time showed enhanced mineralization, which showed that the composites possessed high bioactivity property. Cell viability of the nanofiber composite, studied with osteoblast (MG-63) cells, was observed to be higher in the composites containing higher concentrations of nHA.

Keywords: Polyphenylene ether ether sulfone, Nanohydroxyapatite, Bioactivity, Biomineralization, Osteoblast cells

Background

The term biomaterial, which is closely related to applications that repair or replace a part of or the whole tissue, must have favorable mechanical as well as biological properties and should play a major role in developing successful implantations by means of inducing cell adhesion, proliferation, and differentiation on the surface of the biomaterial (Sista et al. 2011). In developing new biomaterials for tissue replacement, the structure and properties of the tissue which is to be replaced, i.e., the biological template, must be taken into consideration. This is because, if properties of the new material are significantly different from those of the host tissue, the material under development will cause dynamic changes to the host tissue after implantation and thus will not achieve the goals embedded in the original conceptual design (Wang 2003). Different materials have been investigated for applications in bone tissue engineering —metals, ceramics, and polymers. As with all materials implanted into the body, the polymers for bone regeneration must be biocompatible. In addition, they should be moldable, shapeable, or polymerizable *in situ* to ensure good integration in the defective area (Seal et al. 2001). However, the polymer materials used for orthopedic application do not exhibit adequate mechanical properties and bioactive behavior, which are the main disadvantages for bone tissue engineering. In order to overcome these problems, polymer/bioactive ceramic composites have been developed for bone tissue engineering, which ensure the achievement of the above-mentioned properties and performance of the material (Ryszkowska et al. 2010).

It is well known that the two fundamental factors to be considered in producing polymer nanocomposites with bone-like properties are (1) good interfacial adhesion between organic polymers and inorganic hydroxyl apatite (HA) and (2) uniform dispersion of HA at the nanolevel

---

* Correspondence: sangeetha@annauniv.edu
Department of Chemistry, Anna University, Sardar Patel Road, Chennai, Tamil Nadu 600025, India

in the polymer nanofiber (Lee et al. 2007). When such a composite is immersed in simulated body fluid (SBF), biologically active HA layers are formed on the implant due to the ion-exchange reaction between the bioactive implant and the surrounding body fluids which is chemically and crystallographically equivalent to the mineral phase of the bone (Pielichowska and Blazewicz 2010). In addition, HA is known to smartly utilize the apatite that is mineralized on their surfaces as an interface to integrate spontaneously with the living tissue (Kim et al. 2004). Nanofiber composites have certain favorable characteristics and properties, such as porosity, the surface area-to-volume ratio, pore size, pore interconnectivity, structural strength, and biocompatibility, which play a major role in the design and fabrication of polymeric materials for bone tissue engineering (Tan et al. 2003; Teoh 2004).

Although existing bioactive materials possess high compressive strength, they are unfortunately very brittle and have inherently poor tensile and torsional properties. Material selection is especially important in bone tissue engineering because a supporting substrate is critical in maintaining mechanical strength and structural support as well as providing the optimal culturing environment for bone formation during the early stages of the regenerative process (Lu et al. 2003). The large surface area-to-weight ratio of the composite material offered by electrospinning (Reneker and Chun 1996; Li and Xia 2004; Darrell et al. 2006; Ramakrishna et al. 2006; Greiner and Wendorff 2007) is achieved by means of decreasing the diameter of the fiber from the micrometer (10–100 μm) to submicron or nanometer level (10×10^{-3} to 100×10^{-3}), resulting in the appearance of several amazing characteristics such as flexibility in surface functionalities and superior mechanical properties (stiffness and tensile strength) compared with any other known form of material (Huang et al. 2003).

The present study is focused on designing and developing the polymer—polyphenylene ether ether sulfone (PPEES) nanofiber composites reinforced with nanohydroxyapatite (nHA)—and evaluating its potential application as an orthopedic biomaterial. The prepared nanofiber composite was subjected to characterization and morphology studies using Fourier transform infrared (FTIR)-attenuated total reflectance spectroscopy (ATR) and scanning electron microscopy (SEM)-energy dispersive X-ray spectroscopy (EDX) to identify the presence of a structural group and morphology of the composites, respectively. Inverted fluorescence microscopy was used to identify the viability of bone-like cells over the nanofiber composite. The composite was then investigated *in vitro* for its multifunctional properties (mechanical and biological properties) before and after incubation in SBF in order to evaluate the compatibility of the biomaterial for orthopedic application.

Results and discussion
FTIR-ATR

The FTIR-ATR spectra of nHA reinforced PPEES nanofiber composites and bare nanofiber mat are shown in Figure 1. From the spectra, the intense broad band (Figure 1b–d) observed at 3500 cm^{-1} was assigned to the OH stretching vibration which was mainly observed when steric hindrance prevents polymeric association and also confirmed the interaction of nHA with PPEES. The intensity of the peak at 3068 cm^{-1} (Figure 1a) was due to the fact that the adsorption peak of C-H stretching vibration was overlapped by the O-H stretching vibration peaks of nHA. The C=C aromatic ring vibrations were attributed to the peaks occurring at 1578 cm^{-1}; intensity of the peak decreased with the addition of nHA due to the interaction of HA with the polymer backbone. The peak at 1375 cm^{-1} corresponds to the ester linkage of the polymer chain. The strong absorption peaks just above 1250 and 1100 cm^{-1} correspond to the diaryl sulfone (Ar-SO$_2$-Ar) and diaryl ether (Ar-O-Ar) groups, respectively (Dahe et al. 2011). It was observed that the phosphate vibrations were merged with the S=O vibrations just above 1000 cm^{-1}. The aromatic ring CH bending vibration occurred just above 800 cm^{-1}.

Morphology

The surface morphology of PPEES and its nanofiber composite with different weight percentages are shown in Figure 2. The SEM image showed the nanofiber in the range of approximately 100–150 nm in diameter, which would provide high surface area-to-weight ratio than any other form of material such as films or membranes. The composite material with high surface area supported apatite formation, cell adhesion, proliferation, and differentiation of bone-like cells. Non-reinforced PPEES

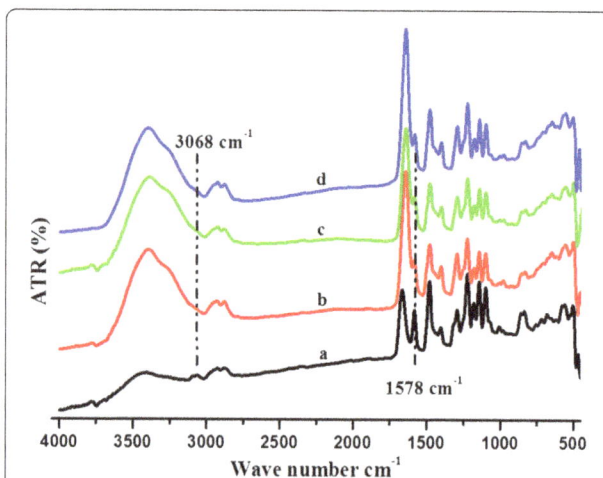

Figure 1 FTIR-ATR spectra of PPEES nanofiber and its composite: (*curve a*) PPEES nanofiber, (*curve b*) PPEES 1, (*curve c*) PPEES 2, and (*curve d*) PPEES 3.

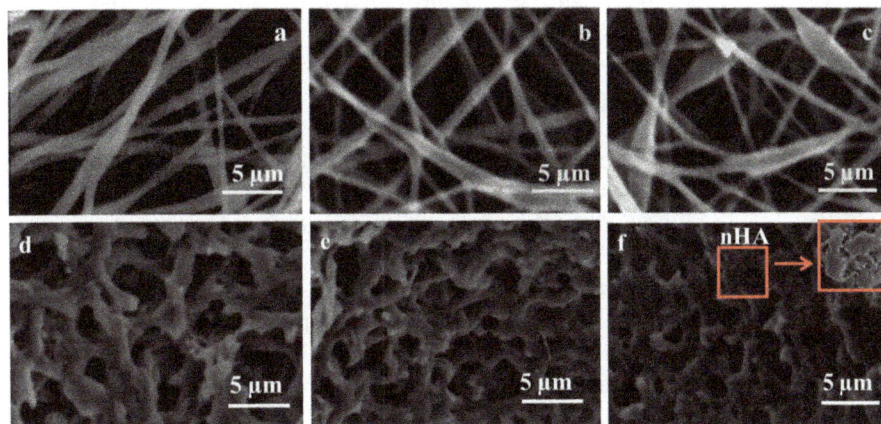

Figure 2 SEM images showing surface morphology (a–c) and cross-sectional images (d–f) of PPEES and its composite. (**a**) PPEES, (**b**) PPEES 2, (**c**) PPEES 3, (**d**) PPEES, (**e**) PPEES 2, and (**f**) PPEES 3.

showed good fiber formation (Figure 2a) without the formation of beads. SEM analysis of the nHA reinforced composites (PPEES 1, 2, and 3) revealed that at higher filler concentrations, the bead formation increased. In comparison, PPEES 1 (figure not shown) and PPEES 2 (Figure 2b) showed better fiber formation than PPEES 3 (Figure 2c), and random dispersion of nHA in the PPEES nanofiber matrix was observed. In PPEES 3, some beads were observed in between the nanofibers with respect to the increased weight percentage of nHA (Figure 2c). In addition, further evidence for nHA encapsulation within the polymer nanofiber matrix was confirmed with the cross-sectional image of the nanofiber composite (Figure 2d–f). The inset in Figure 2f shows evidence for the presence of nHA. Though both PPEES 1 and PPEES 2 showed good fiber mat formation, PPEES 2 was chosen over PPEES 1 for the rest of the *in vitro* studies since it contained the higher percentage of nHA.

Mechanical properties

Tensile strength of the PPEES and PPEES 2 nanofiber composites are represented in Table 1. While it was noted that the PPEES nanofiber mat possessed low levels of stiffness, as inferred from Table 1, the incorporation of the inorganic filler into the PPEES nanofiber matrix enhanced the percentage of elongation with slight reduction in tensile strength, which was concurrent with earlier studies (Salerno et al. 2010). The increase in percentage of elongation was explained by considering that the HA particles in the continuous phase of the polymer matrix constituted a second phase. The interface between the two phases (PPEES and nHA) acts like a grain boundary, resisting the propagation of a crack when subjected to tensile forces. Rather, a friction force that is opposite to the direction of the crack-initiating force will be present at the edges of the crack, leading to the observed increase in elongation

(Zebarjad et al. 2011). When developing a biomaterial for orthopedic application, due importance shall be given to the mechanical properties of the material because most of the polymeric materials, even if they possess good biocompatibility, fail to withstand *in vivo* stresses due to their brittle nature. The fabrication of composite materials reinforced with fillers is one way of avoiding the problems of brittleness. The property transition from being brittle to ductile of the composite was achieved by the incorporation of the inorganic filler, resulting in an enhanced mechanical strength of the composite (Broza et al. 2005; Bunsell and Renard 2005; Bismarck et al. 2001).

Furthermore, the mechanical properties of the composite materials were studied with the percentage elongation obtained from the tensile tests (Table 1). Substantial increased elongation on PPEES 2 nanofiber composite was observed when compared with the bare PPEES nanofiber mat. Also, the percentage elongation of PPEES 2 was increased significantly with respect to time when the composite material was incubated with SBF. In addition, it was observed that the nHA-reinforced polymer material facilitated an enhanced growth of apatite on the surface, resulting in a decrease in the pore sizes in between the nanofibers, which in turn improved the mechanical

Table 1 Tensile strength and percentage elongation of the nanofiber composites

Composites Studied	Tensile strength (MPa)	Elongation (%)
PPEES	56.32	33.25
PPEES2	52.26	42.05
PPEES2[a]	49.17	44.74
PPEES2[b]	48.26	46.21
PPEES2[c]	47.78	52.31

[a]10 days of incubation with SBF; [b]15 days of incubation with SBF; [c]30 days of incubation with SBF.

strength of the composite with more elongation than the bare PPEES nanofiber. From the results, it was found that the incorporation of nHA considerably influenced the stiffness of the composite material.

Biomineralization

The surface morphology of mineralized PPEES and PPEES 2 composites are shown in Figure 3. The PPEES nanofiber showed no signs of apatite formation even after 30 days of incubation in SBF (Figure 3a). Comparatively, there was significant apatite layer formation on the composite material containing nHA with respect to the incubation time, implying the vital role of nHA in the bioactivity. In addition, it was perceived that in bioactive ceramics, nHA acted as a nucleation site and enhanced the growth of apatite by utilizing the existing ions in the SBF solution. Moderate increase in density of the minerals on PPEES 2 composite after 5, 15, and 30 days of incubation in SBF solution was evident (Figure 3b–e) due to the presence of nHA, and it clearly elucidated that the increased incubation time improved the enhanced apatite formation.

Moreover, the mineralization of HA on the polymer nanofiber composite after incubation in SBF was analyzed using EDX, and the spectra were shown in Figure 4. PPEES 2 showed apatite formation over the surface of the nanofiber after 15 days of incubation in SBF (Figure 4b), and interestingly, significant mineralization with dense apatite layer was observed on day 30 (Figure 4c). However, in the case of the bare polymer, the absence of mineralization was obvious. From the results, it was inferred that the polymer nanocomposite reinforced with HA acted as a stimulus for the formation of the apatite layer and was found to be a major factor for mineralization.

Furthermore, the analysis showed that cauliflower-like morphology (Shanmuga Sundar and Sangeetha 2012) of the apatite layer composed mainly of HA, as visualized from SEM identified with peaks of Ca and P elements, with the intensity of the peaks increasing with increase in the incubation time in SBF-endorsed biomineralization (Kim et al. 2004). The Ca and P peaks observed from the EDX analysis were typical of HA. Specifically, in vitro, an acellular SBF with ion concentrations nearly equal to those in the blood plasma could reproduce the formation of apatite layer on the polymer nanofiber composite (Kim et al. 2004). From these in vitro studies, it was confirmed that the polymer composite reinforced with HA would potentially offer enhanced biomineralization after implantation in vivo.

Cell viability

Figure 5 shows the viability of osteoblast-like cells after being cultured with PPEES 2 composites at different time intervals. Good cell adherence on the electrospun nanofibers might be due to the large surface area available for cell attachment (Bhattarai et al. 2005). It is well known that the rough surface formed by the incorporation of nHA on the polymer composite considerably favors cell adherence. In the present study, the optical density value obtained for the control was taken as 100%. The percentage viability of cells on the PPEES 2 composite was extensively similar to the control with different periods of time, as shown in the results of the test in Figure 5. In addition, it was noted that the percentage viability of cells observed with PPEES did not differ significantly from that observed for PPEES 2. With the bioactivity of the composite, rendered by the

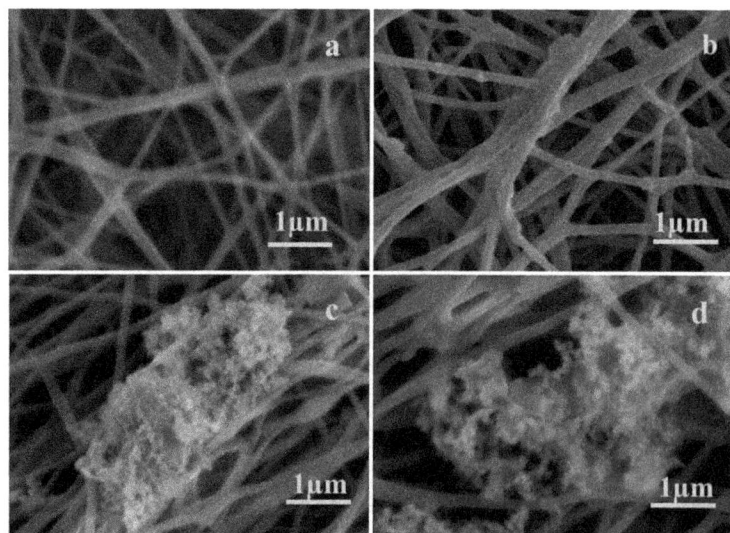

Figure 3 SEM image showing the surface morphology of PPEES and its composites after incubation in SBF. (**a**) PPEES after 30 days, (**b**) PPEES 2 after 5 days, (**c**) PPEES 2 after 15 days, and (**d**) PPEES 2 after 30 days of incubation in SBF.

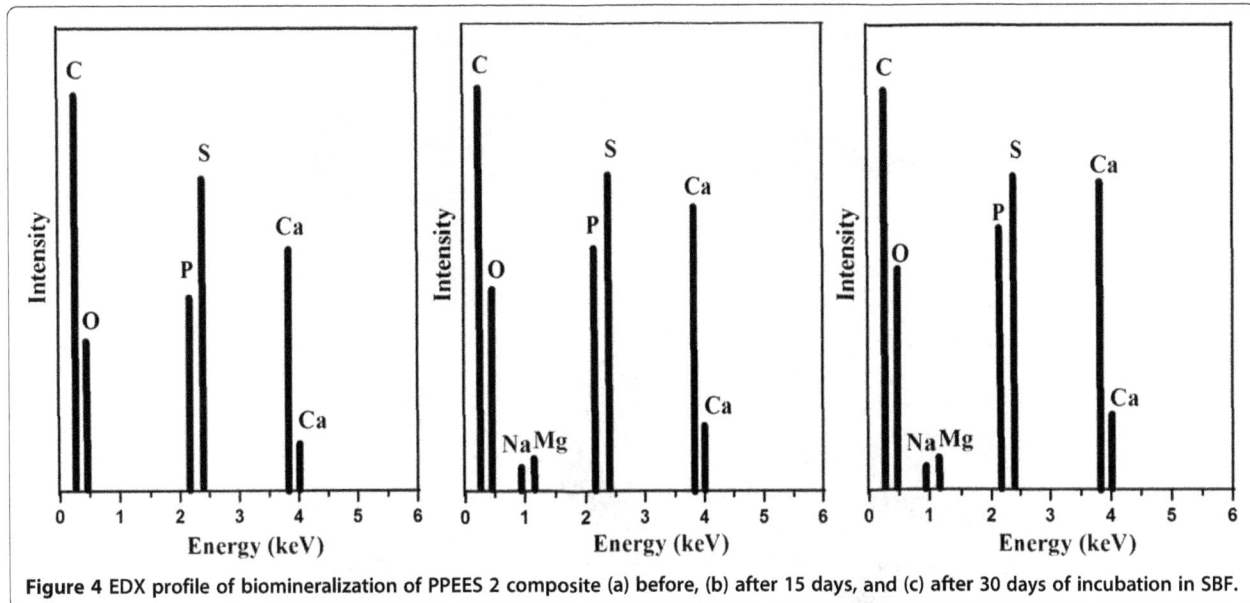

Figure 4 EDX profile of biomineralization of PPEES 2 composite (a) before, (b) after 15 days, and (c) after 30 days of incubation in SBF.

bioactive component, nHA would facilitate tissue (bone) growth adjacent to the implant and thus helping in the osseointegration of the implant *in vivo* (Wang 2003).

Cell morphology

The inverted fluorescence microscopy observation shows the adherence morphology of osteoblast-like cells over the PPEES/nHA nanofiber composites after different culture periods. During the culture period, seeded cells get adhered and proliferated on the fiber composite with apatite formation than with bare nanofiber composite. This may likely be due to higher cell adhesion on apatite-formed PPEES nanofiber composite. This is in agreement with earlier studies (Kang et al. 2008). PPEES 2 nanofiber composite showed more adherences with

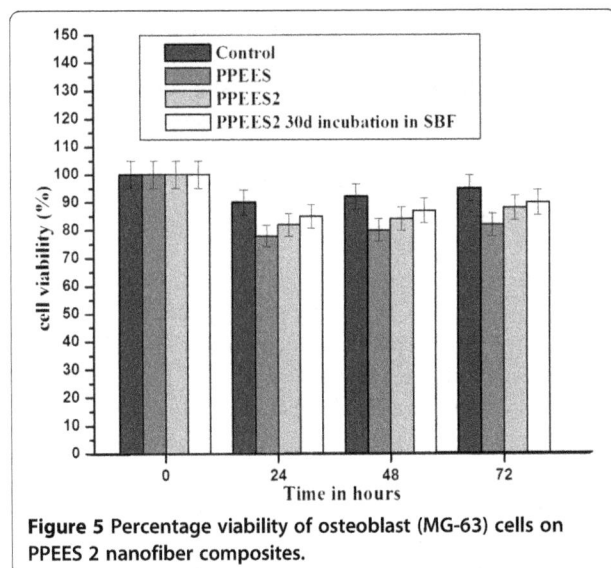

Figure 5 Percentage viability of osteoblast (MG-63) cells on PPEES 2 nanofiber composites.

enthusiastic migration of cells as inferred with fluorescein dye penetration through the osteoblast cell membrane (Figure 6d–f). The intensity of fluorescence on the nanofiber composite increased with the increase in culture time representing the enhanced proliferation of MG-63 cells. From the figure, it is further inferred that the void space in between the nanofibers in the composite was packed with bone-like cells as evidenced by greater cytocompatibility of the apatite-reinforced nanofiber composite. These findings were in line with those observed by Peter et al. (2010) in their studies on chitosan-gelatin/nanohydroxyapatite composites. In the unreinforced chitosan-gelatin (CG) scaffolds, only few cells (osteoblasts) were observed, and the cell morphology was described as rounded. In the case of CG/nHA composite scaffolds, greater cell attachment and spreading were noted while the cell morphology was described as flattened and sheetlike with filopodial extensions. This change in morphology was observed because nHA apparently improved the formation of focal adhesion and allowed for substantial cell spreading. This is quite likely related to the enhanced protein adsorption on the surface in the presence of nHA.

Cell differentiation: ALPase activity

The ability of cells to differentiate on the surface of the composite after implantation was identified using alkaline phosphatase (ALP)ase activity. Figure 7 showed the osteogenic activity of the polymer composite at different culture time intervals. The differentiation of cells on both bare and nHA-incorporated PPEES nanofibers was increased significantly after 1 day of culture. Surprisingly, the rate of differentiation on bare PPEES nanofiber composite was reduced with respect to different periods

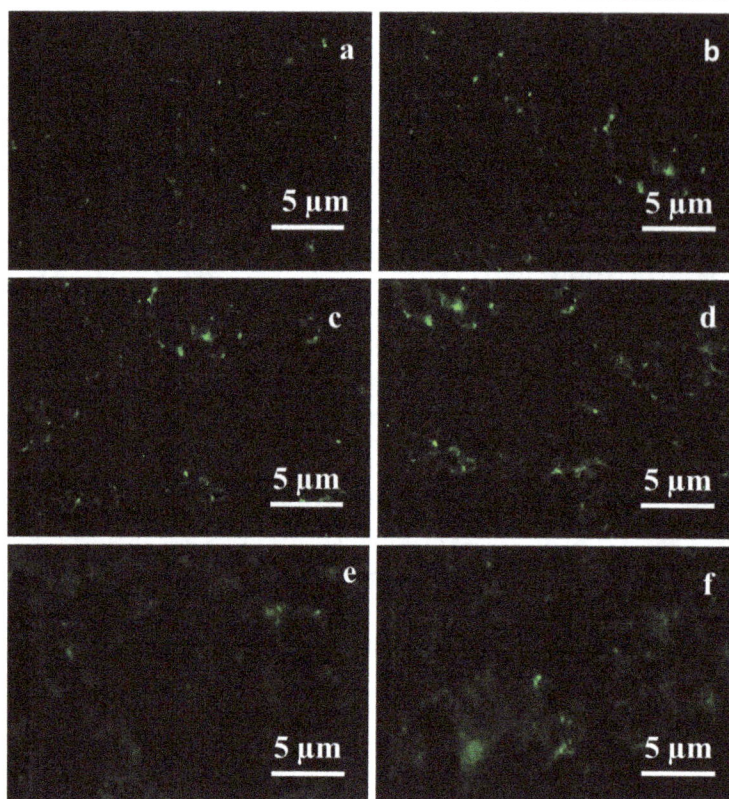

Figure 6 Inverted fluorescence microscopy images after 3-, 7-, and 10-day cultures of PPEES nanofiber (a–c) and PPEES2 nanofiber (d–f) composites.

of culturing time when compared with PPEES nanofiber reinforced with nHA. Rough surface offered by the apatite layer formed over the nanofiber provided the opportunities for cell adhesion and resulted in an increased rate of proliferation and differentiation. After 3 and 5 days of culture, the color intensity of the cell suspension

was found to increase with the addition of alkaline phosphatase which led to the inference of higher cell differentiation in the PPEES 2 nanofiber composite. In the case of bare PPEES nanofibers, the color change was mild, reflecting the low differentiation of MG-63 cells. From the results, it was found that progress in the biomineralization of nanofiber composites significantly encouraged the differentiation of osteoblast cells, which led to bone formation (Dong et al. 2010). Similar studies performed by Srinivasan et al. (2012) using biocompatible alginate/nanobioactive glass ceramic composite scaffolds showed that ALP activity increased up to 7 days and then decreased, indicating the completion of osteoblastic differentiation.

Conclusions

The PPEES nanofiber composite was successfully fabricated using electrospinning technique. The FTIR-ATR study revealed the presence of HA in the PPEES polymer composite. The SEM images of the nanofiber composites confirmed the formation of beads when nHA was above 5 wt. %. The incorporation of nHA in PPEES showed substantially improved biomineralization and osseointegration with greater bone-forming ability *in vitro*. In addition, the apatite formation on the surface

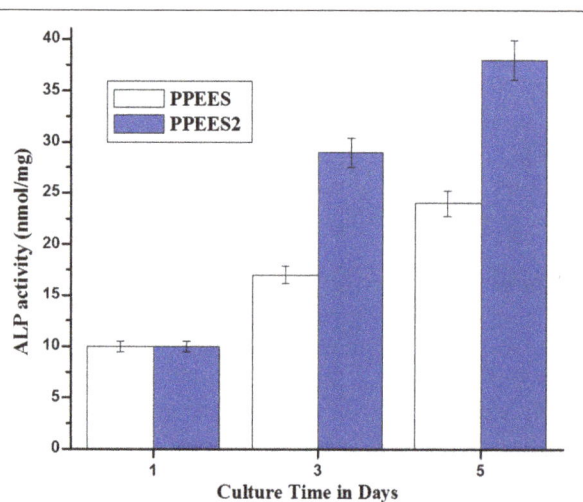

Figure 7 Cell differentiation of osteoblast on PPEES nanofiber and its composite.

of the nanofiber mat after incubation in SBF solution revealed better bioactivity and mechanical property of the composite when compared with the PPEES nanofiber mat. Moreover, elongation of the PPEES 2 composite significantly amplified with respect to apatite formation *in vitro* evidences adequate mechanical property of the composite when implanted *in vivo*. Furthermore, the viability of cells was observed to be higher in the composite with apatite formation, suggesting the affinity of the composite to the natural hard tissues. Hence, it was concluded that the distinctive features of the composite material would play as an ideal candidate for orthopedic application, furthermore in the replacement of hard tissues.

Methods

Materials

Organic polymer, PPEES (CAS number 28212-68-2), and inorganic filler, HA nanopowder (CAS number 12167-74-7), were procured from Sigma-Aldrich Corporation, St. Louis, MO, USA. The solvent used for this study was reagent grade *N*-methyl pyrrolidone (NMP), which was purchased from Merck-India Ltd, Mumbai, India. The bone-like cells (MG-63) used to study the viability of composite materials were acquired from National Center for Cell Sciences, Pune, India. The ingredients utilized for culturing cells such as Dulbecco's modified Eagle's medium, fetal bovine serum (FBS), and penicillin-streptomycin were purchased from HiMedia, Mumbai, India. The diagnostic kits such as MTT and ALP were obtained from Sigma-Aldrich Corporation.

Preparation of nanofiber composite

Various concentrations of PPEES and nHA were used to fabricate the composites to identify their potential for orthopedic applications. The viscous solution was prepared by dissolving PPEES in NMP, and the filler, nHA, was incorporated into it in different weight percentages mentioned in Table 2. The whole content was kept under magnetic stirring overnight and then subjected to ultrasonication for 30 min in order to disperse the nHA uniformly in the solution prior to the start of the electrospinning process. The size of the nHA incorporated into the polymer matrix was around 80 nm as reported

Table 2 Different composition of nHA-reinforced PPEES nanofiber composite

Named composites	Prepared composites	PPEES (g)	nHA (g)
PPEES	PPEES	10	0
PPEES 1	PPEES/nHA (2.5 wt.%)	9.75	0.25
PPEES 2	PPEES/nHA (5.0 wt.%)	9.5	0.5
PPEES 3	PPEES/nHA (7.5 wt.%)	9.25	0.75

through TEM analysis in our previous study (Kalambettu et al. 2012).

The polymer solution containing different concentrations of nHA was loaded into a 2-ml syringe which was linked to a power supply that was capable of generating high voltage up to 50 kV. The flow rate of the syringe pump was regulated using the PICO Espin 2.0 version software. Electrospinning was performed with an electric voltage supplied at 25 kV with a needle tip to a collector distance of 20 cm. The flow rate was adjusted to 0.2 ml/h, and the collecting drum was regulated to rotate at a speed of 1,000 rpm.

A good fiber mat is one which is either devoid or has minimal bead formation (Huang et al. 2003). From the different weight percentages of prepared nanofiber composites, PPEES/5 wt.% nHA (PPEES 2) composite, showed better fiber formation and were used for further investigations of compatibility and mechanical properties for orthopedic applications.

Characterization studies

FTIR-ATR

The composite samples were subjected to FTIR analysis using Alpha T Bruker Optics FTIR spectrophotometer (BRUKER, Billerica, MA, USA). The functional groups present in the polymer and the interaction between the polymer/nHA nanofiber composites were measured using ATR. The frequency of each sample was recorded at a resolution within the scanning range of 4000–500 cm^{-1}.

SEM

The morphology and dispersion of particles in the polymer matrix were observed using HITACHI S-3400 model SEM (Hitachi High-Tech, Minato-ku, Tokyo, Japan). The surface of the materials was sputter-coated with gold before being subjected to SEM in order to make them electroconductive. SEM analysis of the cross section of the composites was also done to better visualize the presence of nHA in the fiber matrix.

Mechanical properties

The ability to resist breaking under tensile stress is one of the most important and widely measured properties of materials used in structural applications. The mechanical properties of the nanofiber and its composite were observed using a universal testing machine (UTM). Tensile testing was done according to the ASTM D638 type 5 standard using Hounsfield UTM with a crosshead speed of 2 mm/min and maximum load of 500 N. The percentage elongation of the nanofiber composites was calculated using the following formula:

$$Percentage\ elongation = \frac{L_x - L_0}{L_0} \times 100, \qquad (1)$$

where L_x = final gage length and L_0 = initial gage length.

Tensile strength can be calculated based on the following formula:

$$Tensil\ Strength = \frac{F}{A}\ (N/mm^2), \qquad (2)$$

where F = force (N) and A = cross-sectional area (mm^2).

Biomineralization

Biomineralization of the polymer nanofiber and its composite with HA was evaluated by analyzing the HA layer formed on the surface of the samples after 15 and 30 days of incubation in SBF solution, which was prepared in the laboratory according to the procedure developed by Kokubo (Leonor et al. 2007). The samples were then retrieved, dried in an oven at 40°C for 4 h, and then examined under SEM-EDX.

Cell viability

Cytotoxicity studies using the composite nanofiber were carried out on 96-well plates using osteoblast cell lines (MG-63) by MTT assay (Mossman 1983; Sgouras and Duncan 1990). MG-63 cell lines were cultured using Dulbecco's modified Eagle's medium (HiMedia), supplemented with 5% FBS and 1% penicillin-streptomycin, and then seeded into the 96-well plate. The wells were sterilized with 70% ethanol followed by UV treatment for 4 h and were neutralized with a phosphate buffered saline (PBS) (pH 7). The wells without the polymer samples were the control groups for the experiment. The MG-63 cell lines were seeded at a density of $6-7 \times 10^3$ cells per well and incubated at 37°C in a humidified atmosphere containing 5% CO_2. In all culture conditions, the medium was renewed every 48 h. After 3 days of incubation, the supernatant of each well was removed and washed with PBS. MTT, diluted in serum-free medium, was added to each well, and the plates were incubated at 37°C for 3 h. After aspirating the MTT solution, acidified isopropanol (0.04 N HCl in isopropanol), was added to each well and pipetted up and down to dissolve the dark blue formazan crystals and then left at room temperature for a few minutes to ensure the dissolution of all crystals. Finally, the absorbance was measured at 570 nm using an ELISA reader. The viability of cells on the composites was visualized using an inverted fluorescence microscope (Nikon Eclipse, TE 300, Nikon Co., Shinjuku, Tokyo, Japan) after different culture periods (Liu and Tirrell 2008). The composites were transferred to a fresh medium containing 50 ng/ml fluorescein diacetate, incubated for 5 min, and then washed with PBS. The membranes of viable cells, penetrated with the dye solution, were then excited at 488 nm under the inverted fluorescence microscope.

Cell differentiation: ALPase activity

Differentiation of osteoblast cells on the nanofiber composite reinforced with nHA was assessed *in vitro* by ALP activity. Many scientists rely upon the ALP activity of osteoblasts as an indicator of their degree of differentiation (Sila-Asna et al. 2007; Thangakumaran et al. 2009). In the present study, a calorimetric assay was used to find the ALPase activity, where the release of yellow-colored *p*-nitrophenol from *p*-nitrophenol phosphate substrate was monitored by measuring the optical density at 405 nm. Sigma diagnostic kit number 104 was used to estimate the ALPase activity of the cells.

Statistical analysis

Quantitative data were expressed as mean ± standard deviation. Statistical analysis was carried out using ANOVA test. Statistical significance was set at 0.05, and Origin version 6.0 software was used.

Competing interests

The authors declare that they have no competing interests with the work published in this manuscript.

Authors' contribution

The author MA carried out all the studies reported in this manuscript under the supervision of the second author DS. Both the authors contributed equally in writing the manuscript. All authors read and approved the final manuscript.

Acknowledgments

The authors would like to thank the Indian Council of Medical Research (ICMR), New Delhi, India for funding the study (Vide letter no. 5/20/5(Bio)/09-NCD letter dated 26.02.2010). Instrumentation facility provided under FIST-DST and DRS-UGC of the Department of Chemistry, Anna University is sincerely acknowledged.

References

Bhattarai N, Edmondson D, Veiseh O, Matsen FA, Zhang M (2005) Electrospun chitosan-based nanofibers and their cellular compatibility. Biomaterials 26:6176–6184

Bismarck A, Mohanty AK, Aranberri-Askargorta I, Czapla S, Misra M, Hinrichsen G, Springer J (2001) Surface characterization of natural fibers: surface properties and water up-take behavior of modified sisal and coir fibers. Green Chemistry 3:100–107

Broza G, Kwiatkowska M, Roslaniec Z, Schulte K (2005) Processing and assessment of poly(butylene terephthalate) nanocomposites reinforced with oxidized single wall carbon nanotubes. Polymer 46:5860–5867

Bunsell AR, Renard J (2005) Fiber reinforced composite materials. In: Goringe MJ, Ma E, Cantor B (eds) Fundamentals of fibre reinforced composite materials. CRC Press, Boca Raton, FL

Darrell H, Reneker DH, Frong H (2006) Polymeric nanofibers: introduction. In: Darrell H, Reneker DH, Frong H (eds) Polymeric nanofibers. ACS Symposium Series, Washington

Dong J-L, Li L-X, Mu W-D, Wang Y-H, Zhou D-S (2010) Bone regeneration with BMP-2 gene-modified mesenchymal stem cells seeded on nano hydroxyapatite/collagen/poly(L-lactic acid) scaffolds. J Bioact Compat Polym 25:547–566

Dahe GJ, Teotia RS, Kadam SS, Bellare JR (2011) The biocompatibility and separation performance of antioxidative polysulfone/vitamin E TPGS composite hollow fiber membranes. Biomaterials 32:352–365

Greiner A, Wendorff JH (2007) Electrospinning: a fascinating method for the preparation of ultrathin fibers. Angew Chem Int Ed 46:5670–5703

Evaluation of polyphenylene ether ether sulfone/nanohydroxyapatite nanofiber composite as a biomaterial...

153

Huang Z-M, Zhang YZ, Kotaki M, Ramakrishna S (2003) A review on polymer nanofibers by electrospinning and their applications in nanocomposites. Compos Sci Technol 63:2223–2253

Kalambettu AB, Rajangam P, Dharmalingam S (2012) The effect of chlorotrimethylsilane on bonding of nano hydroxyapatite with a chitosan-polyacrylamide matrix. Carbohydr Res 352:143–150

Kang S-W, Yang HS, Seo S-W, Han DK, Kim B-S (2008) Apatite-coated poly (lactic-co-glycolic acid) microspheres as an injectable scaffold for bone tissue engineering. J Biomed Mater Res A 85A:747–756

Kim H-M, Himeno T, Kawashita M, Kokubo T, Nakamura T (2004) The mechanism of biomineralization of bone-like apatite on synthetic hydroxyapatite: an in vitro assessment. J R Soc Interface 1:17–22

Lee HJ, Kim SE, Choi HW, Kim CW, Kim KJ, Lee SC (2007) The effect of surface-modified nano-hydroxyapatite on biocompatibility of poly(ε-caprolactone)/hydroxyapatite nanocomposites. Eur Polym J 43:1602–1608

Leonor IB, Kim HM, Balas F, Kawashita M, Reis RL, Kokubo T, Nakamura T (2007) Functionalization of different polymers with sulfonic groups as a way to coat them with biomimetic apatite layer. J Mater Sci Mater Med 18:1923–1930

Li D, Xia Y (2004) Electrospinning of nanofibers: reinventing the wheel? Adv Mater 16:1151–1170

Liu JC, Tirrell DA (2008) Cell response to RGD density in crosslinked artificial extracellular matrix protein films. Biomacromolecules 9(11):2984–2988

Lu HH, El-Amin SF, Scott KD, Laurencin CT (2003) Three-dimensional, bioactive, biodegradable, polymer–bioactive glass composite scaffolds with improved mechanical properties support collagen synthesis and mineralization of human osteoblast-like cells in vitro. J Biomed Mater Res 64A:465–474

Mossman T (1983) Rapid colorimetric assay for cellular growth and survival, application to proliferation and cytotoxicity assays. J Immunol Methods 65:55–63

Peter M, Ganesh N, Selvamurugan N, Nair SV, Furuike T, Tamura H, Jayakumar R (2010) Preparation and characterization of chitosan–gelatin/nanohydroxyapatite composite scaffolds for tissue engineering applications. Carbohydr Polym 80:687–694

Pielichowska K, Blazewicz S (2010) Bioactive polymer/hydroxyapatite (nano) composites for bone tissue regeneration. Adv Polym Sci 232:97–207

Ramakrishna S, Fujihara K, Teo W-E, Yong T, Ma Z, Ramaseshan R (2006) Electrospun nanofibers: solving global issues. Mater Today 9:40–50

Reneker DH, Chun I (1996) Nanometer diameter fibers of polymer, produced by electrospinning. Nanotechnology 7:216–223

Ryszkowska JL, Auguscik M, Sheikh A, Boccaccini AR (2010) Biodegradable polyurethane composite scaffolds containing Bioglass for bone tissue engineering. Compos Sci Technol 70:1894–1908

Salerno A, Zeppetelli S, Maio ED, Iannace S, Netti PA (2010) Novel 3D porous multi-phase composite scaffolds based on PCL, thermoplastic zein and ha prepared via supercritical CO_2 foaming for bone regeneration. Compos Sci Technol 70:1838–1846

Seal BL, Otero TC, Panitch A (2001) Polymeric biomaterials for tissue and organ regeneration. Mater Sci Eng 34:147–230

Sgouras D, Duncan R (1990) Methods for the evaluation of biocompatibility of soluble synthetic polymers which have potential for biomedical use: 1—use of the tetrazolium-based colorimetric assay (MTT) as a preliminary screen for evaluation of in vitro cytotoxicity. J Mater Sci Mater Med 1(2):61–68

Shanmuga Sundar S, Sangeetha D (2012) Investigation on sulphonated PEEK beads for drug delivery, bioactivity and tissue engineering applications. J Mater Sci 47:2736–2742

Sila-Asna M, Bunyaratvej A, Maeda S, Kitaguchi H, Bunyaratavej N (2007) Osteoblast differentiation and bone formation gene expression in strontium-inducing bone marrow mesenchymal stem cell. Kobe J Med Sci 53(1):25–35

Sista S, Wen C, Hogson PD, Pande G (2011) The influence of surface energy of titanium-zirconium alloy on osteoblast cell functions in vitro. J Biomed Mater Res A. doi:10.1002/jbm.a.33013

Srinivasan S, Jayasree R, Chennazhi KP, Nair SV, Jayakumar R (2012) Biocompatible alginate/nano bioactive glass ceramic composite scaffolds for periodontal tissue regeneration. Carbohydr Polym 87(1):274–283

Tan KH, Chua CK, Leong KF, Cheah CM, Cheang P, Abu Bakar MS, Cha SW (2003) Scaffold development using selective laser sintering of polyetheretherketone-hydroxyapatite biocomposite blends. Biomaterials 24:3115–3123

Teoh SH (2004) Introduction to biomaterials engineering and processing—an overview. In: Teoh SH (ed) Engineering materials for biomedical applications. World Scientific, Singapore

Thangakumaran S, Sudarsan S, Arun KV, Talwar A, James JR (2009) Osteoblast response (initial adhesion and alkaline phosphatase activity) following exposure to a barrier membrane/enamel matrix derivative combination. Indian J Dent Res 20(1):7–12

Wang M (2003) Developing bioactive composite materials for tissue replacement. Biomaterials 24:2133–2151

Zebarjad SM, Sajjadi SA, Sdrabadi TE, Yaghmaei A, Naderi B (2011) A study on mechanical properties of PMMA/hydroxyapatite nanocomposite. Engineering 3:795–801

12

Evaluation of silk sericin as a biomaterial: in vitro growth of human corneal limbal epithelial cells on Bombyx mori sericin membranes

Traian V Chirila[1,2,3,4,5*], Shuko Suzuki[1], Laura J Bray[1,6,7], Nigel L Barnett[1,6,8] and Damien G Harkin[1,6,9]

Abstract

Sericin and fibroin are the two major proteins in the silk fibre produced by the domesticated silkworm, *Bombyx mori*. Fibroin has been extensively investigated as a biomaterial. We have previously shown that fibroin can function successfully as a substratum for growing cells of the eye. Sericin has been so far neglected as a biomaterial because of suspected allergenic activity. However, this misconception has now been dispelled, and sericin's biocompatibility is currently indisputable. Aiming at promoting sericin as a possible substratum for the growth of corneal cells in order to make tissue-engineered constructs for the restoration of the ocular surface, in this study we investigated the attachment and growth *in vitro* of human corneal limbal epithelial cells (HLECs) on sericin-based membranes. Sericin was isolated and regenerated from the silkworm cocoons by an aqueous procedure, manufactured into membranes, and characterized (mechanical properties, structural analysis, contact angles). Primary cell cultures from two donors were established in serum-supplemented media in the presence of murine feeder cells. Membranes made of sericin and fibroin-sericin blends were assessed *in vitro* as substrata for HLECs in a serum-free medium, in a cell attachment assay and in a 3-day cell growth experiment. While the mechanical characteristics of sericin were found to be inferior to those of fibroin, its ability to enhance the attachment of HLECs was significantly superior to fibroin, as revealed by the PicoGreen® assay. Evidence was also obtained that cells can grow and differentiate on these substrata.

Keywords: Silk; Silk proteins; Sericin; Corneal limbal epithelial cells; Cell attachment

Introduction

Silks are natural polypeptide composite materials commonly included in the group of fibrous proteins. They are produced by certain species, most notably by spiders (order Araneae) and by silkmoths (order Lepidoptera). The silk produced by the larvae of the domesticated silkmoth (*Bombyx mori*) had an important role in the textile industry for millennia, and is, by far, the most extensively studied silk.

The two constitutive proteinaceous molecular complexes in the silks produced by insects are fibroin and sericin. Following seminal studies by Minoura's group (Minoura et al. 1990, 1995a, b), the silk proteins, especially

the fibroin produced by the *B. mori* silkworm, have been widely investigated as potential biomaterials and considered for tissue engineering applications (Altman et al. 2003; Wang et al. 2006; Vepari and Kaplan 2007; Hakimi et al. 2007; Kearns et al. 2008; Kundu et al. 2008; Wang et al. 2009; Murphy and Kaplan 2009; Hardy and Scheibel 2010; Numata and Kaplan 2010; Ghassemifar et al. 2010; Harkin et al. 2011; Pritchard and Kaplan 2011; Wenk et al. 2011; Gil et al. 2013).

The assembly of polypeptides known as 'sericin' represents about one quarter of the total protein content of *B. mori* cocoons. Being soluble in hot water or alkaline aqueous solutions, sericin can be easily removed in a process known as 'degumming' and isolated as a pure product if needed.

Sericin has been traditionally associated with the immune responses attributed to silk in general (Altman et al. 2003). Sensitization to silk leads to allergic reactions affecting the

* Correspondence: traian.chirila@qei.org.au
[1]Queensland Eye Institute, South Brisbane, Queensland 4101, Australia
[2]Faculty of Science and Engineering, Queensland University of Technology, Brisbane, Queensland 4001, Australia
Full list of author information is available at the end of the article

skin or respiratory tract, while silk as such can also cause an inflammatory response when in direct contact with living tissues. As a result, due to biocompatibility concerns, sericin has been largely neglected as a potential biomaterial. Although still underutilized, there has been a change over the last decade in the attitude regarding the potential applications of sericin in various fields of human activity. This was mainly prompted by environmental and economical concerns with respect to the large amounts of sericin, estimated to exceed 50,000 t annually (Mondal et al. 2007), which have to be discarded as waste from silk processing factories, and also by the realization that sericin is a biocompatible material despite being traditionally regarded as a pathologic allergen. A comprehensive review (Kundu et al. 2008) presents a detailed account of current uses of sericin, chiefly from *B. mori* silkworms, in cosmetics, pharmaceuticals, dietary foods, controlled drug administration, wound dressing, and media for cell culture. Some time ago, it was suggested (Minoura et al. 1995b) that sericin on its own could be a biomaterial, which is obviously different from using it as a supplement in the cell culture medium. However, there have been very few reports on this issue.

We are currently developing and evaluating silk protein substrata or scaffolds for the tissue engineering of the eye (Harkin et al. 2011; Harkin and Chirila 2012), having been the first to assess *B. mori* silk fibroin (henceforward BMSF) as a potential substratum for the restoration of the ocular surface (Chirila et al. 2007, 2008, 2010). We have further evaluated BMSF membranes as substrata for corneal epithelial constructs (Bray et al. 2011, 2012, 2013; George et al. 2013), for corneal endothelial constructs (Madden et al. 2011), and for transplantation of retinal cells (Kwan et al. 2010; Shadforth et al. 2012). The potential advantages of BMSF over other materials (e.g. collagen) in tissue engineering applications have been discussed elsewhere (Harkin and Chirila 2012).

In the present report, we investigated the capacity of sericin regenerated from *B. mori* cocoons to function as a substratum for cell growth, either as such or blended with regenerated fibroin. In particular, human corneal limbal epithelial cells (HLECs) were assessed for the first time on sericin substrata with an aim of furthering the use of sericin in tissue-engineered constructs for ocular surface restoration.

Methods

Materials

B. mori silkworm cocoons were supplied by Tajima Shoji Co Ltd (Yokohama, Japan), with the pupae removed. All chemical reagents were supplied by Sigma-Aldrich (St Louis, MA, USA) with the exceptions mentioned here. Genipin (purity 98%) was purchased from Erica Co Ltd (Xi'an, Shaanxi, China). Water of high purity

(Milli-Q or equivalent, Millipore, Billerica, MA, USA) was used in all experiments. Sartorius Stedim Biotech (Göttingen, Germany) supplied the Minisart®-GF prefilters (0.7 μm) and Minisart® filters (0.2 μm). The dialysis cassettes Slide-A-Lyzer® (MWCO 3.5 kDa) were supplied by Thermo Scientific (Rockford, IL, USA) and dialysis tubes with MWCO 12.4 kDa by Sigma-Aldrich. The olefin copolymer Topas® 8007S-04 was purchased from Advanced Polymers (Frankfurt, Germany).

All cell culture reagents and supplements were purchased from Life Technologies (Mulgrave, Victoria, Australia) with the following exceptions: foetal bovine serum (FBS) from Thermo Scientific (Australia); 3,3,5-triiodo-L-thyronine sodium salt (T3), adenine, transferrin, hydrocortisone, insulin, tris(hydroxymethyl)aminomethane, and EDTA from Sigma-Aldrich; and isoproterenol from Merck (Sydney, Australia). Sterile ethanol 70% *v/v* for sterilization was supplied by ORION Laboratories (Balcatta, Australia).

Unless specified otherwise, all per cent concentrations or compositions in this report are expressed in percentage by weight.

Preparation of fibroin and sericin solutions

The solution of regenerated BMSF was prepared according to a protocol we have previously established (Chirila et al. 2008). The concentration of solution used in experiments was 1.78% (by gravimetric analysis).

To prepare the *B. mori* silk sericin (henceforward BMSS) solution, a published protocol (Nagura et al. 2001) was followed with some modifications. Cocoons (6 g) were cut into 1 cm × 1 cm pieces and washed with water in a 5-L beaker with vigorous stirring for 30 min at room temperature. The washed material was transferred in a screw-capped bottle, with 80 mL water, and autoclaved at 121°C for 30 min. The resulting solution was passed through a paper filter to retain the undissolved fibroin material, then aspirated in a syringe and injected into dialysis tubing (MWCO 12.4 kDa) that was pre-treated by soaking in water for 4 h and changing the water five times. The tube was then placed in a 2-L beaker with hot water (80°C) and dialysed with stirring for 4 h and three water exchanges. The final solution was removed from the tube, filtered through the 0.7-μm Minisart® filter, and collected in a glass container that was kept at 80°C (to prevent gelation) until use. The concentration of BMSS in the solutions resulting from various batches was between 1.12% and 1.24% (by gravimetric analysis).

Preparation of membranes

The BMSF and BMSS membranes were cast from their respective solutions. To produce the blended membranes, the two solutions were mixed before casting at BMSF/BMSS weight ratios of 90/10, 50/50, and 10/90, respectively. For casting, 45-mm glass Petri dishes were first coated with a film of Topas® polymer by the evaporation from a solution in cyclohexane. The solutions were prepared starting with 2 mL of BMSF solution and adding the calculated volumes of the other components. Each solution (BMSF, BMSS, and three mixtures) was poured into the dish, and all dishes were placed in a fan-driven oven and kept for 12 h at room temperature (22°C to 23°C). The dishes with dried membranes were placed in a vacuum enclosure to be annealed in the presence of a container with water, under a vacuum of −80 kPa at room temperature for 24 h. The membranes were then peeled off the Topas® supporting films, and further used in experiments. The thickness of all prepared membranes was approximately 10 μm.

Crosslinking of the BMSF/BMSS 50/50 blend

The blended membrane with the composition BMSF/BMSS 50/50 was prepared also as a crosslinked material for comparative assessment. The crosslinking was carried out with genipin, employing a slightly modified published procedure (Motta et al. 2011). An amount of genipin equivalent to 12% of the final protein content was mixed with 3 mL solution of 1.78% BMSF and stirred for 5 h at 40°C, when 4.6 mL solution of 1.16% BMSS were added. The reaction mixture acquired a light blue hue, which is the hallmark of the reaction between genipin and amino acids (Djerassi et al. 1960). After additionally stirring for 1 h at 40°C, the product was filtered successively through the 0.7-μm and 0.2-μm Minisart® filters. The resulting solution was processed into a membrane as described above, except for the water-annealing treatment. Following the first drying, the membrane was washed thoroughly in water, dried again, and peeled off the Topas® supporting film.

Gel electrophoresis analysis of sericin

The molar mass distribution of BMSS was investigated by sodium dodecyl sulphate-polyacrylamide gel electrophoresis (SDS-PAGE), using a Novex® XCell Sure Lock™ Mini-Cell system (Life Technologies Inc, Carlsbad, CA, USA) and an EPS-250 Series II Power Supply unit (CBS Scientific Company Inc, San Diego, CA, USA). The sericin solution was mixed with both NuPAGE® sample preparation reagent and NuPAGE® sample reducing agent, and heated at 70°C for 10 min. A volume of BMSS solution containing approximately 50 μg protein was loaded into a 1-mm thick 3% to 8% NuPAGE® Novex® Tris-acetate gel in NuPAGE® Tris-acetate SDS running buffer. The gels were run at a voltage of 150 V for 1 h together with a HiMark™ Pre-stained Protein Standard (Life Technologies). The resulting gel was washed in three 5-min stages with distilled water and soaked in SimplyBlue™ SafeStain solution (Life Technologies) containing Coomassie® G-250 stain for 1 h under gentle stirring. The gel was washed in distilled water for 1 h and then photographed.

Mechanical testing of membranes

From each membrane, strips (1 × 3 cm) were cut out and subjected to tensile measurements (for stress, modulus, and elongation) in an Instron 5848 microtester (Instron, Wycombe Buckinghamshire, UK), equipped with a 5 N load cell and a set gauge distance of 14 mm. The samples were loaded by pneumatic grips and submersed in phosphate buffer solution (pre-heated to 37°C ± 3°C) in a BioPuls™ unit for 5 min prior to stretching. Stress–strain plots were recorded, the Young's moduli were computed in the linear region, and elongation at break was also measured. The mean values were calculated from results generated by six measurements for each specimen, which were cut out from at least three different membranes for the same composition.

Analysis by Fourier transform infrared-attenuated total reflectance spectroscopy

The Fourier transform infrared-attenuated total reflectance (FTIR-ATR) spectra of the films (fibroin, sericin, and blends) were collected using a Nicolet Nexus 5700 FTIR spectrometer (Thermo Electron Corp, Marietta, OH, USA), equipped with a Nicolet Smart Endurance diamond ATR accessory. Each spectrum was obtained by co-adding 64 scans in the 4,000 to 525 cm^{-1} range at a resolution of 8 cm^{-1}. Spectra were recorded and analyzed using OMNIC 7 software package (Thermo Electron Corp).

Contact angle analysis

Membranes of BMSF, BMSS, and their blends were cast as films and annealed directly onto microscope glass slides, and they were not removed prior to measurements. The genipin-crosslinked 50/50 blended film was not annealed. The contact angles were measured after placing a water droplet onto the dry films. A water droplet of about 5 μL was applied onto each surface, and photographs were taken with a Sanyo VCB-3512 T CCD camera (Sanyo Electric Co, Moriguchi, Osaka, Japan) at an interval of 5 s after the droplet was dispensed. The resulting contact angle was measured in a goniometer (FTÅ200, First Ten Ångstroms Inc, Portsmouth, VA, USA) using the FTÅ Drop Shape Analysis Software Version 2.0 (2002). The results reported are the average values of 16 measurements for each composition.

Establishment of primary human corneal limbal epithelial cell cultures

The protocol for this stage was detailed in a previous paper (Bray et al. 2013). The ocular tissue was harvested as corneoscleral rims or caps from two different donors, provided by the Queensland Eye Bank (Brisbane, Australia), with donor consent and regulatory ethics approval. In brief, after washing, sectioning, and incubating with 0.25% dispase, the dissociated human limbal epithelial (HLE) sheets were collected, pooled, centrifuged, and re-suspended in 0.25% trypsin in EDTA. After further washing and centrifugation, the HLECs were suspended in serum-supplemented culture medium and propagated in the presence of irradiated 3T3 murine fibroblast feeder cells as described elsewhere (Bray et al. 2011). The passage 1 cultures were further used in the cell attachment assay and in a 3-day cell growth experiment.

Cell attachment assay

The attachment of cells was evaluated according to the manufacturer's instructions for the Quant-iT™ PicoGreen® dsDNA assay (Life Technologies), as detailed in a previous report (Bray et al. 2013). In brief, 2×10^4 HLECs/cm^2 were seeded onto BMSF, BMSS, and their blends, which had been deposited each as coatings onto the bottom of wells of a sterilized 24-well plate. The cells were incubated for 4 h in serum-free medium and washed in PBS. After adding 1 mL of 0.1% Triton-X100, the plates were incubated at room temperature for 1 h. Each well was then triturated and the supernatants were centrifuged. To carry out the PicoGreen analysis, 25 µl of each sample were aliquoted into a 96-well plate with 75 µl of TE buffer. PicoGreen dye was then added at a dilution of 1:200 in the same buffer to each well in 100 µl portions. The plate was read on a fluorescent microplate reader (FLUOstar OPTIMA, BMG Labtech Pty Ltd, Mornington, Victoria, Australia), at the wavelengths 480 nm (excitation) and 520 nm (emission). Cell attachment was assessed based on the DNA content (calculated and plotted in ng/mL), which relates directly to the number of cells.

These experiments were conducted in triplicate for each series of experiments with cells obtained from two different tissue donors. The results were statistically processed by the one-way analysis of variance (ANOVA) in conjunction with Tukey-Kramer multiple comparisons, using the GraphPad Prism® version 6.0. Bright-field microscopy was used to visualize and photograph the 3-day cell cultures in a Nikon TE2000-U (Chiyoda-ku, Tokyo, Japan) instrument.

Results and discussion

The primary and secondary structures of sericin

Historically, sericin in *B. mori* silk triggered the scholars' interest quite early in the development of the modern scientific investigation. The earliest report on sericin was published in 1785 by l'Abbé ('Father') Collomb (Collomb 1785), who was the first to demonstrate its solubility in hot water. A rigorous study of the effects of water, alcohol, alkaline solutions, soap, and acids, and exposure to light on the silk yarn was published two decades later (Roard 1808), showing that the portion of the silk soluble in aqueous or alcoholic media contained at least four fractions, and two of them were assumed to be constituents of the gum coating, i.e. what we call now sericin.

Throughout the next century, research has been focused on improved methods to remove and isolate sericin and to identify its constituents, as overviewed in some earlier publications (Shelton and Johnson 1925; Lucas et al. 1958). The extraction in boiling water or by autoclaving became the techniques of choice for extracting sericin from raw silk or cocoons.

We have investigated by SDS-PAGE the electrophoretic mobility of the sericin components in order to visualize the distribution of their molar masses. Staining revealed a smear pattern (Figure 1), which strongly suggests that the polypeptides in sericin were degraded during processing by autoclaving owing to hydrolytic reactions. Our finding is similar to the smear patterns reported by many investigators for both fibroin and sericin after being processed in aqueous media at elevated temperatures. It is known that the sericin harvested directly from the silkworm's middle silk gland (MSG) display electrophoretic patterns with distinct protein bands, unlike the sericin processed from cocoons in conditions of high temperature. Confirmation of the effect of autoclaving on sericin has been recently provided (Teramoto et al. 2005, 2006) in experiments where pure sericin was collected from the silkworms belonging to a new race known as 'Sericin Hope' that produces exclusively sericin. This race was developed in Japan through the breeding of mutant silkworms *Nd* or *Nd-s*, which practically synthesize sericin only (Julien et al. 2005), with the strain KCS83, a high cocoon-producing species (Mase et al. 2006). Sericin collected directly from the cocoons (Teramoto et al. 2005) or glands (Teramoto et al. 2006) of the Sericin Hope silkworms exhibited an electrophoretic pattern comprising at least five distinct polypeptides; however, these bands could no longer be identified following autoclaving of the samples as they merged into a smear pattern practically identical to that shown in Figure 1. Indeed, temperature is an important factor in the degradation of sericin proteins: upon isolation of sericin from cocoons at room temperature using ethylenediamine/cupric hydroxide (Tokutake 1980) or lithium thiocyanate (Takasu et al. 2002), distinct polypeptide bands could be identified in the electrophoretograms.

Unlike BMSF, which is known to be composed of three well-defined polypeptidic subunits, with the approximate molar masses of 360, 26, and 30 kDa (Inoue et al. 2000; Julien et al. 2005), the number and molar mass of the subunits in BMSS are still disputed. Long time ago, it was noticed that BMSS is a complex mixture of polypeptides in a number that can be as high as 15 (or even higher) according to some investigators (Sprague 1975). Throughout a century of research, different investigators found various numbers of polypeptide fractions in BMSS, such as 2 (Kondo 1921; Shelton and Johnson 1925; Kodama 1926), 3 (Mosher 1934; Rutherford and Harris 1940; Bryant 1948; Kikkawa 1953; Tashiro and Otsuki 1970; Takasu et al. 2002),

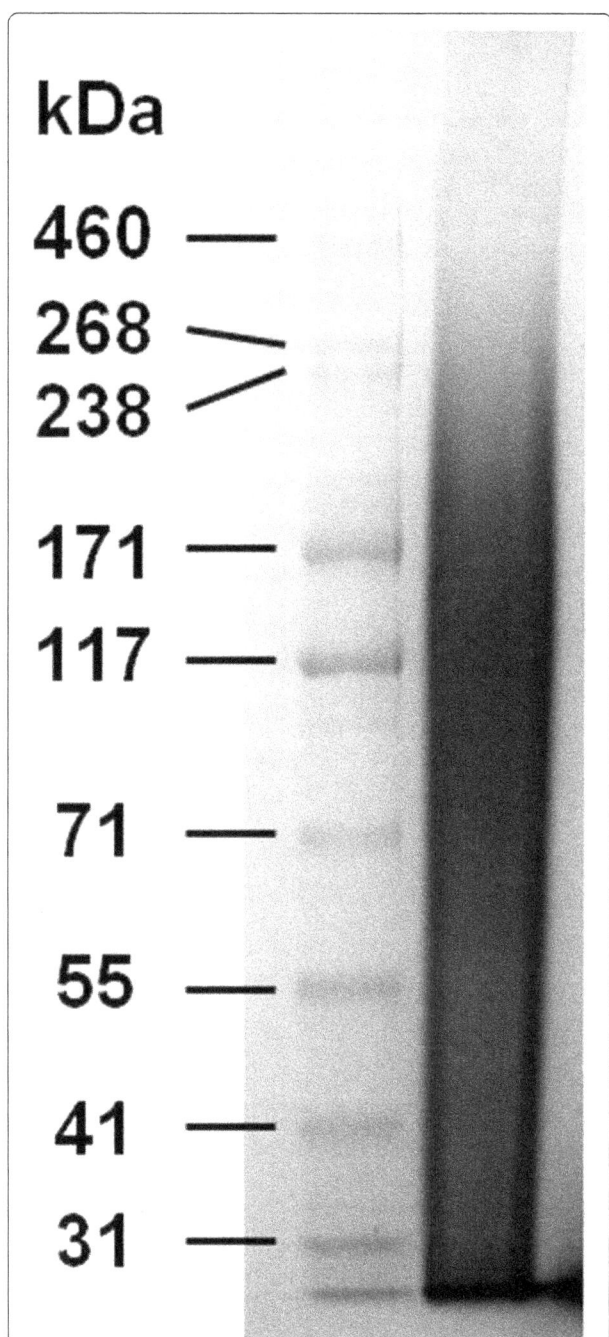

Figure 1 Analysis of sericin by SDS-PAGE. Electrophoretic pattern of sericin regenerated from *B. mori* cocoons (right lane). Left lane shows the molar mass marker positions (see text for details of the analysis).

the genes *Ser1*, *Ser2*, *Ser3*, *MSG-3*, *MSG-4*, and *MSG-5* encode for sericin in *B. mori* silkworm's middle gland (Gamo 1982; Michaille et al. 1986, 1989; Grzelak 1995; Julien et al. 2005; Kundu et al. 2008), and therefore at least six major polypeptides are expected to be synthesized in the gland, such providing compelling justification for its heterogeneous composition.

The FTIR-ATR spectra in the 'Amide I' region (1,590 to 1,720 cm^{-1}) of films of BMSF, BMSS and their blends after water annealing (for 24 h), as well as the BMSF/BMSS 50/50 blend crosslinked with genipin (but not annealed), are all shown in Figure 2. This spectral region is traditionally used for the characterization of the secondary structure of silk proteins. For BMSF, the absorption bands in this region can be assigned as follows: 1,610 to 1,630 and 1,695 to 1,700 cm^{-1} as β-sheet structure; 1,640 to 1,650 cm^{-1} as random-coil structure; 1,650 to 1,660 cm^{-1} as α-helix; and 1,660 to 1,695 cm^{-1} as β turns (Hu et al. 2006, 2011). The spectrum of BMSF film (sample 1) showed peaks at 1,621 and 1,700 cm^{-1}, indicating that the sample had a significant amount of β-sheet crystal component. For the BMSF/SS 90/10 blend film (sample 2), the peak at 1,621 cm^{-1} was smaller than that of sample 1 in relation to the band at 1,645 cm^{-1} corresponding to the random-coil structure. In the spectrum, we can see a decrease of the β-sheet band that indicates lower crystallinity of sample 2 as compared to sample 1. This is due to the presence of sericin that has been found to retard the crystallization of fibroin (Lee 2004). In the spectra of samples 3 and 4, we can see that while the β-sheet band decreases in intensity, a band at a lower wavenumber (1,615 cm^{-1}) increases in intensity, such indicating the predominance of sericin. The spectrum of

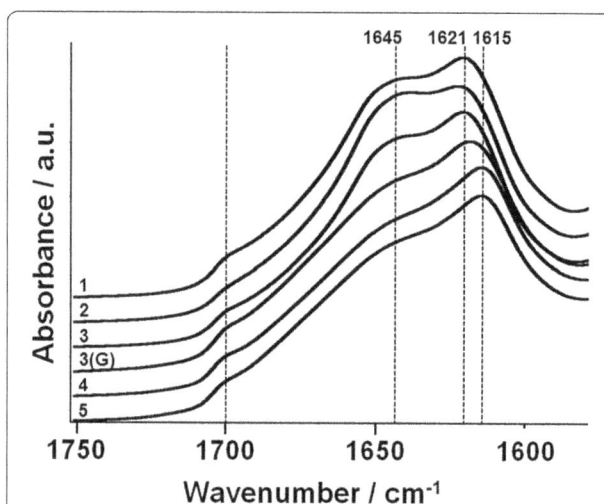

Figure 2 The Amide I region in the FTIR-ATR spectra of membranes. The numbers on the recorded spectra represent the membrane compositions (in% by weight): 1, BMSF 100; 2, BMSF/BMSS 90/10; 3, BMSF/BMSS 50/50; 3(G), BMSF/BMSS 50/50/crosslinked with genipin, not annealed; 4, BMSF/BMSS 10/90; 5, BMSS 100.

5 (Gamo et al. 1977), 6 (Tokutake 1980), or more (Sprague 1975; Fedič et al. 2002), although it was also proposed (Rutherford and Harris 1940) that the existence of fractions is an artefact caused by degradation and that sericin is a homogeneous material. In terms of molecular mass, the range encompassing all reported values is between 20 and 400 kDa. Genomic analysis of sericin showed that

BMSS film (sample 5) displays an intensive band at $1,615$ cm^{-1} as well as a shoulder at $1,700$ cm^{-1}, both corresponding to β-sheet structures and their aggregates (Teramoto and Miyazawa 2005; Teramoto et al. 2005; Teramoto et al. 2008) and signifying a higher crystallinity of BMSS when compared to that of BMSF. The spectrum of the crosslinked sample 3(G) suggests slightly larger β-sheet content than that in the uncrosslinked sample.

Tensile properties of membranes

The results of mechanical testing (Figure 3) indicate poor tensile characteristics for BMSS. Its strength, stiffness (elastic modulus), and elasticity are much lower than those of fibroin. This is expected, as sericin is a globular protein rather than fibrous. It may be also due to the hydrolytic degradation of sericin during processing. For instance, mechanical properties that were significantly superior to those measured in the present study were reported (Teramoto et al. 2008) for the sericin isolated from Sericin Hope cocoons under mild conditions (35°C). On the other hand, sericin extracted by autoclaving could not even be subjected to tensile measurements, as the films were too fragile and broke 'merely by lifting by pincette' (Nagura et al. 2001).

The higher the content of sericin in the blends, the lower their tensile characteristics. When comparing the blends to fibroin, as a confirmation of our results in Figure 3 (sample 2), it was found (Motta et al. 2011) that a BMSF/BMSS 90/10 blend displayed lower tensile strength and modulus, but a slightly higher elongation at break. Crosslinking with genipin of the BMSF/BMSS 50/50 blend was carried out to assess possible improvement of the mechanical properties. Indeed, it enhanced the strength and elasticity, but not the modulus (Figure 3, sample 3(G)), which basically remained the same.

A slight reduction of stiffness has been reported upon crosslinking sericin with dimethylolurea (Nagura et al. 2001) or upon crosslinking the blend BMSF/BMSS 90/10 with either genipin or poly(ethylene glycol) diglycidyl ether (PEG-DE) (Motta et al. 2011), which has been explained, respectively, either by an effect of the change in the water content of the films (which in its turn depends on the crosslinking degree) or as a consequence of pore formation due to the removal of sericin by alcohol treatment. In our study, however, the membranes were not subjected to alcohol treatment prior to testing. With regards to the water content (WC) of BMSF/BMSS 50/50 samples, we measured $58.51\% \pm 0.92\%$ for the crosslinked one and $52.57\% \pm 3.54\%$ for the uncrosslinked water-annealed sample, respectively. However, this difference did not affect the stiffness of the membranes. For comparison, the measured WCs of pure fibroin (sample 1) and of pure sericin (sample 5) were $36.01\% \pm 1.70\%$ and $53.37\% \pm 2.53\%$, respectively.

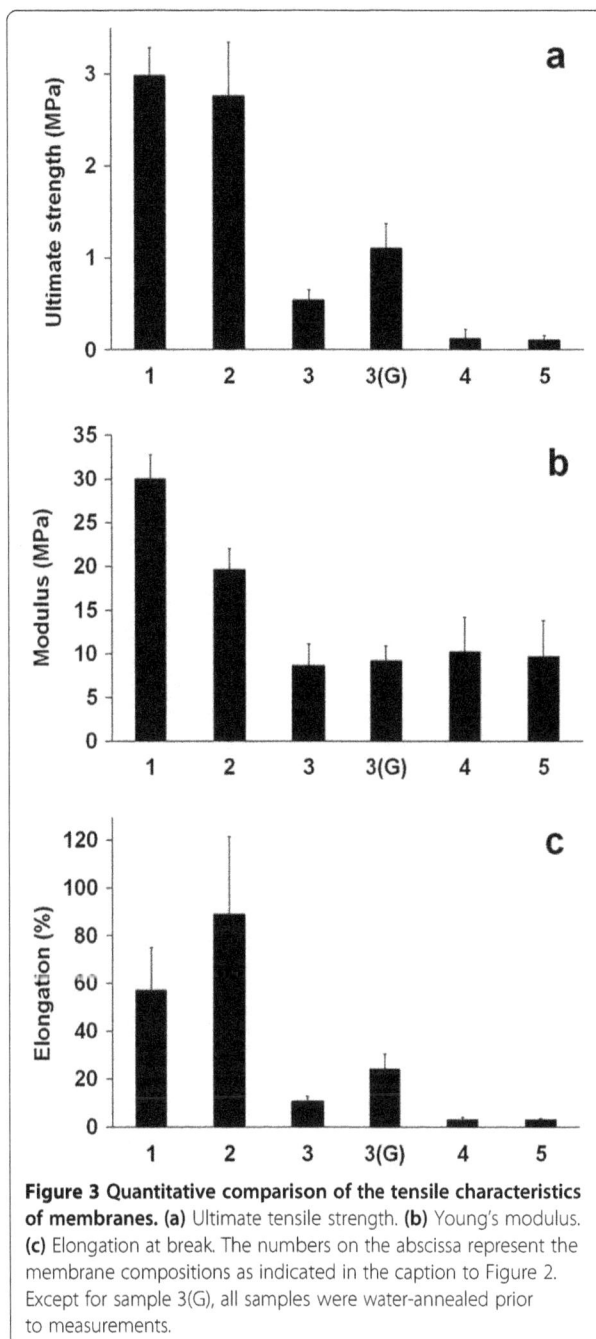

Figure 3 Quantitative comparison of the tensile characteristics of membranes. (a) Ultimate tensile strength. (b) Young's modulus. (c) Elongation at break. The numbers on the abscissa represent the membrane compositions as indicated in the caption to Figure 2. Except for sample 3(G), all samples were water-annealed prior to measurements.

When compared to the tensile strength of the natural human cornea, reported to be around 3.8 MPa (Zeng et al. 2001), sericin-based membranes cannot compete, and even fibroin is less strong. However, the values of Young's moduli for BMSF, BMSS, or blends are higher than those reported (Hjortdal 1996) for the human cornea (9 MPa for meridional loading and 13 MPa for circumferential loading). The potential ways to improve the mechanical properties of BMSS include chemical crosslinking, mild processing conditions to prevent degradation, or blending with other macromolecular materials. To achieve these aims, work is in progress in our laboratories.

Contact angles

Table 1 shows the contact angle values measured for films of BMSF, BMSS, and their blends. The values may reflect in part the vagaries associated with this type of measurement, although we carried out an inordinately large number ($N = 16$) of measurements for each composition. BMSF is clearly more hydrophilic than BMSS, and there is an obvious trend in blends of decreasing hydrophilicity (i.e. higher contact angles) with the increase of BMSS content, although it is difficult to explain why the blends 50/50 and 10/90 displayed contact angles greater than BMSS as such. The crosslinking with genipin induced enhanced hydrophilicity of the film surface.

It may be that the measured values are related to the content of β-sheet in the conformation of sericin leading to indirect effects on the hydrophilicity; but irrespective of possible explanations, the important fact is that the values in Table 1 place all compositions in the category of materials that promote cell attachment (Horbett and Klumb 1996).

Biocompatibility of sericin and its blends

Our rigorous examination of the available literature suggested that blaming exclusively sericin for sensitization to silk was rather speculative and that, in many instances, publications on this topic have been misinterpreted or misquoted. In some early reports on human allergic sensitivity to silk (Clarke and Meyer 1923; Figley and Parkhurst 1933), sericin was indeed suggested as a cause, but only because its solubility in water. Most of the studies published at the same time or later (Taub 1930; Friedman et al. 1957; Brown and Coleman 1957; Kino and Oshima 1979; Häcki et al. 1982; Johansson et al. 1985; Harindranath et al. 1985; Nakazawa and Umegae 1990; Suzuki et al. 1995; Celedón et al. 2001) could not confirm this suggestion. Other investigators also have initially proposed sericin as a pathogen (Zhaoming et al. 1990) only to play down later this assertion (Zhaoming et al. 1996).

There are, however, arguments that appear to support the sensitizing effect of sericin, like in a study (Soong and Kenyon 1984) on the post-operative complications in patients who underwent cataract surgery and were sutured with virgin silk (i.e. containing sericin). However, a previous *in vivo* animal study (Salthouse et al. 1977) had shown that the absence of sericin in degummed sutures did not actually prevent tissue responses. In the reference most frequently cited as a confirmation that sericin is a pathogenic allergen (Dewair et al. 1985), the investigators concluded that the allergenic polypeptides 'are more likely to belong to the sericin group of silk proteins,' a statement that was rather tenuously based on the molecular weights of 12 polypeptides isolated by gel electrophoresis.

In recent papers (Panilaitis et al. 2003; Aramwit et al. 2009a; Hakimi et al. 2010), convincing evidence was provided that sericin does not induce inflammatory or cytotoxic effects. Sericin-mediated activation of a macrophage cell line has been shown to be related *not* to sericin itself but to a physical association between sericin and fibroin, and an inflammatory response could be triggered only in the presence of another pro-inflammatory agent (Panilaitis et al. 2003). Moreover, in experiments involving the growth of human microvascular endothelial cells in the presence of sericin isolated from domesticated or wild silkworm cocoons, no cytotoxic/cytostatic effects have been seen; on the contrary, supplementation with sericin has led to significant enhancement of cell proliferation (Hakimi et al. 2010). These studies alone would exonerate sericin of its potential cytopathologic activities. Moreover, a large number of studies demonstrated the general biocompatibility of sericin and led to many applications involving its contact with body tissues (Kundu et al. 2008; Sehnal 2011).

Considerably fewer reports are available on the use of sericin as a substratum or scaffold for growing cells and tissues. In an early study (Minoura et al. 1995b), BMSS harvested directly from the silk gland was evaluated as a substratum for mouse fibroblasts (line L-929), alone or blended with fibroin. Cell attachment and proliferation on the two silk proteins were similar to those on collagen and higher than on tissue culture plastic. The ability of BMSS, either directly extracted from MSG or regenerated from cocoons, to function as a substratum for cell growth was latter confirmed by others (Terada et al. 2002; Tsubouchi et al. 2005; Xie et al. 2007; Aramwit et al. 2009b, 2010).

We aim at extending the range of sericin's potential use as a substratum or scaffold for cells in ophthalmic tissue engineering. As mentioned in the introductory section, research in this field has been limited so far to the BMSF membranes. In the present study, the *in vitro* attachment and growth of HLECs on BMSS membranes was investigated to gauge the cell response to sericin. The primary cell cultures established from donor human

Table 1 The contact angles of water on the film surfaces

Sample	BMSF/BMSS (% by weight)	Contact angle (deg)[a]
1	100/0	41.7 ± 0.4
2	90/10	51.3 ± 0.6
3	50/50	53.7 ± 1.1
3(G)[b]	50/50	58.4 ± 0.6
4	10/90	59.0 ± 1.2
5	0/100	53.5 ± 1.4

[a]The results are given as mean values ± SEM for $N = 16$; [b]Sample was crosslinked but not annealed.

eye tissue displayed characteristic cobblestone morphology and reached confluence within 2 weeks or so. After being passaged and then re-suspended in serum-free culture medium in the presence of BMSS membranes, the cells attached to the substratum and maintained their viability and morphology, as seen after 3 days of growth (Figure 4). There is no difference, from one substratum to another, regarding the morphology of cells. However, the level of cell attachment was variable, as shown in Figure 5. A dose-dependent increase with the BMSS content is evident with respect to the level of cell attachment. The level of attachment to pure sericin (sample 5) is higher than that to fibroin (sample 1) or a fibroin-rich blend (sample 2), and even to 50/50 blends (samples 3 and 3(G)). These differences are statistically significant and provide compelling evidence of the suitability of BMSS as a substratum for the growth of HLECs. Although the level of attachment to a sericin-rich blend (sample 4) is also higher than that to samples 1, 2, 3, and 3(G), these differences are not statistically significant. An enhancement of cell attachment to sericin and to BMSF/BMSS 10/90 blend when compared to the attachment to the tissue culture plastic (TCP) control sample is also evident, although the differences were not statistically significant.

In terms of contact angle values, both BMSF and BMSS are expected to promote cell attachment, but the differences between BMSS (sample 5) and the blend samples 2, 3, and 3(G) (Table 1) are too small to justify the differences in cell attachment shown in Figure 5. Nevertheless, the higher hydrophilicity of BMSF (sample 1) may be related to a lower level of attachment. However, the mechanism of cell attachment to BMSF or BMSS appears to be more complex and has not been elucidated so far.

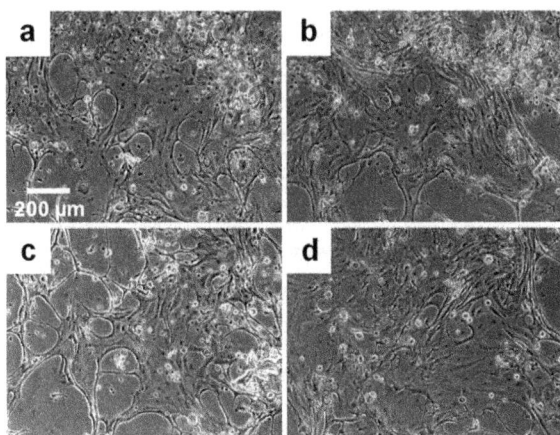

Figure 5 Bar chart showing a quantitative comparison of the HLE cell attachment to BMSF, BMSS, and their blends. The cells were seeded in the absence of serum. The bars represent mean values ± SEM for the total number of viable cells after 4 h estimated from the DNA content using the PicoGreen® dsDNA assay (see text for details). The numbers on the abscissa represent the membrane compositions as indicated in the caption to Figure 2. TCP denotes 'tissue culture plastic' (control sample). The differences between the sample marked with a single asterisk and each of those marked with double asterisk are statistically significant ($p < 0.05$). Except for sample 3(G), all samples were water-annealed prior to use as substrata.

As none of these proteins contains any of the identifiable cell-binding peptide motifs, such as the arginine-glycine-aspartic acid (RGD) sequence, the process of cell attachment has been suggested to be non-specific, likely based on electrostatic interactions (Minoura et al. 1995b). In sericin, according to other investigators, a 170-kDa polypeptide fraction encoded by *Ser1* gene may have an active role in the cell attachment process (Tsubouchi et al. 2005). Our analysis (Figure 1) indicated the presence of polypeptide fractions around this molar mass. However, at this stage it can be only surmised that the cell-adhesive properties of BMSF or BMSS arise from a favourable combination of non-specific interactions promoted by certain characteristics of the substratum's surface such as charge, wettability, and topography. Further investigations are needed to explain the enhanced cell attachment on BMSS as compared to BMSF.

Conclusions

The *B. mori* sericin-based materials are not as strong as those based on *B. mori* fibroin, and they may need improvement in this respect for potential use as surgical implants. However, the attachment of human corneal limbal epithelial cells to sericin or sericin-rich blends is superior to that to fibroin or fibroin-rich blends, with no evidence of any cytopathologic response. These findings would strongly recommend sericin as an appropriate material for making substrata or scaffolds for cellular invasion and tissue formation in applications pertaining to the restoration of damaged ocular surface.

Figure 4 Growth of primary human corneal limbal epithelial (HLE) cells after 3 days in culture. The micrographs show the morphology of cells growing on **(a)** tissue culture plastic (control), **(b)** BMSF, **(c)** BMSF/BMSS 10/90 blend, and **(d)** BMSS. The scale bar is the same for all panels.

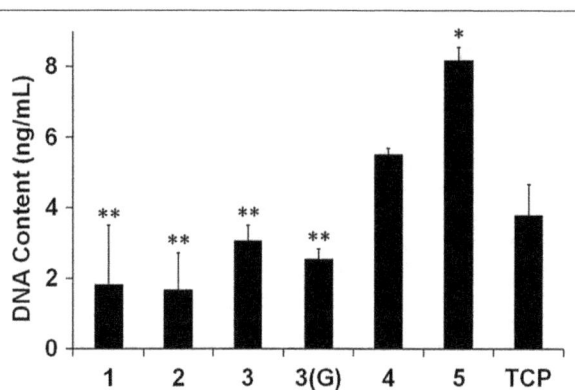

Abbreviations

BMSF: *Bombyx mori* silk fibroin; BMSS: *Bombyx mori* silk sericin; EDTA: Ethylenediaminetetraacetic acid; FBS: Foetal bovine serum; FTIR-ATR: Fourier transform infrared attenuated total reflectance; HLEC: Human corneal limbal epithelial cell; MSG: Middle silk gland; MWCO: Molecular weight cut-off; PEG-DE: Poly(ethylene glycol) diglycidyl ether; SEM: Standard error of the mean; SDS-PAGE: Sodium dodecyl sulphate-polyacrylamide gel electrophoresis; TCP: Tissue culture plastic; WC: Water content.

Competing interests

The authors declare that they have no competing interests.

Authors' contributions

TVC designed the study, coordinated the characterization of silk materials, and wrote the manuscript. SS carried out the production and analysis of silk materials, contributed to the interpretation of results, drafted sections of the manuscript, executed and organized the graphic matter in the manuscript. LJB carried out all cell culture experiments and interpreted the results. NLB contributed to the electrophoretic analysis and to the interpretation of results, and revised critically the manuscript for general content. DGH contributed to the interpretation of cell culture results, and revised critically the manuscript for general content. All authors read and approved the final manuscript.

Acknowledgements

This work was supported by the Prevent Blindness Foundation, Queensland, Australia, through the Viertel's Vision program. LJB and DGH obtained also supplementary funding from Queensland University of Technology, Brisbane, Australia.

Author details

[1]Queensland Eye Institute, South Brisbane, Queensland 4101, Australia. [2]Faculty of Science and Engineering, Queensland University of Technology, Brisbane, Queensland 4001, Australia. [3]Faculty of Health Sciences, The University of Queensland, Herston, Queensland 4029, Australia. [4]Australian Institute of Bioengineering & Nanotechnology, The University of Queensland, St Lucia, Queensland 4072, Australia. [5]Faculty of Science, The University of Western Australia, Crawley, Western Australia 6009, Australia. [6]Faculty of Health, Queensland University of Technology, Brisbane, Queensland 4001, Australia. [7]Max Bergmann Center of Biomaterials, Leibniz Institute for Polymer Research, Dresden, Saxony 01069, Germany. [8]UQ Centre for Clinical Research, The University of Queensland, Herston, Queensland 4029, Australia. [9]Institute of Health and Biomedical Innovation, Kelvin Grove, Queensland 4059, Australia.

References

Altman GH, Diaz F, Jakuba C, Calabro T, Horan RL, Chen J, Lu H, Richmond J, Kaplan DL (2003) Silk-based biomaterials. Biomaterials 24:401–416

Aramwit P, Kanokpanont S, De-Eknamkul W, Srichana T (2009a) Monitoring of inflammatory mediators induced by silk sericin. J Biosci Bioeng 107:556–561

Aramwit P, Kanokpanont S, De-Eknamkul W, Kamei K, Srichana T (2009b) The effect of sericin with variable amino-acid content from different silk strains on the production of collagen and nitric oxide. J Biomater Sci Polym Ed 20:1295–1306

Aramwit P, Kanokpanont S, Nakpheng T, Srichana T (2010) The effect of sericin from various extraction methods on cell viability and collagen production. Int J Mol Sci 11:2200–2211

Bray LJ, George KA, Ainscough SL, Hutmacher DW, Chirila TV, Harkin DG (2011) Human corneal epithelial equivalents constructed on Bombyx mori silk fibroin membranes. Biomaterials 32:5086–5091

Bray LJ, George KA, Hutmacher DW, Chirila TV, Harkin DG (2012) A dual-layer silk fibroin scaffold for reconstructing the human corneal limbus. Biomaterials 33:3529–3538

Bray LJ, Suzuki S, Harkin DG, Chirila TV (2013) Incorporation of exogenous RGD peptide and inter-species blending as strategies for enhancing human corneal limbal epithelial cell growth on Bombyx mori silk fibroin membranes. J Funct Biomater 4:74–88

Brown SF, Coleman M (1957) Severe immediate reactions to biologicals caused by silk allergy. J Am Med Assoc 165:2178–2180

Bryant F (1948) The sericin fractions of raw silk. Textile J Aust 23:190

Celedón JC, Palmer LJ, Xu X, Wang B, Fang Z, Weiss ST (2001) Sensitization to silk and childhood asthma in rural China. Pediatrics 107:e80. doi:10.1542/peds.107.5.e80

Chirila T, Barnard Z, Zainuddin HD (2007) Silk as substratum for cell attachment and proliferation. Mater Sci Forum 561–565:1549–1552

Chirila TV, Barnard Z, Zainuddin HDG, Schwab IR, Hirst LW (2008) Bombyx mori silk fibroin membranes as potential substrata for epithelial constructs used in the management of ocular surface disorders. Tissue Eng A 14:1203–1211

Chirila TV, Harkin DG, Hirst LW, Schwab IR, Barnard Z, Zainuddin Z (2010) Reconstruction of the ocular surface using biomaterials. In: Chirila TV (ed) Biomaterials and regenerative medicine in ophthalmology. Woodhead/CRC, Cambridge/Boca Raton, pp 213–242

Clarke AJ, Meyer GP (1923) A case of hypersensiveness to silk. J Am Med Assoc 80:11–12

Collomb (l'Abbé) (1785) Observations sur la dissolution du vernis de la soie [Observations on the dissolution of the varnish of the silk]. Obs Phys Hist Nat Arts 27(II):95–107

Dewair M, Baur X, Ziegler K (1985) Use of immunoblot technique for detection of human IgE and IgG antibodies to individual silk proteins. J Allergy Clin Immunol 76:537–542

Djerassi C, Gray JD, Kincl FA (1960) Naturally occurring oxygen heterocycles: IX. Isolation and characterization of genipin. J Org Chem 25:2174–2177

Fedič R, Žurovec M, Sehnal F (2002) The silk of Lepidoptera. J Insect Biotechnol Sericol 71:1–15

Figley KD, Parkhurst HJ (1933) Silk sensitivity. J Allergy 5:60–69

Friedman HJ, Bowman K, Fried R, Weitz M (1957) Severe allergic reaction caused by silk as a contaminant in typhoid-parathyphoid vaccine. J Allergy 28:489–493

Gamo T (1982) Genetic variants of the Bombyx mori silkworm encoding sericin proteins of different lengths. Biochem Genet 20:165–177

Gamo T, Inokuchi T, Laufer H (1977) Polypeptides of fibroin and sericin secreted from the different sections of the silk gland. Insect Biochem 7:285–295

George KA, Shadforth AMA, Chirila TV, Laurent MJ, Stephenson S-A, Edwards GA, Madden PW, Hutmacher DW, Harkin DG (2013) Effect of sterilization method on the properties of Bombyx mori silk fibroin films. Mater Sci Eng C 33:668–674

Ghassemifar R, Redmond S, Zainuddin, Chirila TV (2010) Advancing towards a tissue-engineered tympanic membrane: silk fibroin as a substratum for growing human eardrum keratinocytes. J Biomater Appl 24:591–606

Gil ES, Panilaitis B, Bellas E, Kaplan DL (2013) Functionalized silk biomaterials for wound healing. Adv Healthcare Mater 2:206–217

Grzelak K (1995) Control of expression of silk protein genes. Comp Biochem Physiol 110B:671–681

Häcki M, Wüthrich B, Hanser M (1982) Widseide: ein aggressives Inhalationsallergen [Wild silk: a strong inhalation allergen]. Dtsch Med Wschr 107:166–169

Hakimi O, Knight DP, Vollrath F, Vadgama P (2007) Spider and mulberry silkworm silks as compatible biomaterials. Composites B 38:324–337

Hakimi O, Gheysens T, Vollrath F, Grahn MF, Knight DP, Vadgama P (2010) Modulation of cell growth on exposure to silkworm and spider silk fibers. J Biomed Mater Res 92A:1366–1372

Hardy JG, Scheibel TR (2010) Composite materials based on silk proteins. Prog Polym Sci 35:1093–1115

Harindranath N, Prakash O, Rao S (1985) Prevalence of occupational asthma in silk filatures. Ann Allergy 55:511–515

Harkin DG, Chirila TV (2012) Silk fibroin in ocular surface reconstruction - what is its potential as a biomaterial in ophthalmics? Future Med Chem 4:2145–2147

Harkin DG, George KA, Madden PW, Schwab IR, Hutmacher DW, Chirila TV (2011) Silk fibroin in ocular tissue reconstruction. Biomaterials 32:2445–2458

Hjortdal JØ (1996) Regional elastic performance of the human cornea. J Biomechanics 29:931–942

Horbett TA, Klumb LA (1996) Cell culturing: surface aspects and considerations. In: Brash JL, Wojciechowski PW (eds) Interfacial phenomena and bioproducts. Marcel Dekker Inc, New York, pp 351–445

Hu X, Kaplan D, Cebe P (2006) Determining beta-sheet crystallinity in fibrous proteins by thermal analysis and infrared spectroscopy. Macromolecules 39:6161–6170

Hu X, Shmelev K, Sun L, Gil E-S, Park S-H, Cebe P, Kaplan DL (2011) Regulation of silk material structure by temperature-controlled water vapour annealing. Biomacromolecules 12:1686–1696

Inoue S, Tanaka K, Arisaka F, Kimura S, Ohtomo K, Mizuno S (2000) Silk fibroin of Bombyx mori is secreted, assembling a high molecular mass elementary unit consisting of H-chain, L-chain, and P25, with a 6:6:1 molar ratio. J Biol Chem 275:40517–40528

Johansson SGO, Wüthrich B, Zortea-Caflisch C (1985) Nightly asthma caused by allergens in silk-filled bed quilts: clinical and immunologic studies. J Allergy Clin Immunol 75:452–459

Julien E, Coulon-Bublex M, Garel A, Royer C, Chavancy G, Prudhomme J-C, Couble P (2005) Silk gland development and regulation of silk protein genes. In: Gilbert LI, Iatrou K, Gill SS (eds) Comprehensive molecular insect science, vol 2. Elsevier BV, Amsterdam, pp 369–384

Kearns V, MacIntosh AC, Crawford A, Hatton PV (2008) Silk-based biomaterials for tissue engineering. Topics Tissue Eng 4:1–19

Kikkawa H (1953) Biochemical genetics of Bombyx mori (silkworm). Adv Genetics 5:107–140

Kino T, Oshima S (1979) Allergy to insects in Japan: II. The reaginic sensitivity to silkworm moth in patients with bronchial asthma. J Allergy Clin Immunol 64:131–138

Kodama K (1926) The preparation and physico-chemical properties of sericin. Biochem J 20:1208–1222

Kondo K (1921) Research on some properties of sericin [in Japanese]. J Chem Soc Japan [Nippon Kagaku Kaishi] 42:1054–1065

Kundu SC, Dash BC, Dash R, Kaplan DL (2008) Natural protective glue protein, sericin bioengineered by silkworms: potential for biomedical and biotechnological applications. Prog Polym Sci 33:998–1012

Kwan ASL, Chirila TV, Cheng S (2010) Development of tissue-engineered membranes for the culture and transplantation of retinal pigment epithelial cells. In: Chirila TV (ed) Biomaterials and regenerative medicine in ophthalmology. Woodhead/CRC, Cambridge/Boca Raton, pp 390–408

Lee HK (2004) Silk sericin retards the crystallization of silk fibroin. Macromol Rapid Commun 25:1792–1796

Lucas F, Shaw JTB, Smith SG (1958) The silk fibroins. Adv Protein Chem 13:107–242

Madden PW, Lai JNX, George KA, Giovenco T, Harkin DG, Chirila TV (2011) Human corneal endothelial cell growth on a silk fibroin membrane. Biomaterials 32:4076–4084

Mase K, Iizuka T, Okada E, Miyajima T, Yamamoto Y (2006) A new silkworm race for sericin production, "SERICIN HOPE" and its product, "VIRGIN SERICIN". J Insect Biotechnol Sericol 75:85–88

Michaille JJ, Couble P, Prudhomme J-C, Garel A (1986) A single gene produces multiple sericin messenger RNAs in the silk gland of Bombyx mori. Biochimie 68:1165–1173

Michaille JJ, Garel A, Prudhomme JC (1989) The expression of five middle silk gland specific genes is territorially regulated during the larval development of Bombyx mori. Insect Biochem 19:19–27

Minoura N, Tsukada M, Nagura M (1990) Physico-chemical properties of silk fibroin membrane as a biomaterial. Biomaterials 11:430–434

Minoura N, Aiba S, Higuchi M, Gotoh Y, Tsukada M, Imai Y (1995a) Attachment and growth of fibroblast cells on silk fibroin. Biochem Biophys Res Commun 208:511–516

Minoura N, Aiba S, Gotoh Y, Tsukada M, Imai Y (1995b) Attachment and growth of cultured fibroblast cells on silk protein matrices. J Biomed Mater Res 29:1215–1221

Mondal M, Trivedy K, Kumar SN (2007) The silk proteins, sericin and fibroin in silkworm, Bombyx mori Linn., - a review. Caspian J Env Sci 5:63–76

Mosher HH (1934) The sericin fractions of silk. Am Silk Rayon J 13:43–44

Motta A, Barbato B, Foss C, Torricelli P, Migliaresi C (2011) Stabilization of Bombyx mori silk fibroin/sericin films by crosslinking with PEG-DE 600 and genipin. J Bioact Compat Polym 26:130–143

Murphy AR, Kaplan DL (2009) Biomedical applications of chemically-modified silk fibroin. J Mater Chem 19:6443–6450

Nagura M, Ohnishi R, Gitoh Y, Ohkoshi Y (2001) Structures and physical properties of crosslinked sericin membranes. J Insect Biotechnol Sericol 70:149–153

Nakazawa T, Umegae Y (1990) Sericulturist's lung disease: hypersensitivity pneumonitis related to silk production. Thorax 45:233–234

Numata K, Kaplan DL (2010) Silk-based delivery systems of bioactive molecules. Adv Drug Deliv Rev 62:1497–1508

Panilaitis B, Altman GH, Chen J, Jin H-J, Karageorgiou V, Kaplan DL (2003) Macrophage responses to silk. Biomaterials 24:3079–3085

Pritchard EM, Kaplan DL (2011) Silk fibroin biomaterials for controlled release drug delivery. Expert Opin Drug Deliv 8:797–811

Roard J-L (1808) Mémoire sur le décreusage de la soie [Dissertation on the degumming of the silk]. Ann Chim 65:44–58

Rutherford HA, Harris M (1940) Concerning the existence of fractions of the sericin in raw silk. Textile Res 10:221–228

Salthouse TN, Matlaga BF, Wykoff MH (1977) Comparative tissue response to six suture materials in rabbit cornea, sclera, and ocular muscle. Am J Ophthalmol 84:224–233

Sehnal F (2011) Biotechnologies based on silk. In: Vilcinskas A (ed) Insect biotechnology. Gorb SN (ed) Biologically-inspired systems series, vol 2. Springer, Dordrecht, pp 211–224

Shadforth AMA, George KA, Kwan AS, Chirila TV, Harkin DG (2012) The cultivation of human retinal pigment epithelial cells on Bombyx mori silk fibroin. Biomaterials 33:4110–4117

Shelton EM, Johnson TB (1925) Research on proteins: VII. The preparation of the protein "sericin" from silk. J Am Chem Soc 47:412–418

Soong HK, Kenyon KR (1984) Adverse reactions to virgin silk sutures in cataract surgery. Ophthalmology 91:479–483

Sprague KU (1975) The Bombyx mori silk proteins characterization of large polypeptides. Biochemistry 14:925–931

Suzuki M, Itoh H, Sugiyama K, Takagi I, Nishimura J, Kato K, Mamiya S, Baba S, Ohya Y, Itoh H, Yokota A, Itoh M, Ohta N (1995) Causative allergens of allergic rhinitis in Japan with special reference to silkworm moth allergen. Allergy 50:23–27

Takasu Y, Yamada H, Tsubouchi K (2002) Isolation of three main sericin components from the cocoon of the silkworm, Bombyx mori. Biosci Biotechnol Biochem 66:2715–2718

Tashiro Y, Otsuki E (1970) Studies on the posterior silk gland of the silkworm Bombyx mori. J Cell Biol 46:1–16

Taub SJ (1930) Allergy due to silk. J Allergy 1:539–541

Terada S, Nishimura T, Sasaki M, Yamada H, Miki M (2002) Sericin, a protein derived from silkworms, accelerates the proliferation of several mammalian cell lines including a hybridoma. Cytotechnology 40:3–12

Teramoto H, Miyazawa M (2005) Molecular orientation behaviour of silk sericin as revealed by ATR infrared spectroscopy. Biomacromolecules 6:2049–2057

Teramoto H, Nakajima K, Takabayashi C (2005) Preparation of elastic sericin hydrogel. Biosci Biotechnol Biochem 69:845–847

Teramoto H, Kakazu A, Asakura T (2006) Native structure and degradation pattern of silk sericin studied by ^{13}C NMR spectroscopy. Macromolecules 39:6–8

Teramoto H, Kameda T, Tamada Y (2008) Preparation of gel film from Bombyx mori silk sericin and its characterization as a wound dressing. Biosci Biotechnol Biochem 72:3189–3196

Tokutake S (1980) Isolation of smallest component of silk protein. Biochem J 187:413–417

Tsubouchi K, Igarashi Y, Takasu Y, Yamada H (2005) Sericin enhances attachment of cultured human skin fibroblasts. Biosci Biotechnol Biochem 69:403–405

Vepari C, Kaplan DL (2007) Silk as a biomaterial. Prog Polym Sci 32:991–1007

Wang Y, Kim H-J, Vunjak-Novakovic G, Kaplan DL (2006) Stem cell-based tissue engineering with silk biomaterials. Biomaterials 27:6064–6082

Wang X, Cebe P, Kaplan DL (2009) Silk proteins - biomaterials and engineering. In: Lutz S, Bornscheuer UT (eds) Protein engineering handbook. Wiley, Weinheim, pp 939–959

Wenk E, Merkle HP, Meinel L (2011) Silk fibroin as a vehicle for drug delivery applications. J Control Rel 150:128–141

Xie R, Li M, Lu S, Sheng W, Xie Y (2007) Preparation of sericin film and its cytocompatibility. Key Eng Mater 342–343:241–244

Zeng Y, Yang J, Huang K, Lee Z, Lee X (2001) A comparison of biomechanical properties between human and porcine cornea. J Biomechanics 34:533–537

Zhaoming W, Shitai Y, Lixin Z, Yan Y (1990) Silk-induced asthma in children: a report of 64 cases. Ann Allergy 64:365–378

Zhaoming W, Codina R, Fernández-Caldas E, Lockey RF (1996) Partial characterization of the silk allergens in mulberry silk extract. J Invest Allergol Clin Immunol 6:237–241

13

Thermal, mechanical, and moisture absorption properties of egg white protein bioplastics with natural rubber and glycerol

Alexander Jones[1], Mark Ashton Zeller[2] and Suraj Sharma[1*]

Abstract

Petroleum-based plastics have many drawbacks: the large amount of energy required to produce the plastic, the waste generated as a result of plastic production, and the accumulation of waste due to slow degradation rate. It is because of these negative attributes of conventional plastic use that attention is being focused on environmentally friendly plastics from alternative sources. Albumin protein provides one possible source of raw material, with inherent antimicrobial properties that may make it suitable for medical applications. We conducted this study to investigate the various bioplastic properties of the albumin with the use of three plasticizers - water, glycerol, and natural rubber latex. Based on results, 75:25 albumin-water, 75:25 albumin-glycerol, and 80:20 albumin-natural rubber were the best blending ratios for each plasticizer for a subsequent time study to determine water stability, with the 80:20 albumin-natural rubber blend ratio having possessed the best thermal, tensile, and viscoelastic properties overall.

Keywords: Bioplastics; Albumin; Sustainability; Plasticizers

Introduction

Using conventional plastics comes with a multitude of drawbacks: the large amount of energy that is required to produce the plastic, the waste that is a result of plastic production, and the use of materials that do not biodegrade readily. In order to shift the production of plastics towards a more sustainable path, research is being conducted to determine the types of renewable bioplastic resources that could be converted into plastic form. For instance, polylactide biopolymer, one of the few resourceful polymers, is naturally produced on a large scale (Mukerjee 2011). A common theme for various bioplastics that will replace conventional plastics is their tendency to be biodegraded, compared to petroleum-based plastics that are resistant to chemical and biological attacks. According to a review study by Flieger et al. (2003), there are three groups of biodegradable polymers that can be utilized in the production of bioplastics: biopolymers by chemical synthesis,

biopolymers through fermentation process by microorganisms, and biopolymers from chemically modified natural products. Under the classification of chemically modified natural products falls the use of protein in producing bioplastics as protein must be modified chemically by the addition of plasticizers and the use of thermal treatments. The main initiative for the use of modified natural products in bioplastics is the continual drive to find more uses for agricultural commodities (Flieger et al. 2003).

It is necessary to determine the thermal and mechanical properties of bioplastics produced from protein as this will help identify the process by which the bioplastic should be made as well as what applications the resulting plastic will be suitable for. In a study by Sharma et al. (2008), they determined that the albumin from chicken egg white denatures at a temperature of 136.5°C ± 3°C. This indicates that in order to produce plastic from the chicken egg white albumin, the material must be molded at 136.5°C ± 3°C to ensure that the protein will be denatured and able to orient and form a bioplastic. When the tensile properties of the protein-based plastics were measured, it was determined that

* Correspondence: ssharma@uga.edu
[1]Department of Textiles, Merchandising and Interiors, University of Georgia, Athens, GA 30602, USA
Full list of author information is available at the end of the article

the breaking of hydrophobic interactions and hydrogen bonds of the bioplastics initiated a reversible yield point (Sharma et al. 2008). This reversal of the yield point allows tensile stress to be placed onto the bioplastic multiple times as long as the breaking point is not reached.

In order to determine the potential uses of protein-based bioplastics, thermal and mechanical properties must be examined. Since the protein-based bioplastics require a lower processing temperature and possess tensile properties similar to plastics like high-density polyethylene (HDPE), it is possible to manufacture a bioplastic at a lower production cost (Jerez et al. 2007). However, one potential drawback of using protein in plastics is its hygroscopic properties as it was determined that the water absorption of various bioplastics ranged from 40% to 320% (Jerez et al. 2007). This tendency for bioplastics to absorb water may result in plastic with lower elasticity as the moisture content may alter the elastic modulus of the resulting plastic. Another potential drawback that arises when using protein-based (or any polymer) materials is the lack of knowledge about how the materials will react to bacteria in their environment. In a study by Hook et al. (2012), they determined that certain polymers were actually better at preventing microbial growth than either the silicone or silver hydrogel coatings that are commonly applied to medical devices before being implanted into humans. One other potential drawback with bioplastics is permanent deformation when stress is applied (Widiastuti et al. 2013). Although researchers have utilized composites to decrease the amount of deformation and creep, research must still be conducted in order to examine both the positive and negative properties of protein bioplastics in order to determine if they are suitable for certain applications (Dorigato and Pegoretti 2012).

Certain constituents of albumin pose a potential advantage of possessing inherent antibacterial properties allowing for potential pharmaceutical and medicinal uses. Albumin possesses this property most notably from its lysozyme enzyme constituent, utilizing a lysis reaction in which it will break down the peptidoglycan barrier of bacteria consisting of the glycosidic (1–4) β-linkage between the N-acetylglucosamine and the N-acetylmuramic acid (Baron and Rehault 2007). It is also possible to improve the antimicrobial properties of the lysozyme enzyme through the use of chemical modifications. There are various preservatives such as nisin and sodium lactate as well as substances such as ethylenediaminetetraacetic acid, butylparaben, and trisodium phosphate that can be added to the lysozyme to enhance its properties (Cegielska-Radziejewska et al. 2010). The utilization of albumin in medical plastic production would not be an extension on what chicken albumin is being used for today. It is because of inherent antibacterial enzymes and improvement

through chemical modification that albumin is already used in various medical applications, such as circulatory support, drug delivery, and the removal of toxins from the body (Peters 1996). Another area in which albumin bioplastics or its thermoplastic blends (with petroleum-based polymers/biopolymers/biodegradable polymers) could be utilized is in drug elution as well as orthopedic implants and sutures as it would serve as a material that could release a low dose of medicine over a period of time in the body while limiting the risk of infection (Zilberman and Elsner 2008). Our objectives when conducting this study were to modify the properties of albumin plastics through the use of glycerol and natural rubber latex and to evaluate the physical performance of these plastics as time progressed.

Methods
Materials
The albumin (purity ≥99%) utilized in the production of bioplastics was obtained from Sigma-Aldrich Corporation (St. Louis, MO, USA). The plasticizers used to form the bioplastics were obtained through various sources: deionized water was obtained through filtering water in the lab, glycerol was obtained from Sigma-Aldrich with a purity ≥99%, and natural rubber latex (70% solid, 30% water mixture with a pH of 10.8) was obtained from the Chemionics Corporation (Tallmadge, OH, USA).

Preparation of compression-molded samples
Molding of the albumin-based bioplastic blends were performed on a 24-ton bench-top press (Carver Model 3850, Wabash, IN, USA) with electrically heated and water-cooled platens. The stainless steel molds were custom made to form either dog bone-shaped bioplastics for mechanical analysis or two small rectangular flex bars for various property analyses. Data presented in this study was generated from compression-molded samples using a 5-min cook time at 136.5°C followed by a 10-min cooling period, under a pressure of at least 40 MPa as a certain minimum amount of pressure was required in order to mold plastic (Sue et al. 1997). The bioplastic blends were prepared in small batches of ≤6 g and then poured into the molds at a constant weight, with dynamic mechanical analysis (DMA) flex bars made of 2 g and dog bones made of 6 g of albumin powder. After the samples were cooled for 10 min under pressure, the pressure was released and the samples were removed. The samples were then placed in a conditioning chamber for at least 24 h, unless otherwise noted. The conditioning chamber was set to 21.1°C and 65% relative humidity.

Weight change and moisture content analysis

The bioplastic samples were placed in conditioning chamber settings to determine moisture content over time - initial, 1, 2, 3, 4, 5, 6, 24, 48, 72, and 96 h after molding. In order to ensure accurate measurements, four DMA flex bars were prepared and analyzed during this process. Moisture content of plastics were analyzed by cryocrushing bioplastics with liquid nitrogen for each blend type ($n = 4$) and heated at 80°C for 1 h, with 10 min of cooling afterwards. The equation used to determine the moisture content was as follows:

$$MC = [(W_0 - W_{0d})/W_0] \times 100,$$

where W_0 = initial weight of specimen and W_{0d} = weight of specimen after drying.

Dynamic mechanical analysis

After conditioning, DMA flex bars were analyzed for their viscoelastic properties through the use of dynamic mechanical analysis (Menard 1999) using a DMA 8000 dynamic mechanical analyzer from PerkinElmer (Branford, CT, USA) starting at a temperature of 25°C and ending at a temperature of 160°C, with a temperature ramp of 2°C min^{-1}. The settings of the analyzer were set to dimensions of $9 \times 2.5 \times 12.5$ mm^3 using a dual cantilever setup at a frequency of 1 Hz with a displacement of 0.05 mm. Each sample type was analyzed in duplicate ($n = 2$) to ensure accuracy. The DMA flex bars were also tested in intervals of immediate, 24 h, and 5 days after molding in order to determine the viscoelastic properties over time.

Thermal analysis

Thermal gravimetric analysis (TGA) was performed using a Mettler Toledo TGA/SDTA851e (Columbus, OH, USA), and differential scanning calorimetry (DSC) was performed using a Mettler Toledo DSC821e. TGA was performed from 25°C to 800°C under N$_2$ atmosphere with a heating rate of 10°C min^{-1}. DSC was performed from −50°C to −250°C under N$_2$ atmosphere with a heating rate of 20°C min^{-1}. All samples ($n = 2$) were prepared with weights between 2.0 and 4.0 mg as the samples were cut from DMA flex bars for each blend. TGA and DSC tests were conducted in intervals of immediate, 24 h, and 5 days after molding.

Scanning electron microscopy

Albumin scanning electron microscopy (SEM) samples ($n = 2$ for each plastic type) were prepared from cryogenic DMA flex bar fracture surfaces after being placed in a conditioning chamber setting for at least 24 h. DMA flex bars were submerged in liquid nitrogen for 20 s; after that, they were immediately broken. The samples were mounted, then sputter-coated for 60 s with an Au/Pt mix. SEM images were recorded on a Zeiss 1450EP (Carl Zeiss AG, Oberkochen, Germany) variable pressure scanning electron microscope. Coated samples were analyzed at × 20, ×100, and × 500 for each blend type.

Mechanical properties

The mechanical properties of the conditioned albumin bioplastics were measured using the Instron testing system (Model 3343, Instron Corporation, Norwood, MA, USA) interfaced with the Blue Hill software. The test was performed according to the standard test method for tensile properties of plastics (ASTM D 638–10, type I) with a 5-mm min^{-1} crosshead speed, a static load cell of 1,000 N, and a gauge length of 4 cm. Samples were run in quintuplicate ($n = 5$) for each blend type in order to ensure accurate measurement.

Statistical methods

Statistical analyses of data were generated for moisture content analysis and mechanical property analysis through the use of power analysis. For each plastic type tested, statistical values based on the mean and standard deviation were generated, with p values (0.05 or less) compared to plastic types based on properties being tested generated from Student's t test distribution. For moisture content analysis, correlation analysis was also conducted (1 = perfect positive correlation, 0 = no correlation, −1 = perfect negative correlation).

Results and discussion
Initial material analysis
Thermal properties of albumin

An initial degradation peak was shown between 220°C and 230°C, with a much larger peak starting from 245 to 250°C, and 93% of the albumin powder degraded by the end of the TGA run (Figure 1). These results were similar to the results obtained in the work conducted by Sharma and Luzinov (2012). For DSC data, the endothermic dip began at 75°C with a broad peak between 120°C and 125°C. This indicated that the material had fully passed its transition phase - denaturation. An endothermic decomposition or pyrolysis peak occurred at 250°C, which exhibited the onset of degradation. Therefore, the albumin-based bioplastics were molded at 136.5°C as this was the safe temperature of processing albumin into the plastics with as little degradation occurring as possible. Based on the albumin being fully denatured between 120°C and 125°C without degradation, it was determined that the plastics were to be molded higher than this temperature but below temperatures where degradation occurs (Figure 1).

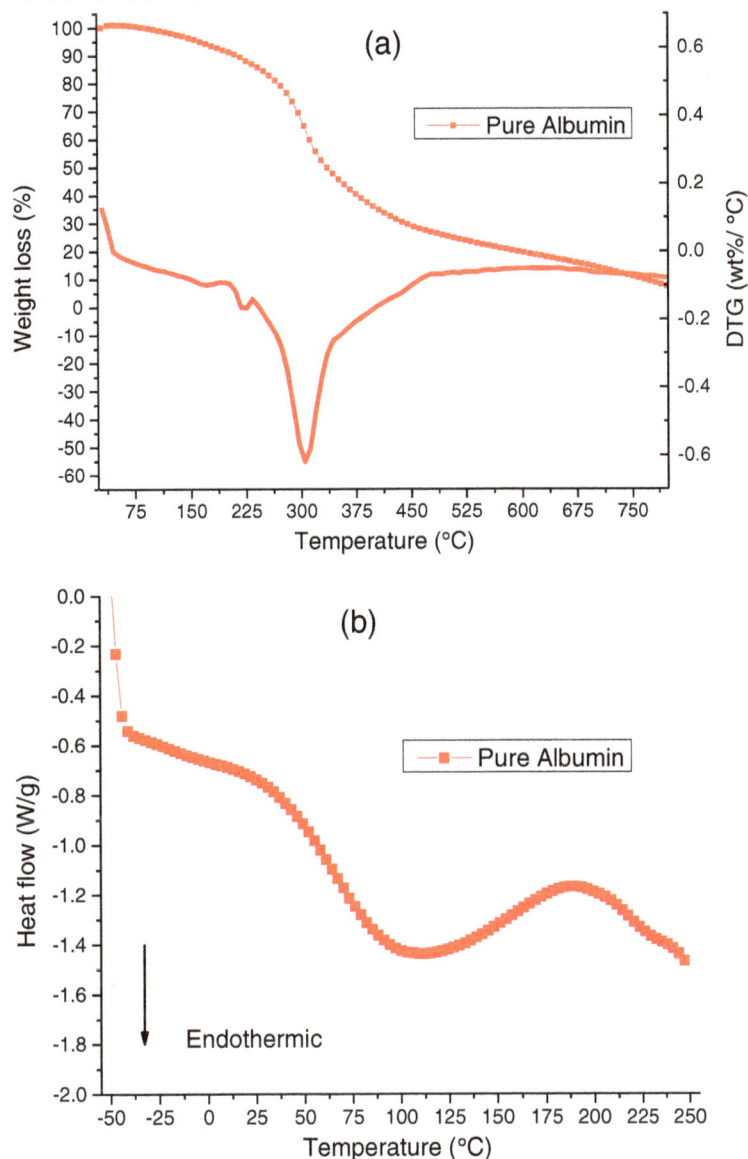

Figure 1 Thermographs of pure albumin powder. (a) TGA and **(b)** DSC.

Dynamic mechanical analysis

In plastics with water as a plasticizer, we found that as the amount of water was increased, the initial modulus of the resulting plastics decreased, with the tanδ peak occurring at 70°C (Figure 2a). This was consistent with the research conducted by González-Gutiérrez et al. (2011). The increased water content caused an increase in the initial tanδ values as well as caused the tanδ peaks to shift to the left (or lowered glass transition temperature) and occurred at lower temperatures, which indicated increased viscous heat dissipation. The shifted curves indicated that the 75:25 albumin-water formulation was the most desirable of the blends examined as this formulation possessed the mix of a modulus that was comparable to the other water plasticized samples

(and higher than the 70:30 albumin-water formulation), while possessing an elasticity (tanδ) that was much higher than the other formulations (and equal to the 70:30 albumin-water formulation), as shown in Figure 2a. The same trends occurred in the albumin plastics that had glycerol as a plasticizer - the higher percentage led to the higher initial tanδ and lower modulus as well as the shifting of the tanδ peaks to the left (Figure 2b). However, at lower content of both water and glycerol, the bioplastics showed anti-plasticization and plasticization phenomena (Galdeano et al. 2009). Based on the results, we determined that the 75:25 albumin-glycerol ratio was the composition with the highest overall tanδ peak as well as moderate modulus values (Figure 2b).

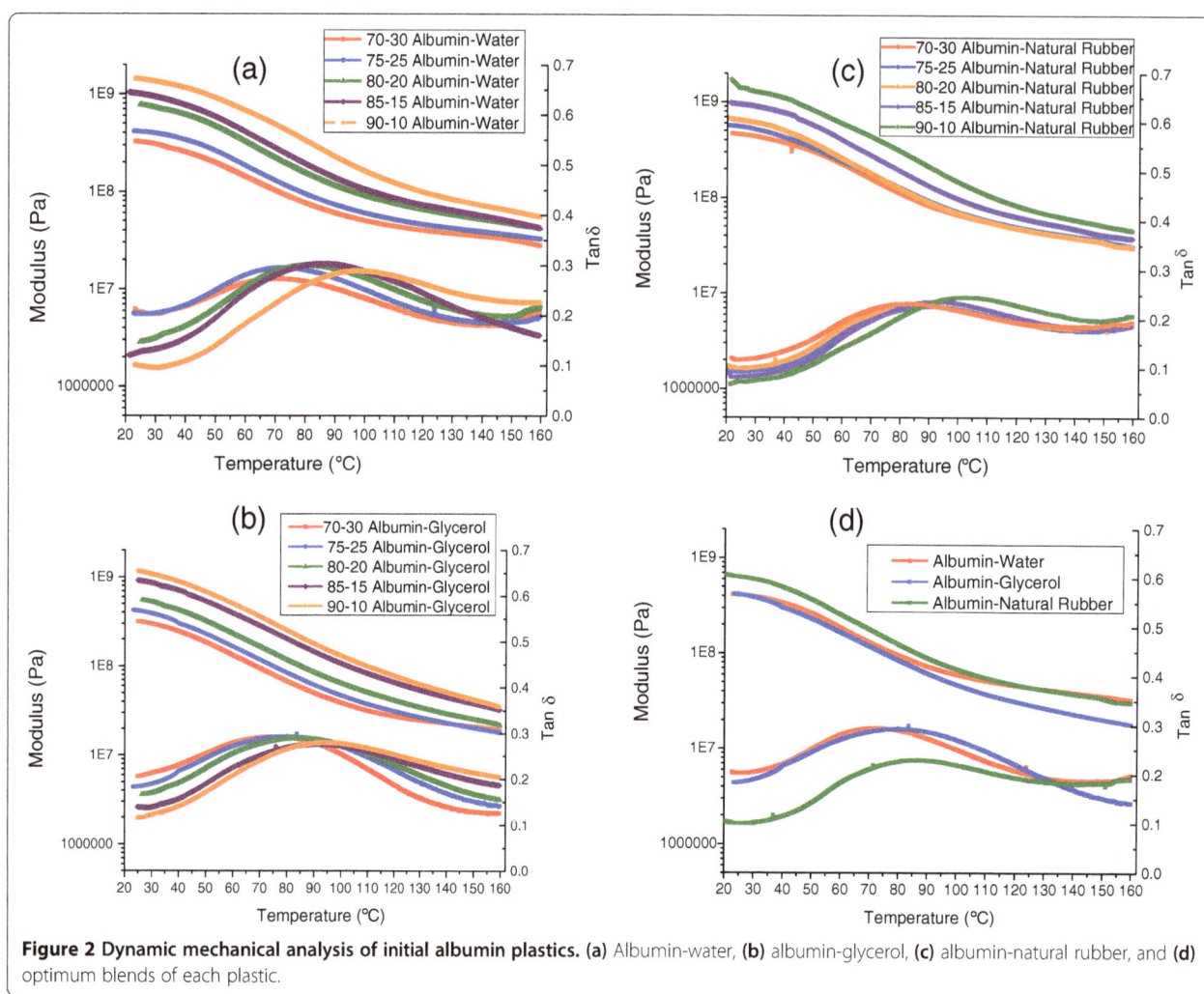

Figure 2 Dynamic mechanical analysis of initial albumin plastics. (a) Albumin-water, (b) albumin-glycerol, (c) albumin-natural rubber, and (d) optimum blends of each plastic.

For the albumin plastics with natural rubber latex as the plasticizer, we observed the same trends, although there was very little difference in the initial tanδ values (Figure 2c). The 80:20 albumin-rubber formulation possessed the optimal mix of high initial modulus and tanδ as its tanδ values were comparable to the 70:30 and 75:25 albumin-rubber ratios. However, the 80:20 albumin-rubber bioplastics possessed a higher initial modulus while having a tanδ peak at a lower temperature than the bioplastics that contained lower weights of rubber (Figure 2c). When we compared the plastics based on the types of plasticizer used, we found that the initial modulus was similar for all three plasticizers, but the natural rubber-based bioplastics exhibited the lowest initial tanδ values, whereas other plasticizers (water and glycerol) showed highest viscous heat dissipation (Pommet et al. 2005). With this analysis completed, it was determined that the optimum blends for albumin plastic production were 75:25 albumin-water, 75:25 albumin-glycerol, and 80:20 albumin-rubber (Figure 2d).

Time study
Bioplastic moisture content analysis
The 75:25 albumin-water bioplastics demonstrated a large decrease in moisture content within 48 h of molding, losing an average between 15% and 17% of its initial moisture content (Figure 3). However, after the initial loss of moisture content, the plastics maintained a stable moisture content after 96 h of molding. This loss in moisture content was due to the loss of water from the bioplastic as time progressed, producing a stiffer and more brittle plastic (Van Soest and Knooren 1997). While the water-based bioplastics lost moisture content, the 75:25 albumin-glycerol formulation steadily increased in moisture content after molding, reaching 10% in moisture content 96 h after molding as the time progression correlated (0.978) with moisture content growth. The gain in moisture content was due to a combination of glycerol leaching from the bioplastics and the plastic absorbing ambient moisture. This could render this plastic unsuitable for most applications as the resulting water absorption would alter the properties of

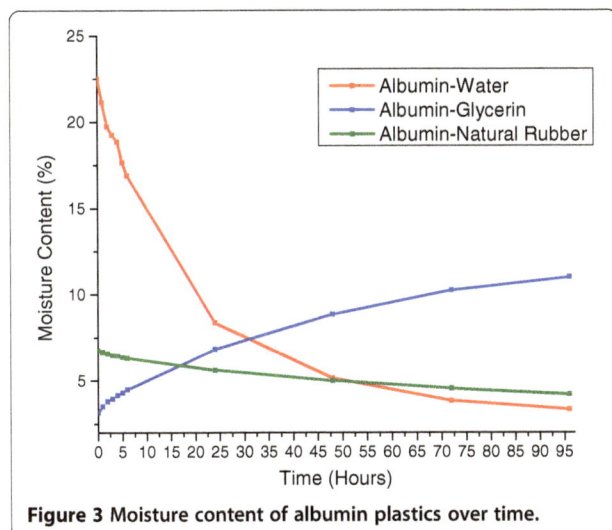

Figure 3 Moisture content of albumin plastics over time.

the plastic, reducing its tensile strength. While the water- and glycerol-based plastics underwent a comparatively large change in moisture content during the study ($p = 0.002$), the 80:20 albumin-rubber bioplastics were more stable in terms of moisture content, maintaining a moisture content between 5% and 7% on average throughout the study, as it was significantly equivalent to the albumin-glycerol values ($p = 0.472$). This moisture content stability was due to natural rubber being stable and nonreactive in the environment, with the only moisture loss due to the loss of water that was in the natural rubber latex molding base (Carvalho et al. 2003).

Based on our findings in the study, it was determined that the natural rubber provided the most stability in terms of moisture content as the other plasticizers either lost moisture over time (water-based bioplastic) or gained moisture (glycerol-based bioplastic) (Figure 3).

Bioplastic dynamic mechanical analysis

The 75:25 albumin-water plastics showed the most significant amount of change: the modulus drastically increased after 5 days of conditioning (initial = 2.5E8 Pa, 5 days = 1.8E9 Pa) with lower initial tanδ values (initial = 0.23, 5 days = 0.09), and the tanδ peak shifted to the right (Figure 4a; Van Soest and Knooren 1997). These characteristics pointed to the unbound water being released over time when it was placed in ambient conditions, which reduced the ability of water to plasticize, as shown in past studies (Verbeek and van den Berg 2010). The change in properties of these plastics over time was most likely due to the amount of moisture loss that occurred over time, which resulted in a stiff plastic. This drastic change in the properties of albumin-water plastics over time pointed to a lack of usability in the material as materials must maintain consistent properties for more than a short period of

time (Figure 4a). The 75:25 albumin-glycerol plastics demonstrated almost completely opposite results, though the amount of change over time was not as drastic (Figure 4b). Conditioning for the glycerol-based plastics led to the lowering of initial modulus (initial = 5.4E8 Pa, 5 days = 2.6E8 Pa) and a slight increase in tanδ (initial = 0.17, 5 days = 0.21) as well as the general lowering and shifting to the left of the tanδ peak. These changes in viscoelastic properties were most likely due to the gradual leaching of glycerol from the plastic, with ambient moisture taken in to replace it, weakening the hydrogen bonds within the plastic in the process (Lodha and Netravali 2005). When it came to the 80:20 albumin-rubber plastics, there was very little change that occurred in the plastic after it was allowed to condition for 24 h as the tanδ and modulus values were essentially identical after the plastics were given time to condition (Figure 4c). This lack of change in properties may have been due to natural rubber lacking the ability to react to the environment; if given enough time, it will remain bonded to the albumin and maintain its basic properties (Carvalho et al. 2003).

When the plasticizers are compared, the 75:25 albumin-water plastics gradually became the most brittle over time as their modulus increased and tanδ decreased. The 75:25 albumin-glycerol plastics underwent the exact opposite in properties as they initially started as the most brittle of the plastics but at the end of the experiment became the most ductile due to its loss of modulus and maintaining its initial tanδ value. As for the 80:20 albumin-natural rubber plastics, they maintained the most consistent among the plastics as after the decrease of tanδ and increase of modulus after 24 h, their viscoelastic properties normalized (Figure 4d,e,f).

Bioplastic thermal analysis

In the 75:25 albumin-water plastics, the glass transition temperature range of 40°C to 60°C was more evident in the 24-h and 5-day samples (Figure 5a), while for all three samples, an endothermic peak was seen around 225°C. This peak in the albumin-water bioplastics could have been attributed to the degradation or decomposition of protein polymers. For the 75:25 albumin-glycerol plastics, the glass transition phase of 50°C to 110°C was also very noticeable, with a small dip beginning at180°C (due to glycerol) and a larger endothermic peak between 215°C and 220°C for bioplastics that had been molded on the same day (Figure 5b). The larger endothermic peaks were at 250°C for the 24-h and 5-day samples. This shift in protein decomposition to higher temperature could have been attributed to the absorption of moisture and reorganized polymer chains due to the displacement of unbound glycerol molecules (Chen et al. 2005). As for the 80:20 albumin-rubber samples, a much clearer glass transition of 40°C to 80°C occurred

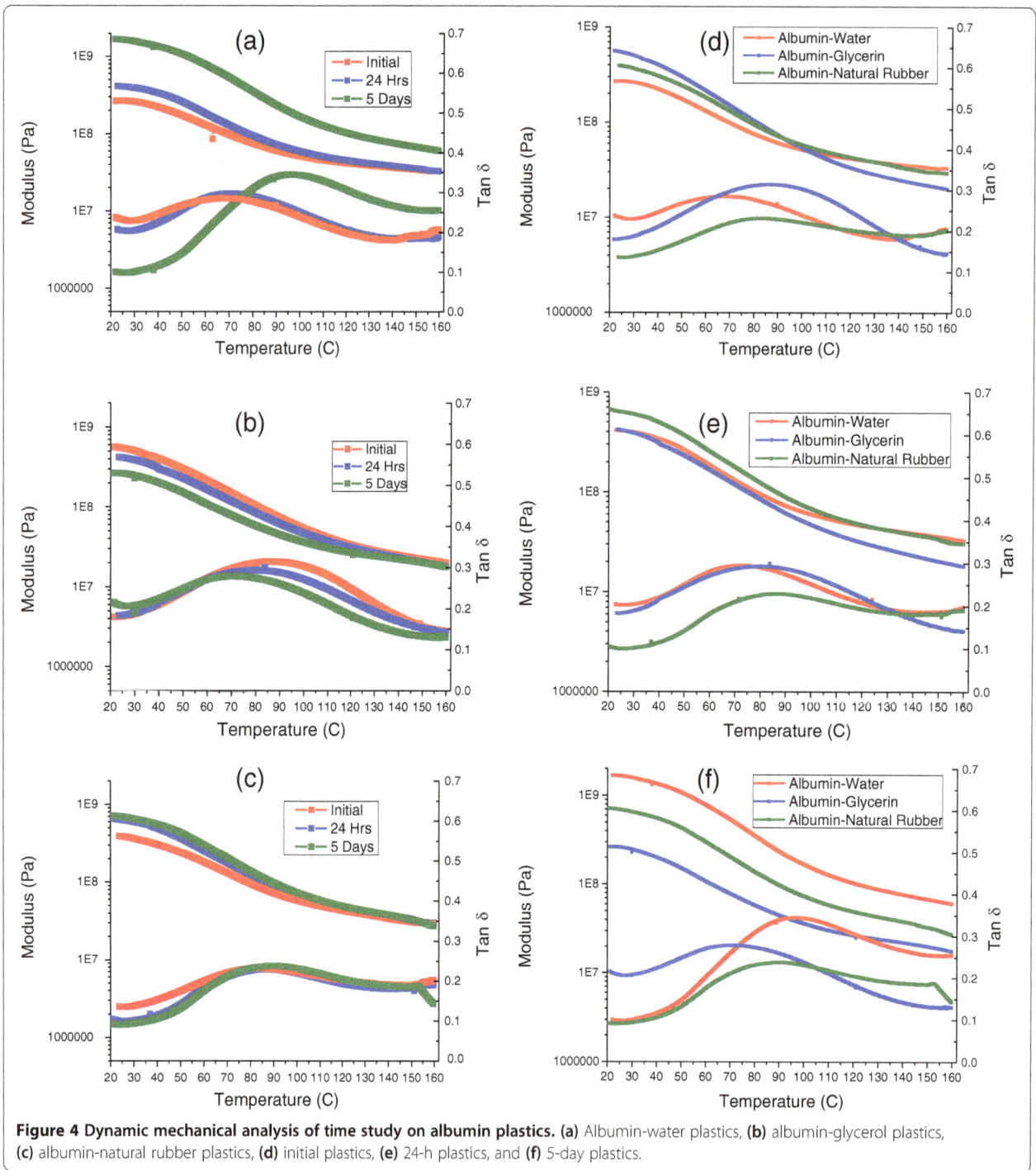

Figure 4 Dynamic mechanical analysis of time study on albumin plastics. (a) Albumin-water plastics, **(b)** albumin-glycerol plastics, **(c)** albumin-natural rubber plastics, **(d)** initial plastics, **(e)** 24-h plastics, and **(f)** 5-day plastics.

with the 24-h and 5-day samples in comparison with the initial sample, with all three samples beginning to have an endothermic decomposition peak at 225°C (Figure 5c).

When the plastics were compared with each other, it was found that the natural rubber-based bioplastics underwent a much more noticeable endothermic enthalpy change initially, but after conditioning, it recovered to normal glass

transition phase, similar to that in other plasticizers used (Figure 5d,e,f).

In terms of the thermogravimetric analysis, we found that the amount of time after molding did not have an effect on the amount of mass loss at higher temperatures as all of the curves were similar depending on the type of plasticizer used. The 75:25 albumin-water plastics possessed one degradation peak at 300°C, where the

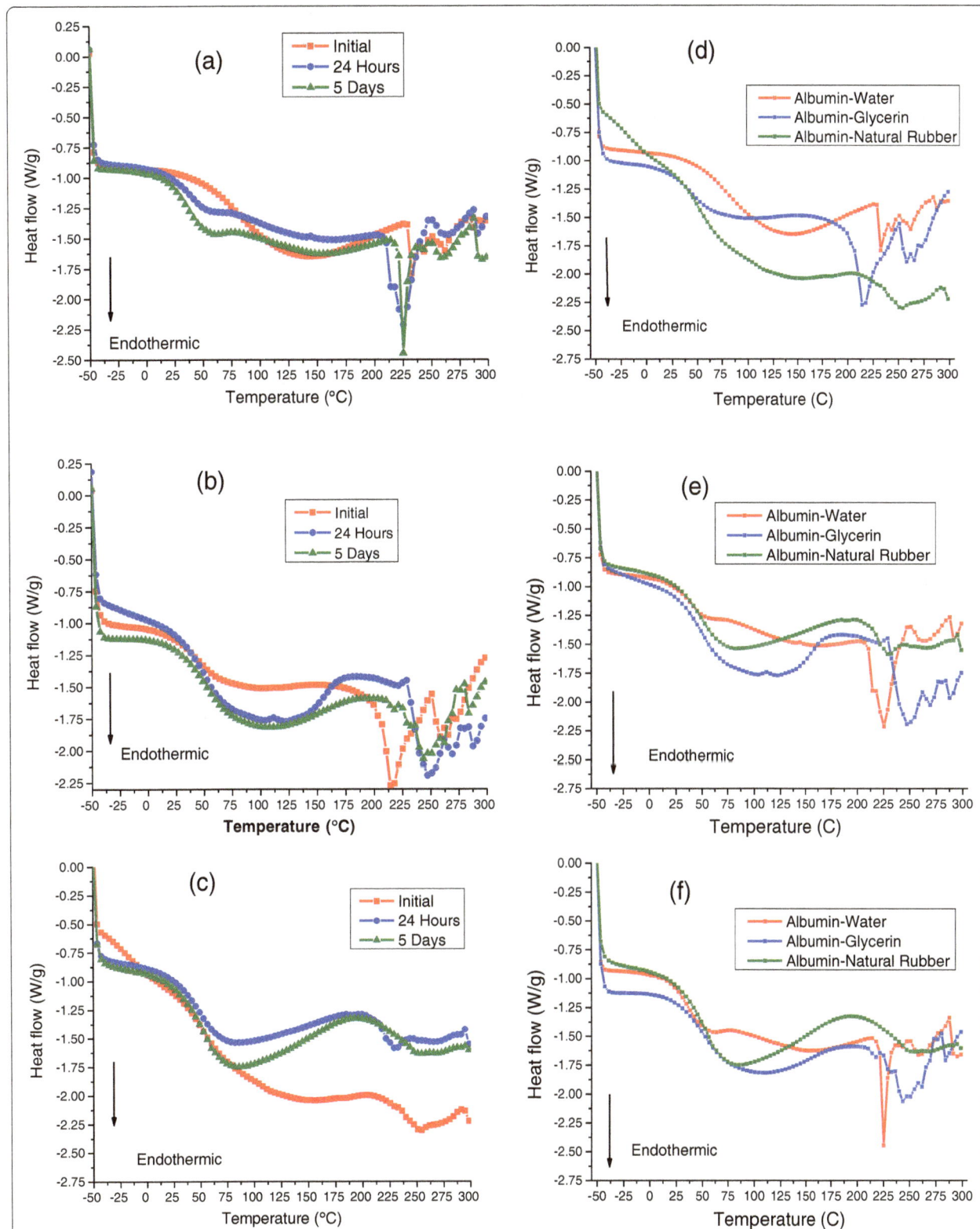

Figure 5 Differential scanning calorimetry of time study on albumin plastics. (a) albumin-water plastics, **(b)** albumin-glycerol plastics, **(c)** albumin-natural rubber plastics, **(d)** initial plastics, **(e)** plastics after 24 h, and **(f)** plastics after 5 days.

protein within the plastic begins to degrade, while a small drop at the beginning of each curve was due to moisture loss (Figure 6a). This moisture loss was more evident in the initial plastic, as this sample had the highest amount of moisture, resulting in a mass drop between 50°C and 100°C. For the 75:25 albumin-glycerol

plastics, there was one bimodal degradation peak for all of the samples (Figure 6b). The first peak began at 225° C, which was most likely due to glycerol degradation (flashpoint of glycerol is around 180°C, with mass loss occurring in nitrogen gas environments at 199°C (Castelló et al. 2009)) within the bioplastic; the peak was

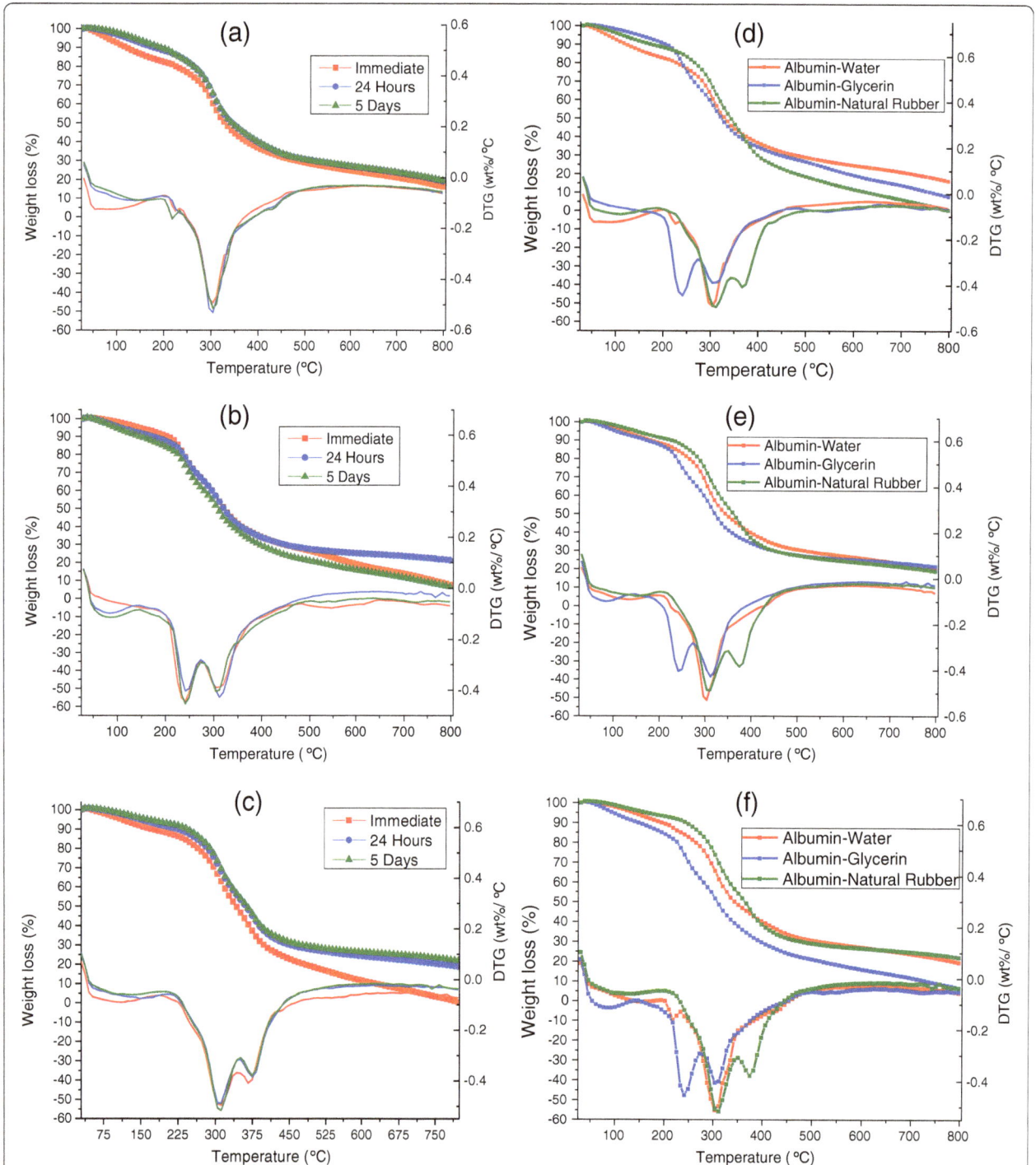

Figure 6 Thermogravimetric analysis of time study on albumin plastics. (a) Albumin-water plastics, (b) albumin-glycerol plastics, (c) albumin-natural rubber plastics, (d) initial plastics, (e) plastics after 24 h, and (f) plastics after 5 days.

right-shifted, most likely due to stabilization in the albumin matrix. The second peak between 300°C and 325°C was most likely due to the albumin protein itself degrading. The 80:20 albumin-rubber samples also possessed a bimodal degradation peak, although we found that the peaks were in different temperature ranges compared to one in glycerol-plasticized bioplastics (Figure 6c). For instance, the first peak seen at 300°C was the initial degradation of the protein in the plastic. However, the second peak seen at a higher temperature at 375°C was most likely due to natural rubber latex degradation, with the rubber degrading between 350°C and 360°C (Mathew et al. 2001). The initial sample of natural rubber latex sample also degraded at a higher rate and a slightly lower temperature due to the water contained in the natural rubber latex molding base still being present in the plastic.

When all three plasticizers are compared, we found that as time progressed, the albumin-glycerol plastics possessed the lowest onset degradation temperature, which may have been due to the increase of moisture in the plastic leading to destructurization (Figure 6d,e,f). The albumin-water and albumin-natural rubber plastics showed higher onset degradation temperatures. This was most likely due to the natural latex making cross-links with plastic matrix and the water-based plastic becoming stiffer over conditioning time due to restructured polymer chains. Once the degradation of the plasticizer had occurred, however, it was found that the water and natural rubber plastics would reach about the same rate of mass loss, with the glycerol plastics losing almost all of their mass over the time of the experiment.

Overall, based on the DSC data, it was found that after conditioning for 24 h, the plastics would maintain consistent values. As for the TGA data, it could have been determined that the albumin-natural latex plastics had the highest onset degradation temperatures, while the 75:25 albumin-glycerol plastics had the lowest temperature for degradation onset.

Tensile properties of bioplastics

In terms of the amount of extension that occurred before breaking, the 75:25 albumin-water plastics possessed by far the highest amount of extension, extending nearly 200% on average before a ductile break; the other plastics extended around only 75% before breaking (Figure 7) as the glycerol and natural rubber plastics were statistically undistinguishable ($p = 0.943$). One possible reason why water was able to facilitate a higher extension (but not load bearing) could have been due to the bonding that would occur within the structure as plasticization occurred, which had been found in previous research that had been conducted (Pommet et al. 2005; Verbeek and van den Berg 2010). As for the load that was required to

break the bioplastics and the modulus of the bioplastic itself, we found that the 80:20 albumin-rubber plastics required a much higher load to break the samples (around 12 MPa) and inherently had a much higher modulus near 60 MPa. For the 75:25 albumin-water and 75:25 albumin-glycerol plastics, the maximum loads that we observed were undistinguishable from each other as a p value of 0.757 illustrates. This ability for the albumin-rubber plastics to undergo a high load before plastic deformation may have been due to the natural rubber providing a more load-bearing material in the structure of the plastic, counteracting any potential losses due to long-range plasticization prevention. When the rubber serves as a load-bearing constituent of the plastic, it was possible for the plastic to undergo a higher amount of stress before breaking (Carvalho et al. 2003). Comparing each of the plastic types overall, it was evident that the ductile 75:25 albumin-water and the 80:20 albumin-rubber plastics were the best types of bioplastic to use, depending on the intended use, as the water-based samples allowed large amounts of extension before breaking, while the rubber-based samples were stiff and required the highest amount of load needed to break the samples. As for the brittle 75:25 albumin-glycerol plastics, there was very little benefit in terms of tensile properties as it possessed neither the extension nor the strength that the other plastics possess. The weak tensile properties of the albumin-glycerol plastics could have been explained through disordered conformations as the relatively large chemical structure of glycerol prevented any long-range plasticization to occur (Aman Ullah et al. 2011). When long-range plasticization was prevented, there was a limit on how much force a plastic could have undergone; as when the short polymer chains were broken under stress, it resulted in a violent break at a lower stress than a long chain polymer.

When we examined the tensile properties obtained through this study, we found that the results were similar to the results obtained in the study by Jerez et al. (2007). In their study, they determined that the lower processing temperature of the albumin bioplastic molding process resulted in modulus values that were similar to those of polymers, such as low-density polyethylene (LDPE) and HDPE (Jerez et al. 2007). This discovery made it possible for the use of albumin plastics in place of LDPE and HDPE in certain applications as the reduction of protein as a total percentage of the blend and the lower processing temperature would lower the costs of plastic production.

Based on the results, we determined that the 80:20 albumin-natural rubber plastics were able to undergo the greatest amount of stress, while the 75:25 albumin-water plastics were able to undergo the greatest amount of strain before breaking (Figure 7).

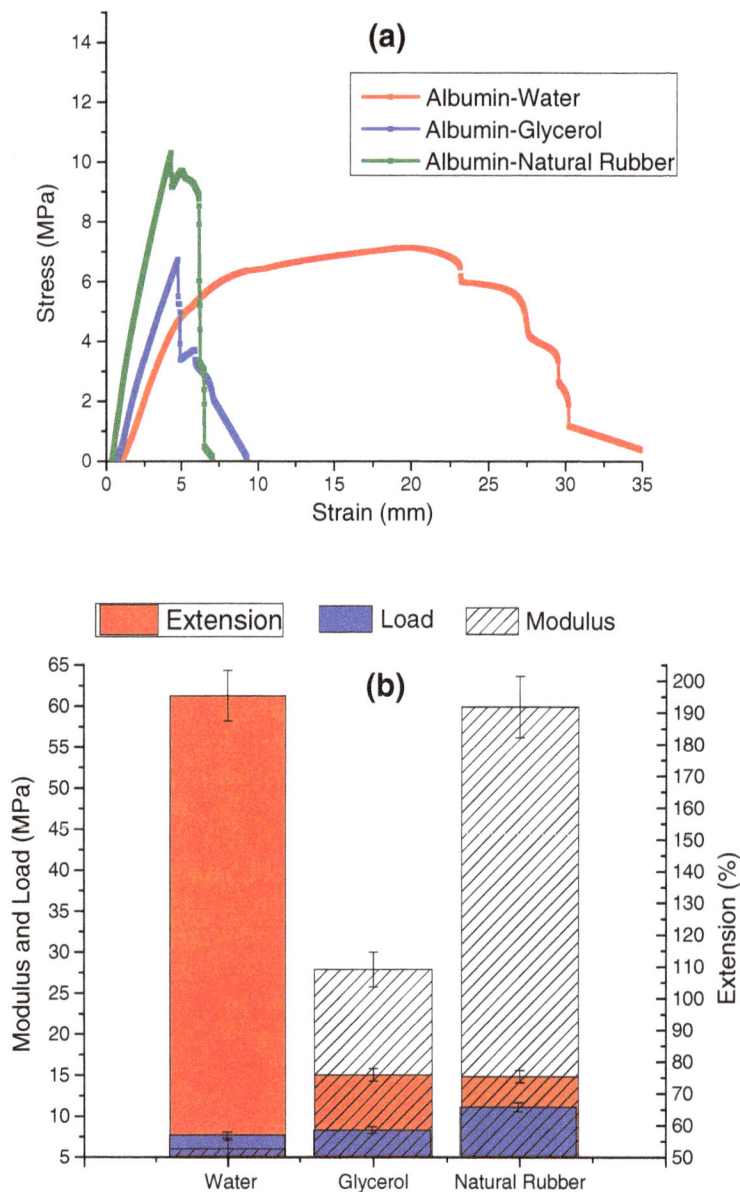

Figure 7 Tensile properties of time study on albumin plastics after 24 h of conditioning. (a) Stress–strain curve. **(b)** Modulus, load, and extension chart.

Scanning electron microscopy images of bioplastics

When the 75:25 albumin-water plastics were analyzed through scanning electron microscopy, the pictures illustrated that when the plastic was broken, a scratched and pitted surface was the result, pointing to the fact that the break was not clean and that mixture of the albumin and the plasticizer was fairly homogenous (Figure 8a). This scratching indicated the high level of toughness needed to break the sample, which could also have been seen in the high extension of the albumin-water bioplastics (Figure 7). For the 75:25 albumin-glycerol bioplastics, the spotted surface of the broken plastic was the aspect of highest interest as these images could have

been the evidence of glycerol leaching from the plastic on a much smaller scale, with the moisture being removed from the sample when the SEM chamber was sealed under high vacuum (Figure 8b). As the glycerol slowly leached into the environment, pores would form inside the plastic, with moisture from the environment causing the pores to absorb water. When the SEM chamber was sealed and moisture was vacuumed from the testing chamber, cracks formed in the plastic because the moisture was being removed from the plastic. With the 80:20 albumin-rubber bioplastics, what was most evident with these pictures was the lack of homogenous mixture of the albumin and the rubber as there were

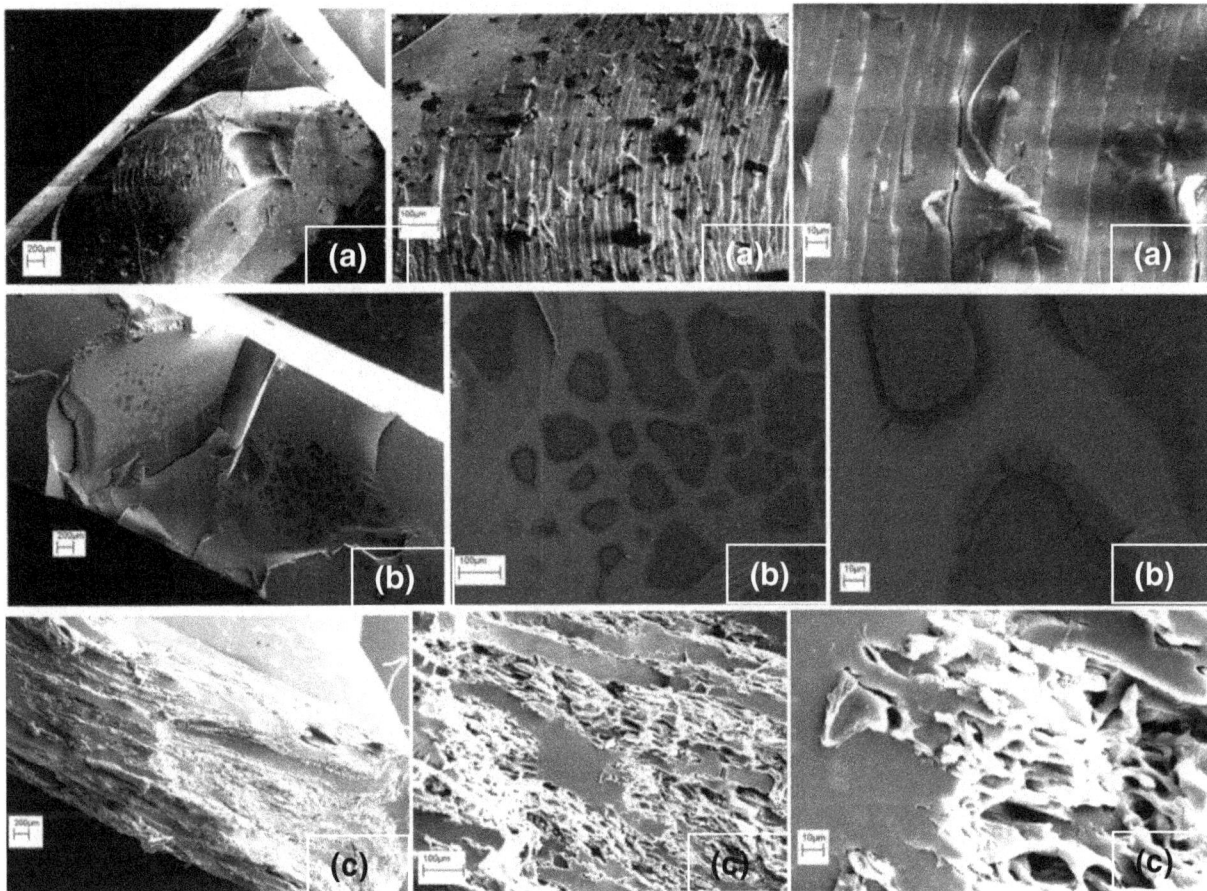

Figure 8 Scanning electron Microscopy images of albumin bioplastics. (a) Albumin-water, **(b)** albumin-glycerol, and **(c)** albumin-natural rubber bioplastics. Magnification × 20, ×100, and × 500.

pockets of albumin and pockets of rubber throughout the whole plastic sample (Figure 8c). The jagged surface of the plastic also illustrated the amount of force that was required to break the plastic as the break would have been sudden and drastic (Carvalho et al. 2003). This abrupt breaking point was also illustrated in the tensile strength and modulus results (Figure 7); the latex was able to hold the plastic together until giving way under a high load. Based on the results of this analysis, we determined that it was glycerol leaching in the 75:25 albumin-glycerol plastics that altered the properties of the plastic, while multiple phases of material in the 80:20 albumin-natural rubber plastics could have been observed.

Conclusions

When comparing the amounts of plasticizer used in molding bioplastics from albumin, we found that the blending ratios that had the best combination of initial modulus and elasticity were the 75:25 albumin-water, 75:25 albumin-glycerol, and 80:20 albumin-natural Rubber. Of these blends, we found that the 80:20 albumin-

natural rubber bioplastic provided the best thermal, viscoelastic, and tensile properties, while properties of the 75:25 albumin-water and 75:25 albumin-glycerol plastics were not comparable due to moisture loss in the water-based bioplastics over time, while glycerol leaching occurred in the glycerol-based bioplastics as time passed. There are multiple avenues of interest that should be examined in the light of the knowledge gained in this study. In order to determine whether the albumin-natural rubber (as well as albumin-water for short-term uses) plastics would be suitable for medical applications or not, further studies are needed to ensure that the plastics will inhibit bacterial growth in order to prevent post-operation infection as well as potentially aid in the application of drugs through elution. Another area of interest requiring further research is the continued modification of the blending by altering the percentages of each component used as well as adding different materials into the blends in order to reach the optimum properties for other applications. As for the aspect of moisture content, it would be beneficial to examine the possible uses of other materials that would limit the amount of moisture content change that would occur with albumin plastics as it has been demonstrated that this area has a significant effect on the overall properties of the plastic.

Competing interests

The authors declare that they have no competing interests.

Authors' contributions

AJ conducted all experimental studies and statistical analyses as well as helped the corresponding author, SS, in preparing the manuscript draft. SS was the lead researcher in conceptualizing the idea of initiating this research followed by preparing the discussion section. MAZ helped in conducting DMA experiments and analyzing data from mechanical and thermal analyses. All authors read and approved the final manuscript.

Acknowledgements

The authors are grateful to the Georgia Research Alliance (GRA), USA, as well as the Department of Textiles, Merchandizing, and Interiors at the University of Georgia for the financial support to complete this work. Also, the authors would like to thank Dr. John Shields from the Center for Ultrastructural Research at the University of Georgia for providing the facilities and knowledge to aid in the morphological analysis of the albumin plastics.

Author details

[1]Department of Textiles, Merchandising and Interiors, University of Georgia, Athens, GA 30602, USA. [2]ALGIX, LLC, Athens, GA 30602, USA.

References

Aman Ullah TV, Bressler D, Elias AL, Jianping W (2011) Bioplastics from feather quill. Biomacromolecules 12:3826–3832

Baron F, Rehault S (2007) Compounds with antibacterial activity. In: Huopalahti R, Lopez-Fandiño R, Anton M, Schade R (eds) Bioactive egg compounds. Springer-Verlag, Berlin, pp 191–198

Carvalho AJF, Job AE, Alves N, Curvelo AAS, Gandini A (2003) Thermoplastic starch/natural rubber blends. Carbohydr Polym 53(1):95–99

Castelló M, Dweck J, Aranda DAG (2009) Thermal stability and water content determination of glycerol by thermogravimetry. J Therm Anal Calorim 97(2):627–630

Cegielska-Radziejewska RLG, Szablewski T, Kijowski J (2010) Physico-chemical properties and antibacterial activity of modified egg white—lysozyme. Eur Food Res Technol 231:959–964

Chen P, Zhang L, Cao F (2005) Effects of moisture on glass transition and microstructure of glycerol-plasticized soy protein. Macromol Biosci 5(9):872–880

Dorigato A, Pegoretti A (2012) Biodegradable single-polymer composites from polyvinyl alcohol. Colloid Polym Sci 290(4):359–370

Flieger M, Kantorova M, Prell A, Řezanka T, Votruba J (2003) Biodegradable plastics from renewable sources. Folia Microbiology 48(1):27–44

Galdeano MC, Mali S, Grossmann MVE, Yamashita F, Garcia MA (2009) Effects of plasticizers on the properties of oat starch films. Mater Sci Eng C 29(2):532–538

González-Gutiérrez J, Partal P, García-Morales M (2011) Effect of processing on the viscoelastic, tensile and optical properties of albumen/starch-based bioplastics. Carbohydr Polym 84(11):308–315

Hook A, Chang CY, Yang J, Luckett J, Cockayne A, Atkinson S, Mei Y, Bayston R, Irvine DR, Langer R, Anderson DG, Williams P, Davies MC, Alexander MR (2012) Combinatorial discovery of polymers resistant to bacterial attachment. Nat Biotechnol 30:868–875

Jerez A, Partal P, Martínez I, Gallegos C, Guerrero A (2007) Protein-based bioplastics: effect of thermo-mechanical processing. Rheologica Acta 46:711–720

Lodha P, Netravali AN (2005) Thermal and mechanical properties of environment-friendly 'green' plastics from stearic acid modified-soy protein isolate. Ind Crop Prod 21(1):49–64

Mathew A, Packirisamy S, Thomas S (2001) Studies on the thermal stability of natural rubber/polystyrene interpenetrating polymer networks: thermogravimetric analysis. Polym Degrad Stab 72:423–429

Menard K (1999) Dynamic mechanical analysis: a practical introduction. CRC Press, Boca Raton

Mukerjee TKN (2011) PLA based biopolymer reinforced with natural fibre: a review. J Polym Environ 19:714–725

Peters T Jr (1996) All about albumin: biochemistry, genetics, and medical applications. Academic Press, San Diego

Pommet M, Redl A, Guilbert S, Morel M-H (2005) Intrinsic influence of various plasticizers on functional properties and reactivity of wheat gluten thermoplastic materials. Journal of Cereal Science 42(1):81–91

Sharma S, Luzinov I (2012) Water aided fabrication of whey and albumin plastics. J Polym Environ 20:681–689

Sharma SHJ, Hodges JN, Luzinov I (2008) Biodegradable plastics from animal protein coproducts: feathermeal. J Appl Polymer Sci 110:459–467

Sue HJ, Wang S, Lane JL (1997) Morphology and mechanical behaviour of engineering soy plastics. Polymer 38(20):5035–5040

Van Soest JJG, Knooren N (1997) Influence of glycerol and water content on the structure and properties of extruded starch plastic sheets during aging. J Appl Polymer Sci 64(7):1411–1422

Verbeek CJR, van den Berg LE (2010) Extrusion processing and properties of protein-based thermoplastics. Macromol Mater Eng 295(1):10–21

Widiastuti I, Sbarski I, Masood SH (2013) Creep behavior of PLA-based biodegradable plastic exposed to a hydrocarbon liquid. J Appl Polymer Sci 127(4):2654–2660

Zilberman A, Elsner JJ (2008) Antibiotic-eluting medical devices for various applications. J Control Release 130:202–215

14

Engineering of chitosan and collagen macromolecules using sebacic acid for clinical applications

G Sailakshmi, Tapas Mitra and A Gnanamani[*]

Abstract

Transformation of natural polymers to three-dimensional (3D) scaffolds for biomedical applications faces a number of challenges, viz., solubility, stability (mechanical and thermal), strength, biocompatibility, and biodegradability. Hence, intensive research on suitable agents to provide the requisite properties has been initiated at the global level. In the present study, an attempt was made to engineer chitosan and collagen macromolecules using sebacic acid, and further evaluation of the mechanical stability and biocompatible property of the engineered scaffold material was done. A 3D scaffold material was prepared using chitosan at 1.0% (w/v) and sebacic acid at 0.2% (w/v); similarly, collagen at 0.5% (w/v) and sebacic acid at 0.2% (w/v) were prepared individually by freeze-drying technique. Analysis revealed that the engineered scaffolds displayed an appreciable mechanical strength and, in addition, were found to be biocompatible to NIH 3T3 fibroblast cells. Studies on the chemistry behind the interaction and the characteristics of the cross-linked scaffold materials suggested that non-covalent interactions play a major role in deciding the property of the said polymer materials. The prepared scaffold was suitable for tissue engineering application as a wound dressing material.

Keywords: Chitosan, Collagen, Sebacic acid, Mechanical strength, Biocompatible

Background

Polymers, in general, (whether natural or synthetic) play major roles in biomaterial preparations. However, natural polymers or polymers derived from living creatures are of greater interest, and most research publications majorly discussed chitosan and collagen Parenteau-Bareil et al. (2010; Iwasaki et al. 2011). Further, Ko et al. (2010) suggested the use of naturally derived polymers as three-dimensional (3D) scaffolds, which received much attention due to their low cost, ease of processing, and biocompatibility.

With regard to chitosan and collagen, after extraction, both these materials did not have much stability to act as a biomaterial for clinical applications and demanded stabilizers in the form of cross-linkers Friess (1998; Austero et al. 2012). Diisocyanates, Resimene Ligler et al. (2001), N, N-disuccinimidyl suberate Schauer et al. (2003), epichlorohydrin Wei et al. (1992), Genipin Jin

et al. (2004), and glutaraldehyde Tual et al. (2000) were the stabilizers studied for chitosan and chromium Usha and Ramasami (2000), while aldehydes Sheu et al. (2001), hexamethylene diisocyanate Miles et al. (2005), carbodiimide Nam et al. (2008), acyl azides Petite et al. (1990), citric acid, maleic acid derivatives Saito et al. (2008), and various other physical treatments, such as UV Weadock et al. (1995) and gamma irradiation Olde Damink et al. (1995a), were the stabilizers studied for type I collagen. All the said exogenous cross-linkers cross-linked with chitosan or with type I collagen through (1) covalent amide/imine linkage, (2) metal-protein complex formation (chromium cross-linking with type I collagen), and (3) H-bond formation (between poly-phenolic -OH group with different types of amino acids of type I collagen molecule and amino group of chitosan). Although the resultant biomaterial upon cross-linking with these agents was acceptable, the complete utilization of these materials was restricted because of low mechanical strength Schiffman and Schauer (2007). The biocompatibility of the scaffold

* Correspondence: gnanamani3@gmail.com
Microbiology Division, Central Leather Research Institute (CSIR, New Delhi), Adyar, Chennai 20, Tamil Nadu, India

material is also questionable because of the release of toxic components from some of the cross-linkers upon usage Gough et al. (2002).

In general, the mechanical property of any biomaterial depends on the interaction between the cross-linkers/ stabilizers and the parent molecule (here, it is chitosan and collagen type I) Rinaudo (2010). As summarized above, reports on bonding interaction of the said cross-linkers suggest the predominance of covalent interactions which ultimately restrict the molecule to attain the desired mechanical strength.

Hence, in order to engineer the macromolecules and also to obviate the problems associated with the mechanical property and biocompatibility of biomaterials, we attempted to cross-link the parent molecule with a suitable cross-linker through non-covalent interactions. In our previous study, we detailed the cross-linking chemistry between malonic acid (MA) with chitosan/collagen Mitra et al. (2012). Observations with short-chain dicarboxylic acid (MA, three carbons) cross-linked chitosan/collagen scaffold gave impetus to carry out further work with long-chain dicarboxylic acids, namely sebacic acid (SA) (ten carbons).

Thus, in the present study, SA was chosen to cross-link with natural polymers, and to this day, no reports are available on the non-covalent interaction of SA with chitosan and type I collagen. Sebacic acid (decanedioic acid), a C-8 dicarboxylic acid ($HOOC-(CH_2)_8-COOH$), is a white flake or powdered crystal generally used as a component in metalworking fluids, surfactants, lubricants, detergents, oiling agents, emulsifiers, etc. Sebacic acid is the natural metabolic intermediate in ω-oxidation of medium- to long-chain fatty acids Liu et al. (1996). According to Tamada and Langer (1992), it is safe under *in vivo* condition.

The present study emphasizes the detailed chemistry behind the engineering of chitosan and collagen using SA and the evaluation of the thermal, mechanical, and biocompatible properties of the engineered chitosan and collagen scaffolds.

Results and discussion
Scaffold preparation using sebacic acid: understanding the cross-linking chemistry

For the preparation of any scaffold materials, the solution form of the parent compound/polymer is required to proceed further. However, in the case of chitosan and collagen, acetic and formic acids were generally used for dissolution Ohkawa et al. (2004; El-Tahlawy et al. 2006). The 'proton exchange' between the -COOH groups of acid molecule and free -NH₂ groups of chitosan and collagen could be the reason for the dissolution in the said acids.

Therefore, it has been expected that like acetic acid, sebacic acid is also able to donate protons to dissolve chitosan and type I collagen. Further, similar to the

interaction of TPP Bhumkar and Pokharkar (2006) with chitosan, sebacic acid may also interact with both natural polymers through ionic interaction. Because of the said proton exchange, chitosan and type I collagen dissolve in the presence of sebacic acid in water; the following schematic representation (Scheme 1a,b) illustrates the nature of proton exchange between sebacic acid with chitosan and with collagen for better understanding.

Because of the said interaction, both natural polymers were completely dissolved in water in the presence of sebacic acid. With the resulting solution, scaffolds were prepared and subjected to characterization studies. Figure 1 shows the morphological features of the cross-linked scaffolds, namely sebacic acid cross-linked chitosan (SACCH) and sebacic acid cross-linked collagen (SACC). The 3D scaffold material was highly porous, and the pore structures of the membranes were well distributed and interconnected. It was obvious that most of the membrane volume was taken up by interconnecting pore space. The high porosity suggests the suitability of this scaffold for biomedical applications, including serving as absorption sponges and matrices for cell proliferation.

Fourier transform infrared spectroscopy (FT-IR) studies were conducted to monitor chemical modifications in the chitosan and collagen structures upon cross-linking with SA. Figure 2 illustrates the FT-IR spectral details of SA, chitosan, collagen, SACCH, and SACC. Table 1 demonstrates the FT-IR peak assignments of SA, chitosan, and collagen. In the SACCH spectrum, few significant changes were observed. A broad, strong absorption peak in the region of 3,433 to 2,928 cm^{-1} resulted from the superimposed -OH and $-NH_3^+$ stretching bands. Absorption in 1,640 and 1,557 cm^{-1} corresponded to the presence of asymmetric N-H ($-NH_3^+$) bends and asymmetric $-COO^-$ stretching, respectively. A peak observed at 1,403 cm^{-1} was due to symmetric $-COO^-$ stretching. Other absorption peaks around 1,257, 1,157, and 899 cm^{-1} observed in the SACCH spectrum were similar to the native chitosan spectrum which exhibits that there was no change in the main backbone of the chitosan structure Lopez et al. (2008).

In the SACC spectrum, few changes were observed when compared with native type I collagen. A broad, strong absorption peak in the region of 3,551 to 3,101 cm^{-1} resulted from the superimposed -OH and $-NH_3^+$ stretching bands. In the type I collagen spectrum, a sharp intense amide I band observed around 1,658 cm^{-1} disappeared with the appearance of two new bands in 1,681 and 1,625 cm^{-1} in the SACC spectrum; these bands were supposed to be caused by $-NH_3^+$ and $-COO^-$, respectively. Moreover, when compared with native type I collagen spectrum, there was a reduction in the region of 1,557 cm^{-1} (overlapped band of amide II and free primary amines) in the SACC spectrum, which may be due to the reduction of free -NH₂ group in

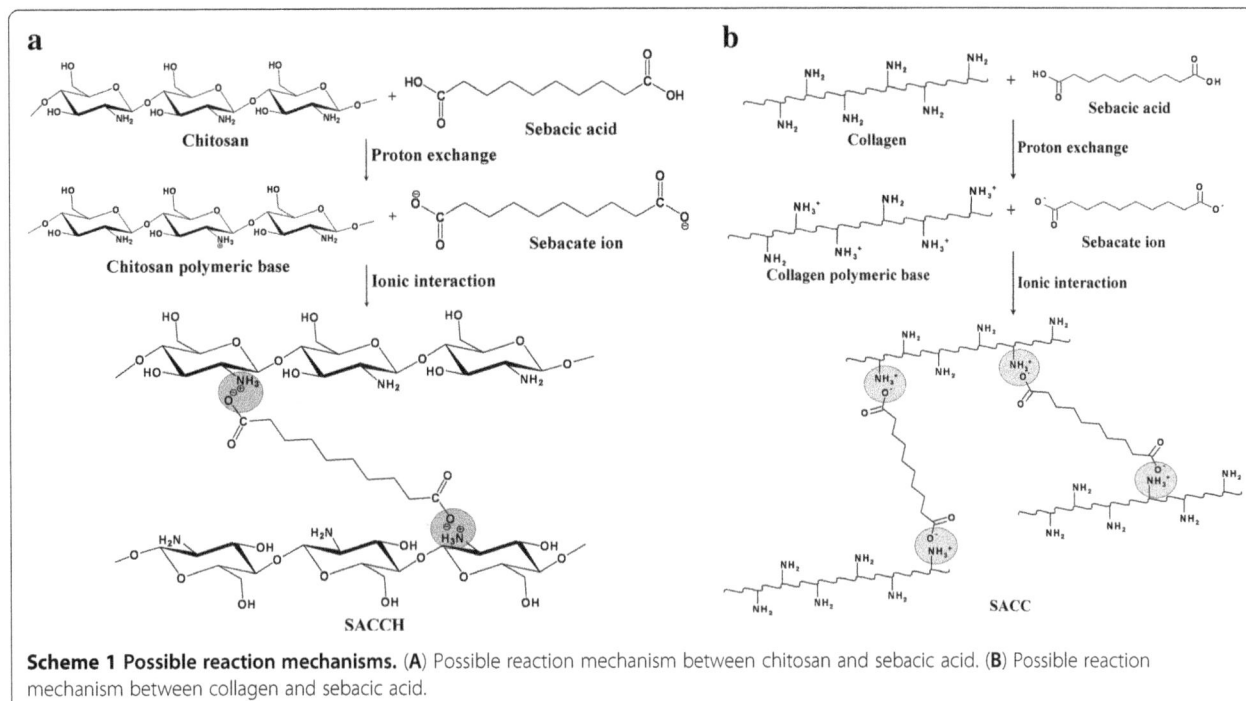

Scheme 1 Possible reaction mechanisms. (**A**) Possible reaction mechanism between chitosan and sebacic acid. (**B**) Possible reaction mechanism between collagen and sebacic acid.

the SACC. In the SACC spectrum, the observed band around 525 cm^{-1} was ascribed to the N-H oscillation of -NH$_3^+$. Results from FT-IR analysis reflected that SA was ionically cross-linked with chitosan and type I collagen Pavia et al. (2001; Lawrie et al. 2007).

Though FT-IR analysis displayed the ionic interaction between the cross-linker and the natural polymers, results on the percentage of cross-linking degree calculations suggested that increasing the concentration of SA increases the degree of cross-linking up to 0.4% and confirmed the interaction. About 60% to 65% cross-linking was observed with 0.2% SA with chitosan and collagen. However, in the case of experiments with

glutaraldehyde, about 88% to 93% of cross-linking was observed with 0.2% concentration.

With regard to the mechanical property of the scaffold materials, it is a fundamental property for any scaffold material in the application point of view. From the results, we observed that the mechanical strength of the scaffold increased with the increase in sebacic acid concentration up to 0.2%. Further increase in SA concentration leads to the decrease in mechanical strength (results not shown). Table 2 illustrates the tensile strength, Young's modulus, and stiffness of native and sebacic acid (0.2%) cross-linked scaffolds. High tensile strength (MPa) values were observed for both the cross-

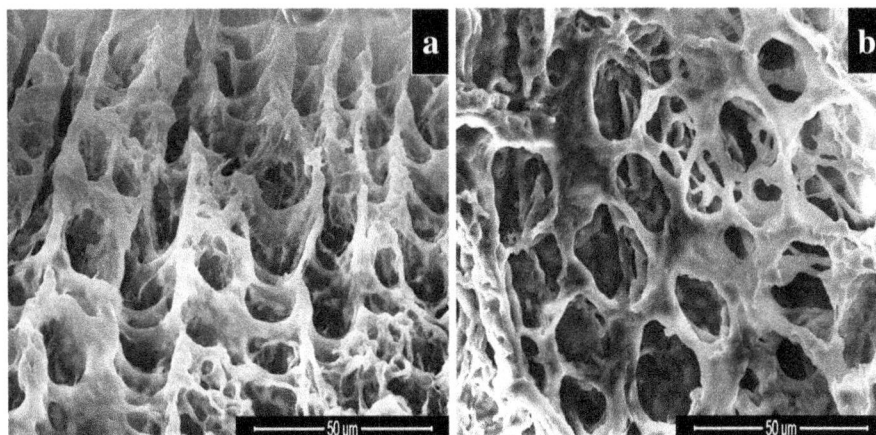

Figure 1 SEM micrographs of (a) sebacic acid cross-linked chitosan and (b) sebacic acid cross-linked collagen scaffolds.

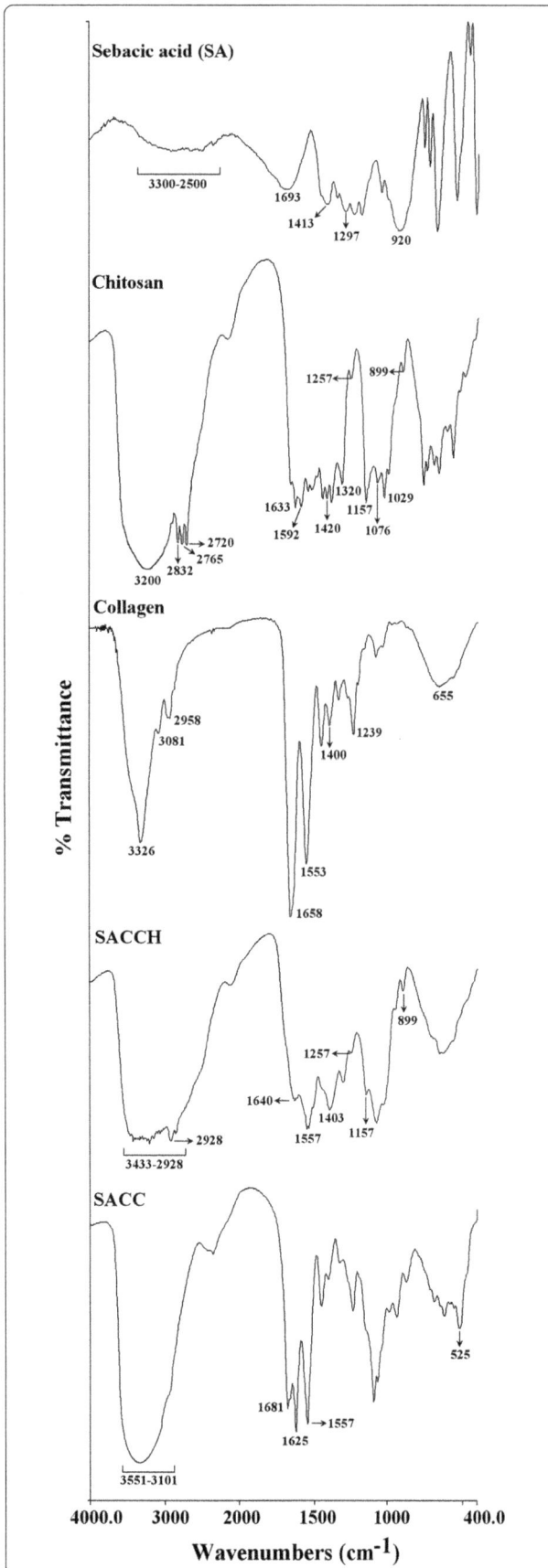

Figure 2 FT-IR spectra of SA, chitosan, type I collagen, SACCH, and SACC scaffolds. SA, sebacic acid; SACCH, sebacic acid cross-linked chitosan; SACC, sebacic acid cross-linked collagen.

linked scaffolds (SACCH 8.94; SACC 2.96) than for the native polymers (chitosan 0.37; type I collagen 0.13). Moreover, the Young's modulus of SACCH and SACC were 200.8 and 18.61, respectively. The stiffness values (SACCH 9.53 N/mm; SACC 0.69 N/mm) were also greater than those of the native polymers (chitosan 0.79 N/mm; type I collagen 0.3 N/mm).

All these observations on mechanical properties suggest that sebacic acid cross-linked materials demonstrated appreciable mechanical strength compared to glutaraldehyde, wherein we observed brittleness similar to the observation made by Schiffman and Schauer (2007). Further, when the concentration of SA was increased > 0.2%, a decrease in mechanical strength was observed, and this could be reasoned to the high degree of

Table 1 FT-IR analysis of SA, chitosan, and collagen

Material	Wave number (cm^{-1})	Peak assignment
SA	3,300 to 2,500	Overlapping stretching vibration of C-H and O-H groups (v_{C-H}-v_{OH})
	1,693	-C=O group ($v_{C=O}$)
	1,413	C-O-H in-plane bending (δ_{C-O-H})
	1,297	C–O stretching vibration (v_{C-O})
	920	Out-of-plane bending of the bonded O-H (δ_{O-H})
Chitosan	3,200	-NH$_2$ stretching vibration (v_{NH})
	2,832, 2,765, 2,720	Symmetric or asymmetric -CH$_2$ stretching vibration attributed to the pyranose ring (v_{C-H})
	1,633	-C=O in acetamide group (amide I band)
	1,592	-NH$_2$ bending vibration in amino group (δ_{NH})
	1,420, 1,320	Vibrations of OH, CH in the ring
	1,257	C-O group
	1,157	-C-O-C in the glycosidic linkage
	1,076, 1,029	C-O stretching in acetamide (v_{C-O})
	899	Corresponds to the saccharide structure
Collagen	3,326	-NH$_2$ stretching vibration (v_{NH})
	3,081	Fermi resonance overtone of 1,553 band
	2,958	C-H stretching (v_{C-H})
	1,658	Amide I band ($v_{C=O}$)
	1,553	Amide II band (δ_{NH})
	1,400	Amide III band (v_{C-N})
	1,239	C-N stretching of amine (v_{C-N})
	655	Out-of-plane N-H wagging of amide and amine (δ_{NH})

v, stretching; δ, bending.

Table 2 Assessment of mechanical properties of chitosan, SACCH, collagen and SACC

Samples	Maximum load[a]	Tensile strength[a]	Elongation at break[a]	Extension at maximum load[a]	Young's modulus/ tensile modulus[a]	Stiffness (κ)[a]
	(N)	(MPa)	(%)	(mm)	(MPa)	(N/mm)
Chitosan	1.32 ± 0.08	0.37 ± 0.03	8.33 ± 0.6	1.67 ± 0.04	4.43 ± 0.5	0.790 ± 0.07
SACCH	5.61 ± 0.5	3.21 ± 0.2	8.63 ± 1	1.73 ± 0.04	37.1 ± 2.2	3.24 ± 0.4
Collagen	0.37 ± 0.03	0.13 ± 0.02	6.17 ± 0.8	1.23 ± 0.05	2.1 ± 0.7	0.3 ± 0.01
SACC	2.22 ± 0.5	2.96 ± 0.5	31.84 ± 2	3.18 ± 0.4	18.61 ± 1.5	0.698 ± 0.02

The assessment is in terms of tensile strength, elongation at break, Young's modulus, and stiffness. [a]Mean ± SD

cross-linking of SA with the polymers, as evidenced from the 2,4,6-trinitrobenzenesulfonic acid (TNBS) assay Wang et al. (2002).

Further, the reason for the brittle nature of the glutaraldehyde cross-linked material is that glutaraldehyde could covalently cross-link with chitosan and collagen through the formation of a double bond (C=N, imine bond) between the -CHO group of glutaraldehyde and the -NH$_2$ group of natural polymers (chitosan and collagen), which results in the large energy barrier for the rotation of associated groups linked by a double bond (C=N), and finally, this provided the brittle nature to the material.

Thermogravimetric analyses for the experimental samples SA, chitosan, type I collagen, SACCH, SACC, glutaraldehyde cross-linked chitosan (GACCH), and glutaraldehyde cross-linked collagen (GACC) were illustrated in Figure 3a,b, and the corresponding thermal degradation values were displayed in Table 3. From the results, we observed that the incorporation of SA with chitosan and type I collagen tended to shift the thermal region into higher temperature, and such a shift is attributed to an increase in thermal stability.

Differential scanning calorimetry (DSC) studies were performed to understand the behavior of SACCH and SACC on the application of thermal energy. The thermogram values of SA, chitosan, type I collagen, SACCH, SACC, GACCH, and GACC were shown in Figure 4a,b. DSC studies recorded the melting temperature of SA (135°C) and the degradation temperature differences among chitosan (107°C), type I collagen (92°C), SACCH (119°C), and SACC (125°C), whereas in GACCH and GACC, it was observed at 149°C and 151°C, respectively. The higher transition temperature suggests that SACCH and SACC have high stability at high-temperature environment. Thermal stability also influences on the durability of the scaffolds. A similar kind of observation was reported by Bhumkar and Pokharkar (2006).

Results on binding energy calculations based on the bioinformatics tool for the cross-linking of SA with chitosan and type I collagen using AutoDock software (The Scripps Research Institute La Jolla, CA 92037, USA) proved that chitosan and type I collagen cross-links with sebacic acid not only through ionic interaction, but also through

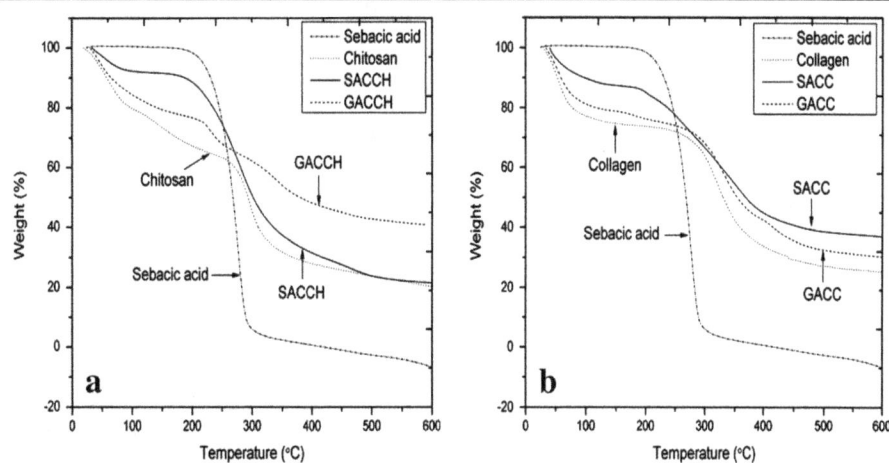

Figure 3 Thermogravimetric analyses for experimental samples. (a) Thermogravimetric analysis of sebacic acid (SA), chitosan, sebacic acid cross-linked chitosan (SACCH), and glutaraldehyde cross-linked chitosan (GACCH) scaffolds. (b) Thermogravimetric analysis of SA, collagen, sebacic acid cross-linked collagen (SACC), and glutaraldehyde cross-linked collagen (GACC) scaffolds.

Table 3 Thermal analyses of SA, chitosan, SACCH, GACCH, collagen, SACC, and GACC under N$_2$ air atmosphere

Temperature (°C)	Percentage of weight loss (heating rate 20°C/min)						
	SA	Chitosan	SACCH	GACCH	Collagen	SACC	GACC
100	0	21	8	16	23	11	20
200	1	33	12	24	27	15	24
300	95	56	50	38	37	34	33
400	100	73	69	52	67	55	58
500	100	77	77	58	73	62	68
600	100	81	79	60	75	63	70

SA, sebacic acid; SACCH, sebacic acid cross-linked chitosan; GACCH, glutaraldehyde cross-linked chitosan; SACC, sebacic acid cross-linked collagen; GACC, glutaraldehyde cross-linked collagen.

multiple intermolecular hydrogen bonding. AutoDock is an automated procedure for predicting the interaction of ligands with biomacromolecular targets. Hundred runs were given for docking SA with chitosan and type I collagen. The best binding energy values and their corresponding rank and run numbers were depicted in Table 4. The binding energy of −4.21 and −4.49 (kcal/mol) was observed when SA interacted with chitosan and type I collagen, respectively. These interactions were made by multiple intermolecular hydrogen bonds between the -COOH group of SA and the -NH$_2$ group of chitosan and the free ε-NH$_2$ group of lysine from type I collagen (Figure 5a,b). In addition to ionic cross-linking, hydrogen bonding interaction also improves the mechanical property of the scaffold. The ionic interaction and hydrogen bonding between the -COOH group of cross-linker and the -NH$_2$ group of natural polymers (chitosan/collagen) has already been reported Milosavljevic et al. (2010; Vijayaraghavan et al. 2010). The details of these intermolecular hydrogen bonding sites are given in the following sections.

Intermolecular hydrogen bond details between SA and chitosan

H (3) of chitosan is linked to O (14) of SA with a bond distance of 1.91531, and H (3) of chitosan is linked to O (13) of SA with a bond distance of 1.99012.

Intermolecular hydrogen bond details between SA and type I collagen

Lysine amino acid (LYS) (12) H2 of type I collagen is linked with O (7) of SA with a bond distance of 2.06994, LYS (12) H2 of type I collagen is linked with O (8) of SA with a bond distance of 2.09251, LYS (12) H3 of type I collagen is linked with O (13) of SA with a bond distance of 2.11339, and LYS (12) H3 of type I collagen is linked with O (14) of SA with a bond distance of 1.91386.

Concerning the biocompatibility of the resulting polymers, cell attachment and viability assays were carried out. 3-[4,5-Dimethylthiazol-2-yl]-2,5-dephenyltetrazolium bromide (MTT) assay was done to check the toxicity profile of the prepared scaffolds (SACCH, SACC, GACCH,

and GACC). Only cells that are metabolically normal can turn the tetrazolium salts into purple crystals. Compared with the native chitosan and type I collagen, SACCH and SACC showed no significant differences in absorbance (Figure 6) and inferred that scaffolds being in direct contact with fibroblast did not lead to apoptosis or necrosis. MTT results clearly indicated that NIH 3T3 cells are viable on the surface of the SA cross-linked scaffolds (SACCH and SACC). However, only 10% cells were viable with the samples of GACCH and GACC. Though glutaraldehyde (GA) has been widely used as chemical cross-linking agent Jorge-Herrero et al. (1999) because of stabilizing the collagen efficiently, and the cross-linking is thought to involve the formation of Schiff bases Olde Damink et al. (1995b), the poor biocompatibility of GA-cross-linked biomaterials with some other cell lines, including human fibroblasts, osteoblasts, Chang cells, and endothelial cells, had been reported Gough et al. (2002). The side effects of GA treatment were attributed to the degradation of the GA-derived cross-links and the continuous release of aldehydes contributing to prolong toxic effects Schmidt and Baier (2000).

In the cell viability assay, we observed intense fluorescence of the cells on the surface of the native and SA cross-linked scaffolds and suggest the viability of the cells as illustrated in Figure 7.

The SEM images of the cell seeded SACCH and SACC scaffolds displayed in Figure 8a,b demonstrated that after being cultured for a prolonged time (12 days), fibroblasts were detected in the scaffolds (SACCH and SACC) with typical spindle-shaped morphology, suggesting that the cells were infiltrated into the scaffolds.

When comparing the present results with the previous work carried out with malonic acid (three carbons) Mitra et al. (2012), no significant differences in mechanical and thermal stability were observed. However, when comparing the pore size (16 to 18 μm) determination from SEM results of malonic acid cross-linked chitosan/collagen, an increased pore size (27 to 30 μm) was observed in SACCH and SACC scaffolds. This could be due to the long chain length of sebacic acid (ten carbons) compared

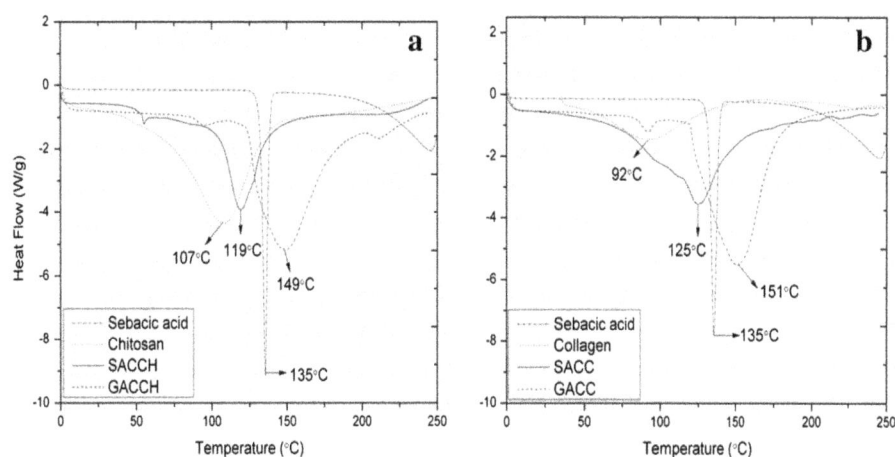

Figure 4 DSC analyses of SA, chitosan, type I collagen, SACCH, SACC, GACCH, and GACC. (a) DSC analysis of sebacic acid (SA), chitosan, sebacic acid cross-linked chitosan (SACCH), and glutaraldehyde cross-linked chitosan (GACCH) scaffolds. **(b)** DSC analysis of sebacic acid (SA), collagen, sebacic acid cross-linked collagen (SACC), and glutaraldehyde cross-linked collagen (GACC) scaffolds.

with malonic acid (three carbons). In addition, the docking studies suggested that there was stronger intermolecular hydrogen bonding interaction between sebacic acid and chitosan/collagen compared with the interaction between malonic acid and chitosan/collagen.

Conclusions

The present study explicitly demonstrates sebacic acid acting as suitable cross-linker for the preparation of biocompatible scaffolds from natural polymers (chitosan and collagen) with appreciable mechanical properties compared to short-chain dicarboxylic acids. The interactions between SA and chitosan or collagen were identified as non-covalent, i.e., both for ionic and multiple intermolecular hydrogen bonding interactions. These non-covalent interactions offered high mechanical strength to the resultant material. All the instrumental analyses and bioinformatics tool authenticated the non-covalent interactions. The scaffold material prepared upon cross-linking of SA with chitosan or collagen was the green method of preparation. Sebacic acid not only dissolves the above-said natural polymers, but also improves the property of the material through its non-covalent interactions with natural polymers. No toxic compounds were involved in this preparation, and the resultant material can be used as a wound dressing material or as an implant in clinical applications.

Methods

Materials

Chitosan from shrimp shells (≥75% deacetylated), sebacic acid, and Picrylsulfonic acid (TNBS) were obtained from Sigma-Aldrich Corporation (St. Louis, MO, USA). MTT and dexamethasone were purchased from HiMedia Laboratories (Mumbai, India). All the other reagents were of analytical grade and used without further purification. Type I collagen from bovine skin was extracted according to the procedure followed by Mitra et al. (2011).

Preparation of a 3D scaffold

In the powder form of chitosan (1%), sebacic acid at different concentrations was mixed, and 20 ml of water was added to the mixture which was stirred at room temperature until the solution became homogenous. Similarly, collagen (0.5%) and sebacic acid at different concentrations were mixed in the presence of 20 ml of water and stirred at 4°C to obtain a clear homogenous

Table 4 Binding energy values of SACCH and SACC scaffolds

	Rank	Binding energy (Kcal/mol)	Number of runs
Binding energy calculation between chitosan and sebacic acid based on AutoDock tool software	1	−4.21	2
	2	−4.17	24
	3	−3.92	65
	4	−3.73	29
Binding energy calculation between type I collagen and sebacic acid based on AutoDock tool software	1	−4.49	49
	2	−4.36	20
	3	−4.04	58
	4	−3.94	21
	5	−3.79	51
	6	−3.71	63
	7	−3.63	15
	8	−3.20	85
	9	−3.13	53

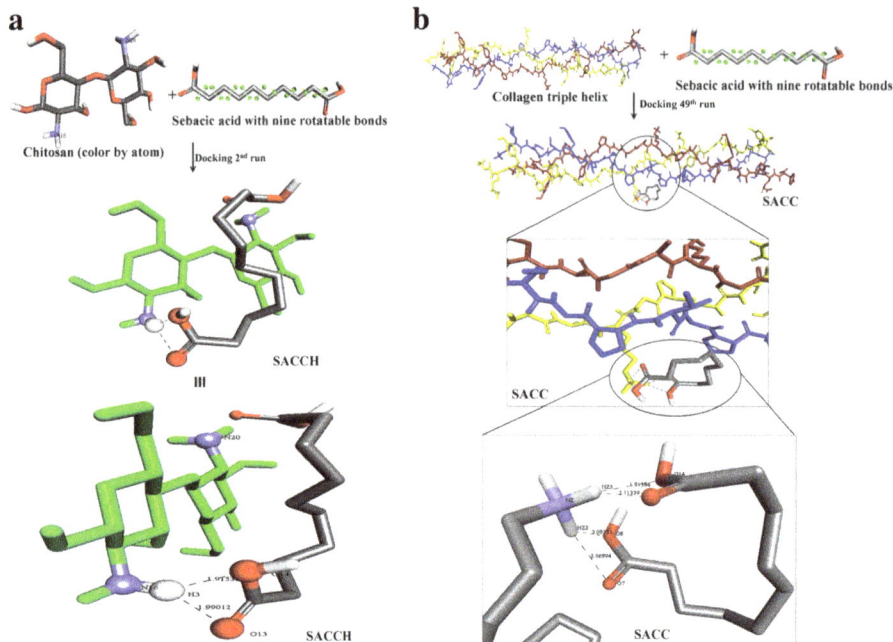

Figure 5 Multiple hydrogen bonding interactions. (**a**) Multiple hydrogen bonding interaction between chitosan and sebacic acid. Images of SACCH are displayed in two different orientations. The first orientation demonstrates the interaction, and the second illustrates the details on hydrogen bonding. (**b**) Multiple hydrogen bonding interaction between collagen and sebacic acid. The black dotted line indicates the hydrogen bond. Various colors denote the different atoms to be recognized: white color is for hydrogen atom (H), red color indicates oxygen atom (O), gray is for carbon atom (C), and blue corresponds to nitrogen atom (N).

solution. The concentration of SA varied between 0.05% and 0.5% (w/v). The samples were subjected to centrifugation in order to remove any nonreactive molecules, and a clear solution obtained from centrifugation at 5,000 rpm for 10 min was poured in Tarsons vials (Tarsons Product Pvt. Ltd., Kolkata, India), which have an inner diameter of 4.5 cm and frozen at –4°C for 2 h, –20°C for 12 h,

and –80°C for another 12 h according to Peng et al. (2006). The frozen samples were lyophilized for 48 h at a vacuum of 7.5 mTorr (1 Pa) and a condenser temperature of –70°C (PENQU CLASSIC PLUS, Lark, India). The resultant 3D scaffold material was neutralized with repeated washings with 0.05 N of NaOH and ethanol mixture, followed by washings with water and ethanol

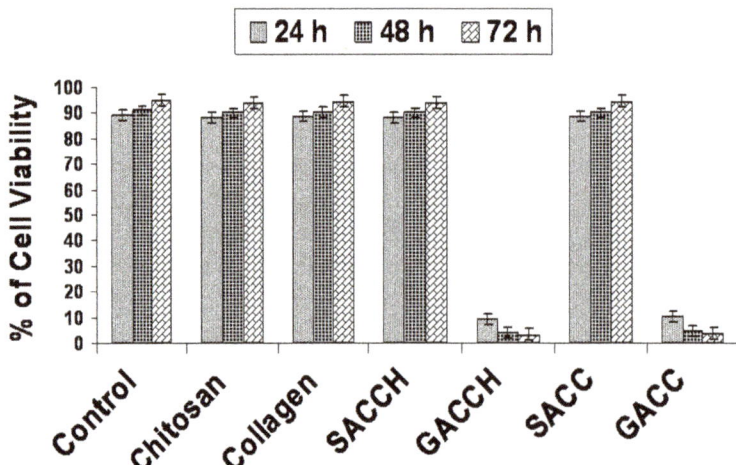

Figure 6 MTT analysis at 24-, 48-, and 72-h time interval. The analysis was for control, chitosan, collagen, sebacic acid cross-linked chitosan (SACCH), glutaraldehyde cross-linked chitosan (GACCH), sebacic acid cross-linked collagen (SACC), and glutaraldehyde cross-linked collagen (GACC) scaffolds.

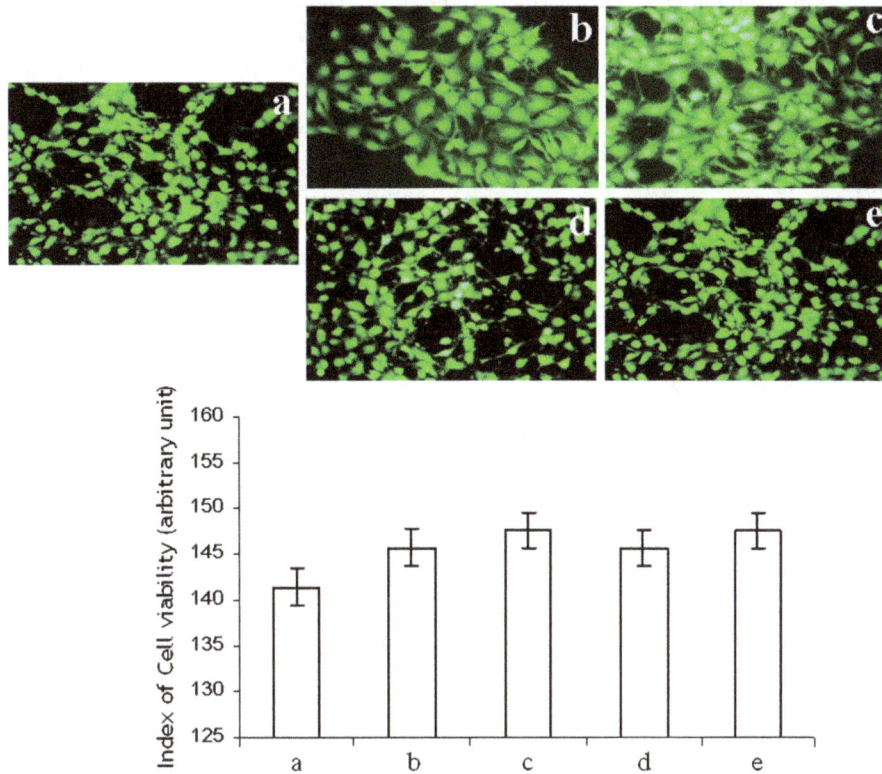

Figure 7 Cell viability index (arbitrary unit) assessed in SACCH and SACC compared with parent molecules and control. (a) Control, (b) chitosan, (c) collagen, (d) sebacic acid cross-linked chitosan (SACCH), and (e) sebacic acid cross-linked collagen (SACC). The assay was carried out using a cell tracker kit. NIH3T3 cells were treated on the surface of native and cross-linked scaffolds for 6 h followed by incubation with cell viable dye cell tracker for 30 min. Fluorescence images of the cells were acquired using DP71 camera adapted to an Olympus IX71 microscope (Olympus Corporation, Shinjuku, Tokyo, Japan). Intensity of green positive cells were counted and plotted. Next, fluorescence intensities of the images were calculated using Adobe Photoshop version 7.0.

mixture; finally, it was again lyophilized for 24 h. The scaffolds obtained during this procedure were designated as SACCH and SACC. For comparative analysis, GACCH and GACC were prepared according to the method described previously using 0.2% glutaraldehyde.

SEM analysis of SACCH and SACC

The physical texture and the morphology of the scaffold of SACCH and SACC were assessed using a scanning electron micrograph. SEM micrograph analysis was made using FEI Quanta (FEI Company, Hillsboro, OR,

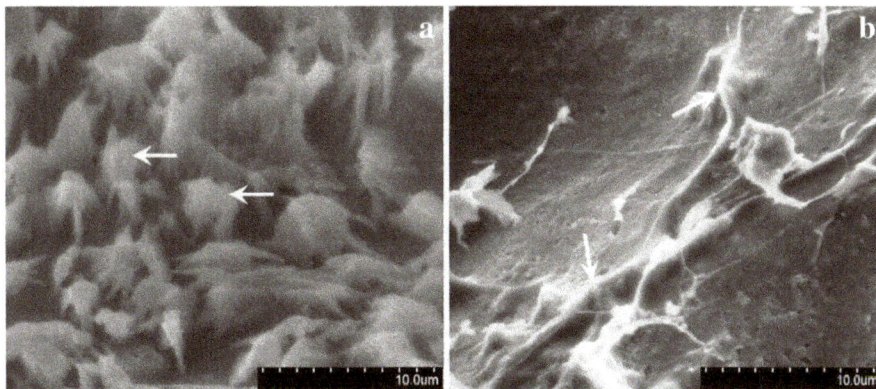

Figure 8 Attachment of fibroblast cells on the (a) SACCH and (b) SACC scaffolds. In (a), white arrows indicate the adhered cells on the scaffold. In (b), porous SACC scaffold was completely covered by fibroblast cells which are indicated by the white arrow.

USA) FEG 200 high-resolution scanning electron microscope under a high voltage at 20 kV.

FT-IR analysis

Functional group analysis for SA, chitosan, type I collagen, SACCH, and SACC scaffolds were made using Spectrum One FT-IR (PerkinElmer Instruments, Branford, CT, USA). All spectra were recorded with the resolution of 4 cm^{-1} in the range of 400 to 4,000 cm^{-1} with 20 scans.

Cross-linking degree (2,4,6-trinitrobenzenesulfonic acid assay) determination

Degree of cross-linking was quantified using TNBS assay according to the procedure summarized by Bubnis and Ofner (1992). In brief, native and cross-linked (SACCH and SACC) biopolymer materials were cut into small pieces at 4.5 mm. Cut pieces (6 mg) were immersed in a 2-ml solution (1 ml of 4% disodium hydrogen orthophosphate (w/v) and 1 ml of 0.5% TNBS (v/v)) and incubated at 40°C for 2 h. Pure sebacic acid at respective percentages was also treated with a 2-ml solution in separate test tubes. Termination of reaction was by the addition of 3 ml of 6 M HCl (v/v), and the incubation was continued at 60°C for 90 min. The absorbance of the resulting solution was measured at 345 nm using a UV-visible spectrophotometer (UV-2450, Shimadzu Corporation, Nakagyo-ku, Kyoto, Japan), and the percentage of cross-linking was calculated from the difference in the absorbance divided by the absorbance of the native material and then multiplied by 100.

Mechanical properties of SACCH, SACC, GACCH, and GACC scaffolds

Mechanical properties, *viz.*, Young's modulus, ultimate tensile strength, stiffness, and percentage of elongation of the dried scaffold materials, were measured using a universal testing machine (model 1405, Instron Corporation, Norwood, MA, USA) at a crosshead speed of 5 mm min^{-1} at 25°C and 65% relative humidity. Length and width of the dumbbell-shaped test samples were maintained at 20 and 5 mm, respectively, according to Shanmugasundaram et al. (2004). All the mechanical tests were performed with dried samples and were examined in triplicates.

Thermogravimetric analysis

Thermal decomposition analysis of SA, native, and cross-linked scaffolds (chitosan, type I collagen, SACCH, SACC, GACCH, and GACC) was carried out under nitrogen flow (40 and 60 ml min^{-1}) with ramp at 20°C min^{-1} using TGA Q 50 (TA Instruments, New Castle, DE, USA) with an isothermal temperature accuracy of ±1°C.

Differential scanning calorimetry

DSC analysis for SA, native, and cross-linked scaffolds (chitosan, type I collagen, SACCH, SACC, GACCH, and GACC) was analyzed using a differential scanning calorimeter (model DSC Q 200, TA Instruments) with standard mode at nitrogen (50 ml min^{-1}) atmosphere with ramp at 10°C min^{-1}.

Docking and binding energy calculations

For the docking study, chemical structures of chitosan and SA were generated using ACD/ChemSketch ACD (2009), and the 3D structure of type I collagen was generated using the gencollagen program. The docking technique is useful to find the binding efficiency with a ligand and a chemical compound. To find out the interaction between chitosan and type I collagen with SA, AutoDock 4.2 was used to calculate Morris et al. (2009) the free energy of binding of SA with chitosan and type I collagen.

Biocompatibility of SACCH and SACC scaffold an *in vitro* assessment

Cell viability study (MTT assay)

NIH 3T3 fibroblast cells were grown in Dulbecco's modified Eagle's medium supplemented with 10% fetal bovine serum (v/v) and 1% antibiotics and were incubated at 37°C in 5% CO_2 humidified atmosphere. Polystyrene 96-well culture plates (Tarsons) were coated with native chitosan, type I collagen, SA cross-linked chitosan (SACCH), GA cross-linked chitosan, SA cross-linked type I collagen (SACC), and GA cross-linked collagen samples. The plates were dried using a laminar air flow hood, followed by 30 min of UV sterilization. The cells were seeded at the density of 0.5×10^6 per well and incubated at 37°C in humidified atmosphere containing 5% CO_2. At scheduled time points of 24, 48, and 72 h, the supernatant of each well was replaced with MTT diluted in serum-free medium, and the plates incubated at 37°C for 4 h. After removing the MTT solution, acid isopropanol (0.04 N HCl in isopropanol) was added to each well, was then pipetted up and down to dissolve all the dark-blue crystals, and was left at room temperature for a few minutes to ensure the dissolution of all crystals. Finally, absorbance was measured at 570 nm using a UV spectrophotometer Mossmann (1983). Each experiment was performed at least three times. The sets of three wells for the MTT assay were used for each experimental variable.

Cell tracker assay to detect live cells

Cell viability was measured using 5-chloromethylfluorescein diacetate probe (CMFDA) (Invitrogen Life Technologies, Carlsbad, CA, USA). NIH 3T3 cells were subjected to respective treatment conditions. Cells were probed with 5 μM CMFDA and incubated for 2 h. Cells were then washed with sterile PBS, and images were taken using DP71 camera adapted to an Olympus IX71 (Olympus Corporation) microscope Majumder et al. (2011). The assay was carried out using a cell tracker kit. NIH3T3 Cells were treated on the surface of native and cross-linked scaffolds for 6 h, followed by the incubation with cell viable dye cell tracker for 30 min. Fluorescence images of the cells were acquired using a DP71 camera adapted to an Olympus IX71 microscope. The intensity of green positive cells was counted and plotted. Next, fluorescence intensities of the images were calculated using Adobe Photoshop version 7.0. No cell tracker assay was carried out for GACCH and GACC based on the observations made with the MTT assay.

Cell morphology of NIH 3T3 cells in SACCH and SACC scaffolds

SACCH and SACC scaffolds (2 × 2 × 1 cm) were placed individually in six-well culture plates (Tarsons, India) and sterilized ETO. Culture media were added to the scaffolds overnight. NIH 3T3 fibroblast cells were seeded onto the scaffolds at a density of 5×10^4 cells and incubated in an atmosphere of 5% CO_2 at 37°C. The medium was changed every 24 h. The morphology of cells was examined after 12 days according to the following procedure:

1. The cells-and-scaffold constructs were fixed in 2.5% glutaraldehyde and dehydrated through a graded ethanol series.
2. The dried cells-and-scaffold were coated with gold (E-1010 Ion sputter, Hitachi High-Tech, Minato-ku, Tokyo, Japan) and examined under SEM (S-3400 N Hitachi High-Tech).

Experiments for GACCH and GACC were not conducted.

Statistical analysis

When necessary, the experimental results were expressed as the mean ± SD values of triplicates.

Competing interests
The authors declare that they have no competing interests.

Authors' contributions
The work was designed by TM, GS and AG and performed by TM and GS. Manuscript was drafted by TM & GS and written and corrected by AG. All authors read and approved the final manuscript.

Acknowledgments
One of the authors Mr. Tapas Mitra acknowledges CSIR, New Delhi for the financial assistance provided in the form of CSIR-SRF. The authors thank Dr. Swaraj Sinha and Dr. Suvro Chatterjee, AU-KBC Research Centre, MIT Campus, Anna University, Chennai for their help and lab facility to perform cell line studies.

References
ACD (2009) ChemSketch Version 12. Advanced Chemistry Development, Inc., Toronto

Austero MS, Donius AE, Wegst UGK, Schauer CL (2012) New crosslinkers for electrospun chitosan fibre mats. I. Chemical analysis. J R Soc Interface 9:2551–2562

Bhumkar DR, Pokharkar VB (2006) Studies on effect of pH on crosslinking of chitosan with sodium tripolyphosphate: a technical note. AAPS Pharm Sci Tech 7:1–6

Bubnis WA, Ofner CM (1992) The determination of ε-amino groups in soluble and poorly soluble proteinaceous materials by a spectrophotometric method using trinitrobenzenesulfonic acid. Anal Biochem 207:129–133

El-Tahlawy K, Gaffar MA, El-Rafie S (2006) Novel method for preparation of cyclodextrin/grafted chitosan and its application. Carbohydr Polym 63:385–392

Friess W (1998) Collagen—biomaterial for drug delivery. Eur J Pharm Biopharm 45:113–136

Gough JE, Scotchford CA, Downes S (2002) Cytotoxicity of glutaraldehyde crosslinked collagen/poly (vinyl alcohol) films is by the mechanism of apoptosis. J Biomed Mater Res 61:121–130

Iwasaki N, Kasahara Y, Yamane S, Igarashi T, Minami A, Nisimura SI (2011) Chitosan-based hyaluronic acid hybrid polymer fibers as a scaffold biomaterial for cartilage tissue engineering. Polymers 3:100–113

Jin J, Song M, Hourston DJ (2004) Novel chitosan-based films cross-linked by Genipin with improved physical properties. Biomacromolecules 5:162–168

Jorge-Herrero E, Fernandez P, Turnay J, Olmo N, Calero P, Garcia L, Freile I, Olivares JL (1999) Influence of different chemical cross-linking treatments on the properties of bovine pericardium and collagen. Biomaterials 20:539–545

Ko HF, Feir CS, Kumta PN (2010) Novel synthesis strategies for natural polymer and composite biomaterials as potential scaffolds for tissue engineering. Phil Trans R Soc A 368:1981–1997

Lawrie G, Keen I, Drew B, Temple AC, Rintoul L, Fredericks P, Grondahl L (2007) Interactions between alginate and chitosan biopolymers characterized using FTIR and XPS. Biomacromolecules 8:2533–2541

Ligler FS, Lingerfelt BM, Price RP, Schoen PE (2001) Development of uniform chitosan thin-film layers on silicon chips. Langmuir 17:5082–5084

Liu G, Hinch B, Beavis AD (1996) Mechanisms for the transport of α, β-dicarboxylates through the mitochondrial inner membrane. J Biol Chem 271:25338–25344

Lopez FA, Merce ALR, Alguacil FJ, Delgado AL (2008) A kinetic study on the thermal behaviour of chitosan. J Therm Anal Cal 91:633–639

Majumder S, Siamwala JH, Srinivasan S, Sinha S, Sridhara SRC, Soundararajan G, Seerapu HR, Chatterjee S (2011) Simulated microgravity promoted differentiation of bipotential murine oval liver stem cells by modulating BMP4/Notch1 signaling. J Cell Biochem 112:1898–1908

Miles CA, Avery NC, Rodin VV, Bailey AJ (2005) The increase in denaturation temperature following cross-linking of type-I collagen is caused by dehydration of the fibres. Mol Biol 346:551–556

Milosavljevic NB, Kljajevic LM, Popovic IG, Filipovic JM, Krusic MTK (2010) Chitosan, itaconic acid and poly(vinyl alcohol) hybrid polymer networks of high degree of swelling and good mechanical strength. Polym Int 59:686–694

Mitra T, Sailakshmi G, Gnanamani A, Raja ST, Thiruselvi T, Mangala Gowri V, Selvaraj NV, Ramesh G, Mandal AB (2011) Preparation and characterization of a thermostable and biodegradable biopolymers using natural cross-linker. Int J Biol Macromol 48:276–285

Mitra T, Sailakshmi G, Gnanamani A, Mandal AB (2012) Preparation and characterization of malonic acid cross-linked chitosan and collagen 3D scaffolds: an approach on non-covalent interactions. J Mater Sci Mater Med 23:1309–1321

Morris GM, Huey R, Lindstrom W, Sanner MF, Belew RK, Goodsell DS, Olson AJ (2009) Autodock4 and AutoDockTools4: automated docking with selective receptor flexibility. J Comput Chem 30:2785–2791

Mossmann T (1983) Rapid colorimetric assay for cellular growth and survival: application to proliferation and cytotoxicity assays. Immunol Met 65:55–63

Nam K, Kimura T, Kishida A (2008) Controlling coupling reaction of EDC and NHS for preparation of type-I collagen gels using ethanol/water co-solvents. Macromol Biosci 8:32–37

Ohkawa K, Cha DI, Kim H, Nishida A, Yamamoto H (2004) Electrospinning of chitosan. Macromol Rapid Commun 25:1600–1605

Olde Damink LHH, Dijkstra M, Van Luyn MJA, Van Wachem PB, Nieuwenhuis P, Feijen J (1995a) Glutaraldehyde as a cross-linking agent for collagen-based biomaterials. J Mater Sci Mater Med 6:460–472

Olde Damink LHH, Dijkstra PJ, van Luyn MJA, Van Wachem PB, Nieuwenhuis P, Feijen J (1995b) Influence of ethylene oxide gas treatment on the *in vitro* degradation behavior of dermal sheep type-I collagen. J Biomed Mater Res 29:149–155

Parenteau-Bareil R, Gauvin R, Berthod F (2010) Collagen-based biomaterials for tissue engineering applications. Materials 3:1863–1887

Pavia DL, Lampman GM, Kriz GS (2001) Introduction to spectroscopy, 3rd edition. Thomson Learning, Inc, USA

Peng L, Cheng XR, Wang JW, Xu DX, Wang GE (2006) Preparation and evaluation of porous chitosan/collagen scaffolds for periodontal tissue engineering. J Bioact Compat Polym 21:207–220

Petite H, Rault I, Huc A, Menasche P, Herbage D (1990) Use of the acyl azide method for cross-linking type-I collagen-rich tissues such as pericardium. J Biomed Mater Res 24:179–187

Rinaudo M (2010) New way to crosslink chitosan in aqueous solution. Eur Polym J 46:1537–1544

Saito H, Murabayashi S, Mitamura Y, Taguchi T (2008) Characterization of alkali-treated type-I collagen gels prepared by different crosslinkers. J Mater Sci Mater Med 19:1297–1305

Schauer CL, Chen MS, Chatterley M, Eisemann K, Welsh ER, Price R, Schoen PE, Ligler FS (2003) Color changes in chitosan and poly(allyl amine) films upon metal binding. Thin Solid Films 434:250–257

Schiffman JD, Schauer CL (2007) Cross-linking chitosan nanofibers. Biomacromolecules 8:594–601

Schmidt CE, Baier JM (2000) Acellular vascular tissues: natural biomaterials for tissue repair and tissue engineering. Biomaterials 21:2215–2231

Shanmugasundaram N, Ravikumar T, Babu M (2004) Comparative physico-chemical and in vitro properties of fibrillated collagen scaffolds from different sources. J Biomater Appl 18:247–264

Sheu MT, Huang JC, Yeh GC, Ho HO (2001) Characterization of type-I collagen gel solutions and type-I collagen matrices for cell culture. Biomaterials 22:1713–1719

Tamada J, Langer R (1992) The development of polyanhydrides for drug delivery applications. J Biomater Sci Polym Ed 3:315–353

Tual C, Espuche E, Escoubes M, Domard AJ (2000) Transport properties of chitosan membranes: influence of crosslinking. Polym Sci Part B: Polym Phys 38:1521–1529

Usha R, Ramasami T (2000) Effect of crosslinking agents (basic chromium sulfate and formaldehyde) on the thermal and thermomechanical stability of rat tail tendon type-I collagen fibre. Thermochim Acta 356:59–66

Vijayaraghavan R, Thompson BC, MacFarlane DR, Ramadhar K, Surianarayanan M, Aishwarya S, Sehgal PK (2010) Biocompatibility of choline salts as crosslinking agents for collagen based biomaterials. Chem Comm 46:294–296

Wang Y, Ameer GA, Sheppard BJ, Langer R (2002) A tough biodegradable elastomer. Nat Biotechnol 20:602–606

Weadock KS, Miller EJ, Bellincampi LD, Zawadsky JP, Dunn MG (1995) Physical crosslinking of type-I collagen fibers: comparison of ultraviolet irradiation and dehydrothermal treatment. J Biomed Mater Res 29:1373–1379

Wei YC, Hudson SM, Mayer JM, Kaplan DL (1992) The cross linking of chitosan fibers. Polym Sci Part A: Polym Chem 30:2187–2193

15

Comparative studies on osteogenic potential of micro- and nanofibre scaffolds prepared by electrospinning of poly(ε-caprolactone)

Ting-Ting Li[1], Katrin Ebert[2], Jürgen Vogel[1] and Thomas Groth[1*]

Abstract

The biocompatibility and osteogenic potential of four fibrous scaffolds prepared by electrospinning of poly(ε-caprolactone) (PCL) was studied with MG-63 osteoblast cells. Two different kinds of scaffolds were obtained by adjustment of spinning conditions, which were characterized as nano- or microfibrous. In addition of one nanofibrous, scaffold was made more hydrophilic by blending PCL with Pluronics F 68. Scaffolds were characterized by scanning electron microscopy and water contact angle measurements. Morphology and growth of MG63 cells seeded on the different scaffolds were investigated by confocal laser scanning microscopy after vital staining with fluorescein diacetate and by colorimetric assays. It was found that scaffolds composed of microfibres stipulated better growth conditions for osteoblasts probably by providing a real three-dimensional culture substratum, while nanofibre scaffolds restricted cell growth predominantly to surface regions. Osteogenic activity of cells was determined by alkaline phosphatase (ALP) and o-cresolphthalein complexone assay. It was observed that osteogenic activity of cells cultured in microfibre scaffolds was significantly higher than in nanofibre scaffolds regarding ALP activity. Overall, one can conclude that nanofibre scaffold provides better conditions for initial attachment of cells but does not provide advantages in terms of scaffold colonization and support of osteogenic activity compared to scaffolds prepared from microfibres.

Keywords: Poly(ε-caprolactone); Electrospinning; Microfibres; Nanofibres; Osteoblasts; Cell growth; Osteogenic activity

Introduction

The traditional clinical methods of bone defect management, such as autografts and allografts of cancellous bone, can be limited by the large size of bone defects, poor viability of host environment and unpredictable graft resorption (Burg et al. 2000). Transplanting bone generated by autologous cells and tissue engineering could eliminate the problems of defect size, donor site scarcity, immune rejection and pathogen transfer. Cells, scaffolds and signals are the triad of tissue engineering, which is related to the main components of biologic tissues (Bell 2000). In order to grow into functional three-dimensional (3D) tissues or organs, osteoblasts progenitor cells require external signals including mechanical, structural and chemical

cues. In actual bone tissue, the inhabitancy and the structural cue of cells are the extracellular matrix (ECM), which is composed of a composite meshwork of collagen fibres encased within a hard matrix of calcium phosphate (Burg et al. 2000). Hence, it would be desirable in tissue engineering of bone to apply biomaterials, which are organized as a three-dimensional scaffold-mimicking specific feature of the bone matrix, such as the fibrous elements of ECM (Guillame-Gentil et al. 2010).

Electrospinning is a versatile method for the production of fibres with diameters from micrometre to nanometre scale (Szentivanyi et al. 2011). It has gained serious interest for mimicking the structure of fibrillar extracellular matrix components and was proposed for applications in tissue engineering of bone as well (Yoshimatoa et al. 2003). Principally, electrospinning can be performed either with polymeric solutions or melts. In this process, the polymer solution or melt moves under the influence of a strong electric field between two electrodes. One of the electrodes

---

* Correspondence: thomas.groth@pharmazie.uni-halle.de
[1]Department Pharmaceutics and Biopharmaceutics, Biomedical Materials Group, Martin Luther University Halle-Wittenberg, Institute of Pharmacy, Heinrich-Damerow-Strasse 4, Halle (Saale) 06120, Germany
Full list of author information is available at the end of the article

is located in the spinning nozzle, while opposite to it, a grounded collector is located. As the polymer solution leaves the nozzle first, a hemispherical droplet is formed. If a critical strength of the electrical field is achieved, this form becomes conical, which is known as the Taylor cone. The polymer solution is ejected from this cone as the applied electric field is strong enough to overcome the surface tension of the fluid forming a jet that moves in a whipping mode to the grounded electrode. Due to the simultaneously occurring stretching of the jet and evaporation of the solvent, thin and solid fibres are deposited on the collecting electrode. The dimension and the morphology of the electrospun polymeric nanofibres can be influenced by a number of parameters such as composition of solution or blend, strength of the electric field, spinning distance, nozzle diameter, polymer concentration and the surface tension and electric conductivity of the polymer solution (Szentivanyi et al. 2011; Yoshimatoa et al. 2003; Gugutkov et al. 2013; Andiapann et al. 2013). Recent work has shown that electrospinning can be used to produce aligned fibres or combine fibres of different diameter in a sequential manner to achieve structures that guide cell adhesion, movement and differentiation (Gugutkov et al. 2013; Kim et al. 2010; Pham et al. 2006). This allows the generation of structures that resemble the composition of different tissues like the arterial wall, skin and other tissues (Gugutkov et al. 2013; Andiapann et al. 2013; Asran et al. 2010; Nottelet et al. 2009).

Poly(ε-caprolactone) (PCL) is an interesting material for tissue engineering application because of its non-toxic degradation products, controlled degradability and useful mechanical properties (Cruz et al. 2008). Electrospinning of biodegradable PCL and its application in tissue engineering has been reported recently (Pham et al. 2006; Nottelet et al. 2009; Kweon et al. 2003; Venugopal et al. 2006). It was also shown that interaction of cells with electrospun scaffolds could be significantly improved by modification of the surface chemistry and consequently, the hydrophobic/hydrophilic character of the polymer (Szentivanyi et al. 2011; Asran et al. 2010; Cruz et al. 2008). Indeed, as a weakness of nanofibre scaffolds, one can consider that usually a rather tight network is formed, which prevents penetration of cells. In addition, the mechanical properties of the obtained scaffold are rather weak. On the other hand, thicker fibres (in micrometre scale) provide better mechanical support but are less similar to ECM structure. Therefore, it was aimed here to study the influence of fibre dimensions, surface properties of fibres and the three-dimensional structure of scaffolds on adhesion, proliferation and differentiation of MG-63 human osteosarcoma cells. To achieve this goal, we varied the viscosity and conductivity of PCL solutions by additives like phosphate-buffered saline solution and change of voltage obtaining fibres of different diameter and scaffolds

of lower density from microfibres and higher density from nanofibres. As additional parameter that could affect initial adhesion of cells, the wetting properties of the rather hydrophobic PCL fibres were improved by the addition of amphiphilic Pluronics F 68 to generate one type of hydrophilic nanofibrous scaffold. The results shall help to provide a better understanding how the scaffold architecture affects the behaviour of osteoblasts as a prerequisite for tissue engineering of bone.

Methods
Scaffold manufacturing conditions
PCL with a molecular weight of 80,000 was obtained from Sigma-Aldrich, Steinheim, Germany. Pluronics F-68 from Sigma, Steinheim, Germany was used as a non-ionic fluorescein diacetate (FDA) approved surfactant. All solvents were received from Merck, Darmstadt, Germany in p. a. quality. The PCL was dissolved in tetrahydrofurane (THF) and dimethylformamide (DMF) at room temperature until a homogeneous solution was obtained. The weight ratio of THF/DMF of 9:1 was equal for all solutions. Phosphate-buffered saline (PBS) tablets (Aldrich, Germany) were dissolved in 200 ml deionised water (0.01 M phosphate buffer, 0.0027 M KCl, 0.137 M NaCl). For one set of scaffolds, this PBS solution was added to the PCL solution to obtain a concentration of 1.44 wt.% PBS solution in the spinning solution to increase the conductivity of spinning solution. Prior to the spinning experiments, all solutions were heated for at least 15 min at 30°C under stirring and then filtrated through a 20-μm metal filter. Solution preparation and electrospinning were always performed within 1 day.

Electric conductivity of the spinning solutions was measured in a cell equipped with two oppositely deposited platinum electrodes. The cell was connected to a commercial conductivity meter LF 530 (WTW, Weilheim, Germany). About 2 to 3 ml of solution was used for conductivity measurements. Viscosities of the spinning solutions were measured with an Ubbelohde viscosimeter at 25°C (capillary IIIc for 16 wt.% PCL solutions, capillary IV for 18 wt.% PCL solutions). The viscosities given in Table 1 represent an average of three measurements.

Electrospinning was performed with a set-up designed and constructed at GKSS Research Centre. The polymer solution was fed through a glassy capillary with an inner diameter of 0.2 mm. The spinning solution was in contact with a platinum electrode. The solution flow was adjusted with an infusion pump (Medipan Typ 610 BS, Medipan, Warsaw, Poland). The experiments were performed between 25 and 35 kV and a spinning distance between 25 and 35 cm to have another parameter to vary the fibre diameter. The electrospun fibres were collected on aluminium foils. The abbreviations used in this paper for the four PCL scaffolds electrospun from 18 wt.% PCL

Table 1 Composition of solutions and spinning parameters

Scaffold	Composition spinning solution					Distance [cm]	Voltage [kV]
	C_{PCL} [%]	$C_{PluronicF68}$ [%]	C_{PBS} [%]	Solvent	Viscosity [mm^2/s]		
18%PCL	18	/	/	THF/DMF 9:1	1,598	25	35
18%PCL + Pluronic F68	18	1% of PCL	/	THF/DMF 9:1	1,585	27	25
16%PCL + PBS	16	/	1.44% of PCL	THF/DMF 9:1	1,202	25	25
16%PCL	16	/	/	THF/DMF 9:1	1,132	35	27

solution, 16 wt.% PCL solution, 18 wt.% PCL solution with an addition of Pluronic F 68, and 16 wt.% PCL/ 1.44 wt.% PBS are *18%*, *16%*, *18% + F 68*, and *16% + PBS*, respectively.

Characterization of scaffolds

Static contact angle measurements were performed with a DSA 100 device from Kruess, Hamburg, Germany. A droplet of deionised water with a volume of 5 μl was deposited on the scaffolds on which the measurements were performed (sessile drop mode). The contact angles given in the paper are an average of 5 values measured within the first 5 s. The morphology of scaffolds was performed with a scanning electron microscope (Leo 1550 VP Gemini® field emission scanning electron microscope from Carl Zeiss Company, Jena, Germany). Samples were sputtered in a magnetron with a 2.5-nm-thick Au/Pd layer. The average distance of fibres was obtained by evaluation of scanning electron micrographs with Image J software measuring the largest distance of fibres on the surface region of scaffolds and naming this as mesh size as a parameter that should classify qualitatively the density of fibre network. The fibre diameter was also obtained from scanning electron micrographs by image analysis.

Sample preparation and sterilization

The scaffolds on aluminium foil were cut into discs with a diameter of 14 mm. Replicates of each scaffold type were placed in 24-well culture plates. To avoid floating of scaffolds, glass rings were placed on the samples. Scaffolds were sterilized in 70% ethanol for 1 h and washed three times with distilled water. Scaffolds were kept in distilled water over night. The next day, distilled water was aspirated and substituted by culture medium. Before seeding, medium was removed and then replaced by the cell suspension.

MG63 cell culture

MG63 cells were routinely cultured in 75- or 25-cm^2 flasks at 37°C in a humidified incubator with 5% CO_2. Cells were fed by standard Dulbecco's modified Eagle's medium (DMEM, Biochrom AG, Berlin, Germany) supplemented with 10% *v/v* fetal calf serum (FCS, PromoCell, Heidelberg,

Germany), 1% Pen/Strep/Fungizone (PromoCell, Heidelberg, Germany) and 1% sodium pyruvate (Biochrom AG, Berlin, Germany). MG63 cells were seeded at a density of 10^4 cells/ml in wells containing the sterilized scaffolds. Empty wells of tissue culture polystyrene plates were used as control. All plates were incubated at 37°C in a humidified incubator with 5% CO_2. During the culture period, measurements were done on days 1, 3, 7, 10 and 14. For the measurement of calcium deposition MG-63 cells were cultured for 3 and 4 weeks. During these experiments osteogenic factors (L-ascorbic acid 2-phosphate (0.2 mM, Fluka, Steinheim, Germany), β-glycerophosphate (10 mM, Fluka, Steinheim, Germany) and dexamethasone (0.1 μM, Sigma, Steinheim, Germany) were added to the standard culture medium to support the osteogenic activity of MG-63 cells.

Morphology and distribution of cells

Cell morphology was investigated by confocal laser scanning microscopy (CLSM) (LEICA DM IREZ TCS SP2 AOBS spectral confocal microscope, Leica Microsystems, Singapore) by staining MG-63 cells with FDA (Sigma, Steinheim, Germany). This allowed also studying the distribution of viable cells within the scaffolds. First, culture medium was aspirated and replaced by 1 ml fresh medium. Then, 5 μl FDA solutions (5 mg FDA/ml in acetone) was added to each well. After 5-min incubation at 37°C, the scaffold was transferred to a glass support slide, and the morphology and distribution of cells were evaluated by CLSM (excitation wavelength 485 nm, emission wavelength 520 nm).

Cell viability and growth

Viability of cells was measured by QBlue assay (QBlue Cell Viability Assay Kits, BioChain, Newark, CA, USA). On the measuring day, the scaffolds were transferred to a new 24-well plate containing 500 μl of fresh medium. Fifty microlitres Qblue assay reagent was added to each well. Fresh medium without scaffold represented a blank value. After incubation for 2 h at 37°C, 100 μl medium was transferred from each well to a new black 96-well plate. Fluorescent intensity (excitation wavelength 544 nm, emission wavelength 590 nm) was measured with a fluorescence plate reader (BMG LABTECH, Fluostar OPTIMA, Offenburg, Germany).

Cell growth was measured by modification of the protocol of LDH Cytotoxicity Assay (WST-8 mix reagents, BioCat, Mountain View, CA, USA). On the measuring day, the scaffolds were transferred to a new 24-well plate. Cells were lysed with 0.5% TritonX-100 in distilled water for 30 min at 37°C. After that, the whole plate was centrifuged at $250 \times g$ (1,300 rpm) for 10 min to remove cell debris. Ten microlitres cell lysis solutions was transferred from each well to a 96-well plate. Ten microlitres lysis solutions (0.5% Triton × 100) was used as blank reference. One hundred microlitres LDH reaction mixture was added into each well. The plate was incubated 30 min in room temperature without light. The absorbance was measured at 492 nm with a plate reader (BMG LABTECH, Fluostar OPTIMA, Offenburg, Germany).

Measurement of osteogenic activity

Alkaline phosphatase (ALP) is a typical marker of early stage osteoblastic differentiation (Owen et al. 1990). Here, the quantification of ALP was determined by the hydrolysis of p-nitrophenylphosphate (pNPP, Roth, Karlsruhe, Germany) to p-nitrophenol (p-NP) at pH 10.2. p-NPP solution was prepared in bicarbonate buffer (NaHCO$_3$) at pH 10.2 to obtain a concentration of 0.3 mg/ml. Fifty microlitres of residual cell lysis solutions (prepared for LDH assay) was transferred to a 96-well plate. Fifty microlitres lysis solution (0.5% TritonX-100 only) was used as blank reference. Then, the samples were incubated with 100 µl 0.3 mg/ml p-NPP solution for 1.5 h at 37°C. After incubation, ALP activity was determined by the absorbance at 405 nm using a plate reader (BMG LABTECH, Fluostar OPTIMA, Offenburg, Germany).

Deposited calcium phosphates form a purple-coloured complex with o-cresolphthalein complexone in an alkaline medium. 1.5 M AMP-Buffer (2-amino-2-methyl-3-propanol, Applichem) at pH 10.7 provides the proper alkaline medium for the colour reaction. After confluence (7-day culture), cells in one part of the 24-well plate were fed with inductor medium. Another half was cultured in DMEM without inductors. After 3 or 4 weeks, scaffolds were transferred into a new 24-well plate, 0.5 ml 0.6 M HCl was added into each well, including two wells without scaffolds as reference. After incubation over night at 37°C, 0.5 ml HCl suspension of each well was mixed with 0.5 solution of 0.16 mM o-cresolphthalein (Sigma, Steinheim, Germany) and 5 mM 8-Hydroxyquinoline-5-sulfate (HQS, VWR). HQS avoids an interference of the assay with by magnesium ions. The intensity of the colour was measured at 575 nm with a spectrophotometer SPECORD 200 (Analytik, Jena, Germany).

Results

Electrospinning and scaffold characterization

Representative scanning electron micrographs of the different scaffolds are shown in Figure 1. From scanning electron micrographs, the average fibre diameter as well as their density and distribution was estimated quantitatively and qualitatively as shown in Table 2 by image analysis with Image J. The fibre density was quantified as 'mesh size' indicating the average distance of fibres in horizontal direction. Scaffolds electrospun from 18 wt.% solutions showed rather thick fibres with diameters above 2 µm and only very few thin fibres with diameters of about 0.5 µm, which was related to the higher viscosity of the polymer solution. The mesh size of the fibre network was rather high with mean inter-fibre distances between 20 and 50 µm. Electrospun scaffolds obtained from a solution containing 18 wt.% PCL and Pluronic F 68 showed a rather non-uniform distribution of diameters with thin fibres of about 0.4 µm and a few very thick spindle-shaped fibres of about 6 µm in diameter. The addition of Pluronics F 68, which was done for making fibres more hydrophilic, had obviously also a huge effect on fibre diameter because neither viscosity nor spinning conditions were largely changed. The mesh sizes of the scaffolds prepared from 18%PCL spinning solutions with Pluronic F 68 was below 20 µm. Fibres obtained from the 16 wt.% solution were more uniform, no beads could be detected. However, with diameters above 5 µm, these fibres were much thicker than those obtained from the 18 wt.% solution. The mesh size of this scaffold was larger than 20 up to 100 µm. The reason for such rather unexpected finding may be the change of distance between capillary and collector from 25 cm for PCL 18% to 35 cm for PCL 16%. The addition of 1.44 wt.% PBS to a 16 wt.% PCL solution caused a significant increase in electric conductivity from 0.029 to 0.56 mS/m, which had a significant effect on fibre diameter. The resulting fibres of this solution showed a distribution of diameters between 0.6 and 2.8 µm. No beads were detected. The mesh sizes of the scaffolds obtained from the spinning solutions containing PBS were below 20 µm. According to the results of the diameter and mesh size measurements, meshes prepared from 16 and 18%PCL are called here microfibrous while 16% + PBS and 18% + F68 are denominated as nanofibrous scaffolds.

Table 2 displays also the wetting properties obtained by static water contact angle (WCA) measurements. WCA of the scaffolds electrospun from 16 wt.%, 16 wt.% + PBS and 18%PCL solutions did not differ significantly and were above 100°, which indicated the rather hydrophobic nature of all scaffolds spun from PCL. By contrast, the addition of Pluronics F 68 led a scaffold with much better wetting properties because the WCA of 18%PCL + F68 scaffold was about 50°. For all samples, a decrease

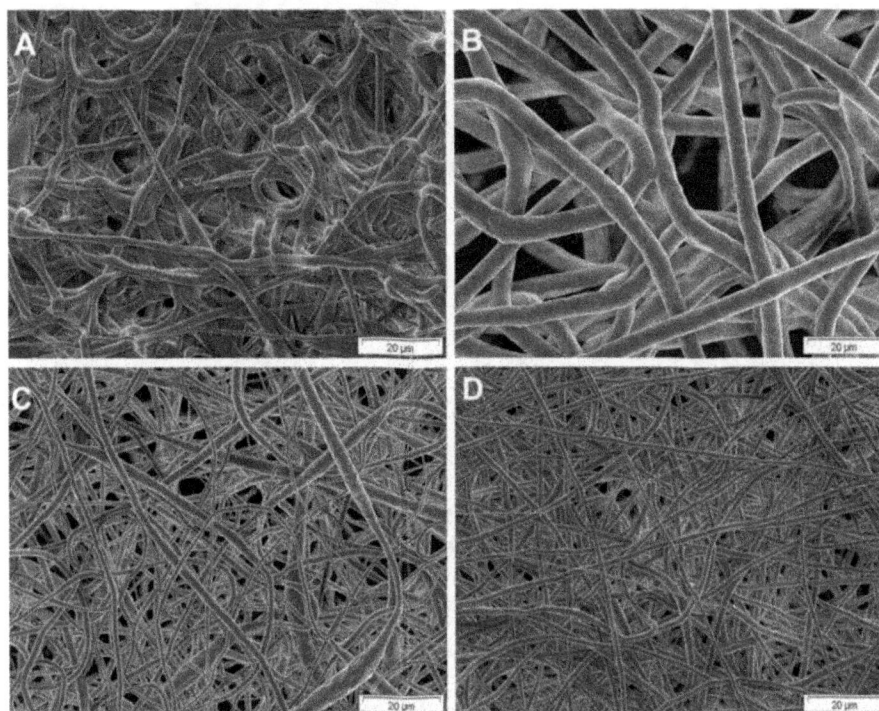

Figure 1 Representative scanning electron micrographs of the different scaffolds. (A) PCL scaffold electrospun from 18 wt.% PCL in THF/DMF 9:1 (voltage, 35 kV); spinning distance, 25 cm; nozzle, 0.2 mm. **(B)** PCL scaffold electrospun from 16 wt.% PCL in THF/DMF 9:1 (voltage, 27 kV); spinning distance, 35 cm; nozzle, 0.2 mm. **(C)** PCL scaffold electrospun from 18 wt.% PCL/0.18 wt.% Pluronics F 68 (voltage. 35 kV); spinning distance, 25 cm; nozzle, 0.2 mm. **(D)** PCL scaffold electrospun from 18 wt.% PCL/1.44 wt.% PBS (voltage, 25 kV); spinning distance, 25 cm; nozzle, 0.2 mm.

in contact angle with time was observed (data not shown here), which is related to the porous character of the scaffold.

Morphology and distribution of cells in/on the scaffold

The morphology of viable MG 63 cells in the scaffold was characterized by confocal laser scanning microscopy (CLSM) after FDA staining. Cells were visualized along z-sections of confocal images in the scaffolds. In Figures 2 and 3, micrographs are shown as apparent two-dimensional (2D) images composed of all z-sections made, which was arranged by confocal microscopy software. This was done to visualize the total number of cells colonizing the scaffold within the range of visibility of confocal microscopy and presence of cells within the scaffold. Basically, all scaffolds supported MG 63 cell attachment and growth, which is demonstrated in Figure 2. It was observed that only a

few adhering cells were found on all scaffolds with minor difference in their morphology after 24 h. An increase in cell numbers was observed after 3 days with small differences in morphology of cells between the scaffolds. Particularly MG-63 cells on the hydrophilic, nanofibrous 16% + F68 scaffolds developed a more elongated phenotype. These differences in morphology became more evident after 7 days of culture, which can be also seen in the higher magnification micrographs shown in Figure 3. MG 63 cells cultured on scaffolds with lower fibre diameters below the micrometre scale such as 16% + PBS and 18% + F68 promoted a more spread and elongated phenotype of cells, while cells growing on fibres with larger diameter like 16% and 18% were still round or growing in aggregates (see Figure 3 as well). After 10 days, cell density had increased greatly. Again, there was a notable difference between both major types of

Table 2 Characterization of scaffolds

| Parameter | Spinning solution | | | |
	18%	18% + F68	16% + PBS	16%
Fibre diameter, mean ± SD, $n = 20$	2.7 ± 1.0 µm	939 ± 1,030 nm	937 ± 360 nm	5.0 ± 0.6 µm
Mesh size, M	20 < M < 50 µm	M < 20 µm	M < 20 µm	20 < M < 100 µm
Fibre density	++	+++	++++	+
Water contact angle [deg]	104.8 ± 4.1	46.6 ± 17.9	112.9 ± 4.9	101.1 ± 6.7

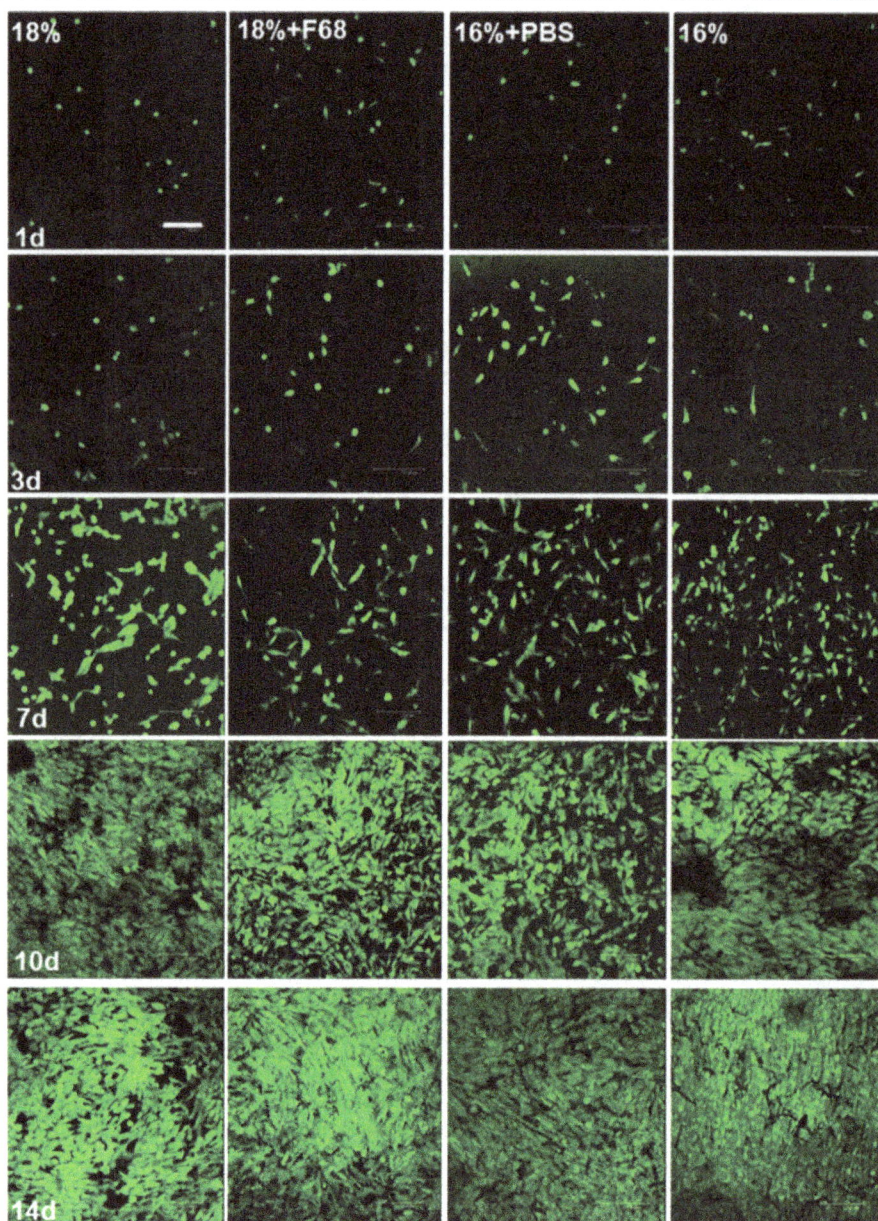

Figure 2 Confocal micrographs of MG63 cells in scaffolds stained with FDA after different cultivated times. The confocal images were taken along z-sections into the scaffolds and shown as 2D pictures composed by Leica Confocal Software. Scale bar, 150 μm.

scaffolds. Those with lower fibre diameter (16% + PBS; 18% + F68) seemed to host less cells as indicated by the slightly lower density of cells. No major differences were observed after 14 days. Cell layers on all scaffolds appeared to be quite dense, though one should note that the images shown in Figures 2 and 3 represent a merge of different z-sections down to 70 μm inside the scaffold.

To show the distribution of cells within the scaffold, z-sections were made at different positions after 14 days of culture. Results are shown in Figure 4. Again, major differences were found between scaffolds composed of larger and smaller fibres. MG-63 cells cultured on nano-fibre scaffolds (16% + PBS; 18% + F68) were growing predominantly in regions very close to the surface up to a depth of about 20 μm. No further viable cells were detected in lower regions of nanofibrous scaffolds. In opposite to this finding, cells cultured on 16% micro-fibre scaffold were growing predominantly in deeper regions down to 70 μm (data not completely shown in Figure 4), while MG 63 cells growing well in 18% micro-fibre scaffolds showed an apparent enrichment in central regions. In lower regions of 18% scaffolds only, some viable cells were found.

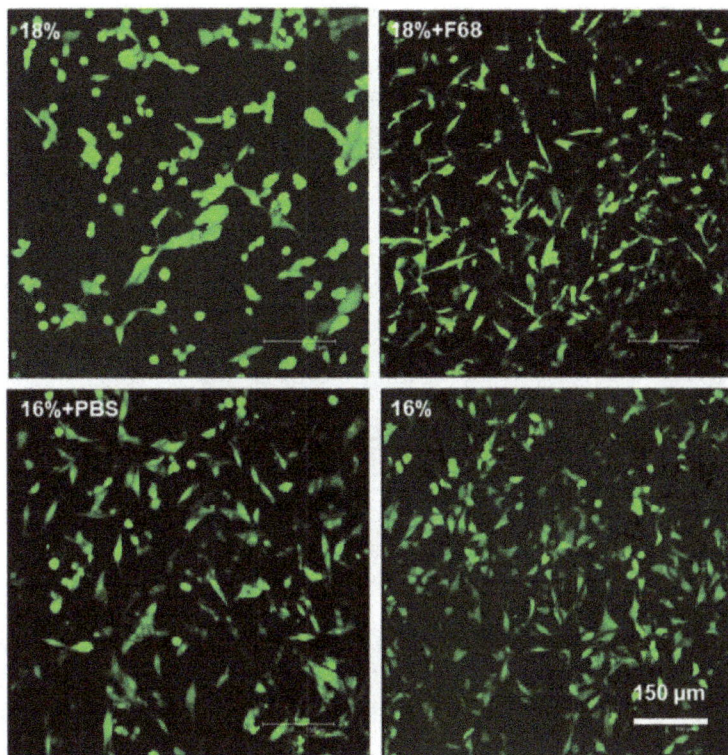

Figure 3 CLSM pictures of MG63 cells in scaffolds stained with FDA after 7 days of culture. Scale bar, 150 μm.

Metabolic activity and growth of cells

In addition to the morphological studies with CLSM, also quantitative data on metabolic activity and growth of cells were obtained. Figure 5A shows the results of experiments with QBlue to measure the metabolic activity of cells. Cells grown on tissue culture polystyrene (TCP) were considered as 2D control system here. It was observed that MG 63 cells cultured on tissue culture polystyrene (TCP) possessed always the highest metabolic activity during all days of measurement. It was also detected that metabolic activity increased during the culture period of 14 days with small and mostly non-significant differences between the different nanofibre scaffolds. This was also expected from the results of studies with vital cell staining and CLSM. There was one exception, namely 16%PCL where metabolic activity of cells was significantly lower than in all other samples. LDH assay was used here in a modified version (after induced lysis of cells) to estimate number of viable cells to quantify cell growth (see Figure 5B). It is interesting to note that the quantity of cells on TCP was only significantly higher up to 7 days and remained then comparable to cell quantities on/in the fibre scaffolds. By contrast, no such significant differences in cell quantities were observed with LDH proliferation assay among the different scaffolds. Only 16%PCL scaffold cells seemed to support cell grow more until day 7

measured by LDH assay in comparison to the other scaffolds, which was significant on day 7 only ($p \leq 0.05$). It is also interesting to note that cell quantities in/on scaffolds measured LDH assay on days 7 and 14 were equal or slightly higher than on the 2D substratum TCP, while metabolic activity measured by QBlue assay was always highest on the control substratum.

Osteogenic activity

The activity of alkaline phosphatase (ALP) was measured also after induced lysis of cells as an early marker of osteoblasts differentiation. Figure 6A shows ALP activity of MG 63 cells seeded in scaffolds and TCP over 14 days. The ALP activity of MG 63 cells on scaffolds increased until day 10 although an initial lag period was observed until day 3 for most of the scaffolds but not TCP and 18%. It was also found that ALP activity was always significantly greater in cells on TCP compared to the scaffolds. Beginning from day 7 on, microfibrous scaffolds 18% and 16% showed higher values than the nanofibrous scaffolds 18% + F68 and 16% + PBS. These differences were also significant ($p \leq 0.05$).

The mineralization of calcium phosphate by MG 63 cells seeded on TCP and scaffolds was measured by quantification of o-cresolphthalein complexone. As shown in Figure 6B, after 3 and 4 weeks, the cells cultured with

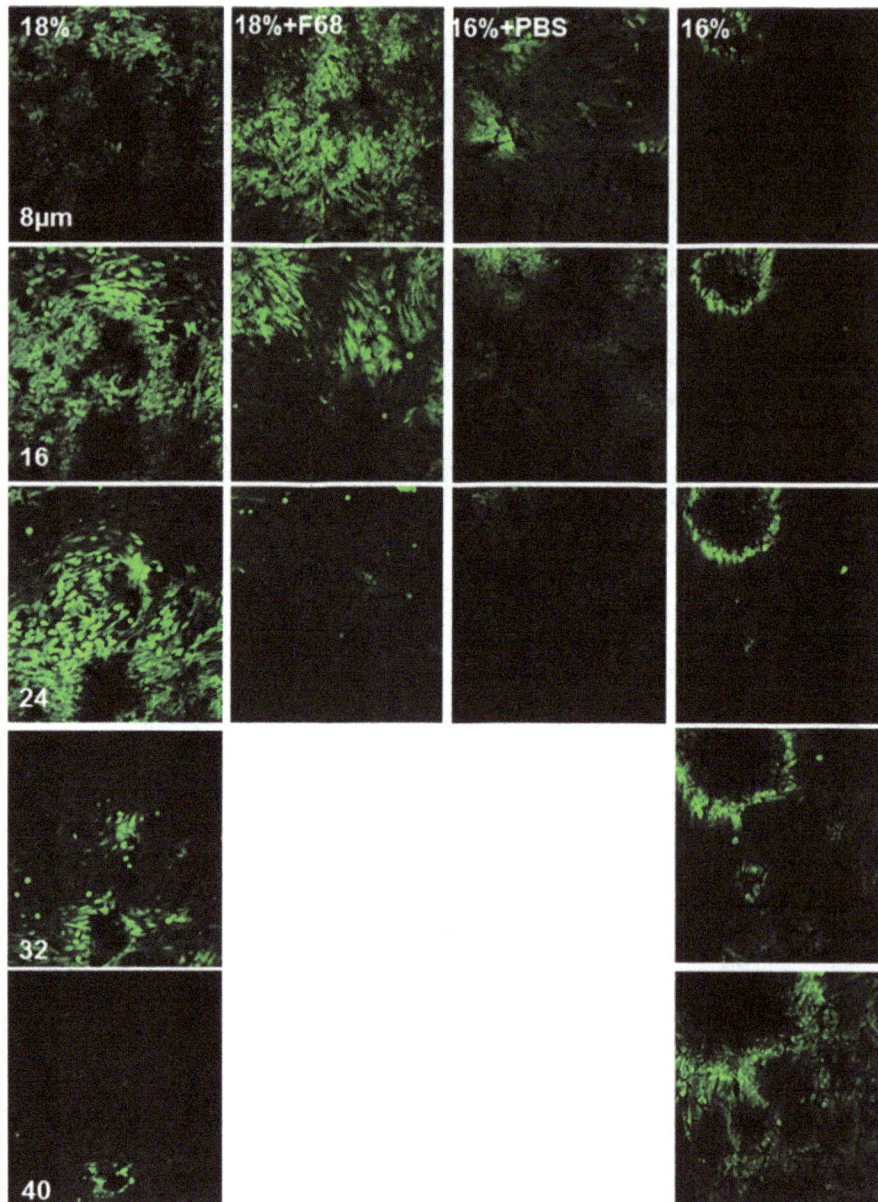

Figure 4 Confocal micrographs of MG63 cells in scaffolds stained with FDA after 14 days of culture. Five layers of confocal images were presented along z-sections from the surface to the inner part of scaffolds. Scale bar, 150 μm.

inductor medium always displayed more calcium deposition than the MG 63 cells in normal medium, which was also expected. It was also found that cells on TCP, cells treated by osteogenic medium showed almost five times more calcium deposition than the cells fed by standard medium both after 3 and 4 weeks culture. Under all conditions, mineralization of cells on TCP was much higher than in scaffolds. As the culture period prolonged, calcium deposition increased in all cases. Comparing the scaffolds, no remarkable differences were found after 3 weeks of culture. However, after 4 weeks, an obvious increase in

calcium content was seen for cells cultured in 18%PCL and 16%PCL + PBS scaffolds.

Discussion

In this study, the effect of fibre diameter and corresponding mesh sizes of electrospun PCL scaffolds on viability, growth and osteogenic activity of MG 63 osteoblast cell line were studied. By the adjustment of the polymer solution composition and spinning conditions, two PCL scaffolds were prepared that can be designated as microfibrous scaffolds (18%, 16%) with mesh sizes from 20 to

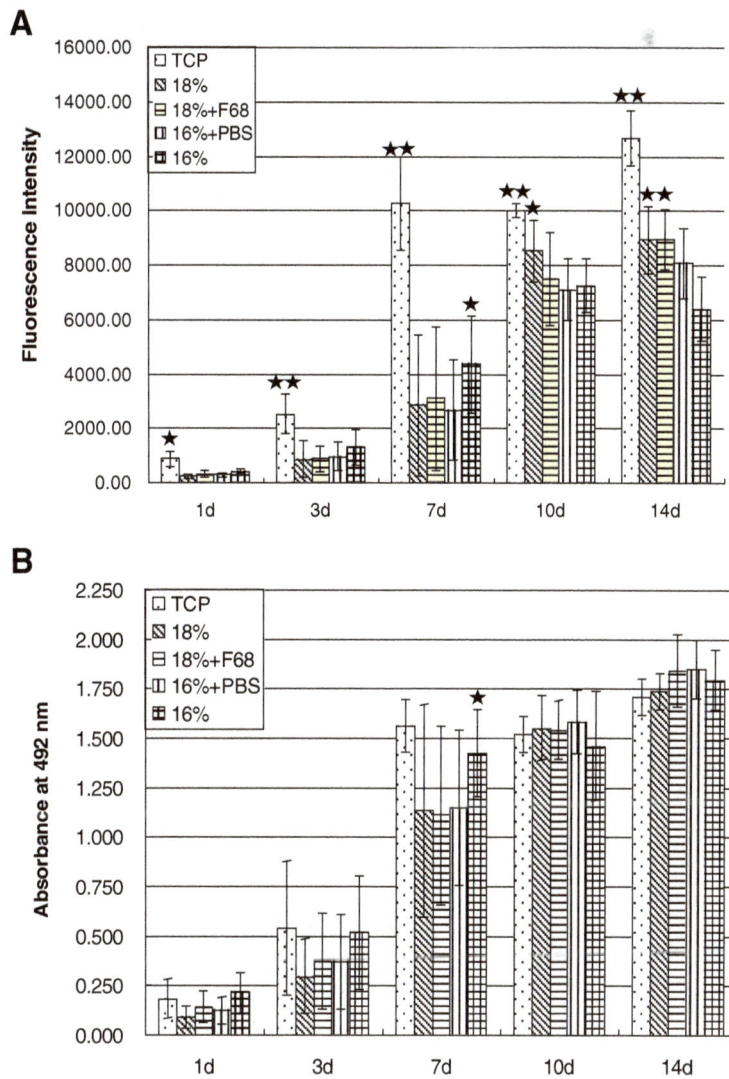

Figure 5 Cell viability and cell growth. (A) QBlue test; cells were cultured in standard medium in scaffolds and TCP as control fluorescence intensity was measured at excitation of 544 nm and emission of 590 nm. Values are mean ± SD, $n = 8$. The data were statistically analyzed with t test. $**p < 0.001$, there was a significant difference of cell viability between MG 63 cells seeded on TCP and scaffolds at each time point. $^*p < 0.05$, there was a significant difference of cell viability between MG 63 cells seeded on 16%PCL and the other three scaffold on day 7, and between 18%PCL and the other three scaffold on day 10, and between 18%PCL, 18%PCL + Pluronic and 16%PCL + PBS, 16%PCL on day 14. **(B)** Growth of MG63 cells cultured by standard medium in scaffolds and TCP as control. Absorbance was measured at 492 nm. Values are mean ± SD, $n = 8$. The data were statistically analyzed with a t test. $^*p < 0.05$, there was a significant difference of cell viability between MG 63 cells seeded on 16%PCL and the other three scaffold on day 7.

100 μm. By contrast, the two other scaffolds had a nanofibrous character (16% + PBS, 18% + F68) with mesh sizes below 20 μm (see also Table 2). As a general conclusion, one can state that nanofibrous scaffolds restricted the growth of cells predominantly to the surface, while microfibrous scaffolds allowed growth in deeper regions. The latter seemed also to enhance the osteogenic activity with regard to alkaline phosphatase activity, but this was also accompanied by a reduced metabolic activity of cells in one case (16% scaffold).

One observation during electrospinning of nanofibres from 18 wt.% PCL solutions was that the majority of fibres seemed to stick together at the points of contact. Most probably residuals of the lower volatile solvent DMF were still entrapped in the fibre at the time during deposition on the collector. This can usually be prevented by avoiding lower volatile solvents. However, because of the low dielectric constant of THF ($\varepsilon = 7.52$), the addition of DMF ($\varepsilon = 38$) was required to render the polymer solution electrically more conductive. Surprisingly, the scaffold made

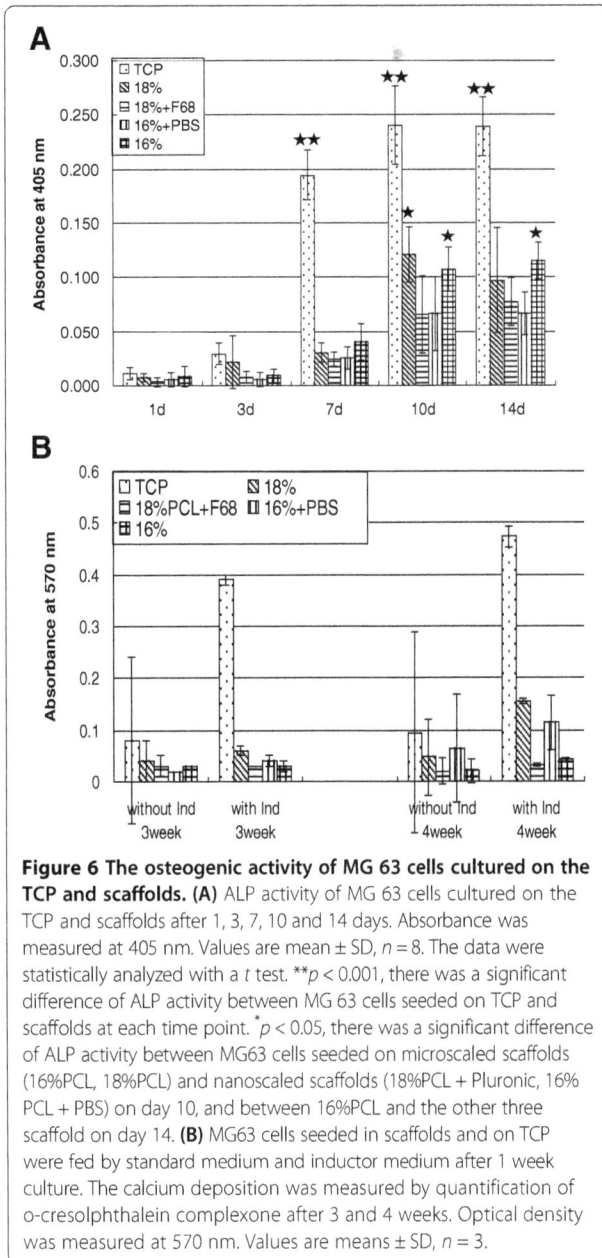

Figure 6 The osteogenic activity of MG 63 cells cultured on the TCP and scaffolds. (A) ALP activity of MG 63 cells cultured on the TCP and scaffolds after 1, 3, 7, 10 and 14 days. Absorbance was measured at 405 nm. Values are mean ± SD, $n = 8$. The data were statistically analyzed with a t test. **$p < 0.001$, there was a significant difference of ALP activity between MG 63 cells seeded on TCP and scaffolds at each time point. *$p < 0.05$, there was a significant difference of ALP activity between MG63 cells seeded on microscaled scaffolds (16%PCL, 18%PCL) and nanoscaled scaffolds (18%PCL + Pluronic, 16% PCL + PBS) on day 10, and between 16%PCL and the other three scaffold on day 14. **(B)** MG63 cells seeded in scaffolds and on TCP were fed by standard medium and inductor medium after 1 week culture. The calcium deposition was measured by quantification of o-cresolphthalein complexone after 3 and 4 weeks. Optical density was measured at 570 nm. Values are means ± SD, $n = 3$.

of 16%PCL solution showed significantly thicker fibres than those obtained from the higher concentrated solution, which was an unexpected finding. Normally, a decrease in polymer concentration usually causes also a decrease in fibre diameter (Szentivanyi et al. 2011; Yoshimatoa et al. 2003; Nottelet et al. 2009; Kweon et al. 2003). However, during the making of 16%PCL fibres, processing conditions were changed by reduction of voltage and increase of distance, which had a significant effect on fibre diameter increase. An improvement of uniformity of electrospun fibres can be achieved by the increase in electric conductivity of the spinning solution (Szentivanyi et al. 2011; Kweon et al. 2003).

Therefore, usually organic salts or organic acids are added. However, in case of scaffolds applied in tissue engineering, biocompatibility and toxicity of additives have to be considered. For this reason, phosphate-buffered saline was used to increase the conductivity of 16 wt.% PCL solution. The resulting scaffolds showed a relatively uniform distribution of thinner fibres in the nanometer range. For rendering the hydrophobic PCL fibres, more hydrophilic Pluronic F 68 was added to the spinning solution as well. The resulting fibres were characterized by the lowest contact angle measured within this study (~47°), while all other scaffolds showed water contact angles above 100°.

Previous investigation showed that cell adhesion is generally improved on hydrophilic surfaces (Groth et al. 2010). However, in this study, 18% + F68 as the most hydrophilic scaffolds did not show a strikingly different response regarding cell attachment and proliferation. This confirms a previous report that wetting properties of 3D scaffolds do not strongly affect cell attachment and proliferation (Spasova et al. 2007). On the other hand, it was reported that the spreading of mouse fibroblasts was more supported on hydrophilic than on the hydrophobic electrospun surfaces (Bhattarai et al. 2006). Therefore, although hydrophilic scaffolds do not significantly promote cell proliferation, they might support cell spreading and increase viability of cells (Bacakova et al. 2004).

A noticeable difference was observed between the two microfibre scaffolds 18% and 16%. The viability and growth of cells were significantly higher in 16% scaffolds than in all other scaffolds before 10 days of culture. After that, there was no remarkably higher cell growth and the cell viability decreased in this scaffold after 14 days. In comparison to that, cell viability was significantly higher in 18% fibre networks after 10 days of culture. The characteristics of scaffolds (Table 2) show that 16% possessed the largest mesh size (sizes from 20 to 100 μm); followed by the mesh size of 18% nanofibre (sizes from 20 to 50 μm). Z-sections of confocal images revealed that the depth of cell growth inside scaffolds was consistent with the order of mesh size. Cells were growing deepest in 16% followed by 18% scaffolds. Indeed, MG63 cells in 18% were well-distributed in the 3D structure of the scaffold (Figure 5). The different behaviour of cells in the two micro-fibre scaffolds is probably due to the different pore structure of the scaffolds. Boudriot et al. suggested that the lack or at least reduced supply of oxygen and nutrients in the core of scaffolds diminishes metabolic activity of cells (Boudriot et al. 2006). In this study, the most porous scaffold 16% allowed adhesion of MG63 cells on the bottom area of the scaffolds, where they proliferated initially quite well. However, an increasing number of cells inside the scaffold increases the oxygen consumption and leads probably to hypoxic conditions after 10 days.

By contrast, the well-distributed cells in 3D structure and the free space remained inside 18%PCL scaffolds provide better oxygen supply to the cells. This is also reflected by the reduced metabolic activity of cells in 16% scaffolds measured by QBlue assays, while cell quantities measured by the modified LDH assay were not different.

Previous studies suggested that nanofibre scaffolds with diameters like collagen fibres and 3D structures mimicking the natural extracelluar matrix promote cell adhesion, proliferation and differentiation (Gugutkov et al. 2013; Dzenis 2004). Also, Boudriot et al. showed with nanofibres from identical polymer solutions that the biocompatibility increased as the fibre diameter decreased (Boudriot et al. 2006). However, in this study, 18% + F68 and 16% + PBS scaffolds representing nanoscaled fibres did not showed significantly better biocompatibility than the microscaled fibre scaffolds 18% and 16%. The only advantage was more initial cell spreading during the early culture periods shown by CLSM (Figure 4). After 14 days of culture, the growth of most cells was still restricted to the surface region of the two nanofibre scaffolds (18% + F68, 16% + PBS). Obviously, the pore size of nanofibre scaffolds was not large enough to allow penetration of cells into the 3D structure. Therefore, the viability of cells could was affected by a reduced supply of oxygen. This result points to the important role of porosity of fibrous scaffolds as a major determinant of cell behaviour (Hutmacher 2000).

ALP is an early marker of osteogenic differentiation. ALP activity is present at high level in cells, which mineralized their matrix such as osteoblasts (Owen et al. 1990). The production of a mineralized matrix is considered as the final process of osteoblast phenotype in differentiation by quantification of calcium deposition (Owen et al. 1990; Bancroft et al. 2002). During osteoblast differentiation, L-ascorbic acid 2-phosphate, β-glycerophosphate and dexamethasone are used as osteogenic factors to effectively stimulate bone formation (Cheng et al. 1996). In this study, the cells treated by osteogenic inductor medium produced almost five times more calcium than the cells fed by standard medium both after 3 and 4 weeks culture. ALP expression follows immediately the down-regulation of cell proliferation. Later, the increasing expression of osteocalcein and osteopontin reflects the beginning of the mineralization step by deposition of the calciumphosphate hydroxyapatite (Owen et al. 1990). In this study, cells cultured on tissue culture polystyrene as control material (TCP) reduced their growth from day 7 on indicated by stagnant signal for LDH assay, while a dramatic increase of ALP activity was found on the same day. This shows that cells on TCP were confluent and started down-regulation of cell proliferation. On the four nanofibre scaffolds, ALP expression remained at a basal level while cells were still proliferating until day 7 (Figure 6A). That is the reason why the ALP activity decreased on day 7. This phenomenon indicates that the cell proliferation was lasting longer inside the scaffolds, which went along with delayed expression of ALP. It is also interesting to note that Koegler et al. modified PLGA scaffolds with different percentage of polyethylene oxide (PEO) - the hydrophilic block of the Pluronics copolymer. He found an increased ALP activity of MG63 cells on PLGA scaffolds coated with higher PEO concentrations (Koegler and Griffith 2004). However, in this study here, the most hydrophilic 18%PCL + Pluronic scaffold (1% Pluronic inside PCL polymer solution) did not express any stronger effect on MG63 cells differentiation. Moreover, it was observed during the differentiation studies that MG63 cells seeded in both microfibre scaffolds possessed a significantly higher ALP activity than in nanofibre scaffolds. However, calcium deposition after 4 weeks was higher only in one of the in microfibre 18%PCL scaffold, while also one of nanofibre 16%PCL + PBS scaffold showed also more calcium deposition. Here, the pore structure of electrospun scaffolds may play a key role in controlling the osteogenic response. Compared to 18%PCL scaffolds, the microfibre scaffold 16%PCL expressed high ALP activity but low calcium deposition, which could be related due to the lower metabolic activity of cells that grow on the bottom of the scaffold and suffer obviously from lack of oxygen. On the other hand, as the only scaffolds causing stronger aggregation of MG63 cells, high calcium deposition in 16%PCL + PBS scaffolds may demonstrate that nanofibre scaffolds can promote mineralization of matrix as well.

In comparison with TCP, all scaffolds showed much lower ALP expression and calcium deposition of MG63 cells. Van den Dolder et al. suggested that seeding at high cell density (8×10^5 cells/ml) initially could further enhance calcification (Van der Dolder et al. 2002). High cell density is known to enhance cell-cell contacts and communication between osteoblasts and promote their differentiation (Cheng et al. 1998). Therefore, the lower ALP activity and calcium deposition on all scaffold materials might be due to the lower cell density during the culture compared to TCP. Previous work has also shown that even though biomaterial scaffolds support the growth and division of osteoblasts cells (MG63), they often alter their phenotypic expression, leading to a loss of their key characteristics, such as bone matrix formation (Zhang and Zhang 2004).

Conclusions

In summary, all types of 3D scaffolds prepared in this study by electrospinning of poly(ε-caprolactone) were well biocompatible and supported proliferation of MG63 osteoblasts. The different hydrophilicity of 3D scaffolds did not affect cell adhesion and proliferation significantly as evident by the behaviour of cells on 18% + F68 scaffold. In general, MG63 cells seeded in microfibre scaffolds presented good viability and increased osteogenic activity regarding activity of ALP. Hence, it can be concluded that one important characteristic of scaffolds prepared by electrospinning is to promote osteoblasts function in 3D structures to provide an adequate porosity for cell growth inside the scaffold by arrangement of fibres to allow a homogenous distribution of cells. However, under conditions of oxygen transfer by diffusion only, the metabolic activity of cells can be hampered by insufficient oxygen supply due to the growth inside the scaffolds. Compared to the cells in microfibre scaffolds, the growth of cells seeded on nanofibre scaffolds expressed better initial cell spreading, which may affect growth and differentiation of osteoblasts in a positive manner. Although the ALP expression of osteoblasts in nanofibre scaffolds was remarkably lower, high calcium deposition was observed in one of them (16%PCL + PBS) scaffolds after long-term culture. This indicates at least in line with previous studies that nanofibre scaffolds are also quite useful for promoting osteogenic activity (Kweon et al. 2003; Zhang and Zhang 2004). The obvious disadvantage of nanofibre scaffolds to block entry of cells due to too close deposition of fibres can be overcome by a combination of nano- and microfibres to a suitable scaffold architecture with sufficient mechanical strength that allows penetration of cells into the structure but provides ale specific topographical stimuli to cells due to interaction with nanofibres as shown by some recent work of different groups (Kim et al. 2010; Pham et al. 2006; Tuzlakoglu et al. 2005).

Competing interests
The authors declare that they have no competing interest.

Authors' contributions
TTL performed all cell culture work, assessment of cell adhesion and proliferation by colorimetric measurements and also confocal laser scanning microscopy. She was supported by JV doing some of the confocal laser scanning microscopy work. KE carried out fabrication and characterization of nanofibre meshes. TG contributed to conceptual work, discussions and writing of the manuscript. All authors read and approved the final manuscript.

Acknowledgements
M. Aderhold and K. Prause are kindly acknowledged for making the scanning electron micrographs.

Author details
[1]Department Pharmaceutics and Biopharmaceutics, Biomedical Materials Group, Martin Luther University Halle-Wittenberg, Institute of Pharmacy, Heinrich-Damerow-Strasse 4, Halle (Saale) 06120, Germany. [2]GKSS Research Centre Geesthacht GmbH, Institute of Polymer Research, Max-Planck-Str.1, Geesthacht 21502, Germany.

References
Andiapann M, Sundaramoorthy S, Panda N, Meiyazhaban G, Winfred SB, Venkatamaran G, Krishna P (2013) Electrospun eri silk fibroin scaffold coated with hydroxyapatite for bone tissue engineering applications. Progress Biomater 2:6

Asran AS, Razghandi KH, Aggarwal N, Michler GH, Groth T (2010) Nanofibers from blends of polyvinyl alcohol and polyhydroxybutyrate as potential scaffold material for tissue engineering of skin. Biomacromolecules 11:3413–3421

Bacakova L, Filova E, Rypacek F, Svorcik V, Stary Y (2004) Cell adhesion on artificial materials for tissue engineering. Physiol Res 53(suppl I):S35–S45

Bancroft GN, Sikavitsas VI, van der Dolder J, Sheffield TL, Ambrose CG, Jansen JA, Mikos A (2002) Fluid flow increases mineralized matrix deposition in 3 D perfusion culture of marrow stromal osteoblasts in a dose dependent manner. Proc Natl Acad Sci U S A 99:12600–12605

Bell E (2000) Tissue Engineering in perspective. In: Vacanti JP (ed) Principles of tissue engineering, 2nd edn. Academic Press, San Diego, pp 35–61

Bhattarai SR, Bhattarai N, Viswanathamurthi P, Yi HK, Hwang PH, Kim HY (2006) Hydrophilic nanofibrous structure of polylactide; fabrication and cell affinity. J Biomed Mat Res 78A:247–257

Boudriot U, Dersch R, Greiner A, Wendorff JH (2006) Electrospinning approaches toward scaffold engineering - a brief overview. Artif Organs 30:785–792

Burg KJL, Porter S, Kellam JF (2000) Biomaterial development for bone tissue engineering. Biomaterials 21:2347–2359

Cheng SL, Zhang SF, Avioli LV (1996) Expression of bone matrix proteins during dexamethasone-induced mineralization of human bone marrow stromal cells. J Cell Biochem 61:182–193

Cheng SL, Lecanda F, Davidson MK, Warlow PM, Zhang SF, Zhang L, Suzuki S, St. John T, Civitelli R (1998) Human osteoblasts express a repertoire of cadherins, which are crucial from BMP-2 induced osteogenic differentiation. J Bone Miner Res 13:633–644

Cruz DMG, Coutinho DF, Martinez EC, Ribelles JLG, Sanchez MS (2008) Blending polysaccharides with biodegradable polymers II: structure and biological response of chitosan/polycaprolactone blends. J Biomed Mat Res B: App Biomater 87B:544–554

Dzenis Y (2004) Spinning continuous fibers for nanotechnology. Science 304:1917–1919

Groth T, Liu Z-M, Niepel M, Peschel D, Kirchhof K, Altankov G, Faucheux N (2010) Chemical and physical modifications of biomaterials surfaces to control adhesion of cells. In: Shastri VP, Altankov G, Lendlein A (eds) Advances in regenerative medicine: role of nanotechnology and engineering principles. Nato Science Series Springer, Netherlands, pp 253–284

Gugutkov D, Gustavsson J, Ginebra MP, Altankov G (2013) Fibrinogen nanofibers for guiding endothelial cell behavior. Biomater Sci 1:1065–1073

Guillame-Gentil O, Semenov O, Roca A, Groth T, Zahn R, Vörös J, Zenobi-Wong M (2010) Engineering the extracellular environment: strategies for building 2D and 3D cellular structures. Adv Materials 22:5443–5462

Hutmacher DW (2000) Scaffolds in tissue engineering of bone and cartilage. Biomaterials 21:2529–2543

Kim SJ, Jang DH, Park WH, Min B-M (2010) Fabrication and characterization of 3-dimensional nanofibre/microfiber scaffolds. Polymer 51:1320–1327

Koegler WS, Griffith LG (2004) Osteoblast response to PLGA tissue engineering scaffolds with PEO modified surface chemistries and demonstration of a patterned cell response. Biomaterials 25:2819–2830

Kweon H, Yoo MK, Park IK, Kim TH, Lee HC, Lee HS, Oh JS, Akaike T, Cho CS (2003) A novel degradable polycaprolactone networks for tissue engineering. Biomaterials 24:801–808

Nottelet B, Pektok E, Mandracchia D, Tille J-C, Walpoth B, Gurny R, Möller M (2009) Factorial design optimization and in vivo feasibility of poly(e-caprolactone)-micro- and -nanofiber based small diameter vascular grafts. J Biomed Mat Res 89A:865–875

Owen TA, Aronow M, Shalhoub V, Barone LM, Wilming L, Tassinari MS, Kennedy MB, Pockwinse S, Lian JB, Stein GS (1990) Progressive development of the rat osteoblast phenotype in vitro: reciprocal relationships in expression of genes associated with osteoblast proliferation and differentiation during formation of the bone extracellular matrix. J Cell Physiol 143:420–430

Pham QP, Sharma U, Mikos AG (2006) Electrospun poly(ε-caprolactone) microfiber and multilayer nanofiber/microfiber scaffolds: Characterization of scaffolds and measurements of cellular infiltration. Biomacromolecules 7:2796–2805

Spasova M, Stoilova O, Manolova N, Raskov I (2007) Preparation of PLLA/PEG nanofibers by electrospinning and potential applications. J Bioact Compat Polymers 22:62–76

Szentivanyi AL, Zernetsch H, Menzel H, Glasmacher B (2011) A review of developments in electrospinning technology: new opportunities for the design of artificial tissue structures. Int J Artif Organs 10:986–997

Tuzlakoglu K, Bolgen N, Salgado AJ, Gomes ME, Piskin E, Reis RL (2005) Nano and microfiber combined scaffolds: a new architecture for bone tissue engineering. J Mat Sci Mat Med 16:1099–1104

Van der Dolder J, Vehof JWM, Soauwen PHM, Jansen JA (2002) Bone formation by rat bone marrow cells cultured on titanium fiber mesh: effect of culture time. J Biomed Mat Res 62A:350–358

Venugopal JR, Zhang Y, Ramakrishna S (2006) In vitro culture of human dermal fibroblasts on electrospun polycaprolactone collagen nanofibrous membrane. Artif Organs 30:440–446

Yoshimatoa H, Shina YM, Teraia H, Vacanti JP (2003) A biodegradable nanofiber scaffold by electrospinning and its potential for bone tissue engineering. Biomaterials 24:2077–2082

Zhang Y, Zhang M (2004) Cell growth and function on calcium reinforced chitosan scaffolds. J Mat Sci Mat Med 15:255–260

Permissions

List of Contributors

Sergey V Dorozhkin
Kudrinskaja sq. 1-155, Moscow 123242, Russia

Maryam Mobed-Miremadi
Department of Biomedical, Chemical and Materials Engineering, San Jose State University, San Jose, CA 95192-0082, USA

Raki Komarla Nagendra
MSE Biomedical Devices Concentration, San Jose State University, San Jose, CA, USA

Sujana Lakshmi Ramachandruni
MSE Biomedical Devices Concentration, San Jose State University, San Jose, CA, USA

Jason James Rook
MSE Biomedical Engineering, San Jose State University, San Jose, CA, USA

Mallika Keralapura
Department of Electrical Engineering, San Jose State University, San Jose, CA, USA

Michel Goedert
Department of Biomedical, Chemical and Materials Engineering, San Jose State University, San Jose, CA 95192-0082, USA

Olivia Donaldson
Villanova University, 800 East Lancaster Avenue, Villanova, PA 19085, USA

Zuyi Jacky Huang
Villanova University, 800 East Lancaster Avenue, Villanova, PA 19085, USA

Noelle Comolli
Villanova University, 800 East Lancaster Avenue, Villanova, PA 19085, USA

Qizhi Chen
Department of Materials Engineering, Monash University, Clayton, Victoria 3800, Australia

Chenghao Zhu
Department of Materials Engineering, Monash University, Clayton, Victoria 3800, Australia

George A Thouas
Department of Zoology, The University of Melbourne, Parkville, Victoria 3010, Australia

Reem Ajaj
Section of Biomaterials, Division of Conservative Dental Sciences, School of Dentistry, King Abdulaziz University, Jeddah 22254, Saudi Arabia
Department of Oral Diagnostic Sciences, Division of Biomaterials, State University of New York at Buffalo, 355 Squire Hall, 110 Parker Hall, Buffalo NY 14214, USA

Robert Baier
Department of Oral Diagnostic Sciences, Division of Biomaterials, State University of New York at Buffalo, 355 Squire Hall, 110 Parker Hall, Buffalo NY 14214, USA

Jude Fabiano
Department of Restorative Dentistry, State University of New York at Buffalo, Buffalo NY 14214, USA

Peter Bush
Department of Restorative Dentistry, State University of New York at Buffalo, Buffalo NY 14214, USA

Hamideh Aghajani-Lazarjani
Biotechnology Group, Department of Chemical Engineering, Faculty of Engineering, Tarbiat Modares University, P.O. Box 14115-143, Tehran 1411713116, Iran

Ebrahim Vasheghani-Farahani
Biotechnology Group, Department of Chemical Engineering, Faculty of Engineering, Tarbiat Modares University, P.O. Box 14115-143, Tehran 1411713116, Iran

Sameereh Hashemi-Najafabadi
Biotechnology Group, Department of Chemical Engineering, Faculty of Engineering, Tarbiat Modares University, P.O. Box 14115-143, Tehran 1411713116, Iran

Seyed Abbas Shojaosadati
Biotechnology Group, Department of Chemical Engineering, Faculty of Engineering, Tarbiat Modares University, P.O. Box 14115-143, Tehran 1411713116, Iran

Saleh Zahediasl
Endocrine Physiology Laboratory, Endocrine Research Centre, Research Institute for Endocrine Sciences, Shahid Beheshti University of Medical Sciences, Tehran 3197619751, Iran

Taki Tiraihi
Department of Anatomy, School of Medical Sciences, Tarbiat Modares University, Tehran 1411713116, Iran

Fatemeh Atyabi
Faculty of Pharmacy, Tehran University of Medical Sciences, Tehran 1419733171, Iran

Nidhi Puri
Polymer and Soft Material Section, CSIR-National Physical Laboratory, Dr. K.S. Krishnan Road, New Delhi 110012, India

Vikash Sharma
Polymer and Soft Material Section, CSIR-National Physical Laboratory, Dr. K.S. Krishnan Road, New Delhi 110012, India

Vinod K Tanwar
Polymer and Soft Material Section, CSIR-National Physical Laboratory, Dr. K.S. Krishnan Road, New Delhi 110012, India

Nahar Singh
Polymer and Soft Material Section, CSIR-National Physical Laboratory, Dr. K.S. Krishnan Road, New Delhi 110012, India

Ashok M Biradar
Polymer and Soft Material Section, CSIR-National Physical Laboratory, Dr. K.S. Krishnan Road, New Delhi 110012, India

Rajesh
Polymer and Soft Material Section, CSIR-National Physical Laboratory, Dr. K.S. Krishnan Road, New Delhi 110012, India

Lydia G Berezhna
Protista Biotechnology AB, Bjuv SE-26722, Sweden

Alexander E Ivanov
Protista Biotechnology AB, Bjuv SE-26722, Sweden

André Leistne
Polymerics GmbH, Berlin D-12681, Germany

Anke Lehmann
Polymerics GmbH, Berlin D-12681, Germany

Maria Viloria-Cols
Protista Biotechnology AB, Bjuv SE-26722, Sweden

Hans Jungvid
Protista Biotechnology AB, Bjuv SE-26722, Sweden

Moslem Tavakol
Biotechnology Group, Faculty of Chemical Engineering, Tarbiat Modares University, P.O. Box 14115–143, Tehran, Iran

Ebrahim Vasheghani-Farahani
Biotechnology Group, Faculty of Chemical Engineering, Tarbiat Modares University, P.O. Box 14115–143, Tehran, Iran

Sameereh Hashemi-Najafabadi
Biotechnology Group, Faculty of Chemical Engineering, Tarbiat Modares University, P.O. Box 14115–143, Tehran, Iran

Ramesh P Babu
Centre for Research Adoptive Nanostructures and Nano Devices, TrinityCollege, Dublin 2, Ireland
School of Physics, Trinity College Dublin, Dublin 2, Ireland

Kevin O'Connor
School of Biomolecular and Biomedical Sciences, Centre for Synthesis and Chemical Biology, UCD Conway Institute, and Earth Institute, University College Dublin, Belfield, Dublin Ireland

Ramakrishna Seeram
NUSNNI, National University of Singapore, 2 Engineering Drive 3, Singapore 117581, Singapore
Institute of Materials Research and Engineering, Singapore 117602, Singapore Jinan University, Guangzhou, China

Manickam Ashokkumar
Department of Chemistry, Anna University, Sardar Patel Road, Chennai, Tamil Nadu 600025, India
Dharmalingam Sangeetha
Department of Chemistry, Anna University, Sardar Patel Road, Chennai, Tamil Nadu 600025, India

Traian V Chirila
Queensland Eye Institute, South Brisbane, Queensland 4101, Australia
Faculty of Science and Engineering, Queensland University of Technology, Brisbane, Queensland 4001, Australia
Faculty of Health Sciences, The University of Queensland, Herston, Queensland 4029, Australia
Australian Institute of Bioengineering & Nanotechnology, The University of Queensland, St Lucia, Queensland 4072, Australia
Faculty of Science, The University of Western Australia, Crawley, Western Australia 6009, Australia

Shuko Suzuki
Queensland Eye Institute, South Brisbane, Queensland 4101, Australia

Laura J Bray
Queensland Eye Institute, South Brisbane, Queensland 4101, Australia
Faculty of Health, Queensland University of Technology, Brisbane, Queensland 4001, Australia
Max Bergmann Center of Biomaterials, Leibniz Institute for Polymer Research, Dresden, Saxony 01069, Germany

Nigel L Barnett
Queensland Eye Institute, South Brisbane, Queensland 4101, Australia
Faculty of Health, Queensland University of Technology, Brisbane, Queensland 4001, Australia
UQ Centre for Clinical Research, The University of Queensland, Herston, Queensland 4029, Australia

Damien G Harkin
Queensland Eye Institute, South Brisbane, Queensland 4101, Australia
Faculty of Health, Queensland University of Technology, Brisbane, Queensland 4001, Australia
Institute of Health and Biomedical Innovation, Kelvin Grove, Queensland 4059, Australia

Alexander Jones
Department of Textiles, Merchandising and Interiors, University of Georgia, Athens, GA 30602, USA

Mark Ashton Zeller
ALGIX, LLC, Athens, GA 30602, USA

Suraj Sharma
Department of Textiles, Merchandising and Interiors, University of Georgia, Athens, GA 30602, USA

G Sailakshmi
Microbiology Division, Central Leather Research Institute (CSIR, New Delhi), Adyar, Chennai 20, Tamil Nadu, India

Tapas Mitra
Microbiology Division, Central Leather Research Institute (CSIR, New Delhi), Adyar, Chennai 20, Tamil Nadu, India

A Gnanamani
Microbiology Division, Central Leather Research Institute (CSIR, New Delhi), Adyar, Chennai 20, Tamil Nadu, India

Ting-Ting Li
Department Pharmaceutics and Biopharmaceutics, Biomedical Materials Group, Martin Luther University Halle-Wittenberg, Institute of Pharmacy, Heinrich-Damerow-Strasse 4, Halle (Saale) 06120, Germany

Katrin Ebert
GKSS Research Centre Geesthacht GmbH, Institute of Polymer Research, Max-Planck-Str.1, Geesthacht 21502, Germany

Jürgen Vogel
Department Pharmaceutics and Biopharmaceutics, Biomedical Materials Group, Martin Luther University Halle-Wittenberg, Institute of Pharmacy, Heinrich-Damerow-Strasse 4, Halle (Saale) 06120, Germany

Thomas Groth
Department Pharmaceutics and Biopharmaceutics, Biomedical Materials Group, Martin Luther University Halle-Wittenberg, Institute of Pharmacy, Heinrich-Damerow-Strasse 4, Halle (Saale) 06120, Germany

www.ingramcontent.com/pod-product-compliance
Lightning Source LLC
Chambersburg PA
CBHW050437200326
41458CB00014B/4978